AF594932

Plant Proteomic Research 4.0

Plant Proteomic Research 4.0

Editors

Sixue Chen
Setsuko Komatsu

MDPI • Basel • Beijing • Wuhan • Barcelona • Belgrade • Manchester • Tokyo • Cluj • Tianjin

Editors
Sixue Chen
University of Florida
USA

Setsuko Komatsu
Fukui University of Technology
Japan

Editorial Office
MDPI
St. Alban-Anlage 66
4052 Basel, Switzerland

This is a reprint of articles from the Special Issue published online in the open access journal *International Journal of Molecular Sciences* (ISSN 1422-0067) (available at: https://www.mdpi.com/journal/ijms/special_issues/plant-proteomic_4).

For citation purposes, cite each article independently as indicated on the article page online and as indicated below:

LastName, A.A.; LastName, B.B.; LastName, C.C. Article Title. *Journal Name* **Year**, *Volume Number*, Page Range.

ISBN 978-3-0365-2662-1 (Hbk)
ISBN 978-3-0365-2663-8 (PDF)

Cover image courtesy of Sixue Chen

Contents

About the Editors . . . . **vii**

Sixue Chen and Setsuko Komatsu
Plant Proteomic Research 4.0: Frontiers in Stress Resilience
Reprinted from: *Int. J. Mol. Sci.* **2021**, *24*, 13362, doi:10.3390/ijms222413362 . . . . **1**

Božena Klodová and Jan Fíla
A Decade of Pollen Phosphoproteomics
Reprinted from: *Int. J. Mol. Sci.* **2021**, *22*, 12212, doi:10.3390/ijms222212212 . . . . **7**

Piengtawan Tappiban, Yining Ying, Feifei Xu and Jinsong Bao
Proteomics and Post-Translational Modifications of Starch Biosynthesis-Related Proteins in Developing Seeds of Rice
Reprinted from: *Int. J. Mol. Sci.* **2021**, *22*, 5901, doi:10.3390/ijms22115901 . . . . **21**

Tosin Victor Adegoke, Yifeng Wang, Lijuan Chen, Huimei Wang, Wanning Liu, Xingyong Liu, Yi-Chen Cheng, Xiaohong Tong, Jiezheng Ying and Jian Zhang
Posttranslational Modification of Waxy to Genetically Improve Starch Quality in Rice Grain
Reprinted from: *Int. J. Mol. Sci.* **2021**, *22*, 4845, doi:10.3390/ijms22094845 . . . . **47**

Yuehan Pang, Yaqi Hu and Jinsong Bao
Comparative Phosphoproteomic Analysis Reveals the Response of Starch Metabolism to High-Temperature Stress in Rice Endosperm
Reprinted from: *Int. J. Mol. Sci.* **2021**, *22*, 10546, doi:10.3390/ijms221910546 . . . . **71**

Orarat Ginsawaeng, Michal Gorka, Alexander Erban, Carolin Heise, Franziska Brueckner, Rainer Hoefgen, Joachim Kopka, Aleksandra Skirycz, Dirk K. Hincha and Ellen Zuther
Characterization of the Heat-Stable Proteome during Seed Germination in Arabidopsis with Special Focus on LEA Proteins
Reprinted from: *Int. J. Mol. Sci.* **2021**, *22*, 8172, doi:10.3390/ijms22158172 . . . . **95**

Yueyang Sun, Yu Zou, Jing Jin, Hao Chen, Zhiying Liu, Qinru Zi, Zeyang Xiong, Ying Wang, Qian Li, Jing Peng and Yi Ding
DIA-Based Quantitative Proteomics Reveals the Protein Regulatory Networks of Floral Thermogenesis in *Nelumbo nucifera*
Reprinted from: *Int. J. Mol. Sci.* **2021**, *22*, 8251, doi:10.3390/ijms22158251 - **123**

Bonoso San-Eufrasio, Ezequiel Darío Bigatton, Victor M. Guerrero-Sánchez, Palak Chaturvedi, Jesús V. Jorrín-Novo, María-Dolores Rey and María Ángeles Castillejo
Proteomics Data Analysis for the Identification of Proteins and Derived Proteotypic Peptides of Potential Use as Putative Drought Tolerance Markers for *Quercus ilex*
Reprinted from: *Int. J. Mol. Sci.* **2021**, *22*, 3191, doi:10.3390/ijms22063191 . . . . **143**

Setsuko Komatsu, Hisateru Yamaguchi, Keisuke Hitachi, Kunihiro Tsuchida, Yuhi Kono and Minoru Nishimura
Proteomic and Biochemical Analyses of the Mechanism of Tolerance in Mutant Soybean Responding to Flooding Stress
Reprinted from: *Int. J. Mol. Sci.* **2021**, *22*, 9046, doi:10.3390/ijms22169046 . . . . **161**

Junwei Zhang, Dongmiao Liu, Dong Zhu, Nannan Liu and Yueming Yan
Endoplasmic Reticulum Subproteome Analysis Reveals Underlying Defense Mechanisms of Wheat Seedling Leaves under Salt Stress
Reprinted from: *Int. J. Mol. Sci.* **2021**, *22*, 4840, doi:10.3390/ijms22094840 . . . . **175**

Xuemei Zhang, Bowen Tan, Dan Zhu, Daniel Dufresne, Tingbo Jiang and Sixue Chen
Proteomics of Homeobox7 Enhanced Salt Tolerance in *Mesembryanthemum crystallinum*
Reprinted from: *Int. J. Mol. Sci.* **2021**, *22*, 6390, doi:10.3390/ijms22126390 . . . . **191**

Kai-Ting Fan, Yang Hsu, Ching-Fang Yeh, Chi-Hsin Chang, Wei-Hung Chang and Yet-Ran Chen
Quantitative Proteomics Reveals the Dynamic Regulation of the Tomato Proteome in Response to *Phytophthora infestans*
Reprinted from: *Int. J. Mol. Sci.* **2021**, *22*, 4174, doi:10.3390/ijms22084174 . . . . **209**

Kibrom B. Abreha, Erik Alexandersson, Svante Resjö, Åsa Lankinen, Daniela Sueldo, Farnusch Kaschani, Markus Kaiser, Renier A. L. van der Hoorn, Fredrik Levander and Erik Andreasson
Leaf Apoplast of Field-Grown Potato Analyzed by Quantitative Proteomics and Activity-Based Protein Profiling
Reprinted from: *Int. J. Mol. Sci.* **2021**, *22*, 12033, doi:10.3390/ijms222112033 . . . . **237**

About the Editors

Sixue Chen (Professor of Biology and Director of Proteomics) completed his Ph.D. study in plant biochemistry in China, and postdoctoral research in Germany, Denmark and University of Pennsylvania, USA. He is a Professor in Department of Biology, and Director of Proteomics and Mass Spectrometry at Interdisciplinary Center for Biotechnology Research of University of Florida. Dr. Chen has established three major research projects: plant stomatal disease triangle, glucosinolate metabolism, and photosynthesis transition from C3 to CAM. His lab is known for application of proteomics and mass spectrometry technologies in elucidating plant molecular networks. He has been continuously funded by NSF, USDA, NIH and UF internal fund. In addition to his own research program, Dr. Chen has collaborated extensively with faculty at University of Florida and across the country. His research has led to more than 260 publications. Dr. Chen serves as Associate Editors and Board Members of Metabolomics, Frontiers in Plant Proteomics, International Journal of Molecular Sciences, Journal of Proteomics, Archives of Proteomics and Bioinformatics, and other journals.

Setsuko Komatsu (Professor of Biology) is a Professor at Fukui University of Technology, Japan from 2018. Before this position, she was Chief of Field Omics Research Unit at National Institute of Crop Science and Professor at University of Tsukuba, Japan. Since 1990, she started working on plant proteomics using protein sequencer and mass spectrometry. Her main research interests are within the field of crop proteomics, phytohormone, biochemistry, and molecular biology with a special focus on signal transduction in cells. She serves as Section Editor-in-Chief of "Molecular Plant Sciences" in International Journal of Molecular Sciences.

International Journal of
Molecular Sciences

MDPI

Editorial

Plant Proteomic Research 4.0: Frontiers in Stress Resilience

Sixue Chen [1,2,3,*] and Setsuko Komatsu [4,*]

1 Department of Biology, Genetics Institute, University of Florida, Gainesville, FL 32611, USA
2 Plant Molecular and Cellular Biology Program, University of Florida, Gainesville, FL 32610, USA
3 Proteomics and Mass Spectrometry, Interdisciplinary Center for Biotechnology Research, University of Florida, Gainesville, FL 32610, USA
4 Faculty of Environmental and Information Sciences, Fukui University of Technology, Fukui 910-8505, Japan
* Correspondence: schen@ufl.edu (S.C.); skomatsu@fukui-ut.ac.jp (S.K.)

Citation: Chen, S.; Komatsu, S. Plant Proteomic Research 4.0: Frontiers in Stress Resilience. *Int. J. Mol. Sci.* **2021**, *24*, 13362. https://doi.org/10.3390/ijms222413362

Received: 30 November 2021
Accepted: 5 December 2021
Published: 12 December 2021

Publisher's Note: MDPI stays neutral with regard to jurisdictional claims in published maps and institutional affiliations.

Large-scale high-throughput multi-omics technologies are indispensable components of systems biology in terms of discovering and defining parts of the system. Once the parts are functionally characterized, they become foundational for synthetic biology. Proteomics, as one of the essential omics in systems biology, provides the functional analysis of the expressed genome and generates detailed information that can be integrated with those obtained by other traditional and new omics approaches [1–3]. Systems biology and synthetic biology are the current frontiers of many areas of plant biology. As sessile organisms, plants are subject to varying environmental conditions, such as frequent changes in light exposure, humidity, temperature extremities, flooding, and other abiotic and biotic stresses, which can disturb and inhibit their growth and development [4,5]. These challenges to plants and crops are growing more severe as the effects of climate change become worse and increase the likelihood of inflicting extreme stresses on plants [4,6]. For example, the past seven years have been on track to be the warmest on record, according to the World Meteorological Organization (WMO) State of the Global Climate 2021 report [6]. In summer 2021, heatwaves in Canada and southwestern USA have pushed temperatures to almost 50 °C in a village in British Columbia and to 54.4 °C in California's Death Valley. Many parts of the Mediterranean also experienced record temperatures. Accompanying the exceptional heat, drought and natural fires devastated communities and agriculture production. How plants respond to these environmental stresses, how crop production will be affected by climate change, and what strategies will be effective to boost crop stress resilience are fundamental questions for the plant research community.

Over the past two decades, plant proteomics has provided a deep knowledge and understanding of the diverse proteoforms and different plant processes owing to significant advancements in concepts, technologies, approaches, and platforms in plant proteomics. Proteome directly reflects proteotype, which is the proteomic state that uniquely underlies a phenotype [7]. Thus, proteotype connects genotype to phenotype, and it is affected by development and the environment. A proteome contains all the proteoforms, i.e., all the protein species that are expressed and modified, not just the genome-centric proteome. It has become clear that plant proteomics has moved from simply cataloging to profiling posttranslationally modified proteins (i.e., proteoforms), which include proteins with phosphorylation, acetylation, methylation, redox modifications, ubiquitinylation, glycosylation, etc. It has also moved from a 2D-gel-based approach to a gel-free liquid chromatography (LC)-based approach. In addition to traditional data-dependent acquisition (DDA), data-independent acquisition (DIA) has shown great utility and power in potentially achieving high proteome coverage and depth [8–10]. Thus, the amalgamation of diverse analytical techniques, complemented with genome-sequence data, modern bioinformatics tools, and improved sample preparation and fractionation strategies, provides an unprecedented opportunity to characterize novel proteins/proteoforms in spatial and temporal resolution and under different environmental conditions. This special issue of Plant Proteomic

Research 4.0 captures the recent advancements in proteomics and addresses the current challenges of plant stress response and resilience in the ever-changing climate. It contains 12 articles, including 3 reviews and 9 original research articles.

The three reviews deal with pollen phosphoproteomics [11], starch biosynthesis-related proteins and posttranslational modifications (PTMs) in rice developing seeds [12], and PTMs of waxy proteins in rice grain [13]. Klodova and Fila [11] summarized phosphoproteomics studies in the past decade of male gametophyte developmental stages (mostly mature pollen grains) in different species (Arabidopsis, tobacco, maize, and kiwifruit). They highlighted common phosphoproteins in the different species and compared pollen phosphoproteomes with soybean root hair phosphoproteome to deduce shared mechanisms underlying the polarized tip growth of pollen tubes and root hairs. The next review by Tappiban et al. [12] provides a comprehensive analysis of proteins and PTMs involved in starch biosynthesis at different developmental stages of rice developing seeds. Several starch biosynthesis proteins were targeted by phosphorylation, acetylation, succinylation, lysine 2-hydroxyisobutyrylation, and malonylation. One enzyme, phosphoglucomutase, is commonly targeted by five different PTMs. The crosstalk and functions of these PTMs in starch biosynthesis are to be elucidated in future studies. The third review by Adegoke et al. [13] followed up with PTM of waxy protein (Wx), i.e., the granule-bound starch synthase, responsible for amylose biosynthesis. Wx is subject to phosphorylation modification, and a decrease in phosphorylation leads to low Wx activity and low amylose content. Employing PTM regulation of Wx may be effective in improving the yield of low-amylose rice to meet the market demand.

The nine research articles include three related to temperature, two on water stress, two on salt stress, one on fungal pathogens, and the last one on field-grown potato leaf apoplast proteome. Pang et al. [14] reported high-temperature stress on the phosphoproteomes of developing grains of two indica rice varieties with different starch qualities. The results revealed dynamic phosphorylation changes to starch biosynthesis enzymes in response to high-temperature stress, highlighting the importance of phosphorylation-mediated regulation in the biosynthesis of amylose and amylopectin, which determines the rice grain quality. Ginsawaeng et al. [15] employed both proteomics and gas chromatography–mass spectrometry-based metabolomics to reveal a heat-stable proteome and a highly connected late embryogenesis abundant (LEA) protein network during Arabidopsis seed germination. In the same vein of the heat-stable proteome, Sun et al. [10] are interested in understanding the proteomic basis of how sacred lotus maintains a stable floral chamber temperature under varied environmental temperatures between about 8 and 45 °C. They implemented a DIA-based quantitative proteomics approach to identify and quantify a total of 6913 proteins in lotus. A protein module highly related to the thermogenic phenotype was proposed to mainly involve metabolic processes, fatty acid degradation, and ubiquinone synthesis. Although high-temperature stress is different from drought stress, drought is often associated with high temperatures. San-Eufrasio et al. [16] aimed to identify proteins and proteotypic peptides as potential drought tolerance markers for holm oak using a label-free proteomics dataset of 4470 leaf proteins identified from oak seedlings from four different Andalusian populations in southern Spain. A total of 30 proteins and 46 derived peptides were selected as putative markers of drought tolerance in at least two populations. Among them, subtilisin and chaperone GrpE were increased under drought in three populations, thus they have a high potential of being the drought tolerance markers in oak trees. In addition to drought, climate change also brought irregular flooding [4,6]. Komatsu et al. [17] investigated molecular mechanisms of soybean flooding tolerance using gel-free/label-free proteomics with wild-type and flooding-tolerant mutants generated using gamma-ray irradiation. Although flooding stress affected proteins in both wild-type soybean and the mutant, those exhibiting opposite changes in abundance between the wild type and the mutant under flooding stress were interesting (e.g., calreticulin). Further analyses of alcohol dehydrogenase and glycoprotein profiles suggest that the mutant exhibits flooding tolerance through decreasing fermentation and enhancing glycoprotein folding, thereby minimizing cell death under flooding stress. Like the water stresses, salt stress poses a major negative impact on plant growth and agricultural production. Zhang et al. [18] analyzed endoplasmic reticulum (ER) proteomic changes in wheat leaves in response to salt stress using label-free proteomics. Of the 233 ER-localized differential proteins, salt stress significantly increased the levels of protein disulfide isomerase and heat shock proteins and decreased ribosomal proteins. Transcriptional regulation accounts for about half of the differential protein changes, highlighting the potential involvement of transcriptional factors in plant salt stress response. Zhang et al. [19] actually cloned a homeobox 7 (HB7) transcription factor important for salt tolerance of common ice plants. Overexpression of HB7 enhanced plant tolerance to 500 mM NaCl treatment. Label-free proteomics revealed that proteins increased in HB7-overexpression plants are involved in transport, catalytic activity, biosynthesis of specialized metabolites, and response to stimuli. Identifying the downstream targets of HB7 is an interesting direction towards understanding the mechanisms of HB7-mediated salt tolerance and potentially utilizing HB7 in marker-based crop breeding for improving salt tolerance. Tomatoes and potatoes are widely grown and consumed worldwide, especially in Europe, America, and Asian countries. However, late blight disease caused by *Phytophthora infestans* is a major challenge to the production of tomatoes and potatoes. Fan et al. [9] examined tomato leaf proteomic changes at different stages of *P. infestans* pathogenesis using DIA proteomics. A total of 6631 tomato proteins and 678 *P. infestans* proteins were profiled at three different time-points of pathogenesis. Tomato proteins regulated by *P. infestans* during different phases of pathogenesis have functions in immunity, signaling, defense, and metabolism. The results provide an important resource for developing new strategies towards controlling late blight disease. Currently, most of the studies on plant response to abiotic and/or biotic stresses have been conducted under controlled laboratory conditions. Whether the results obtained can be translated into field conditions in agricultural practice is not known. Abreha et al. [20] took a big stride by investigating apoplastic proteomic changes of potato growing at two field sites in June to August across two years. The plants were also divided into fungicide treated groups and untreated control groups. Although no significant differences were observed between fungicide treated and untreated control samples, they did observe many differential apoplastic proteins in response to environmental or developmental factors in the two growing seasons. Using activity-based protein profiling, the differential activities of serine hydrolases and β-glycosidases revealed the seasonal effects. The apoplastic proteomics data obtained from the field-grown plants not only lay an important foundation for understanding the physiological state of crops grown under complex environmental conditions, but also bring up an important question of the translationability of growth chamber/greenhouse experiments. This work paves a new direction for agriculture-focused proteomics research.

In summary, the articles collected in this special issue of Plant Proteomic Research 4.0 reflect the current frontiers of plant proteomics, focusing on development and environmental factors. An obvious theme is environmental stresses, including temperature [10,14,15], drought [16], flooding [17], salt stress [18,19], pathogen [9], and agriculture field conditions [20]. Another topic is proteoforms/PTMs that several articles have focused on [11–14,16,17]. In addition, many studies are on crop species, e.g., tomato [9], maize [11], rice [12–14], soybean [17], wheat [18], and potato [20]. Last but not least, 2D gel-based proteomics seems to have become obsolete, with LC-based proteomics taking its place. It is great to see the application of DIA proteomics [9,10]. It will inspire more and more scientists to use this fairly new technology in addition to traditional DDA proteomics. The guest editors hope that this special issue will provide readers with a framework for understanding the status of plant proteomics and insights into new technologies and directions of proteomics in the frontiers of systems biology and synthetic biology. Due to time and space limitations, there are still many exciting developments in plant proteomics, e.g., PTM crosstalks, protein complexes, multi-omics, and single-cell/single-cell type proteomics that were not fully covered or included here. For sure, these new developments will be the subject of future special issues on plant proteomic research. Finally, the guest editors would like to express gratitude to all the authors for their contributions and the reviewers for their critical assessments of these articles. Additionally, they also want to thank the Assistant Editor, Ms. Dani Wu, for the opportunity to serve as guest editors for "Plant Proteomic Research 4.0". They would highly appreciate your continuous support for future special issues.

Author Contributions: S.C. and S.K. have made substantial, direct, and intellectual contributions to the published articles and approved them for publication. S.C. drafted this editorial with editing from S.K. All authors have read and agreed to the published version of the manuscript.

Funding: This work was partly supported by the Agriculture and Food Research Initiative (AFRI) (Award No. 2020-67013-31615/accession no. 1022409) from the USDA National Institute of Food and Agriculture (S.C.), and the University of Florida (S.C.).

Institutional Review Board Statement: Not applicable.

Informed Consent Statement: Not applicable.

Data Availability Statement: Not applicable.

Acknowledgments: The authors thank Daniel Chen from the Judy Genshaft Honors College and College of Arts and Sciences at the University of South Florida for critical reading and editing of the manuscript. They also acknowledge Brianne' De Los Santos and Rosie Kereston from the University of Florida Genetics Institute for assistance with the cover image.

Conflicts of Interest: The authors have no conflict of interest to declare.

References

1. Chen, S.; Harmon, A.C. Advances in plant proteomics. *Proteomics* **2006**, *6*, 5504–5516. [CrossRef] [PubMed]
2. David, L.; Kang, J.; Dufresne, D.; Zhu, D.; Chen, S. Multi-omics revealed molecular mechanisms underlying guard cell systemic acquired resistance. *Int. J. Mol. Sci.* **2021**, *22*, 191. [CrossRef] [PubMed]
3. Kang, J.; David, L.; Cang, J.; Chen, S. Three-in-one simultaneous extraction of proteins, metabolites and lipids for multi-omics. *Front. Genet.* **2021**, *12*, 635971. [CrossRef] [PubMed]
4. Raza, A.; Razzaq, A.; Mehmood, S.; Zou, X.; Zhang, X.; Lv, Y.; Xu, J. Impact of climate change on crops adaptation and strategies to tackle its outcome: A Review. *Plants* **2019**, *8*, 34. [CrossRef] [PubMed]
5. Giordano, M.; Petropoulos, S.A.; Rouphael, Y. Response and defence mechanisms of vegetable crops against drought, heat and salinity stress. *Agriculture* **2021**, *11*, 463. [CrossRef]
6. World Meteorological Organization (WMO). State of the Global Climate 2021: Extreme Events and Major Impacts. Available online: https://library.wmo.int/doc_num.php?explnum_id=10859 (accessed on 4 December 2021).
7. Walley, J.W.; Shen, Z.; Sartor, R.; Wu, K.J.; Osborn, J.; Smith, L.G.; Briggs, S.P. Reconstruction of protein networks from an atlas of maize seed proteotypes. *Proc. Natl. Acad. Sci. USA* **2013**, *110*, E4808-17. [CrossRef] [PubMed]
8. Zhang, F.; Ge, W.; Ruan, G.; Cai, X.; Guo, T. Data-independent acquisition mass spectrometry-based proteomics and software tools: A glimpse in 2020. *Proteomics* **2020**, *17–18*, e1900276. [CrossRef] [PubMed]
9. Fan, K.-T.; Hsu, Y.; Yeh, C.-F.; Chang, C.-H.; Chang, W.-H.; Chen, Y.-R. Quantitative proteomics reveals the dynamic regulation of the tomato proteome in response to *Phytophthora infestans*. *Int. J. Mol. Sci.* **2021**, *22*, 4174. [CrossRef] [PubMed]
10. Sun, Y.; Zou, Y.; Jin, J.; Chen, H.; Liu, Z.; Zi, Q.; Xiong, Z.; Wang, Y.; Li, Q.; Peng, J.; et al. DIA-based quantitative proteomics reveals the protein regulatory networks of floral thermogenesis in *Nelumbo nucifera*. *Int. J. Mol. Sci.* **2021**, *22*, 8251. [CrossRef] [PubMed]
11. Klodova, B.; Fila, J. A decade of pollen phosphoproteomics. *Int. J. Mol. Sci.* **2021**, *22*, 12212. [CrossRef] [PubMed]
12. Tappiban, P.; Ying, Y.; Xu, F.; Bao, J. Proteomics and post-translational modifications of starch biosynthesis-related proteins in developing seeds of rice. *Int. J. Mol. Sci.* **2021**, *22*, 5901. [CrossRef] [PubMed]
13. Adegoke, T.V.; Wang, Y.; Chen, L.; Wang, H.; Liu, W.; Liu, X.; Cheng, Y.-C.; Tong, X.; Ying, J.; Zhang, J. Posttranslational modification of Waxy to genetically improve starch quality in rice grain. *Int. J. Mol. Sci.* **2021**, *22*, 4845. [CrossRef] [PubMed]
14. Pang, Y.; Hu, Y.; Bao, J. Comparative phosphoproteomic analysis reveals the response of starch metabolism to high-temperature stress in rice endosperm. *Int. J. Mol. Sci.* **2021**, *22*, 10546. [CrossRef] [PubMed]
15. Ginsawaeng, O.; Gorka, M.; Erban, A.; Heise, C.; Brueckner, F.; Hoefgen, R.; Kopka, J.; Skirycz, A.; Hincha, D.K.; Zuther, E. Characterization of the heat-stable proteome during seed germination in Arabidopsis with special focus on LEA proteins. *Int. J. Mol. Sci.* **2021**, *22*, 8172. [CrossRef] [PubMed]

16. San-Eufrasio, B.; Bigatton, E.D.; Guerrero-Sánchez, V.M.; Chaturvedi, P.; Jorrín-Novo, J.V.; Rey, M.-D.; Castillejo, M.Á. Proteomics data analysis for the identification of proteins and derived proteotypic peptides of potential use as putative drought tolerance markers for *Quercus ilex*. *Int. J. Mol. Sci.* **2021**, *22*, 3191. [CrossRef] [PubMed]
17. Komatsu, S.; Yamaguchi, H.; Hitachi, K.; Tsuchida, K.; Kono, Y.; Nishimura, M. Proteomic and biochemical analyses of the mechanism of tolerance in mutant soybean responding to flooding stress. *Int. J. Mol. Sci.* **2021**, *22*, 9046. [CrossRef] [PubMed]
18. Zhang, J.; Liu, D.; Zhu, D.; Liu, N.; Yan, Y. Endoplasmic reticulum subproteome analysis reveals underlying defense mechanisms of wheat seedling leaves under salt stress. *Int. J. Mol. Sci.* **2021**, *22*, 4840. [CrossRef] [PubMed]
19. Zhang, X.; Tan, B.; Zhu, D.; Dufresne, D.; Jiang, T.; Chen, S. Proteomics of a homeobox7 enhanced salt tolerance in *Mesembryanthemum crystallinum*. *Int. J. Mol. Sci.* **2021**, *22*, 6390. [CrossRef] [PubMed]
20. Abreha, K.B.; Alexandersson, E.; Resjö, S.; Lankinen, Å.; Sueldo, D.; Kaschani, F.; Kaiser, M.; van der Hoorn, R.A.L.; Levander, F.; Andreasson, E. Leaf apoplast of field-grown potato analyzed by quantitative proteomics and activity-based protein profiling. *Int. J. Mol. Sci.* **2021**, *22*, 12033. [CrossRef] [PubMed]

International Journal of
Molecular Sciences

MDPI

Review

A Decade of Pollen Phosphoproteomics

Božena Klodová [1,2] and Jan Fíla [1,*]

1 Laboratory of Pollen Biology, Institute of Experimental Botany of the Czech Academy of Sciences, Rozvojová 263, 165 02 Prague, Czech Republic; klodova@ueb.cas.cz
2 Department of Experimental Plant Biology, Faculty of Science, Charles University, Viničná 5, 128 00 Prague, Czech Republic
* Correspondence: fila@ueb.cas.cz; Tel.: +420-225-106-452

Abstract: Angiosperm mature pollen represents a quiescent stage with a desiccated cytoplasm surrounded by a tough cell wall, which is resistant to the suboptimal environmental conditions and carries the genetic information in an intact stage to the female gametophyte. Post pollination, pollen grains are rehydrated, activated, and a rapid pollen tube growth starts, which is accompanied by a notable metabolic activity, synthesis of novel proteins, and a mutual communication with female reproductive tissues. Several angiosperm species (*Arabidopsis thaliana*, tobacco, maize, and kiwifruit) were subjected to phosphoproteomic studies of their male gametophyte developmental stages, mostly mature pollen grains. The aim of this review is to compare the available phosphoproteomic studies and to highlight the common phosphoproteins and regulatory trends in the studied species. Moreover, the pollen phosphoproteome was compared with root hair phosphoproteome to pinpoint the common proteins taking part in their tip growth, which share the same cellular mechanisms.

Keywords: phosphoproteomics; pollen tube; male gametophyte; root hair; signal transduction; kinase motif

Citation: Klodová, B.; Fíla, J. A Decade of Pollen Phosphoproteomics. *Int. J. Mol. Sci.* **2021**, *22*, 12212. https://doi.org/10.3390/ijms222212212

Academic Editors: Sixue Chen and Setsuko Komatsu

Received: 30 September 2021
Accepted: 8 November 2021
Published: 11 November 2021

1. Introduction

Species' existence on Earth is maintained by reproduction. The angiosperm (Angiospermae) life cycle consists of two altering generations—a diploid sporophyte and a haploid gametophyte [1]. The adult plants form the sporophyte, in the flowers of which, the spores of two distinct sexes (female and male) and sizes are produced by meiosis. These heterospores undergo mitotic divisions, by which multicellular gametophytes are formed. The female gametophyte develops within the ovary, where it is protected from any damage and in most species, it is composed of seven cells with eight nuclei [2]. On the other hand, a mature male gametophyte is formed by 2 or 3 cells [3]. The microspores undergo the asymmetrical pollen mitosis I, which gives rise to two distinct cells. The smaller generative cell (composed mainly of a nucleus) is engulfed by the bigger vegetative cell. Mature pollen grains are shed from anthers either in such a bi-cellular stage or alternatively undergo pollen mitosis II that forms two sperm cells out of one generative cell prior to pollen grain shedding, meaning they will be in a mature state tri-cellular [4,5]. Mature pollen aims at delivering the genetic information in an intact state to the pistil and to fulfil this task, it represents a resistant, metabolically quiescent stage with a dehydrated cytoplasm surrounded by a tough cell wall. Upon pollination, the cytoplasm of pollen grains re-hydrates [6] and it becomes metabolically active and later, the rapid pollen tube growth starts. The pollen tube growth through transmitting tissues of a pistil is accompanied by intensive communication between these structures [7]. Pollen mitosis II, that forms two sperm cells out of one generative cell, takes place in the bicellular pollen after pollination, for instance, in the case of tobacco after 10–12 h of pollen tube growth [8]. Finally, the pollen tube delivers two sperm cells, the male gametes, to the mature embryo sac. Both carried sperm cells take part in fertilization. One sperm cell fertilizes the egg cell (representing female gamete) to form the zygote and later the embryo, whereas the second sperm cell fuses with the central nucleus of the embryo sac to form endosperm. Such a phenomenon is called double fertilization and is typical of angiosperms [9].

The change from metabolically quiescent, resistant mature pollen to a metabolically active, rapidly growing pollen tube is precisely regulated both at the level of protein synthesis and posttranslational modifications. The former regulation is mediated by the synthesis of mRNAs for storage in translationally inactive EDTA/puromycine-resistant particles (EPPs [10,11]), later described as monosomes [12], since, for instance, tobacco (*Nicotiana tabacum*) pollen tube growth was reported to be highly dependent on translation, but nearly independent of transcription [13]. The stored transcripts are de-repressed once the rapid pollen tube growth starts. Then, the post-translational modifications during pollen tube growth are most importantly represented by phosphorylation, which represents one of the most dynamic posttranslational modifications that mediates the regulation of numerous cellular processes. A similar re-hydration-related phosphorylation was described in xerophyte *Craterostigma plantagineum* [14,15]. The other post-translational modifications (namely glycosylation, methylation, myristoylation or acetylation) were also reported to play an important role in male gametophyte development [16]. Glycoproteins are on the one hand an important structural part of pollen tube cell walls and on the other hand play their roles in pollen tube perception [17].

This review aims at an analysis of the known phosphoproteomic datasets acquired on male gametophyte stages and compares them with the root hair phosphoproteome, since these structures share the same type of tip growth that relies on common cellular mechanisms [18–20].

2. Male Gametophyte Phosphoproteomic Studies

Various enrichment protocols were applied [21,22] to study protein phosphorylation on a large scale by phosphoproteomic techniques. The enrichment techniques are inevitable since (1) only several percent of cellular proteome are phosphorylated in a cell at a given time; (2) both phosphorylated and native isoforms of the same protein co-exist in the cell [23], sometimes even with a much higher concentration of their non-phosphorylated forms; (3) phosphorylated peptides are hardly detected in a positive ion scan mode during mass spectrometry if they are mixed with their non-phosphorylated counterparts [24].

The first phosphoproteomic dataset acquired from any angiosperm male gametophyte stage was represented by Arabidopsis *(Arabidopsis thaliana)* mature pollen, which was published nearly 10 years ago by Mayank and colleagues [25]. The first phosphoproteomic study relied on three phosphopeptide-enriching methods (immobilized metal affinity chromatography—IMAC, metal oxide affinity chromatography—MOAC, and sequential elution from IMAC—SIMAC) and collectively identified 962 phosphopeptides carrying 609 phosphorylation sites, which belonged to 598 phosphoproteins (Table 1). The total number of identified phosphopeptides could be higher than the number of identified phosphorylation sites. This is caused by the fact that the same phosphorylation site is carried by more than one phosphopeptide. Alternatively, some authors also calculate phosphopeptides, which lack the conclusively positioned phosphorylation site due to the insufficient support from the MS/MS spectra. In Arabidopsis pollen phosphoproteome, there prevailed proteins annotated by TopGO [26] which were involved in the regulation of metabolism and protein function, metabolism, protein fate, protein with a binding function, signal transduction mechanisms, and cellular transport. It is worth mentioning that various protein kinases (including AGC protein kinases, calcium-dependent protein kinases, and sucrose non-fermenting protein kinases 1) were amongst the identified phosphopeptides. Two over-represented phosphorylation motifs in the Arabidopsis pollen phosphoproteome were identified—a prolyl-directed motif xxxxxxS*Pxxxxx, and a basic motif xxxRxxS*xxxxxx (the phosphorylation site here and onwards is represented by an asterisk).

Table 1. Summary of the publications that presented angiosperm male gametophyte phosphoproteomes.

Species	Citation	Enrichment Technique	Studied Stages		Number of Identified Phosphoproteins	Number of Identified Phosphopeptides	Number of Identified Phosphorylation Sites	pSer:pThr: pTyr Ratio	Phosphorylation Motifs
			Mature Pollen	Activated Pollen					
Arabidopsis thaliana	Mayank et al. 2012 [25]	IMAC, TiO_2–MOAC, SIMAC	×		598	962	609	86:14:0.16	1 prolyl-directed (xxxxxxS*Pxxxxx) 1 basic (xxxRxxS*xxxxxx)
Nicotiana tabacum	Fíla et al. 2012 [27]	$Al(OH)_3$–MOAC, TiO_2–MOAC of the already identified peptides	×	×	139	52	52	67.3:32.7:0	not identified, too small data set
Nicotiana tabacum	Fíla et al. 2016 [28]	TiO_2–MOAC	×	×	301	471	432	86.4:13:4:0.2	2 prolyl-directed (xxxxxxS*Pxxxxx; xxxxxxT*Pxxxxx) 2 basic (xxxRxxS*xxxxxx; xxxKxxS*xxxxxx) 2 acidic (xxxxxxS*DxExxx; xxxxxxS*xDDxxx)
Zea mays	Chao et al. 2016 [29]	IMAC	×		2257	4638	5292	81.5:14.5:4	8 prolyl-directed 5 basic 4 acidic 10 other
Actinidia deliciosa	Vannini et al. 2019 [30]	MOAC phosphoprotein enrichment + IMAC–Ti phosphopeptide enrichment		×	711	1299	1572	90.3:9:0.7	6 prolyl-directed 5 basic 8 acidic 20 other

The second angiosperm species that was subjected to male gametophyte phosphoproteomic studies was tobacco (*Nicotiana tabacum*) [27,28]. Tobacco became the first species in which the activated pollen grains were taken into consideration, since it identified phosphoproteins from mature pollen, 30-min activated pollen [27,28], and in the more recent study also from 5-min activated pollen [28]. The former study relied on phosphoprotein enrichment by aluminium hydroxide matrix, the eluate of which was separated both by a conventional two-dimensional gel electrophoresis (2D–GE), and by nano liquid chromatography (nLC) [27]. Although 139 phosphoprotein candidates were identified, the number of exactly matched phosphorylation sites was lower, since it identified only 52 phosphorylation sites (Table 1). The number of phosphorylation sites identified in the tobacco male gametophyte was notably broadened in the second study that applied phosphopeptide enrichment by titanium dioxide to identify phosphopeptides from mature pollen, 5-min activated pollen, and 30-min activated pollen [28]. The study described 301 phosphoproteins, which contained 471 phosphopeptides that carried 432 exactly matched phosphorylation sites (Table 1). Furthermore, several regulated phosphopeptides that changed their abundance between the studied stages were identified. There were seven such categories, including phosphopeptides present exclusively in either studied stage. Like in Arabidopsis, the most abundant functional categories were represented by protein synthesis, together with protein destination and storage, transcription, and signal transduction. The motif search in the second phosphoproteomic study revealed five motifs with a central phosphoserine and one motif with a central phosphothreonine. There were prolyl-directed phosphorylations on both serine and threonine (xxxxxxS*Pxxxxx, and xxxxxxT*Pxxxxx), two basic motifs (xxxRxxS*xxxxxx, and xxxKxxS*xxxxxx), and two acidic motifs (xxxxxxS*DxExxx, and xxxxxxS*xDDxxx).

In 2016, the first monocot, represented by maize (*Zea mays*) mature pollen [29], was subjected to phosphoproteomic studies, but no activated stage of male gametophyte was studied. This study relied solely on gel-free techniques combined with IMAC phosphopeptide enrichment. It led to the identification of 4638 phosphopeptides in 2257 proteins that carried 5292 phosphorylation sites (Table 1). The number of phosphorylation sites identified is roughly 10 times higher, whereas the number of phosphopeptides is approximately 5−10 times higher than in Arabidopsis or tobacco pollen phosphoproteomes. The increase could be caused (1) in case of *A. thaliana* by technical improvements after a few years (Arabidopsis phosphoproteome was published 4 years before), and (2) compared to tobacco, maize represents a sequenced plant with an annotated genome [31,32]. It is likely that several tobacco MS spectra were not coupled with any sequence from the available databases since the tobacco genome was not fully annotated when the analyses were performed [33], and although the annotations were improved then, they are still far from completion [34]. Chao et al. (2016) were notably more successful in identifying the phosphorylation motifs over-represented in the presented phosphoproteome—the dataset comprised of 23 phosphoserine motifs and 4 phosphothreonine motifs, representing a total of 27 motifs. There were 8 prolyl-directed motifs, 5 basic motifs, and 4 acidic motifs, which usually represented more specified versions of the above tobacco and Arabidopsis motifs. However, there appeared also a variety of 10 newly discovered motifs. The phosphoprotein categories in maize phosphoproteome were represented by DNA synthesis/chromatin structure, transcription regulation, protein modification, cell organization, signal transduction, cell cycle, vesicular transport, transport of ions and various metabolic pathways. It is worth mentioning that Chao et al. found 430 protein kinases and 105 phosphatases. Some kinases represented the families, the phosphorylation motifs of which were up-regulated in the present phosphoproteome—for instance, calcium-dependent protein kinases (CDPK), leucine rich repeat kinases (LRRK), SNF1-related protein kinases (SnRK), and mitogen-activated protein kinases (MAPK). Finally, Chao et al. (2016) clearly demonstrated that the enrichment techniques are inevitable for studying protein phosphorylation by high-throughput methods. There were 5146 total proteins without phosphorylation in the maize pollen proteome, 1604 proteins in both datasets (total proteome and phosphoproteome), and then an additional 653 phosphoproteins were identified exclusively upon phosphopeptide enrichment. It is obvious that quite a big part of the phosphoproteome would remain undetectable in case the enrichment was not carried out at all.

The last large-scale phosphoproteomic dataset published was that of kiwifruit (*Actinidia deliciosa*) [30]. However, this study did not aim at the identification of developmentally related phosphopeptides under normal conditions, but rather at the identification of phosphorylation regulation upon inhibition by MG132. The peptide aldehyde MG132, also named *N*-Benzyloxycarbonyl-L-leucyl-L-leucyl-L-leucinal, represents a proteasome inhibitor [35]. Nevertheless, the crosstalk between protein phosphorylation and degradation in the male gametophyte was described by high-throughput methods the first time. Collectively, 1299 unique phosphopeptides from 711 phosphoproteins were identified, which carried 1572 phosphorylation sites (Table 1). They took part in protein metabolism, RNA and DNA processing, signalling and development. Moreover, many of these phosphoproteins had their homologues in *A. thaliana* and many of them were either annotated in the phosphoproteomic databases or were homologous to Mayank's *A. thaliana* pollen phosphoproteome [25]. However, several candidates were identified in pollen grains newly, a role which might be related to the proteasome inhibition. In general, MG132 treatment caused notable changes in protein phosphorylation, but not in overall protein expression, by which it pinpointed the importance of post-translational modifications for the regulation rather than the synthesis of novel proteins.

This review article has mainly focused on angiosperms. However, Chen et al. (2012) conducted a study investigating the pollen proteome of *Picea wilsonii*, the first gymnosperm to be analysed by a phosphoproteomic approach [36]. Like kiwifruit pollen phosphoproteome, it did not aim at developmental phosphoproteomics since it studied phosphoproteins related to pollen tube growth on media with low sucrose and calcium ion concentration and as such represented the study of phosphorylation upon various stresses.

3. Common Phosphoproteins in Angiosperm Male Gametophyte Phosphoproteomes

We compared angiosperm mature pollen phosphoproteomes together (Arabidopsis [25], tobacco [28], and maize [29]) to find the common regulatory trends in male gametophytes of these species (Supplementary Table S1). We did not include kiwifruit pollen in these analyses since it represented a different dataset—activated pollen that was influenced by the addition of dimethyl sulfoxide (DMSO) in case of the negative control or even by MG132-mediated proteasome inhibition [30].

First, we compared Arabidopsis and maize pollen phosphoproteomes. As was mentioned above, the Arabidopsis mature pollen phosphoproteome presented 598 phosphoproteins, whereas in maize pollen, there were identified 2257 phosphoproteins. The maize GRMZM identifiers of genome assembly R73_RefGEN_v3 were converted to Zm identifiers with MaizeMine v 1.3, and the homologue search between maize and Arabidopsis were executed by the engines on the same webpage. The comparison of maize AGI homologues with Arabidopsis pollen phosphoproteome resulted in 323 unique identifiers (527 in total, Table S1). To unravel the biological significance of these phosphoproteins, we carried out enrichment analyses for gene ontology terms and KEGG pathway (Figure 1A). The list included 54 transport proteins, mainly taking part in vesicular transport of all three types—COP1, COP2, and clathrin-coated vesicles. Besides these, there appeared also proteins playing their roles in endocytosis and vesicle movement on actin filaments. Then, in connection with pollen desiccation, 13 genes related to salt stress were present in both datasets. Both phosphoproteomes also shared proteins responsible for pollen tube growth and 14 proteins seem to possess a double function, since they were annotated with functions in root development. It is likely that these candidates are common to root hairs and pollen tubes since these tissues share the same mechanisms of tip growth. Moreover, 26 proteins responsible for protein phosphorylation were present. Amongst them, there were 3 mitogen-activated protein kinases—MAPK (At1g18150, At1g73670, and At3g07980), and a cyclin-dependent protein kinase CDK (At4g28980)—that recognize the prolyl-directed phosphorylation motifs (xxxxxxS*Pxxxxx, and xxxxxxT*Pxxxxx) [37]. Then there were 2 casein kinases—CK (At4g26100, and At5g57015) that target the acidic motifs xxxxxxS*DxExxx, and xxxxxxS*xDDxxx [37]. Finally, the basic motifs (xxxRxxS*xxxxxx, and xxxKxxS*xxxxxx) [37] were recognized by SNF1-related protein kinases—SnRK (At1g09020, At3g01090, and At3g29160), and Ca^{2+}-dependent protein kinases—CDPK (At1g35670, and At4g09570).

All mentioned kinase families were reported to play important roles during pollen tube growth [38]. The CDKs appear in the phosphoproteomic datasets since they are required for cell divisions that are part of male gametophyte development [39] and for their activity, they require to be phosphorylated by CDK-activating kinases [40]. Then, they regulate pre-mRNA splicing of callose synthase in pollen tubes to control the formation of a cell wall [41]. SnRKs were already reported to play a key role in pollen germination, where its mutation resulted in the compromised pollen hydration on the stigma [42]. Moreover, the SnRK-mediated phosphorylation is involved in communication by reactive oxygen species [43]. Then, CPK11 and CPK24 were involved in Ca^{2+}-dependent regulation of the K^+ channels [44] and CPK6 was reported to phosphorylate actin depolymerizing factor 1, by which the dynamics of actin filaments are regulated [45].

All motifs recognized by the mentioned kinase families usually appeared as overrepresented in pollen phosphoproteomes. To test whether the phosphorylation sites in kinases are conserved between Arabidopsis and maize pollen phosphoproteomes, we compared the exact positions of phosphorylation sites in these datasets together. There was one common phosphorylation site, particularly VSFNDTPSAIFWT*DYVATR in mitogen-activated protein kinase 8 (At1g18150, and its maize homologue GRMZM2G062761). Then, several other phosphopeptides carry most likely the conserved phosphorylation site, but the phosphorylation position in the Arabidopsis dataset was unfortunately not identified conclusively. However, these proteins share at least the peptide sequence with maize pollen phosphoproteome: serine/threonine-protein kinase SRK2A (At1g10940), serine/threonine-protein kinase SRK2G (At5g08590), serine/threonine-protein kinase SRK2H (At5g63650), SNF1-related protein kinase catalytic subunit α KIN10 (At3g01090), SNF1-related protein kinase catalytic subunit α KIN11 (At3g29160), and Shaggy-related protein kinase iota (At1g06390). Collectively, most kinases with conserved phosphopeptides between maize and Arabidopsis pollen phosphoproteomes were represented by the kinases, with known phosphorylation motifs in the phosphoproteomic datasets.

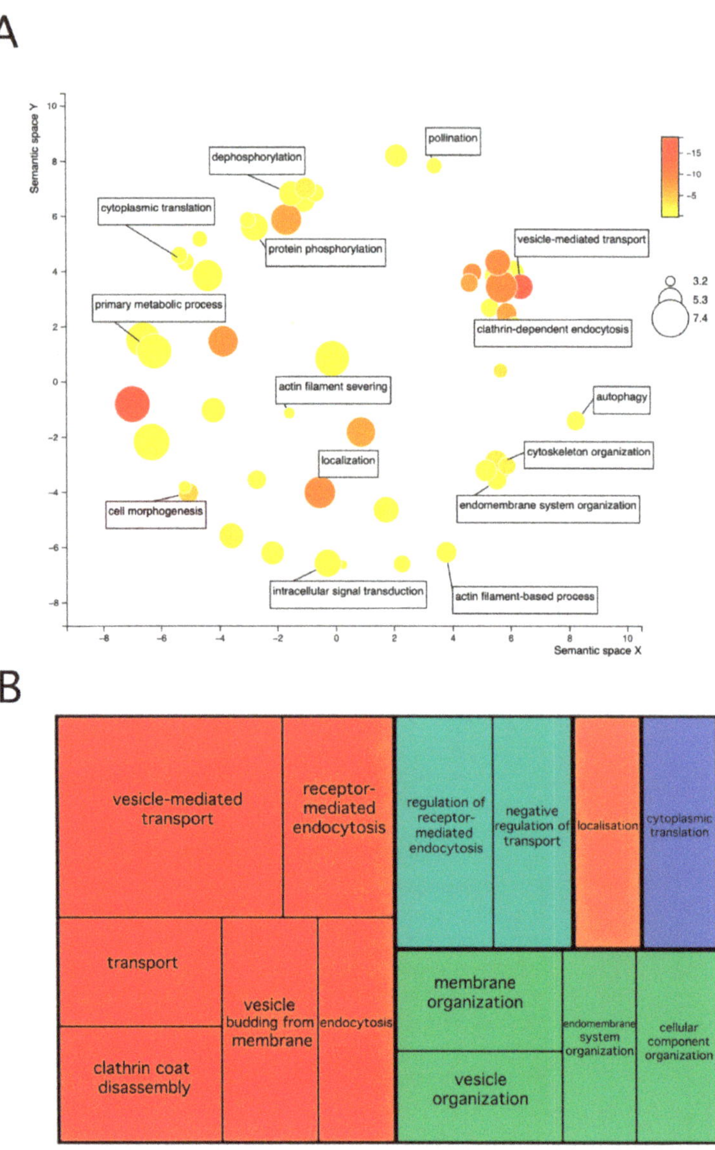

Figure 1. Comparison of pollen phosphoproteomes. (**A**)—GO biological processes enrichment analysis of phosphoproteins common to Arabidopsis and maize. The colours represent the false discovery rate of the enriched term, and the size of the circle represents the relative size of the GO term. (**B**)—A Treemap of enriched GO biological processes among phosphoproteins present in all three pollen samples (Arabidopsis, maize, and tobacco). The plots in (**A**,**B**) were rendered by Revigo [46].

Among the enriched molecular processes, the phosphoproteins shared between Arabidopsis and maize were divided into the following groups: 141 proteins had a binding capacity and were further distinguished as RNA binding, which included, for example, 9 translation initiation factors, cytoskeletal protein binding, phosphatidyl inositol binding or AMP binding. The second distinct group consisted of protein kinases and kinase activators. Finally, three proteins were annotated to localize into the polarized growth, namely *KINKY POLLEN* (At5g49680), putative clathrin assembly protein (At1g03050), and receptor-like kinase lost in pollen tube guidance—LIP1 (At5g16500). These regulatory proteins represent conserved candidates with an important role for pollen tube growth and guidance. The protein *KINKY POLLEN* was reported to play its role in vesicular transport both in pollen tubes and in root hairs, so its mutations led to an aberrant pollen tube [47]. Then, the LIP1 receptor-like kinase was important for pollen tube guidance [48].

The homologues to *Nicotiana tabacum* sequences were retrieved with an NCBI command line blastn tool, with dc-megablast task [49]. The top hits were used for further analysis resulting in a list of 170 unique AGI identifiers. Of these, 55 phosphoproteins were shared exclusively with maize, 21 were common exclusively with Arabidopsis, and 31 were shared with both these datasets, leaving aside 63 unique unshared phosphopeptides (Table S1). The enrichment of biological functions and molecular processes in GO term analysis was similar to the other studied species. The phosphoproteins were involved in endocytosis and vesicular transport. Furthermore, translation and mRNA processing represented the enriched processes, which may correspond to the activated state of pollen and preparation for pollen tube burst. As for the molecular function, 58 proteins were involved in protein binding, from which 30 candidates were annotated as RNA binding. Finally, three proteins, with their functions in chromatin structure, were present.

Considering the overlaps in the pollen datasets, maize and Arabidopsis pollen phosphoproteomes showed a much higher similarity to each other than to the tobacco dataset. This could have been caused by (1) lower total number of phosphoproteins in the tobacco dataset; (2) the type of tissue, since tobacco studies also included 5-min activated and 30-min activated pollen; and (3) pollen type, since both maize and Arabidopsis share tri-cellular pollen, whereas tobacco sheds bi-cellular pollen [5]. Nevertheless, 31 phosphoprotein homologues were present in all 3 datasets. These common proteins include mostly candidates taking part in vesicular transport, suggesting that they represent the basic conserved mechanisms which are important for pollen development (Figure 1B).

To conclude, the phosphoproteins shared at least by some species belonged to similar functional categories. There was usually at least some of the categories typical for tip growth—small GTPase signalling, ion gradient formation, cytoskeleton organization together with vesicular transport [18–20]. Then, the genes, which take part in regulatory mechanisms in protein synthesis, were also amongst the abundant functional categories. Overall, the phosphorylation of specific protein involved mainly in pollen tube growth seems to be conserved in the plant's evolution.

4. Common Trends for Male Gametophyte and Root Hairs

After comparing the three male gametophyte phosphoproteomes (Arabidopsis, tobacco, and maize) together and pinpointing common phosphoproteins, the same male gametophyte datasets were compared with root hair phosphoproteome (Figure 2A, Supplementary Table S2). Root hairs and pollen tubes share the same type of growth—tip growth—and due to this, they rely on the common regulatory mechanisms, such as small GTPase signalling, ion gradient formation, cytoskeleton organization together with regulations of vesicular transport, reactive oxygen species (ROS) signalling and a massive decrease in pH [18–20,50]. The only available root hairs phosphoproteome belongs to soybean (*Glycine max*), the roots of which were studied with respect to the nodule formation that accommodate the symbiotic nitrogen-fixing bacteria, typical for leguminous plants [51]. They presented both root hair phosphoproteome and the phosphoproteome of the corresponding shaved roots (i.e., roots with removed root hairs). Collectively, the study led to the identification of 1625 phosphopeptides carrying 1659 phosphorylation sites, which belonged to 1126 phosphoproteins. These phosphoproteins were assigned to the following functional categories: DNA/RNA-related proteins, signal transduction, miscellaneous group (proteins with multiple functions), and protein trafficking.

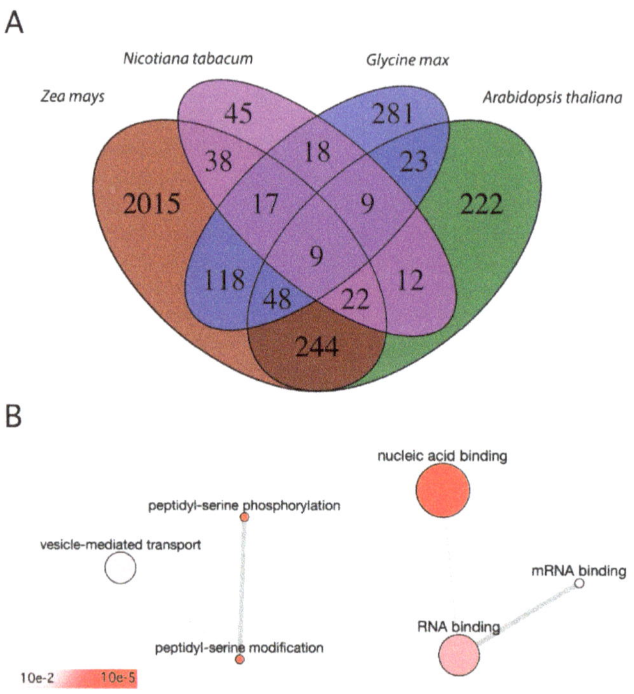

Figure 2. Comparison of pollen phosphoproteomes with root hair phosphoproteome. (**A**)—Venn diagram shows the overlap of several Arabidopsis homologue phosphoproteins discovered in pollen samples of Arabidopsis *(Arabidopsis thaliana)*, maize (*Zea mays*), and tobacco (*Nicotiana tabacum*) together with soybean (*Glycine max*) root hair phosphoproteome. (**B**)—The biological processes (left) and molecular function (right) GO terms of phosphoproteins shared between the Arabidopsis pollen phosphoproteome and soybean root hair phosphoproteome. The colour represents false discovery rate of the enriched term. The plots were rendered by Revigo [46].

To establish the shared regulatory pathways between polarized tip growth of pollen tubes and root hairs, we compared Arabidopsis mature pollen phosphoproteome (since its genome is from the studied species annotated best [52]) with soybean root hair phosphoproteome. Only phosphopeptides that were present in the root hairs were used for further comparative analyses since phosphopeptides identified solely in the shaved roots were removed for being irrelevant to root hairs. There were 825 annotated phosphopeptides present in the root hairs, and the Arabidopsis homologues were retrieved from the Phytozome database [53] using the Wm82.a4.v1 as a reference genome [54,55]. In total, 254 proteins (represented by 89 unique AGI identifiers) were shared between Arabidopsis pollen phosphoproteome and soybean root hair phosphoproteome (Table S2). These included proteins taking part in peptidyl-serine phosphorylation (5 proteins) and vesicle-mediated transport (11 proteins). These 5 kinases were represented by calcium-dependent protein kinase 4 (At4g09570), calcium-dependent protein kinase 11 (At1g35670), 3-phosphoinositide-dependent protein kinase 1 (At5g04510), 3-phosphoinositide-dependent protein kinase 2 (At3g10540), and casein kinase 1-like protein 1 (At4g26100). As mentioned above, CPK11 was involved in the Ca^{2+}-dependent regulation of the K^{+} channels [44], whereas it inhibited (together with CPK4) the root growth by phosphorylation of 1-aminocyclopropane-1-carboxylate synthase, by which its activity during ethylene synthesis was increased [56]. The 3-phosphoinositide-dependent protein kinase 1 was important in several physiological processes where cell proliferation and growth are of key importance [57], and it was proven to be the regulator of AGC1 kinases [58]. The casein kinase 1-like protein 1 regulated cell division by the phosphorylation of Kip-related protein 6 [59] and their other activities throughout plant development were reviewed recently [60]. Then, if biological function was considered, 36 proteins were reported to show the binding capacity (Figure 2B).

Most of these shared proteins were common also to maize pollen phosphoproteome (57 out of 89) or tobacco phosphoproteome (18 out of 89). Collectively, there were 9 common phosphoproteins for all compared datasets (including tobacco, maize and Arabidopsis pollen phosphoproteomes, and soybean root hairs phosphoproteome). These proteins had the following AGI identifiers: At1g11360, At1g20760, At1g21630, At1g59610, At5g57870, At1g20110, At5g41950, At5g52200, and At4g35890. Six of these proteins were functionally annotated; there were candidates working in RNA metabolism (La-related protein 1), translation initiation factor 4G-1, protein phosphatase inhibitor 2, dynamin 2B, and proteins FREE1, and HLB1. In summary, the phosphorylation of tip growth regulators seems to be partially conserved between pollen tip growth and root hair tip growth. However, these modifications can probably maintain a different role in each tissue. In future contexts, it may prove interesting to repeat this comparison within one species.

5. Beyond Pollen Phosphoproteomics

In the previous sections, we considered the published male gametophyte phosphoproteomes. However, it should be mentioned that also whole anthers were subjected to phosphoproteomic studies. Although mature pollen grains are part of anther samples, the surrounding sporophyte tissues usually dominate. Ye and colleagues identified the proteome and phosphoproteome of *Arabidopsis thaliana* anthers [61]. In total, they identified 3908 phosphorylation sites on 1637 phosphoproteins. Amongst these 1637 phosphoproteins, there appeared 493 newly identified ones, whereas the others were already deposited to the public phosphoproteomic database and/or were identified in Mayank's mature pollen phosphoproteome [25]. The other species with known anther phosphoproteome were represented by kenaf [?] and rice [61]. Unfortunately, there are not any pollen phosphoproteomic datasets for these species, so a direct comparison of mature pollen and anther phosphoproteomes is not currently possible.

6. Conclusions

The studies of protein phosphorylation in angiosperm male gametophyte initiated in 2012 by *Arabidopsis thaliana* mature pollen phosphoproteome. After almost 10 years, there appeared more studies, namely on tobacco, maize, and kiwifruit. However, the kiwifruit study was performed with respect to proteasome inhibition by MG132, but not to pollen development under normal conditions. The only activated pollen phosphoproteome is represented so far by the dataset from tobacco. For the future, the activated pollen of more species should be studied and compared to mature pollen since the phosphorylation dynamics is the most interesting aspect of their post-translational modifications.

The comparison of mature pollen phosphoproteomes between different angiosperm species revealed that the common phosphoproteins played their role in the vitally important processes for pollen tube growth—vesicular transport, metabolism, protein phosphorylation, and cytoskeleton dynamics. It seems that the basic cellular processes are conserved even between monocots and dicots, but the number of available datasets remains limited. For the future, the data acquired on more species should enable the comparison of mature pollen from both monocots and dicots with both bicellular and tricellular pollen (recently reviewed in [?]). Such a comparison will most likely highlight the pollen mitosis II-related kinases and other regulatory proteins.

After the decade of pollen phosphoproteomics, the research is surely not finished and deserves our future interest, especially on the emphasis of activated pollen.

Supplementary Materials: The following are available online at https://www.mdpi.com/article/10.3390/ijms222212212/s1. Supplementary Table S1—The comparison of pollen phosphoproteomes from maize (*Zea mays*) [29], tobacco (*Nicotiana tabacum*) [28], and *Arabidopsis thaliana* [25]. Supplementary Table S2—The comparison of pollen phosphoproteomes (from maize [29], tobacco [28], and Arabidopsis [25]) with soybean root hair phosphoproteome [51].

Author Contributions: Conceptualization, J.F.; data analysis, B.K.; writing—original draft preparation, B.K. and J.F.; writing—review and editing, J.F. All authors have read and agreed to the published version of the manuscript.

Funding: This research was funded by the Czech Ministry of Education, Youth and Sports [LTC20050] and the Czech Science Foundation [19-01723S].

Institutional Review Board Statement: Not applicable.

Informed Consent Statement: Not applicable.

Data Availability Statement: Not applicable.

Acknowledgments: We acknowledge David Honys for his valuable comments during revision of the manuscript.

Conflicts of Interest: The authors declare no conflict of interest. The funders had no role in the design of the study; in the collection, analyses, or interpretation of data; in the writing of the manuscript, or in the decision to publish the results.

Abbreviations

2D-GE	two-dimensional gel electrophoresis
AMP	adenosine monophosphate
CDK	cyclin-dependent protein kinase
CDPK	calcium-dependent protein kinase
CK	casein kinase
DMSO	dimethyl sulfoxide
EDTA	2,2′,2′′,2′′′-(Ethane-1,2-diyldinitrilo)tetraacetic acid
EPP	EDTA/puromycine-resistant particles
GO	gene ontology
IMAC	immobilized metal affinity chromatography
LRRK	leucine rich repeat kinase
MAPK	mitogen-activated protein kinase
MOAC	metal oxide affinity chromatography
MS	mass spectrometry
MS/MS	tandem mass spectrometry
NCBI	National Center for Biotechnology Information
nLC	nano liquid chromatography
ROS	reactive oxygen species
SIMAC	sequential elution from IMAC
SnRK	SNF1-related protein kinase

References

1. Friedman, W.E.; Floyd, S.K. Perspective: The origin of flowering plants and their reproductive biology—A tale of two phylogenies. *Evolution* **2001**, *55*, 217–231. [PubMed]
2. Christensen, C.A.; Subramanian, S.; Drews, G.N. Identification of gametophytic mutations affecting female gametophyte development in Arabidopsis. *Dev. Biol.* **1998**, *202*, 136–151. [CrossRef] [PubMed]
3. Borg, M.; Twell, D. Life after meiosis: Patterning the angiosperm male gametophyte. *Biochem. Soc. T.* **2010**, *38*, 577–582. [CrossRef] [PubMed]
4. Williams, J.H.; Taylor, M.L.; O'Meara, B.C. Repeated evolution of tricellular (and bicellular) pollen. *Am. J. Bot.* **2014**, *101*, 559–571. [CrossRef]

5. Brewbaker, J.L. Distribution and phylogenetic significance of binucleate and trinucleate pollen grains in angiosperms. *Am. J. Bot.* **1967**, *54*, 1069–1083. [CrossRef]
6. Vogler, F.; Konrad, S.S.A.; Sprunck, S. Knockin' on pollen's door: Live cell imaging of early polarization events in germinating Arabidopsis pollen. *Front. Plant Sci.* **2015**, *6*, 246. [CrossRef]
7. Hafidh, S.; Potěšil, D.; Fíla, J.; Feciková, J.; Čapková, V.; Zdráhal, Z.; Honys, D. In search of ligands and receptors of the pollen tube: The missing link in pollen tube perception. *Biochem. Soc. Trans.* **2014**, *42*, 388–394. [CrossRef]
8. Hafidh, S.; Breznenová, K.; Růžička, P.; Feciková, J.; Čapková, V.; Honys, D. Comprehensive analysis of tobacco pollen transcriptome unveils common pathways in polar cell expansion and underlying heterochronic shift during spermatogenesis. *BMC Plant Biol.* **2012**, *12*, 24. [CrossRef]
9. Raghavan, V. Some reflections on double fertilization, from its discovery to the present. *New Phytol.* **2003**, *159*, 565–583. [CrossRef]
10. Honys, D.; Reňák, D.; Feciková, J.; Jedelský, P.L.; Nebesářová, J.; Dobrev, P.; Čapková, V. Cytoskeleton-associated large RNP complexes in tobacco male gametophyte (EPPs) are associated with ribosomes and are involved in protein synthesis, processing, and localization. *J. Proteome Res.* **2009**, *8*, 2015–2031. [CrossRef]
11. Honys, D.; Čapková, V. Temporal changes in the RNA distribution between polysomes and postpolysomal ribonucleoprotein particles in tobacco male gametophyte. *Biol. Plant.* **2000**, *43*, 517–522. [CrossRef]
12. Hafidh, S.; Potěšil, D.; Müller, K.; Fíla, J.; Michailidis, C.; Herrmannová, A.; Feciková, J.; Ischebeck, T.; Valášek, L.S.; Zdráhal, Z.; et al. Dynamics of the pollen sequestrome defined by subcellular coupled -omics. *Plant Physiol.* **2018**, *178*, 258–282. [CrossRef]
13. Čapková, V.; Hrabětová, E.; Tupý, J. Protein synthesis in pollen tubes: Preferential formation of new species independent of transcription. *Sex. Plant Reprod.* **1988**, *1*, 150–155. [CrossRef]
14. Röhrig, H.; Colby, T.; Schmidt, J.; Harzen, A.; Facchinelli, F.; Bartels, D. Analysis of desiccation-induced candidate phosphoproteins from *Craterostigma plantagineum* isolated with a modified metal oxide affinity chromatography procedure. *Proteomics* **2008**, *8*, 3548–3560. [CrossRef] [PubMed]
15. Röhrig, H.; Schmidt, J.; Colby, T.; Bräutigam, A.; Hufnagel, P.; Bartels, D. Desiccation of the resurrection plant *Craterostigma plantagineum* induces dynamic changes in protein phosphorylation. *Plant Cell Environ.* **2006**, *29*, 1606–1617. [CrossRef] [PubMed]
16. Hafidh, S.; Fíla, J.; Honys, D. Male gametophyte development and function in angiosperms: A general concept. *Plant Reprod.* **2016**, *29*, 31–51. [CrossRef] [PubMed]
17. Lindner, H.; Kessler, S.A.; Muller, L.M.; Shimosato-Asano, H.; Boisson-Dernier, A.; Grossniklaus, U. TURAN and EVAN mediate pollen tube reception in Arabidopsis synergids through protein glycosylation. *PLoS Biol.* **2015**, *13*, e1002139. [CrossRef] [PubMed]
18. Palanivelu, R.; Preuss, D. Pollen tube targeting and axon guidance: Parallels in tip growth mechanisms. *Trends Cell Biol.* **2000**, *10*, 517–524. [CrossRef]
19. Hepler, P.K.; Winship, L.J. The pollen tube clear zone: Clues to the mechanism of polarized growth. *J. Integr. Plant Biol.* **2015**, *57*, 79–92. [CrossRef] [PubMed]
20. Šamaj, J.; Muller, J.; Beck, M.; Böhm, N.; Menzel, D. Vesicular trafficking, cytoskeleton and signalling in root hairs and pollen tubes. *Trends Plant Sci.* **2006**, *11*, 594–600. [CrossRef]
21. Fíla, J.; Honys, D. Enrichment techniques employed in phosphoproteomics. *Amino Acids* **2012**, *43*, 1025–1047. [CrossRef]
22. Dunn, J.D.; Reid, G.E.; Bruening, M.L. Techniques for phosphopeptide enrichment prior to analysis by mass spectrometry. *Mass Spectrom. Rev.* **2010**, *29*, 29–54. [CrossRef]
23. Obaya, A.J.; Sedivy, J.M. Regulation of cyclin-Cdk activity in mammalian cells. *Cell. Mol. Life Sci.* **2002**, *59*, 126–142. [CrossRef] [PubMed]
24. Janek, K.; Wenschuh, H.; Bienert, M.; Krause, E. Phosphopeptide analysis by positive and negative ion matrix-assisted laser desorption/ionization mass spectrometry. *Rapid Commun. Mass Spectrom.* **2001**, *15*, 1593–1599. [CrossRef] [PubMed]
25. Mayank, P.; Grossman, J.; Wuest, S.; Boisson-Dernier, A.; Roschitzki, B.; Nanni, P.; Nuehse, T.; Grossniklaus, U. Characterization of the phosphoproteome of mature Arabidopsis pollen. *Plant J.* **2012**, *72*, 89–101. [CrossRef]
26. Alexa, A.; Rahnenfuhrer, J.; Lengauer, T. Improved scoring of functional groups from gene expression data by decorrelating GO graph structure. *Bioinformatics* **2006**, *22*, 1600–1607. [CrossRef]
27. Fíla, J.; Matros, A.; Radau, S.; Zahedi, R.P.; Čapková, V.; Mock, H.-P.; Honys, D. Revealing phosphoproteins playing role in tobacco pollen activated in vitro. *Proteomics* **2012**, *12*, 3229–3250. [CrossRef]
28. Fíla, J.; Radau, S.; Matros, A.; Hartmann, A.; Scholz, U.; Feciková, J.; Mock, H.P.; Čapková, V.; Zahedi, R.P.; Honys, D. Phosphoproteomics profiling of tobacco mature pollen and pollen activated *in vitro*. *Mol. Cell. Proteom.* **2016**, *15*, 1338–1350. [CrossRef]
29. Chao, Q.; Gao, Z.F.; Wang, Y.F.; Li, Z.; Huang, X.H.; Wang, Y.C.; Mei, Y.C.; Zhao, B.G.; Li, L.; Jiang, Y.B.; et al. The proteome and phosphoproteome of maize pollen uncovers fertility candidate proteins. *Plant Mol. Biol.* **2016**, *91*, 287–304. [CrossRef]
30. Vannini, C.; Marsoni, M.; Scoccianti, V.; Ceccarini, C.; Domingo, G.; Bracale, M.; Crinelli, R. Proteasome-mediated remodeling of the proteome and phosphoproteome during kiwifruit pollen germination. *J. Proteom.* **2019**, *192*, 334–345. [CrossRef] [PubMed]
31. Jiao, Y.P.; Peluso, P.; Shi, J.H.; Liang, T.; Stitzer, M.C.; Wang, B.; Campbell, M.S.; Stein, J.C.; Wei, X.H.; Chin, C.S.; et al. Improved maize reference genome with single-molecule technologies. *Nature* **2017**, *546*, 524–527. [CrossRef] [PubMed]
32. Schnable, P.S.; Ware, D.; Fulton, R.S.; Stein, J.C.; Wei, F.S.; Pasternak, S.; Liang, C.Z.; Zhang, J.W.; Fulton, L.; Graves, T.A.; et al. The B73 maize genome: Complexity, diversity, and dynamics. *Science* **2009**, *326*, 1112–1115. [CrossRef] [PubMed]
33. Sierro, N.; Battey, J.N.D.; Ouadi, S.; Bakaher, N.; Bovet, L.; Willig, A.; Goepfert, S.; Peitsch, M.C.; Ivanov, N.V. The tobacco genome sequence and its comparison with those of tomato and potato. *Nat. Commun.* **2014**, *5*, 3833. [CrossRef] [PubMed]

34. Edwards, K.D.; Fernandez-Pozo, N.; Drake-Stowe, K.; Humphry, M.; Evans, A.D.; Bombarely, A.; Allen, F.; Hurst, R.; White, B.; Kernodle, S.P.; et al. A reference genome for *Nicotiana tabacum* enables map-based cloning of homeologous loci implicated in nitrogen utilization efficiency. *BMC Genom.* **2017**, *18*, 448. [CrossRef]
35. Lee, D.H.; Goldberg, A.L. Proteasome inhibitors: Valuable new tools for cell biologists. *Trends Cell Biol.* **1998**, *8*, 397–403. [CrossRef]
36. Chen, Y.; Liu, P.; Hoehenwarter, W.; Lin, J. Proteomic and phosphoproteomic analysis of *Picea wilsonii* pollen development under nutrient limitation. *J. Proteome Res.* **2012**, *11*, 4180–4190. [CrossRef]
37. Lee, T.Y.; Lin, Z.Q.; Hsieh, S.J.; Bretana, N.A.; Lu, C.T. Exploiting maximal dependence decomposition to identify conserved motifs from a group of aligned signal sequences. *Bioinformatics* **2011**, *27*, 1780–1787. [CrossRef]
38. Heberle-Bors, E.; Voronin, V.; Touraev, A.; Testillano, P.S.; Risueno, M.C.; Wilson, C. MAP kinase signaling during pollen development. *Sex. Plant Reprod.* **2001**, *14*, 15–19. [CrossRef]
39. Takatsuka, H.; Umeda-Hara, C.; Umeda, M. Cyclin-dependent kinase-activating kinases CDKD;1 and CDKD;3 are essential for preserving mitotic activity in *Arabidopsis thaliana*. *Plant J.* **2015**, *82*, 1004–1017. [CrossRef]
40. Shimotohno, A.; Matsubayashi, S.; Yamaguchi, M.; Uchimiya, H.; Umeda, M. Differential phosphorylation activities of CDK-activating kinases in *Arabidopsis thaliana*. *FEBS Lett.* **2003**, *534*, 69–74. [CrossRef]
41. Huang, X.-Y.; Niu, J.; Sun, M.-X.; Zhu, J.; Gao, J.-F.; Yang, J.; Zhou, Q.; Yang, Z.-N. CYCLIN-DEPENDENT KINASE G1 is associated with the spliceosome to regulate CALLOSE SYNTHASE 5 splicing and pollen wall formation in Arabidopsis. *Plant Cell* **2013**, *25*, 637–648. [CrossRef] [PubMed]
42. Li, D.D.; Guan, H.; Li, F.; Liu, C.Z.; Dong, Y.X.; Zhang, X.S.; Gao, X.Q. Arabidopsis shaker pollen inward K^+ channel SPIK functions in SnRK1 complex-regulated pollen hydration on the stigma. *J. Integr. Plant Biol.* **2017**, *59*, 604–611. [CrossRef] [PubMed]
43. Gao, X.Q.; Liu, C.Z.; Li, D.D.; Zhao, T.T.; Li, F.; Jia, X.N.; Zhao, X.Y.; Zhang, X.S. The Arabidopsis KIN beta gamma subunit of the SnRK1 complex regulates pollen hydration on the stigma by mediating the level of reactive oxygen species in pollen. *PLoS Genet.* **2016**, *12*, e1006228. [CrossRef] [PubMed]
44. Zhao, L.-N.; Shen, L.-K.; Zhang, W.-Z.; Zhang, W.; Wang, Y.; Wu, W.-H. Ca^{2+}-dependent protein kinase 11 and 24 modulate the activity of the inward rectifying K^+ channels in Arabidopsis pollen tubes. *Plant Cell* **2013**, *25*, 649–661. [CrossRef]
45. Dong, C.H.; Hong, Y. Arabidopsis CDPK6 phosphorylates ADF1 at N-terminal serine 6 predominantly. *Plant Cell Rep.* **2013**, *32*, 1715–1728. [CrossRef] [PubMed]
46. Supek, F.; Bošnjak, M.; Škunca, N.; Šmuc, T. REVIGO summarizes and visualizes long lists of gene ontology terms. *PLoS ONE* **2011**, *6*, e21800. [CrossRef]
47. Procissi, A.; Guyon, A.; Pierson, E.S.; Giritch, A.; Knuiman, B.; Grandjean, O.; Tonelli, C.; Derksen, J.; Pelletier, G.; Bonhomme, S. KINKY POLLEN encodes a SABRE-like protein required for tip growth in Arabidopsis and conserved among eukaryotes. *Plant J.* **2003**, *36*, 894–904. [CrossRef]
48. Liu, J.J.; Zhong, S.; Guo, X.Y.; Hao, L.H.; Wei, X.L.; Huang, Q.P.; Hou, Y.N.; Shi, J.; Wang, C.Y.; Gu, H.Y.; et al. Membrane-bound RLCKs LIP1 and LIP2 are essential male factors controlling male–female attraction in Arabidopsis. *Curr. Biol.* **2013**, *23*, 993–998. [CrossRef]
49. Camacho, C.; Coulouris, G.; Avagyan, V.; Ma, N.; Papadopoulos, J.; Bealer, K.; Madden, T.L. BLAST plus: Architecture and applications. *BMC Bioinform.* **2009**, *10*, 421. [CrossRef]
50. Zhang, M.J.; Zhang, X.S.; Gao, X.Q. ROS in the male–female interactions during pollination: Function and regulation. *Front. Plant Sci.* **2020**, *11*, 177. [CrossRef]
51. Tran, H.N.N.; Brechenmacher, L.; Aldrich, J.T.; Clauss, T.R.; Gritsenko, M.A.; Hixson, K.K.; Libault, M.; Tanaka, K.; Yang, F.; Yao, Q.M.; et al. Quantitative phosphoproteomic analysis of soybean root hairs inoculated with *Bradyrhizobium japonicum*. *Mol. Cell. Proteom.* **2012**, *11*, 1140–1155.
52. Cheng, C.Y.; Krishnakumar, V.; Chan, A.P.; Thibaud-Nissen, F.; Schobel, S.; Town, C.D. Araport11: A complete reannotation of the *Arabidopsis thaliana* reference genome. *Plant J.* **2017**, *89*, 789–804. [CrossRef] [PubMed]
53. Goodstein, D.M.; Shu, S.Q.; Howson, R.; Neupane, R.; Hayes, R.D.; Fazo, J.; Mitros, T.; Dirks, W.; Hellsten, U.; Putnam, N.; et al. Phytozome: A comparative platform for green plant genomics. *Nucleic Acids Res.* **2012**, *40*, D1178–D1186. [CrossRef] [PubMed]
54. Valliyodan, B.; Cannon, S.B.; Bayer, P.E.; Shu, S.Q.; Brown, A.V.; Ren, L.H.; Jenkins, J.; Chung, C.Y.L.; Chan, T.F.; Daum, C.G.; et al. Construction and comparison of three reference-quality genome assemblies for soybean. *Plant J.* **2019**, *100*, 1066–1082. [CrossRef] [PubMed]
55. Schmutz, J.; Cannon, S.B.; Schlueter, J.; Ma, J.X.; Mitros, T.; Nelson, W.; Hyten, D.L.; Song, Q.J.; Thelen, J.J.; Cheng, J.L.; et al. Genome sequence of the palaeopolyploid soybean. *Nature* **2010**, *463*, 178–183. [CrossRef]
56. Luo, X.; Chen, Z.; Gao, J.; Gong, Z. Abscisic acid inhibits root growth in Arabidopsis through ethylene biosynthesis. *Plant J.* **2014**, *79*, 44–55. [CrossRef] [PubMed]
57. Otterhag, L.; Gustavsson, N.; Alsterfjord, M.; Pical, C.; Lehrach, H.; Gobom, J.; Sommarin, M. Arabidopsis PDK1: Identification of sites important for activity and downstream phosphorylation of S6 kinase. *Biochimie* **2006**, *88*, 11–21. [CrossRef] [PubMed]
58. Xiao, Y.; Offringa, R. PDK1 regulates auxin transport and Arabidopsis vascular development through AGC1 kinase PAX. *Nat. Plants* **2020**, *6*, 544–555. [CrossRef]
59. Qu, L.; Wei, Z.; Chen, H.H.; Liu, T.; Liao, K.; Xue, H.W. Plant casein kinases phosphorylate and destabilize a cyclin-dependent kinase inhibitor to promote cell division. *Plant Physiol.* **2021**, *187*, 917–930. [CrossRef]

60. Kang, J.M.; Wang, Z. Mut9p-LIKE KINASE family members: New roles of the plant-specific casein kinase I in plant growth and development. *Int. J. Mol. Sci.* **2020**, *21*, 1562. [CrossRef]
61. Ye, J.Y.; Zhang, Z.B.; Long, H.F.; Zhang, Z.M.; Hong, Y.; Zhang, X.M.; You, C.J.; Liang, W.Q.; Ma, H.; Lu, P.L. Proteomic and phosphoproteomic analyses reveal extensive phosphorylation of regulatory proteins in developing rice anthers. *Plant J.* **2015**, *84*, 527–544. [CrossRef] [PubMed]

International Journal of
Molecular Sciences

MDPI

Review

Proteomics and Post-Translational Modifications of Starch Biosynthesis-Related Proteins in Developing Seeds of Rice

Piengtawan Tappiban [1], Yining Ying [1], Feifei Xu [1] and Jinsong Bao [1,2,*]

[1] Key Laboratory of Nuclear Agricultural Sciences of Ministry of Agriculture and Zhejiang Province, Institute of Nuclear Agricultural Sciences, College of Agriculture and Biotechnology, Zhejiang University, Zijingang Campus, Hangzhou 310058, China; Piengtawan.tap@hotmail.com (P.T.); ying_erin@163.com (Y.Y.); xuxufei@zju.edu.cn (F.X.)

[2] Hainan Institute of Zhejiang University, Yazhou Bay Science and Technology City, Yazhou District, Sanya 572025, China

* Correspondence: jsbao@zju.edu.cn; Tel.: +86-571-86971932

Abstract: Rice (*Oryza sativa* L.) is a foremost staple food for approximately half the world's population. The components of rice starch, amylose, and amylopectin are synthesized by a series of enzymes, which are responsible for rice starch properties and functionality, and then affect rice cooking and eating quality. Recently, proteomics technology has been applied to the establishment of the differentially expressed starch biosynthesis-related proteins and the identification of posttranslational modifications (PTMs) target starch biosynthesis proteins as well. It is necessary to summarize the recent studies in proteomics and PTMs in rice endosperm to deepen our understanding of starch biosynthesis protein expression and regulation, which will provide useful information to rice breeding programs and industrial starch applications. The review provides a comprehensive summary of proteins and PTMs involved in starch biosynthesis based on proteomic studies of rice developing seeds. Starch biosynthesis proteins in rice seeds were differentially expressed in the developing seeds at different developmental stages. All the proteins involving in starch biosynthesis were identified using proteomics methods. Most starch biosynthesis-related proteins are basically increased at 6–20 days after flowering (DAF) and decreased upon the high-temperature conditions. A total of 10, 14, 2, 17, and 7 starch biosynthesis related proteins were identified to be targeted by phosphorylation, lysine acetylation, succinylation, lysine 2-hydroxyisobutyrylation, and malonylation, respectively. The phosphoglucomutase is commonly targeted by five PTMs types. Research on the function of phosphorylation in multiple enzyme complex formation in endosperm starch biosynthesis is underway, while the functions of other PTMs in starch biosynthesis are necessary to be conducted in the near future.

Keywords: rice; starch biosynthesis; proteomics; posttranslational modification; starch functionality; cooking and eating quality

Citation: Tappiban, P.; Ying, Y.; Xu, F.; Bao, J. Proteomics and Post-Translational Modifications of Starch Biosynthesis-Related Proteins in Developing Seeds of Rice. *Int. J. Mol. Sci.* **2021**, *22*, 5901. https://doi.org/10.3390/ijms22115901

Academic Editor: Sixue Chen

Received: 13 April 2021
Accepted: 28 May 2021
Published: 31 May 2021

1. Introduction

Rice (*Oryza sativa* L.) is one of the most consumed cereal grains for half of the world's population and ranks as the third-largest crop, after sugarcane and maize [1,2]. The production quantity of rice was an estimated 214.08 million tonnes reported in 2018 [1], of which China, India, and Indonesia were the top-three highest production countries (Figure 1). Cultivated rice consists of two subspecies, *O. sativa japonica* and *O. sativa indica* [3,4], which are highly distinct in terms of geographical distribution, visible morphological traits [5], and physiological characteristics (e.g., biotic and abiotic stress responses, cold tolerance, and seed quality) [6].

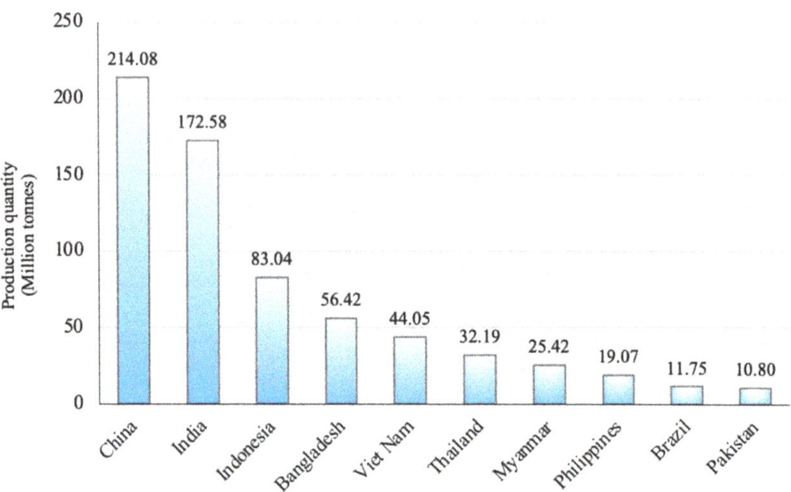

Figure 1. Production quantity of rice in 2018 from the top 10 countries based on FAOSTAT [1].

Rice seed consists of a minuscule embryo containing most of the genetic information and a relatively large endosperm containing most of the nutrient storage [7]. In mature rice endosperm, starch is a primary component with 80–90% of the total dry weight [8] and somewhat considered as a complex carbohydrate including amylose and amylopectin, which are synthesized and packed to form a large semicrystalline granule in amyloplasts through a large suite of enzyme activities [9]. Amylose is a linear chain made up of α-1,4 glycosidic linked glucose molecules with very few α-1,6 branches, whereas amylopectin, the main component of starch granule, is a highly branched chain of glucose units joined by both α-1,4 and α-1,6 glycosidic bonds [9–11]. Starch from different plant origins varies in its physicochemical properties due to the ratio of amylose and amylopectin and differences in branching density of semicrystalline structure [12]. Besides starch properties and functionalities, most studies of rice endosperm have been focused on starch biosynthesis protein expression and regulation [13–15]. Four main classes of starch biosynthesis enzymes including ADP-glucose pyrophosphorylase (AGPase), starch synthase (SS), starch branching enzyme (BE), and starch debranching enzyme (DBE) are presented in rice developing seeds [9,10,16].

Proteomics has become an important tool to study the dynamic and diverse biological processes and analyze expression patterns, variation, function, and interaction of proteins at a given time, in a particular tissue, or among different treatments of biotic and abiotic stresses in plants [16–19]. Using two-dimensional gel electrophoresis (2-DE)-based protein identification method equipped with mass spectrometric techniques, the profile of entire protein expression can be rapidly generated with highly reproducible [17,20], allowing for selection of the reasonable gene(s), which is possibly involved in regulatory mechanisms underlying starch biosynthesis [13,21,22]. In the last two decades, proteomic analyses of rice endosperm have been applied to a broad range of processes including differentially expressed proteins from the specific issues of plant tissues/organs [23], developmental stages [13,21,22,24–26], chalky and translucent parts [19], as well as under high temperature (HT) condition [13,27,28] in which the starch biosynthesis-related proteins were affected and reported.

The posttranslational modifications (PTMs) refer to the chemical modification events resulting from the covalent attachment of chemical groups, such as phosphate, acetyl, succinyl, methyl, and oligosaccharides, to amino acid side chains of the particular proteins [29]. PTM is a crucial step for functional protein maturation and important in signal transduction, apoptosis, transcriptional regulation, etc., by changing the chemical nature of polypeptide chains during or after protein biosynthesis [29,30]. Recently, a number of PTMs have been identified and verified by many highly effective techniques, coupled with database and bioinformatics tools. In rice endosperm, PTMs targeted starch biosynthesis proteins have been identified including phosphorylation [27,28,31], acetylation [30,32–34], succinylation [34], malonylation [35], and lysine 2-hydroxyisobutyrylation [36].

As the global population grows, it is expected that, by 2035, additional demand of approximately 112 million metric tons of rice needs to be produced to keep food security [37]. Understanding the function of proteomics and PTMs in rice seeds will contribute to our understanding of the mechanism and regulation of starch biosynthesis, which is one of the most important topics for high-yield and high-quality rice production [8,28,38,39]. This review summarizes the current knowledge in starch biosynthesis proteins through the studies of proteomics and PTMs, delving into the contribution of starch biosynthesis proteins to starch properties and functionality.

2. Significant Proteins for Starch Biosynthesis in Rice Seeds

2.1. *Amylose and Amylopectin Biosynthesis*

The enzymatic machinery for starch biosynthesis is summarized in Figure 2. AGPase (EC 2.7.7.27), a step limiting for starch biosynthesis [40,41], is a heterotetramer composed of large and small subunits, which catalyzes the glucose-1-phosphate (G1P) to ADPglucose [9,14,42,43]. In rice, seven AGPase subunits were reported [44]. To promote an α-1,4 bond and elongate the glucan chain, starch synthases (SSs) (EC 2.4.1.21) participate in the transfer of a glucosyl moiety from ADPglucose to the nonreducing end of the existing chains [45,46]. SSs are clustered into five classes including soluble SS I (SSI), SSII, SSIII, SSIV, and GBSS [45]. Two GBSS isoforms of rice—GBSSI and GBSSII—are primarily expressed in endosperm and leaves, respectively [44]. GBSSI is the only enzyme responsible for amylose biosynthesis (Figure 2).

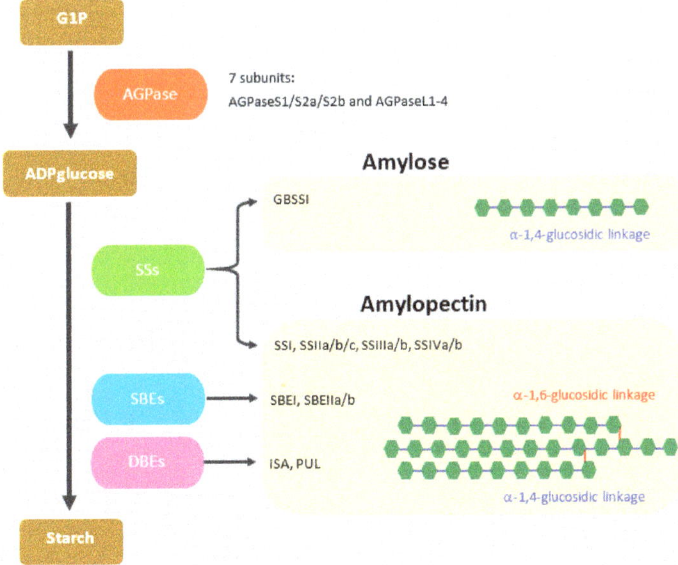

Figure 2. Brief schematic of the major proteins involved in starch biosynthesis in rice endosperm. Amylose and amylopectin are composed of glucose units (green rectangle) with α-1,4-glucosidic linkage (blue line). The branch chains of amylopectin are connected through α-1,6-glucosidic linkage (red line).

In contrast, amylopectin, as compared with amylose, generally has a distinct fine structure and is synthesized by three main enzymes including SS, starch branching enzyme (BE), and starch debranching enzyme (DBE) [47]. The SSs play a crucial role in the elongation of α-1,4-glucosidic chains and also regulating the quality and quantity of starch, especially in defining amylopectin chain-length distribution [11,47]. In rice, SSs contain multiple paralogs including SSI, SSIIa(SSII-3)/IIb(SSII-2)/IIc(SSII-1), SSIIIa(SSIII-2)/IIIb(SSIII-1), and SSIVa(SSIV-1)/IVb(SSIV-2) [45]. Among those isoforms, SSI, SSIIa, and SSIIIa are responsible for priming the short degree of polymerization (DP) 6 to 7 chains [15], short A and B_1 [42], and B_2-B_4 chains (DP $\geq$ 30) [43], respectively.

BEs (EC 2.4.1.18) specifically introduce the branch point, the α-1,6-glucosidic linkage into the glucan chain by cutting the α-1,4-linked glucan and also transferring to another chain at the 6-hydroxyl position and thus is considered as a key enzyme regulating amylopectin structure [48]. Three different isoforms of BEs including BEI, BEIIa, and BEIIb are present in rice endosperm [47]. BEI forms a variety of both short chains and intermediate chains (DP $\leq$ 40) by attacking the branched glucan in both the outer and inner chains of amylopectin [49]. Conversely, BEIIa and BEIIb function only in the outer amylopectin structure and specifically responsible for transferring the short chain of DP 6–15 and DP 6–7, respectively [47,49].

DBEs (EC 3.2.1.10) are composed of two types, isoamylase (ISA; EC 3.2.1.68) and pullulanase (PUL; EC 3.2.1.41), and hydrolyze the improper α-1,6-glucans [47]. Both ISA and PUL debranch the amylopectin as well as other substrates, i.e., glycogen and phytoglycogen, for ISA and pullulan for PUL [47]. However, the improperly located branches are mainly removed by ISA and partially by PUL [50].

Besides those main enzymes, starch phosphorylase (Pho; EC 2.4.1.1) also plays a crucial role in starch biosynthesis and degradation [51–53]. Pho catalyzes the phosphorolytic of the outermost glucose residue to generate G1P, which is reversibly added to the end of α-glucan chains [53], and recently, Pho is reported to play a crucial role in starch biosynthesis at low temperature [54]. Pho composted of a plastidial form or Pho1 (Pho-L) and a cytosolic form known as Pho2 (Pho-H) and both forms are different in terms of structure, kinetic properties, the expression pattern, and subcellular localization [53]. The enzyme activity of Pho1 is observed only in the endosperm of rice, while Pho2 is found in both endosperm and photosynthetic organs [53]. In addition, G1P is also the result of the reaction of plastidic phosphoglucomutase (PGM; EC 5.4.2.2), which catalyzes the conversion of G6P to G1P [55].

2.2. Phosphorylation and Dephosphorylation of Glucan Chains

Phosphorylation is only known in vivo covalent modification of starch. Two important isoforms of glucan water dikinase (GWD), GWD1 (EC 2.7.9.4) and phospho glucan water dikinase (PWD, EC 2.7.9.5) catalyze the phosphorylation at the C6 and C3 positions of glucosyl residues in amylopectin, respectively [56–59]. Both GWDs preferably phosphorylate the longer amylopectin chains at a range of DP 30–100 [57,60] and disrupt the glucan chains to access hydration [61]. The reduction of GWDs contributes to starch degradation by α-amylase (EC 3.2.1.1) [62]. In rice, *OsGWD1* (Os06g0498400) is responsible for leaf starch excess1 (LSE1) [63]. The LSE1 mutant lacked exons 22–32 displayed a higher starch content (5–10-fold) and lower phosphorylation (0.05 $\pm$ 0.02%) in the leaf blades, compared with those of wild type [63]. In addition, the LSE1 mutant had fewer panicles, lower ripened grains, smaller grains, and a lower grain yield, as compared to those of the wild type. Overexpression of potato *GWD1* in rice enhanced the phosphorylation at both C6 (9-fold) and C3 (2-fold) positions, increased amylose content, and displayed a minor change in starch granule morphology of endosperm [64]. Besides the alteration of starch properties and functionality [63,65,66], starch phosphorylation also interrupts the crystalline structure of amylopectin [67].

Dephosphorylation is an essential mechanism required for starch degradation [68]. Starch Excess4 (SEX4, EC 4.3.1.3.48) catalyzes the removal of phosphate groups at both C3 and C6 positions [68,69], while Like Sex Four2 (LSF2) specifically dephosphorylates at C3 position [69]. In Arabidopsis leaves, the deficiency of SEX4 caused an increase in starch accumulation [68,70], but a loss of LSF2 was found to have no obvious effects on starch levels [69]. Even though SEX4 and LSF2 play a crucial role in the dephosphorylation of Arabidopsis, their biological function remains unclear in rice endosperm. In addition, it was found that SEX4 was primarily expressed in the anthers of rice [71]. Recently, *OsSEX4* (LOC_Os03g01750/Os03g0107800)-knockdown rice caused an increase in starch accumulation in suspension-cultured cells, leaves, and rice straw, indicating that the function of *OsSEX4* is conserved with Arabidopsis [65]. The transgenic rice plants also exhibited a chalky grain phenotype and had no effects on vegetative growth and grain yield [65].

2.3. Disproportionation to Nonreducing End of Starch

The disproportionating enzyme (DPE, EC 2.4.1.25), a 4-α-glucanotransferase, cleaves the α-1,4 glucosidic bond, transfers the glucan moiety to the nonreducing end, and finally forms a new 1,4 glucosidic bond [66,72,73]. There are two isoforms of DPEs, plastid-located DPE1 and cytoplasm-located DPE2, which differ in expression profiles, subcellular localization, protein structure, and reaction properties [72,74]. The identified OsDPE1 and OsDPE2 are composed of 594 and 946 amino acids, respectively [72]. Both DPEs showed the conserved domain of glucoside hydrolase of family 77 (GH77). The OsDPE1 contains only one domain of GH77, while the OsDPE2 has two copies of GH77 at the C-terminal and two copies of N-terminal carbohydrate-binding module 20 (CBM20) [72]. The activities of DPEs are different; OsDPE1 catalyzes the maltotriose transfer reaction by using the glucose as its acceptor, whereas OsDPE2 participates in the glucose transfer reaction from maltose to glycogen acceptor [72]. Recently, DPE1 mediates the reaction of transferring maltooligosyl units from amylose and amylopectin to amylopectin [73].

2.4. Starch Granule Initiation

The initiation of starch granules is recently studied in order to understand the mechanisms and factors that influence the number of granules per plastid and the morphogenesis of granules [75]. Several key proteins playing a crucial role in granule initiation were discovered through the homologs of Protein Targeting to Starch (PTST) proteins [75,76]. PTST contains an N-terminal coiled-coil domain and a C-terminal carbohydrate-binding module 48 (CBM48) mediating protein–protein interaction and starch-binding domain, respectively [77,78]. In Arabidopsis chloroplast, PTST1 interacts directly with GBSS via the coiled-coil domain in the stroma and then locates to starch granules by using the CBM48 domain [78,79]. PTST2 interacts with SS4 and soluble maltooligosaccharides (MOS), while PTST3 interacts with PTST2 [80]. In addition, PTST2 is also associated with Mar-Binding Filament Protein (MFP1) to facilitate the normal PTST2 localization [81]. Therefore, the executive functions of PTST1, 2, and 3 are required for amylose biosynthesis, normal granule initiation, and cofunction with PTST2, respectively [80].

In rice, the functions of GBSS-binding protein (OsGBP) and Floury Endosperm6 (FLO6) are similar to PTST1 and PTST2, respectively. A newly identified OsGBP interacts directly with both GBSSI and GBSSII in yeast two-hybrid assay [82]. The coiled-coil domain is responsible for GBSS binding, while the CBM48 is essential for targeting GBSSs to starch granules during amylose biosynthesis [82]. Based on the CRISPR/Cas9 gene editing, mRNA and protein abundance of *osgbp* mutants were significantly decreased in both leaves and grains, compared to wild type, leading to the reduction of starch content and number of starch granules with smaller size in leave and the presence of large chalkiness area in the endosperm [82].

FLO6 plays an essential role in starch granule formation and contains N-terminal transit peptide and C-terminal CBM48 domain for plastid localization and binding to starch, respectively [83]. FLO6 interacts with ISA1 via its N-terminus with no effect on the enzyme activity. As compared to the wild type, the *flo6* mutant showed many smaller granules with irregular shapes and rough surfaces [83].

3. Proteomic Profiling of Starch Biosynthesis-Related Proteins

The proteome of each living cell refers to the entire proteins expressed by a genome, which is highly dynamic and altering in response to intra- and extracellular factors across time points [18]. With different purposes of proteomics analysis, proteins involving starch biosynthesis of rice endosperm have been intensively studied (Tables 1–3).

3.1. Specific Starch Biosynthesis-Related Proteins in Rice Seeds

To identify the tissue-specific expression in rice and understand the mechanisms that regulate starch biosynthesis, a total of 1022, 1350, and 877 unique proteins were identified from leaf, root, and seed tissues of rice, respectively, by using both 2-DE and high-performance liquid chromatography–tandem mass spectrometry (HPLC–MS/MS) coupled with multidimensional protein identification technology (MudPIT) [23]. The unique peptides of starch biosynthesis-related proteins were achieved for 7.43% (162/2180) of those from the root, followed by 2.29% (54/2358) and 0.37% (10/2712) from leaf and root tissues, respectively. Pho, AMY, ISA, and PGM were also observed in root tissue. AGPase and GBSS were detected in the leaf, while only PGM was observed in all three tissues (leaf, root, and seed) [23] (Table 1). At the mature stage of rice endosperm, 14 starch biosynthesis proteins were identified as the starch granule-associated proteins, and additional Hsp70, putative Brittle-1 protein, and PPDK were also identified (Table 1) [32].

Table 1. List of starch biosynthesis proteins of rice endosperm based on proteomics in response to the organ-/tissue-specific differences.

Sample	Aim of Study	Technique	Identified Proteins	Details of Results
leaf, root, and seed of Nipponbare (*Japonica*) [23]	To identify protein expression in leaf, root (49 DAG), and seed (14 DAG).	2-DE HPLC–MS/MS MudPIT	AGPase (id: 7670) leaf, seed AGPase small subunit (id: 44074) leaf, seed AGPase (id: 9904, 34550) seed AGPase (id: 50182) leaf GBSS (id: 31122) leaf, seed GBSS (id: 31130) seed SS precursor (id: 99443, 52528) seed SS (id: 26269) seed SBE isoform rbe3 (id: 36892) seed SBE (id: 20648, 27094, 53238, 55740) seed DBE (id: 14376) PhoH isoenzyme (id: 12500) seed Phol (id: 32714) seed AMY precursor (id: 21708) seed AMY (id: 24707) seed ISA (id: 23091) seed ISA (id: 23496) seed PGM, chloroplast precursor (id: 34039) seed PGM, cytoplasmic (id: 38302) leaf, root, seed	Proteins involving in starch biosynthesis were observed in both leaf and seed tissues. Starch degradation-related proteins were observed only in seed tissue. Two isoforms of small AGPase subunit were detected in both leaf and seed whereas another two isoforms of large AGPase subunit were identified only in seed tissue. The third isoform of large AGPase subunit was observed in leaf.
DY1102 (Wuyujing3 (*Japonica*) treated with 0.5% ethyl methanesulfonate (EMS)) (notched-belly mutant with white belly) [19]	To identify the differentially expressed proteins between the chalky and the translucent parts of DY1102 grains.	iTRAQ LC-MS/MS	AGPase, SSII, SSIII, SBE, Pho1, PGM, AMY, and putative starch synthase DULL1 (SSIIIa)	Downregulation of AMY was observed in chalky part. Downregulation of AMY contributes to starch hydrolysis and the formation of chalkiness. SSIIIa was one of the differentially expressed proteins and increased in chalky part. The increase in SSIIIa expression did not result in the increased proportion of long amylopectin chains (DP > 30).

Table 1. *Cont.*

Sample	Aim of Study	Technique	Identified Proteins	Details of Results
Nipponbare (*Japonica*) [32]	To develop a method for rice starch granule purification from mature endosperm and identify starch granule-associated proteins.	LC-MS/MS	AGPase S2, AGPase L1, AGPase L2 GBSSI, GBSSII SBE 1, SBE3 SSI, SSII-1, SSII-3, SSIIIa PUL Pho1 ISA2	Besides 14 identified starch biosynthesis proteins, the other candidate starch granule-associated proteins involving in starch biosynthesis were also identified by LC-MS/MS including Hsp70, putative Brittle-1 protein, and PPDK. Compared with Tris-HCl buffer extraction method, the proteome extracted by the phenol buffer had more proteins and displayed almost all identified proteins extracted by Tris-HCl buffer.

Lin et al. [19] identified 113 differentially expressed proteins between the translucent and chalky parts of rice Wuyujing3. Among these, proteins in carbohydrate metabolism were the third most abundant (15.0%) after the categories of protein synthesis, folding and degradation (27.4%), and unidentified function (24.8%). AGPase, SSII, SSIII, SSIIIa, SBE, Pho1, PGM, and AMY were identified by using the isobaric tags for relative and absolute quantification (iTRAQ) based on the upper and the bottom half of translucent and chalky grains (Table 1). The AMY was downregulated in the chalky part, which was responsible for the processes of starch hydrolysis and chalk formation [19]. SSIIIa functions in $B_{2\text{-}4}$ chains elongation with the degree of polymerization (DP) ≥ 30 [43]. In contrast, Lin et al. [19] found SSIIIa was increased in the chalky part, which was found the greater amount of short chain (DP ≤ 12) and fewer medium and long chains, compared with the translucent parts.

3.2. Starch Biosynthesis-Related Proteins in Different Developmental Stages of Rice Seeds

In cereal endosperm cells, starch granules are synthesized and increase in number and volume until maturity based upon the synergy of multiple enzymes [9,30,84,85]. No starch accumulation was observed in rice endosperm at 2 days after flowering (DAF), and a small amount of starch was found in the pericarp at 4 DAF [22]. A great accumulation of starch in endosperm was noticed after 8 DAF [22,86]. Endosperm remains equally milk white with no translucent region at 10 DAF [26] and 12 DAF [24]. Since the translucent region indicates the accumulation and packing of starch granule, half of the translucent area in the central endosperm were observed at 15 DAF [24,26], while full translucent in the whole endosperm was noticed at 20 DAF [26]. However, rice seed development varies depending on genotypic and environmental conditions [25].

The identified starch biosynthesis-related proteins involved in rice endosperm development were summarized in Table 2. Over 400 protein spots were identified in Taichung Native 1 (TN 1) at 12 DAF [13]. GBSS (Waxy) was identified and increased the expression after 6 DAF [13].

Table 2. List of starch biosynthesis proteins of rice endosperm based on proteomics in response to different developmental stages.

Sample	Aim of Study	Technique	Identified Proteins	Details of Results
Taichung Native 1 (TN 1, *Indica*) and Tainung 67, (TNG 67, *Japonica*) [13]	To investigate the changes in protein expression patterns during rice caryopsis development (6, 9, 12, 15, and 32 DAF).	2-DE LC-MS/MS	GBSS (Waxy)	The expression of GBSS increased after 6 DAF was coincident with the increase in amylose content. GBSS protein was highly expressed in kernels of rice with high amylose content (TN1).
Nipponbare (*Japonica*) [22]	To study the protein expression profiles related to grain filling during 6–20 DAF.	2-DE MALDI-TOF/TOF	ISA I, AMY, Pho, PGM, AGPaseL2, AGPaseL3, AGPaseS2a/b	All identified proteins were continuously increased from 6 to 20 DAF. Some AGPase isoforms had the highest peak of protein expression at 16 DAF and decrease thereafter.
			ISA3	ISA3 increased at 6 DAF, showed the highest expression at 10 DAF, and decreased thereafter.
			SSI	No result of expression pattern.
Zhonghua 10 (*Japonica*) [24]	To study the cellular features and proteomics of rice endosperm from 12, 15, and 18 DAF.	2D-DIGE MALDI-TOF/TOF-MS	PUL Pho1 AGPase L AGPase S2	Most of the protein expression patterns showed increase in abundance from 12 to 18 DAF. Some isoforms of PUL and AGPase S2 had the highest peak of expression at 15 DAF. Pho1 decreased the expression level form 12–18 DAF. AGPase L showed the highest variation of expression patterns including the expression levels continuously decreased and increased from 12–18 DAF, showed the highest peak and lowest peak at 15 DAF. The completion of starch granule packing was firstly observed in the inner part of endosperm at 15 DAF and showed entire endosperm at 18 DAF. AGPase L and Pho1 were significantly coexpressed with proteins in redox regulation (SOD and APX, respectively)
Ilpumbyeo (*Japonica*) [25]	To identify the differentially expressed proteins of rice grains at 10, 20, 30 DAF and the fully mature grain (45 DAF).	MudPIT	Pho1 PUL AMY SS 2–3 GWD SBE	All identified 6 starch biosynthesis proteins were reproducibly identified and differentially expressed during four stages (10, 20, 30, and 45 DAF). All of these proteins had the highest expression levels at the fully mature grain except SS 2–3 in which its abundance increased until 20 DAF after that decreased at 30 DAF and increased at fully mature grain. The authors suggested that the expression profile of starch biosynthesis proteins was similar to previous research of Xu et al. [22]

Table 2. *Cont.*

Sample	Aim of Study	Technique	Identified Proteins	Details of Results
Jinhui No. 809 (*Indica*) [87]	To identify the differentially expressed proteins between superior (SS) and inferior spikelet (IS) at the early (EGS), mid (MGS), and late (LGS) grain-filling stages.	2-DE MALDI-TOF/MS LC-ESI-MS/MS	AGPase GBSS PUL	AGPase, GBSS, and PUL isoforms were downregulated in inferior spikelets at EGS.
			AGPase S	AGPase S showed downregulation in both MGS and LGS.
Jinhui No. 809 (*Indica*) [88]	To identify the differentially expressed proteins of 10 DAF superior spikelet (SS) and 10 and 20 DAF inferior spikelet (IS).	2-DE MALDI-TOF/MS LC-ESI-MS/MS	AGPase GBSS SBE 1 SBE 3 PUL	AGPase had lower expression level in 10 DAF IS compared with both 10 DAF SS and 20 DAF IS. SBE 3, AGPase, PUL, and SBE 1 were detected as the 14-3-3 interacting proteins. AGPase and GBSS might be involved in the developmental stagnancy stage (DSS) of IS especially at the early grain-filling stage.
Zhonghua 10 (*Japonica*) [26]	To identify the SGAPs of rice at 10, 15, and 20 DAF.	2D-DIGE MALDI-TOF/TOF-MS	Pho1 PUL SSI AGPase L2 GBSSI AGPase S2a	Protein abundance of Pho1, PUL, SSI, and AGPase S2a slowly increased from 10 to 15 DAF and then drastically increased from 15 to 20 DAF. GBSSI showed linearly decreased abundance levels from 10 to 20 DAF. GBSSI and SSI were found only in starch granule-associated (SGA) form. AGPase, Pho1, and PUL were observed in both soluble and SGA forms.

The 396 differentially expressed protein spots were identified in Nipponbare during the early (6–8 DAF), mid (8–12), and late (12–20) stages of rice seed development [22]. ISA I, AMY, DBE, Pho 1, PGM, and AGPase were increased from 6 to 20 DAF, while the expression level of ISA3 protein started to increase at 6 DAF, reached the highest level at 10 DAF, and decreased thereafter. During the late stage, the expression levels of PUL, Pho 1, and AGPase increased in abundance from 12 to 18 DAF [22].

Lee et al. [25] reported 4172 nonredundant proteins of fully mature seeds and 889, 913, 1095, and 899 proteins were identified during 10, 20, 30, and 45 DAF. Pho 1, PUL, AMY, SS 2–3, GWD, and SBE were differentially expressed among each interval stage and the highest protein abundance was observed at the fully mature grains (45 DAF). Among those proteins, PUL, AMY, and SBE increased until 20 DAF and slightly decreased at 30 DAF then rapidly increased at fully mature grain, suggesting that process of starch accumulation was intensive at 20 DAF [25].

Besides starch biosynthesis-related proteins, proteins involving in other metabolic processes such as glycolysis, TCA-cycle, lipid metabolism, and proteolysis were also detected at higher levels in the fully mature grain (desiccation phase), compared to the developing stages, suggesting that the accumulation of these proteins might be for seed germination [25].

Zhang et al. [87] found that AGPase, GBSS, and PUL were differentially expressed between superior and inferior spikelets during the rice grain-filling stage, which was downregulated in the inferior spikelets at the early stage. Western blotting indicated that AGPase was downregulated in the inferior spikelets at all stages of the early, mid, and late grain-filling stage, compared to superior spikelets [87]. In addition, SBE 3, AGPase, PUL, and SBE 1 were detected as the interacting proteins with 14–3-3 [88], which might play a crucial role in the termination of inferior spikelets' development.

Yu et al. [26] identified 115 developmentally changed starch granule-associated proteins (SGAPs), with 39% of which involving in starch biosynthesis. Pho1, PUL, SSI, and AGPase S2a slowly increased in abundance from 10 to 15 DAF and then rapidly increased from 15 to 20 DAF, while the levels of GBSSI abundance showed the linearly decreased from 10 to 20 DAF [26].

Overall, the expression patterns of proteins involving in starch biosynthesis, starch degradation, and starch phosphorylation are particularly responsible for rice seed development. Most of those reported proteins are continuously increased during 6–20 DAF. However, proteins involving in starch debranching, degradation (AMY, Pho1, and PUL), and phosphorylation (GWD) are markedly decreased at 30 DAF and then increased to complete seed development.

3.3. Starch Biosynthesis-Related Proteins Respond to High Temperature (HT)

HT particularly affects the yield and quality of rice [89] but no significant changes were observed for seed morphology and size [90]. The effects of HT on starch biosynthesis-related proteins were summarized in Table 3.

Table 3. List of starch biosynthesis-related proteins of rice endosperm based on proteomics in response to HT.

Sample	Aim of Study	Technique	Identified Proteins	Details of Results
Taichung Native 1 (TN 1, *Indica*) and Tainung 67, (TNG 67, *Japonica*) [13]	To determine the candidate proteins associated with grain quality under HT, 35/30 °C (day/night).	2-DE LC-MS/MS	GBSS (Waxy)	HT caused the reduction of GBSS in TGN67 and decreased the levels of amylose content of TNG67 at 15 DAF (12.3 ± 0.5%) compared with those (15.6 ± 0.4%) under the control temperature (30/25 °C). Protein expression of TN1 showed relatively stable in both HT and control conditions. TNG67 showed more sensitivity to HT than TN1.
9311 (*Indica*) [91]	To identify the differentially accumulated proteins of rice at 5, 10, 15, and 20 DAF under day HT (DHT, 35/27 °C) and night HT (NHT, 27/35 °C).	2-DE MALDI-TOF MS/MS	PGM PUL	One and five isoforms of PUL and PGM were differentially accumulated in response to DHT and NHT and detected in all 5, 10, 15, and 20 DAF with different accumulation patterns. Three PUL isoforms (spot 34, 35, and 36) were increased in parallel abundance from 5 to 20 DAF, while another (spot 37 and 38) showed slowly increase at 5–10 DAF and highly increase in abundance at 15 and 20 DAF.
XN0437T (heat-tolerant) XN0437S (heat-sensitive) [89]	To identify the differentially expressed proteins during rice grain development at 1, 3, and 5 day after HT treatment (38.0 ± 0.5 °C) compared with control (25.0 ± 0.5 °C)	2-DE MALDI-TOF/TOF MS	PUL DBE GBSS AGPase L	All 4 proteins involving in starch biosynthesis showed downregulation in both rice lines under HT stress compared with the control treatment. AGPase L was higher accumulated in the heat-tolerant rice at 1 day after HT and showed lower accumulation at 3 and 5 days after HT compared to heat-sensitive rice. PUL, DBE, and GBSS had lower expression levels in heat-tolerant rice at all three-time points.

Table 3. *Cont.*

Sample	Aim of Study	Technique	Identified Proteins	Details of Results
Perfect and chalky rice grains (Koshihikari (*Japonica*)) [92]	To study the proteomic profile of the translucent and opaque grains under moderate (in 2009, 24.4 °C) and HT (in 2010, 28.0 °C) conditions.	iTRAQ MS/MS	SSI SSII PUL GBSSI BEIIb	All identified proteins showed downregulation in chalky rice compared to perfect grain. Protein expression of SSII, PUL, and BEIIb under moderate temperature was lower than HT condition, while the others, SSI and GBSSI showed higher abundance under HT.
			AMY (AmyII-3)	AMY showed upregulation in chalky rice in both conditions and chalky rice under HT stress had higher AMY abundance than moderate temperature.
			DBE BEI	Both DBE and BEI were increased in HT but downregulated in moderate temperature.
KDML105 (*Indica*) [90]	To identify the differentially changed proteins of rice grains under heat stress (40/26 °C) at the milky, dough, and mature stages.	nanoLC-MS/MS	AMY	AMY showed the highest in abundance at milky then decreased in dough and disappeared at mature stage.
			AGPase	AGPase had the un-change expression in both milky and dough stages and double increased in mature stage.
			SBEI GBSSI	Protein abundance of both SBEI and GBSSI was increased almost three times from milky to dough stages. Both SBEI and GBSSI were not found at mature stage.
			AGPase L2	AGPase L2 was detected only at dough stage.
			SBE3 AGPase L2 SSIIa SSI	All proteins were detected only in milky stage in which the AGPase L2 showed the highest abundance followed by SSI, SBE3, and SS IIa.
			ISA	ISA was detected in both milky and mature stages and showed the highest abundance at mature stage.

HT at 35/30 °C (day/night) decreased the GBSS abundance of japonica rice (TNG 67), leading to the lower amylose content observed at 15 DAF (12.3 ± 0.5%), compared with the control temperature at 30/25 °C (15.6 ± 0.4%), while those of indica rice (TN1) were relatively stable under both temperatures [13]. According to the HT condition during grain filling, chalkiness occurrence is increased, and loosely packed of abnormal starch granules were observed in rice endosperm [92,93]. Li et al. [91] reported that the conditions of both day high temperature (35/27 °C, DHT) and night high temperature (27/35 °C, NHT) caused a higher percentage of chalkiness but lower levels of brown rice rate, milled rice rate, head rice rate, amylose content, and gel consistency, compared to the control condition (28/20 °C). Besides a PGM isoform, five isoforms of PUL were detected and differentially accumulated in response to DHT and NHT [91]. Compared to heat-sensitive rice (XN0437S), the lower abundance of PUL, DBE, and GBSS was observed in heat-tolerant rice (XN0437T) at 1 day (d), 3 d, and 5 d of HT stress (38.0 ± 0.5 °C) [89]. The higher accumulation of AGPase L was found in heat-tolerant rice at 1 d and decreased at 3 d and 5 d of HT stress [89].

Compared to the perfect grain, SSI, SSII, PUL, GBSSI, and BEIIb were downregulated, while AMY (AmyII-3) was increased in chalky grain in both moderate (24.4 °C) and HT (28.0 °C) conditions [92]. DBE and BEI were down- and increased in moderate and HT conditions, respectively [92].

Moreover, Timabud et al. [90] reported that heat stress affected the abundance of starch biosynthesis proteins in milky, dough, and mature stages. Under heat treatment, SBE3, AGPase L2, SSIIa, and SSI were expressed and detected only at the milky stage, while SBEI and GBSSI were increased in abundance from milky to the dough but not detectable at the mature stage. AMY was highly expressed at the milky, compared to the dough stage, whereas AGPase was increased in the highest abundance at the mature stage. AGPase S2 was differentially expressed in both dough and mature stages, while ISA was detected at the milky and mature stages [90]. In addition, grains weight under HT were increased more rapidly than the control, especially from late milky to middle dough stages [90,91].

Altogether, HT activates starch degradation rather than starch biosynthesis through the reduction of GBSSI, SSI, SSII, SBEIIb, DBE, and PUL, while some proteins, such as AGPase L, GBSSI, SBEI, and AMY are increased and contributed to the lower amylose content and the higher the chalkiness rate in rice endosperm.

4. Starch Biosynthesis-Related Proteins Targeted by PTMs

PTMs can alter structure formation, activity, stability, structure, and localization of proteins, which are necessary for cellular functions [84,85]. Five types of PTMs targeting starch biosynthesis proteins have been reported from rice seeds (Figures 3–5).

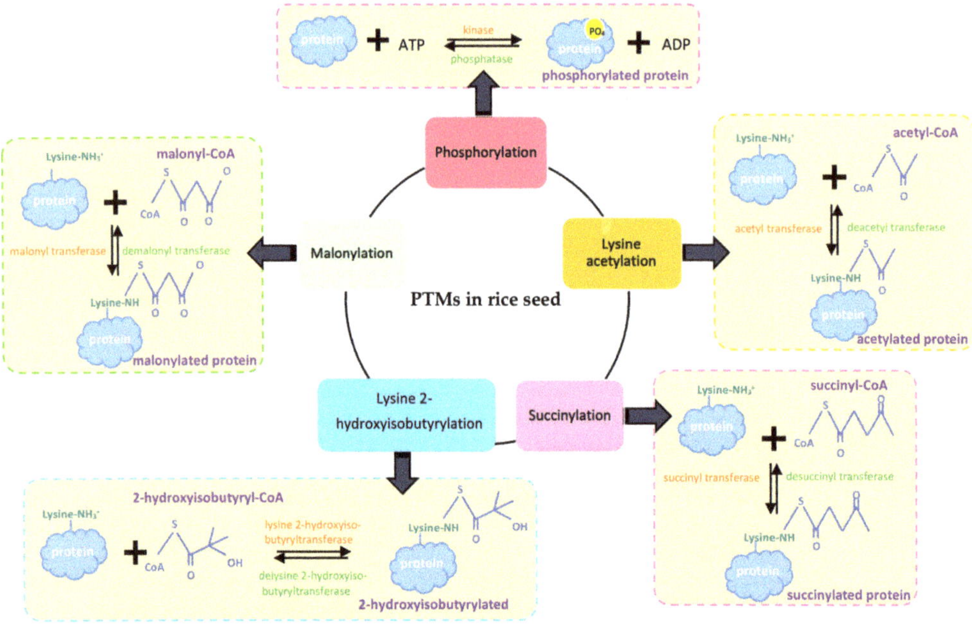

Figure 3. Five types of PTMs targeting starch biosynthesis proteins identified from rice seeds.

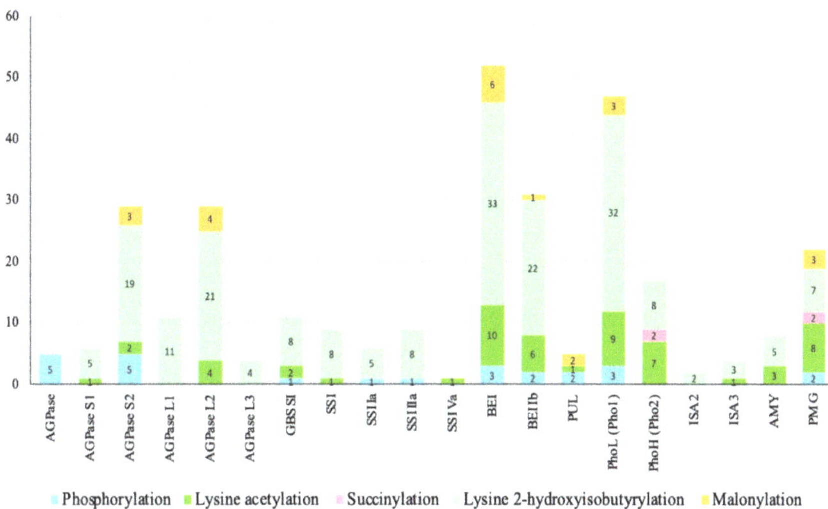

Figure 4. Summary of PTMs targeting starch biosynthesis-related proteins in rice seeds. The number on stacked bar indicates total target sites of PTM(s) identified for each protein.

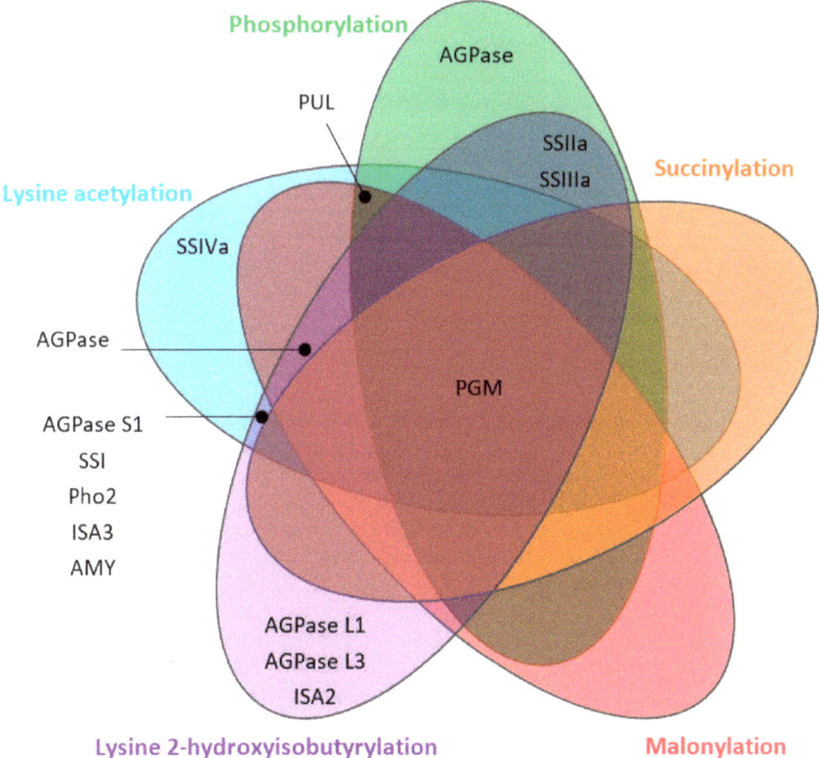

Figure 5. Venn diagram displays the overlap of starch biosynthesis-related proteins targeted by five types of PTMs identified from rice seeds.

4.1. Phosphorylation

4.1.1. Identification of Phosphorylated Protein in Rice Developing Seeds

Protein phosphorylation is a reversible process regulated by kinases and phosphatases and regarded as one of the most important PTMs [94]. The studies on phosphorylation have mainly focused on phosphorylation and/or dephosphorylation of specific proteins or protein families in particular signaling pathways [95]. In eukaryotes, the most common class of phosphorylation is found on serine (S), threonine (T), and tyrosine (Y) residues [96,97] in which the γ-phosphate group covalently reacts with the hydroxyl group of amino acid side chains by potein kinases [98]. A large scale of phosphosites and phosphoproteins has been identified using phosphoproteomic technologies in many cereals such as rice [27,28,31,99–103], wheat [104–108], barley [109,110], and maize [111–115]. Phosphoserine (>90%) is a major phosphorylation type of rice endosperm, followed by phosphothreonine (6–9%) and phosphotyrosine (0.1–0.4%) [28,31]. Nakagami et al. [116] estimated that the proportion of phosphotyrosine in rice is equivalent to that in Arabidopsis and humans.

Recently, the phosphoproteins involved in starch biosynthesis of indica rice cultivars (9311 and Guangluai4) including AGPase (three sites) AGPS2 (five sites), SSIIa (one site), SSIIIa (one site), BEI (four sites), BEIIb (two sites), PUL (three sites) and Pho1 (two sites) were reported by Pang et al. [28] (Table 4).

Table 4. List of identified phosphoproteins involving in starch biosynthesis in rice endosperm based on phosphoproteomics.

Phosphorylated Protein	Uniprot ACCN	Identified Phosphosite(s) [a]	Subspecies	Reference
AGPase	B8XEC3	S62, S381	*indica*	[28]
		T68	*japonica and indica*	[28,31]
	A2Y7W1	S491	*japonica*	[27]
	-	-	*indica*	[87,88]
AGPS2	D4AIA3	S13	*japonica and indica*	[28,31]
		S17, S22, S35, S36	*Indica*	[28]
GBSSI	-	-	*indica*	[87]
SSIIa	P0C586	S126	*indica*	[28]
SSIIIa	Q6Z1D6	S96	*japonica and indica*	[28,31]
BEI	D0TZI4	S562, S620, S814, S815	*indica*	[28]
BEIIb	A2X5K0	S685, S715	*indica*	[28]
PUL	D0TZH1	S154, S155, S869	*indica*	[28]
Pho1	Q9AUV8	S494, S645	*indica*	[28]
PGM	Q9AUQ4	S124	*japonica*	[31]
	Q33AE4	S167	*japonica*	[31]
	-	-	*indica*	[88]

[a] S and T indicate the phosphorylated site on serine and threonine residues, respectively.

Interestingly, one phosphopeptide of both AGPase and SSIIIa showed consistency between japonica and indica rice [28,31]. AGPase and GBSS were detected as downregulated phosphoproteins during grain-filling stages of inferior spikelets, as compared to superior spikelets [87]. Moreover, PGM and AGPase were differentially phosphorylated between the superior and inferior spikelets in which the expression levels of those phosphoproteins of 10 DAF inferior spikelets were lower than both 1 DAF superior spikelets and 20 DAF inferior spikelets [88].

4.1.2. Potential Role of Protein Phosphorylation in Starch Biosynthesis

In amyloplasts, starch biosynthesis isozymes have been demonstrated to display as a complex form (or protein–protein interactions), especially through the regulation of phosphorylation [117–120]. In wheat (*Triticum aestivum*), phosphorylation can activate SBEIIa and SBEIIb enzymes contributed to the protein complex forming of SBEIIa, SBEIIb, and Pho1 at 12–25 days after pollination (DAP) [117]. On the other hand, dephosphorylation reduces the catalytic activities and breaks the complex formation [117]. Furthermore, the phosphorylation-dependent complexes of wheat SSI, SSIIa, and either SBEIIa or SBEIIb were identified in amyloplast at 10–15 DAP [119].

In maize (*Zea mays* L.) endosperm, SSI, SSIIa, and SBEIIb form a trimeric complex in which SBEIIb is phosphorylated [121]. The complex formation is activated by ATP and disassembled by alkaline phosphatase [118,121]. Loss of SBEIIb activity (*amylose extender*, *ae*$^-$ mutant) impacts the protein–protein interactions among SSI, SSIIa, and SBEIIb complex, which is formed in wild type [118]. It was reported that SSI and SSIIa formed the complex possibly through SBEI, SBEIIa, and Pho in the *ae*$^-$ mutant. Since the SBEIIb is replaced by SBEI, a reduction of branch points with longer glucan chains was observed in *ae*$^-$ mutant, as compared to the wild type [118,121]. Recently, the phosphorylated SSIIa is reported that to have interactions with SSI and SBEIIb [120]. In addition, in barley (*Hordeum vulgare*), protein complex formation of SBEIIa, SBEIIb, SSIIa, and SSIIIa was increased by the presence of ATP [122].

In rice endosperm, phosphorylation and dephosphorylation affected the oligomerization and activity of OsGBSSI [123]. The dissociation of OsGBSSI was detected during the phosphatase treatment. The monomer of OsGBSSI increased from 0.07 to 0.86%, and the OsGBSSI activity decreased from 0.17 to 0.11 mol/g/min based upon the increasing phosphatase levels [123]. In addition, GBSSI expression at low temperatures was regulated by phosphorylation [124].

Protein complex formations among starch biosynthesis proteins have been established in rice including the interactions of SSs-SBEs, among SBE isoforms, BEIIa-Pho1, PUL-BEI [125–127], PUL-BEIIb [126], Pho1-Dpe1 [128], and SSI-SSIIa-BEIIb [126]. Recently, an inactive BEIIb forming a complex with SSI, SSIIa, SSIVb, BEI, and BEIIa was reported [126]. Although the evidence of phosphorylation-dependent complex assembly in rice has not been uncovered, the enzymatic activities and enzymatic complexes in rice might be regulated by phosphorylation, as reported in other cereals.

Taken together, phosphorylation is essential for the regulation of starch biosynthesis and has significant effects on enzymatic activities, complex components, and protein–protein interactions. Knockout of one enzyme may lead to changes in protein complex formation, other enzyme activities, and amylopectin structure.

4.2. Lysine Acetylation

Lysine acetylation, a highly conserved PTM in organisms, is well known for the regulation of transcription [129] and reported in a large number of proteins in many biological processes of organisms [130]. Lysine acetylation is a reversible reaction in which an acetyl group (CH_3Co) from acetyl–coenzyme A (CoA) is donated to N^{ε}-terminal amine of a lysine residue through acetyltransferases and removed by deacetylases [131]. The acetylation controls the enzymatic activities of metabolic enzymes and alters the metabolic flux profiles [132]. In rice seeds, starch biosynthesis proteins targeted by lysine acetylation were listed in Table 5.

Table 5. List of lysine acetylation and succinylation on starch biosynthesis proteins from rice seeds.

Acetylated Protein	Uniprot ACCN	Acetylation Position	Modified Peptide [a]	Lysine Motif [b]	Tissue-Specific [b]	Reference
AGPase S1	Q69T99	203	MDYQK(ac)FIQAHR	-	-	[30]
AGPase S2	P15280	217	MDYEK(ac)FIQAHR	-	starch granule/seeds (7 DAP)/seeds (15 DAF)	[32]/[30]/[33]
		261	IVEFAEK(ac)PK	KF	starch granule/seeds (unpollinated pistil and 7 DAP)/seeds (15 DAF)	[32]/[30]/[33]
AGPase L2	Q5VNT5	250	ASDYGLVK(ac)FDDSGR	KF	starch granule/seeds (3 and 7 DAP)/seeds (15 DAF)	[32]/[30]/[33]
		260	VIAFSEK(ac)PK	-	starch granule	[32]
		310	DVLLDILK(ac)SK	-	Seeds (7 DAP)	[30]
		312	SK(ac)YAHLQDFGSEILPR	-	Seeds (7 DAP)	[30]

Table 5. *Cont.*

Acetylated Protein	Uniprot ACCN	Acetylation Position	Modified Peptide [a]	Lysine Motif [b]	Tissue-Specific [b]	Reference
GBSSI	Q0DEV5	444	KFEK(ac)LLK	-	starch granule/seeds (15 DAF)	[32]/[33]
		452	SMEEK(ac)YPGK	KY	starch granule/seeds (15 DAF)	[32]/[33]
SSI	Q0DEC8	193	NFANAFYTEK(ac)HIK	-	seeds (3 and 7 DAP)	[30]
SSIVa	Q5JMA0	589	AQYYGEHDDFK(ac)R	-	seeds (15 DAF)	[33]
SBEI	Q0D9D0	89	LEEFK(ac)DHFNYR	-	starch granule/seeds (15 DAF)	[32]/[33]
		103	YLDQK(ac)CLIEK	-	starch granule/seeds (15 DAF)	[32]/[33]
		118	HEGGLEEFSK(ac)GYLK	KXXXK	starch granule/seeds (15 DAF)	[32]/[33]
		164	DK(ac)FGIWSIK	KF	starch granule/seeds (15 DAF)	[32]/[33]
		236	YVFK(ac)HPR	KH	starch granule/seeds (15 DAF)	[32]/[33]
		372	GYHK(ac)LWDSR	KXXXXR	starch granule/seeds (15 DAF)	[32]/[33]
		614	EGNNWSYDK(ac)CR	-	starch granule/seeds (15 DAF)	[32]/[33]
		662	QIVSDMNEK(ac)DK	-	starch granule/seeds (15 DAF)	[32]/[33]
		697	VGCDLPGK(ac)YR	KY	starch granule/seeds (15 DAF)	[32]/[33]
		809	GM(ox)K(ac)FVFR	KXXXR	starch granule/seeds (15 DAF)	[32]/[33]
SBEIIb	Q6H6P8	134	VVEELAAEQK(ac)PR	-	seeds (15 DAF)	[33]
		303	YIFK(ac)HPQPK	KH	Seed (7 DAP)/seeds (15 DAF)	[30]/[33]
		587	WSEK(ac)CVTYAESHDQALVGDK	-	seeds (unpollinated pistil and 7 DAP)	[30]
		688	FIPGNNNSYDK(ac)CR	-	seeds (7 DAP)	[30]
		738	KHEEDK(ac)MIIFEK	-	starch granule/seeds (15 DAF)	[32]/[33]
		771	VGCLKPGK(ac)YK	KY	starch granule/seeds (15 DAF)	[32]/[33]
ISA3	Q6K4A4	130	K(ac)YFGVAEEK	KY	seeds (15 DAF)	[33]
PUL	Q7X834	805	NEENWHLIK(ac)PR	-	seeds (15 DAF)	[33]
PMG	Q9AUQ4	8	VLFSVTK(su)K	-	embryos (24 HAI)	[34]
		18	ATTPFDGQK(ac)PGTSGLR	-	embryos (24 HAI)/seeds (15 DAF)	[34]/[33]
			ATTPFDGQK(su)PGTSGLR		embryos (24 HAI)	[34]
		69	YFSK(ac)DAVQIITK	-	embryos (24 HAI)	[34]
		206	LMK(ac)TIFDFESIK	-	embryos (24 HAI)	[34]
		215	TIFDFESIK(ac)K	-	seeds (15 DAF)	[33]
		275	EDFGGGHPDPNLTYAK(ac)ELVDR	-	embryos (24 HAI)	[34]
		361	NLNLK(ac)FFEVPTGWK	-	embryos (24 HAI)	[34]
		506	DPVDGSVSK(ac)HQGVR	KH	embryos (24 HAI)/seeds (15 DAF)	[34]/[33]
		543	VYIEQYEK(ac)DSSK	KXXXK	seeds (15 DAF)	[33]
PhoH	Q8LQ33	169	YGLFK(ac)QCITK	-	embryos (24 HAI)	[34]
		409	HMEIIEEIDK(ac)R	-	embryos (24 HAI)	[34]
		412	FK(su)EMVISTR	-	embryos (24 HAI)	[34]
		439	ILDNSNPQK(su)PVVR	-	embryos (24 HAI)	[34]
		645	LVNDVGAVVNNDPDVNK(ac)YLK	-	embryos (24 HAI)	[34]
		747	FEEAK(ac)QLIR	KXXXR	seeds (15 DAF)	[33]
		818	MSILNTAGSGK(ac)FSSDR	-	embryos (24 HAI)	[34]
PhoL	Q9AUV8	216	YK(ac)HGLFK	KH	starch granule/seeds (unpollinated pistil, 3 DAP and 7 DAP)/seeds (15 DAF)	[32]/[30]/[33]
		255	TDVSYPVK(ac)FYGK	KXXXK	starch granule/seeds (15 DAF)	[32]/[33]
		451	YGTEDTSLLK(ac)K	-	starch granule/seeds (15 DAF)	[32]/[33]
		504	SLEPSVVVEEK(ac)TVSK	KXXXK	starch granule/seeds (15 DAF)	[32]/[33]
		594	FQNK(ac)TNGVTPR	-	starch granule/seeds (15 DAF)	[32]/[33]
		734	AFATYVQAK(ac)R	-	seeds (7 DAP)	[30]
		846	AQGK(ac)FVPDPR	KF	starch granule/seeds (15 DAF)	[32]/[33]
		913	DQK(ac)LWTR	KXXXR	starch granule/seeds (15 DAF)	[32]/[33]
		928	MSILNTASSSK(ac)FNSDR	KF	starch granule/seeds (15 DAF)	[32]
AMY	Q0J528	88	LYDLDASK(ac)YGTEAELK	-	embryos (24 HAI)/-	[34]
		123	CADYK(ac)DSR	-	-	[30]
	P27933	88	LYDLDASK(ac)YGTAAELK	-	-	[30]
		215	GYSTDIAK(ac)MYVESCK	-	-	[30]

[a] (ac) and (su) indicate the acetylation and succinylation sites on lysine, respectively. [b] K is the position of the acetylated or succinylated lysine, and X refers to a random amino acid residue.

In total, 2, 2, 2, 10, 2, and 8 sites of lysine acetylation modifications identified from starch granules of mature rice endosperm were reported for AGPase S2, AGPase L2, GBSSI, SBEI, SBEIIb, and Pho L, respectively [32] (Table 5). PGM, Pho L, and AMY were acetylated with six, four, and one sites at the early stage of rice seed germination (24 h after imbibition) [34]. In addition, Wang et al. [30] reported the largest acetylome based on three different developing stages of rice seeds (unpollinated pistil, 3 DAP, and 7 DAP) covering 1817 acetylsites from 972 acetylproteins. Seven starch biosynthesis proteins were targeted by acetyllysine including AGPase S1 (one site), AGPase S2 (two sites), AGPase L2 (three sites), AMY (four sites), SBE3 (three sites), Pho L (two sites), and SSI (one site) (Table 5).

Compared to root, leave, flower, and pollen, the prominent acetylated proteins of rice were observed in seeds at 7, 15, and 21 DAF, as well as in mature dry seeds [33]. A total of 1003 acetylated sites on 692 proteins were identified from rice seeds at 15 DAF [33]. Interestingly, 11 starch biosynthesis proteins were lysine-acetylated including AGPasae S2, AGPasae L2, GBSSI, SSIVa, SBEI, SBEIIb, ISA3, PUL, PMG, Pho H, and Pho L (Table 5).

Among those acetylated proteins, SBEI and Pho L were heavily acetylated on 10 and 8 sites, respectively, which were in a similar result of starch granule acetylome [32]. Out of 7 distinguished motifs of acetylation site reported by Meng et al. [33], 6 motifs displayed amongst the 11 acetylated starch biosynthesis proteins including $\underline{K}$F (5), $\underline{K}$Y (4), $\underline{K}$XXXR (3), $\underline{K}$XXXXR (1), $\underline{K}$XXXK (4) and $\underline{K}$H (4) (Table 5).

Lysine acetylation has a strong impact on the biochemical functions of proteins [133]. For example, Ribulose-15-bisphosphate carboxylase/oxygenase (Rubisco) activity was decreased by lysine acetylation [134], the increased glyceraldehyde-3-phosphate dehydrogenases (GAPDH) acetylation in leaves of *Brachypodium distachyon* L. enhanced the activity in glycolysis and decreased the activity in gluconeogenesis [135]. On the other hand, de-acetylation can increase the pyruvate orthophosphate dikinase (PPDK) activity in maize after 12 h of illumination with white light [134]. It is plausible that lysine acetylation may influence the catalytic activities of starch biosynthesis proteins in rice.

4.3. Succinylation

Succinylation is recently identified as one of the PTMs on lysine residue [136,137] and plays an important role in gene transcription, cellular metabolism, DNA damage response [138], and plant growth [135]. Zhang et al. [136] reported that succinyl–CoA plays a role as a cofactor for lysine succinylation. Among 261 acetylated proteins identified in rice embryos at 24 h after imbibition, two sites from both PGM (8 and 18) and Pho H (412 and 439) were succinylated, which might be involved in the metabolism regulation [34]. Lysine residue at 18 positions on PGM (Uniprot: Q9AUQ4) was targeted for both acetylation and succinylation modifications (Table 5).

Previous studies indicated that succinylation has a potential impact on protein structure and functions [136] as well as cellular processes [139]. Succinylation alters the electrophoretic mobility and isoionic pH of ovalbumin [140], inactivates the canonical carnitine palmitoyltransferase (CPTase) activity, decreases the enolase activity [141], and promotes cell invasion and migration [139]. Succinylation in plants has been reported in rice [34], tomato [142], *B. distachyon* L. [135], patchouli [143], and tea [144]. However, the effect of lysine succinylation in plants is relatively limited.

4.4. Lysine 2-Hydroxyisobutyrylation (K_{hib}) and Malonylation (K_{mal})

K_{hib} is a highly dynamic PTM found on both histone and nonhistone proteins affecting the histone–DNA association and playing role in diverse biological processes [36,145]. K_{hib} introduces a huge change in size, as compared to lysine acetylation, and particularly forms hydrogen bonds with other molecules via its hydroxyl group [145].

In contrast, K_{mal}, a lately identified lysine acylation, is evolutionarily conserved in mammalian and bacteria cells [146,147] and responsible for the regulation of cellular mechanisms and activities [35,146]. K_{mal} triggers more dramatic structural changes than both lysine acetylation and methylation on the substrate proteins [147]. Recently, both

K_{hib} and K_{mal} were identified from the developing rice seeds at 15 DAF reported by Meng et al. [36] and Mujahid [35], respectively (Table 6).

Table 6. Summary of lysine 2-hydroxyisobutyrylation (K_{hib}) and malonylation (K_{mal}) on starch biosynthesis proteins from 15 DAF developing rice seeds identified by Meng et al. [36] and Mujahid et al. [35], respectively.

Protein	Uniprot ACCN	No. of K_{hib} and K_{mal} *	Position
AGPase S1	Q69T99	5	203, 234, 249, 442, 462
AGPase S2	P15280	19	102, 132, 217, 239, 248, 261, 263, 268, 285, 360, 385, 403, 406, 441, 447, 456, 467, 476, 496
		3 *	106, 360, 403
AGPase L1	Q6AVT2	11	100, 194, 196, 247, 299, 331, 326, 369, 446, 456, 470,
AGPase L2	Q5VNT5	21	37, 74, 187, 223, 250, 263, 273, 286, 301, 302, 310, 312, 334, 364, 371, 392, 425, 443, 449, 459, 504,
		4 *	250, 312, 371, 449
AGPase L3	Q688T8	4	202, 228, 315, 376
GBSSI	Q0DEV5	8	181, 192, 309, 381, 385, 530, 538, 549
SSI	Q0DEC8	8	193, 196, 349, 357, 429, 461, 467, 570
SSII-3	Q0DDE3	5	151, 244, 346, 378, 532
SSIIIa	Q6Z1D6	8	228, 649, 761, 794, 808, 961, 1203, 1604
SBEI	Q0D9D0	33	62, 64, 84, 89, 103, 108, 118, 122, 157, 164, 171, 186, 215, 236, 319, 324, 372, 423, 500, 506, 524, 540, 549, 614, 662, 664, 683, 689, 697, 744, 775, 796, 809
		6 *	108, 118, 506, 524, 689, 809
SBEIIb	Q6H6P8	22	134, 146, 158, 191, 231, 268, 299, 328, 386, 466, 558, 564, 571, 587, 603, 612, 636, 677, 688, 719, 738, 773,
		1 *	719
AMY	Q0J528	5	39, 88, 105, 207, 262
	Q0JJV2	1	88
ISA2	Q6AU80	2	319, 369
ISA3	Q6K4A4	2	266, 269
PhoH	Q8LQ33	8	115, 409, 425, 533, 542, 595, 721, 818,
PhoL	Q9AUV8	32	134, 255, 259, 277, 289, 356, 381, 410, 418, 429, 441, 451, 471, 493, 504, 590, 617, 630, 636, 657, 665, 681, 725, 734, 738, 846, 893, 904, 913, 928, 940, 946,
		3 *	259, 493, 657
PUL	Q7X834	2 *	274, 871
PGM	Q33AE4	7	61, 67, 118, 413, 492, 584, 595
	Q9AUQ4	3 *	54, 458, 568

* Indicates the number of K_{mal}.

A total of 2512 K_{hib} proteins were reported in rice [36] and amongst those proteins, 17 proteins involving in starch biosynthesis were targeted (Table 6). The highest modified sites of K_{hib} were observed on SBEI (33 sites), followed by Pho L (32 sites) and AGPase L2 (21 sites). The K_{hib} function on targeted proteins remains unknown in rice. However, there was evidence that K_{hib} modification may introduce the conformational changes of Enolase1 and alter the substrate binding [148]. K_{hib} increased the hydrophobic solvent-accessible surface area and decreased the enzymatic activity of UvSlt2 [149].

For K_{mal}, seven starch biosynthesis proteins out of 247 malonylated proteins were reported [35]. The number of the modified sites of AGPase S2, AGPase L2, SBEI, SBEIIb, Pho L, PUL, and PGM was 3, 4, 6, 1, 3, 2, 3, respectively (Table 6). Although malonylation can occur on several enzymes of the starch biosynthesis mechanism, the potential roles of K_{mal} remain largely unknown. The malonylation has an important role in the enzymatic activity of various proteins. For example, the enzymatic activity of malonylated fructose bisphosphate aldolase B (ALDOB) was decreased by 20%, as compared to nonmalonylated ALDOB [150]. Malonylation increased the enzymatic activity of glyceraldehyde-3-phosphate dehydrogenase (GAPDH) and also interrupted its binding to the target mRNAs [151]. Moreover, cells with elevated K_{mal} had the impaired mitochondrial function and fatty acid oxidation [152].

5. Summary and Future Perspectives

Even though identification of the proteomics and PTMs has been conducted in rice developing seeds, there still remain significant gaps in their regulations in starch biosynthesis. This review presents the significant proteins associated with starch biosynthesis in rice seeds and their expression profiles through different developmental stages and high temperatures. Most of the key proteins in starch biosynthesis are generally increased in the endosperms during 6–20 DAF. SSIIIa and AMY are promising proteins in chalky formation. Hight temperature initiates starch degradation rather than starch biosynthesis and then also results in the reduction of amylose content as well as the increase in chalkiness in rice seeds. Twenty starch biosynthesis proteins are targeted by five types of PTMs including phosphorylation, lysine acetylation, succinylation, lysine 2-hydroxyisobutyrylation, and malonylation. PGM is commonly targeted by all the five PTMs types (Figure 5). Phosphorylation is the most important PTM for starch biosynthesis proteins regarding the regulation of protein complex formation. This information is useful to understand the molecular mechanisms underlying starch biosynthesis, which ultimately affect starch functionality with known and unknown regulatory pathways. Further studies, however, may be focused on the following aspects:

(1) Proteome alteration under climate change environment: Recently, the global population is facing challenging problems caused by global warming and climate change, which have a great impact on rice yield and quality. Further studies are needed to determine the consequences of climate change, e.g., high/low temperatures, carbon dioxide levels, drought stress, etc., on starch biosynthesis mechanism and regulation by using proteomic analysis.

(2) The number and new types of PTMs in rice seeds: Although five types of PTMs were identified from rice seeds, whether there are other PTMs in rice seed has not been fully addressed. For the number of PTMs sites, K_{hib} showed the highest number of targeted starch biosynthesis proteins (17 proteins), while the lowest number was observed in succinylation (2 proteins). Whether more PTMs would be found under the specific genotype or under the specific abiotic conditions such as heat stress, high carbon dioxide levels, etc., is unknown.

(3) The roles and regulation mechanisms of PTMs on starch biosynthesis: Little is known about the roles of individual PTM on the starch biosynthesis proteins and the impact of PTMs on enzymes' activities, protein–protein interaction (protein complex formation), and starch functionality. The phosphorylation is well reported in protein complex formation during the starch biosynthesis process in the endosperm of cereal crops. In-depth regulatory studies on protein–protein interactions are necessary to understand the role of protein complex formation in starch biosynthesis in different crops.

Author Contributions: Conceptualization, J.B. and P.T.; writing—original draft preparation, P.T., Y.Y. and F.X.; writing—review and editing, P.T. and J.B.; funding acquisition, J.B. All authors have read and agreed to the published version of the manuscript.

Funding: This work was financially supported by the National Key Research and Development Program of China (Grant No. 2016YFD0400104), the Natural Science Foundation of China (Grant No. 31871531), and Zhejiang Province (LZ21C130003) (for J.B.), and the Postdoctoral InternationaEx-change Program (for P.T.).

Institutional Review Board Statement: Not applicable.

Informed Consent Statement: Not applicable.

Data Available Statement: All data presented in this review can be found in the references cited in the text.

Conflicts of Interest: The authors declare no conflict of interest.

References

1. FAO. FAOSTAT. Available online: http://www.fao.org/home/en (accessed on 8 February 2020).
2. Jiang, C.; Cheng, Z.; Zhang, C.; Yu, T.; Zhong, Q.; Shen, J.Q.; Huang, X. Proteomic analysis of seed storage proteins in wild rice species of the *Oryza* genus. *Proteome Sci.* **2014**, *12*, 51. [CrossRef]
3. Wang, W.; Mauleon, R.; Hu, Z.; Chebotarov, D.; Tai, S.; Wu, Z.; Li, M.; Zheng, T.; Fuentes, R.R.; Zhang, F.; et al. Genomic variation in 3,010 diverse accessions of Asian cultivated rice. *Nature* **2018**, *557*, 43–49. [CrossRef]
4. Gutaker, R.M.; Groen, S.C.; Bellis, E.S.; Choi, J.Y.; Pires, I.S.; Bocinsky, R.K.; Slayton, E.R.; Wilkins, O.; Castillo, C.C.; Negrão, S.; et al. Genomic history and ecology of the geographic spread of rice. *Nat. Plants* **2020**, *6*, 492–502. [CrossRef]
5. Li, X.; Wu, L.; Wang, J.; Sun, J.; Xia, X.; Geng, X.; Wang, X.; Xu, Z.; Xu, Q. Genome sequencing of rice subspecies and genetic analysis of recombinant lines reveals regional yield- and quality-associated loci. *BMC Biol.* **2018**, *16*, 102. [CrossRef]
6. Khush, G.S. Origin, dispersal, cultivation and variation of rice. *Plant Mol. Biol.* **1997**, *35*, 25–34. [CrossRef] [PubMed]
7. He, D.; Yang, P. Proteomics of rice seed germination. *Front. Plant Sci.* **2013**, *4*, 246. [CrossRef] [PubMed]
8. Kim, Y.J.; Choi, S.H.; Park, B.S.; Song, J.T.; Kim, M.C.; Koh, H.J.; Seo, H.S. Proteomic analysis of the rice seed for quality improvement. *Plant Breed.* **2009**, *128*, 541–550. [CrossRef]
9. Goren, A.; Ashlock, D.; Tetlow, I.J. Starch formation inside plastids of higher plants. *Protoplasma* **2018**, *255*, 1855–1876. [CrossRef]
10. Zeeman, S.C.; Kossmann, J.; Smith, A.M. Starch: Its metabolism, evolution, and biotechnological modification in plants. *Annu. Rev. Plant Biol.* **2010**, *61*, 209–234. [CrossRef]
11. Ball, S.G.; Morell, M.K. From bacterial glycogen to starch: Understanding the biogenesis of the plant starch granule. *Annu. Rev. Plant Biol.* **2003**, *54*, 207–233. [CrossRef] [PubMed]
12. Jobling, S. Improving starch for food and industrial applications. *Curr. Opin. Plant Biol.* **2004**, *7*, 210–218. [CrossRef] [PubMed]
13. Lin, S.K.; Chang, M.C.; Tsai, Y.G.; Lur, H.S. Proteomic analysis of the expression of proteins related to rice quality during caryopsis development and the effect of high temperature on expression. *Proteomics* **2005**, *5*, 2140–2156. [CrossRef]
14. Kawagoe, Y.; Kubo, A.; Satoh, H.; Takaiwa, F.; Nakamura, Y. Roles of isoamylase and ADP-glucose pyrophosphorylase in starch granule synthesis in rice endosperm. *Plant J.* **2005**, *42*, 164–174. [CrossRef]
15. Fujita, N.; Yoshida, M.; Asakura, N.; Ohdan, T.; Miyao, A.; Hirochika, H.; Nakamura, Y. Function and characterization of starch synthase I using mutants in rice. *Plant Physiol.* **2006**, *140*, 1070–1084. [CrossRef]
16. Thurston, G.; Regan, S.; Rampitsch, C.; Xing, T. Proteomic and phosphoproteomic approaches to understand plant–pathogen interactions. *Physiol. Mol. Plant. Pathol.* **2005**, *66*, 3–11. [CrossRef]
17. Cánovas, F.M.; Dumas-Gaudot, E.; Recorbet, G.; Jorrin, J.; Mock, H.-P.; Rossignol, M. Plant proteome analysis. *Proteomics* **2004**, *4*, 285–298. [CrossRef] [PubMed]
18. Newton, R.P.; Brenton, A.G.; Smith, C.J.; Dudley, E. Plant proteome analysis by mass spectrometry: Principles, problems, pitfalls and recent developments. *Phytochemistry* **2004**, *65*, 1449–1485. [CrossRef]
19. Lin, Z.; Zhang, X.; Yang, X.; Li, G.; Tang, S.; Wang, S.; Ding, Y.; Liu, Z. Proteomic analysis of proteins related to rice grain chalkiness using iTRAQ and a novel comparison system based on a notched-belly mutant with white-belly. *BMC Plant Biol.* **2014**, *14*, 163. [CrossRef]
20. Wittmann-Liebold, B.; Graack, H.-R.; Pohl, T. Two-dimensional gel electrophoresis as tool for proteomics studies in combination with protein identification by mass spectrometry. *Proteomics* **2006**, *6*, 4688–4703. [CrossRef]
21. Yang, P.; Li, X.; Wang, X.; Chen, H.; Chen, F.; Shen, S. Proteomic analysis of rice (*Oryza sativa*) seeds during germination. *Proteomics* **2007**, *7*, 3358–3368. [CrossRef]
22. Xu, S.B.; Li, T.; Deng, Z.Y.; Chong, K.; Xue, Y.; Wang, T. Dynamic proteomic analysis reveals a switch between central carbon metabolism and alcoholic fermentation in rice filling grains. *Plant. Physiol.* **2008**, *148*, 908–925. [CrossRef] [PubMed]
23. Koller, A.; Washburn, M.P.; Lange, B.M.; Andon, N.L.; Deciu, C.; Haynes, P.A.; Hays, L.; Schieltz, D.; Ulaszek, R.; Wei, J.; et al. Proteomic survey of metabolic pathways in rice. *Proc. Natl. Acad. Sci. USA* **2002**, *99*, 11969. [CrossRef]
24. Xu, S.B.; Yu, H.T.; Yan, L.F.; Wang, T. Integrated proteomic and cytological study of rice endosperms at the storage phase. *J. Proteome Res.* **2010**, *9*, 4906–4918. [CrossRef] [PubMed]
25. Lee, J.; Koh, H.-J. A label-free quantitative shotgun proteomics analysis of rice grain development. *Proteome Sci.* **2011**, *9*, 61. [CrossRef]

26. Yu, H.; Wang, T. Proteomic dissection of endosperm starch granule associated proteins reveals a network coordinating starch biosynthesis and amino acid metabolism and glycolysis in rice endosperms. *Front. Plant Sci.* **2016**, *7*, 707. [CrossRef]
27. Han, C.; Wang, K.; Yang, P. Gel-based comparative phosphoproteomic analysis on rice embryo during germination. *Plant. Cell Physiol.* **2014**, *55*, 1376–1394. [CrossRef]
28. Pang, Y.; Zhou, X.; Chen, Y.; Bao, J.S. Comparative phosphoproteomic analysis of the developing seeds in two Indica rice (*Oryza sativa* L.) cultivars with different starch quality. *J. Agric. Food Chem.* **2018**, *66*, 3030–3037. [CrossRef]
29. Huang, K.-Y.; Su, M.-G.; Kao, H.-J.; Hsieh, Y.-C.; Jhong, J.-H.; Cheng, K.-H.; Huang, H.-D.; Lee, T.-Y. dbPTM 2016: 10-year anniversary of a resource for post-translational modification of proteins. *Nucleic Acids Res.* **2016**, *44*, D435–D446. [CrossRef]
30. Wang, Y.; Hou, Y.; Qiu, J.; Li, Z.; Zhao, J.; Tong, X.; Zhang, J. A quantitative acetylomic analysis of early seed development in rice (*Oryza sativa* L.). *Int. J. Mol. Sci.* **2017**, *18*, 1376. [CrossRef] [PubMed]
31. Qiu, J.; Hou, Y.; Tong, X.; Wang, Y.; Lin, H.; Liu, Q.; Zhang, W.; Li, Z.; Nallamilli, B.R.; Zhang, J. Quantitative phosphoproteomic analysis of early seed development in rice (*Oryza sativa* L.). *Plant Mol. Biol.* **2016**, *90*, 249–265. [CrossRef]
32. Xing, S.; Meng, X.; Zhou, L.; Mujahid, H.; Zhao, C.; Zhang, Y.; Wang, C.; Peng, Z. Proteome profile of starch granules purified from rice (*Oryza sativa*) endosperm. *PLoS ONE* **2016**, *11*, e0168467. [CrossRef] [PubMed]
33. Meng, X.; Lv, Y.; Mujahid, H.; Edelmann, M.J.; Zhao, H.; Peng, X.; Peng, Z. Proteome-wide lysine acetylation identification in developing rice (*Oryza sativa*) seeds and protein co-modification by acetylation, succinylation, ubiquitination, and phosphorylation. *Biochim. Biophys. Acta (BBA)–Proteins Proteom.* **2018**, *1866*, 451–463. [CrossRef]
34. He, D.; Wang, Q.; Li, M.; Damaris, R.N.; Yi, X.; Cheng, Z.; Yang, P. Global proteome analyses of lysine acetylation and succinylation reveal the widespread involvement of both modification in metabolism in the embryo of germinating rice seed. *J. Proteome Res.* **2016**, *15*, 879–890. [CrossRef]
35. Mujahid, H.; Meng, X.; Xing, S.; Peng, X.; Wang, C.; Peng, Z. Malonylome analysis in developing rice (*Oryza sativa*) seeds suggesting that protein lysine malonylation is well-conserved and overlaps with acetylation and succinylation substantially. *J. Proteom.* **2018**, *170*, 88–98. [CrossRef] [PubMed]
36. Meng, X.; Xing, S.; Perez, L.M.; Peng, X.; Zhao, Q.; Redoña, E.D.; Wang, C.; Peng, Z. Proteome-wide analysis of lysine 2-hydroxyisobutyrylation in developing rice (*Oryza sativa*) seeds. *Sci. Rep.* **2017**, *7*, 17486. [CrossRef]
37. Seck, P.A.; Diagne, A.; Mohanty, S.; Wopereis, M.C.S. Crops that feed the world 7: Rice. *Food Secur.* **2012**, *4*, 7–24. [CrossRef]
38. Chang, T.-S.; Liu, C.-W.; Lin, Y.-L.; Li, C.-Y.; Wang, A.Z.; Chien, M.-W.; Wang, C.-S.; Lai, C.-C. Mapping and comparative proteomic analysis of the starch biosynthetic pathway in rice by 2D PAGE/MS. *Plant Mol. Biol.* **2017**, *95*, 333–343. [CrossRef]
39. Bao, J.S. Biotechnology for rice grain quality improvement. In *Rice Chemistry and Technology*, 4th ed.; Bao, J.S., Ed.; Elsevier: Amsterdam, The Netherlands, 2019; pp. 443–471. [CrossRef]
40. Akihiro, T.; Mizuno, K.; Fujimura, T. Gene expression of ADP-glucose pyrophosphorylase and starch contents in rice cultured cells are cooperatively regulated by sucrose and ABA. *Plant Cell Physiol.* **2005**, *46*, 937–946. [CrossRef]
41. Hannah, L.C.; James, M. The complexities of starch biosynthesis in cereal endosperms. *Curr. Opin. Biotechnol.* **2008**, *19*, 160–165. [CrossRef]
42. Nakamura, Y.; Francisco, P.B.; Hosaka, Y.; Sato, A.; Sawada, T.; Kubo, A.; Fujita, N. Essential amino acids of starch synthase IIa differentiate amylopectin structure and starch quality between japonica and indica rice varieties. *Plant Mol. Biol.* **2005**, *58*, 213–227. [CrossRef]
43. Fujita, N.; Yoshida, M.; Kondo, T.; Saito, K.; Utsumi, Y.; Tokunaga, T.; Nishi, A.; Satoh, H.; Park, J.-H.; Jane, J.-L.; et al. Characterization of SSIIIa-deficient mutants of rice: The function of SSIIIa and pleiotropic effects by SSIIIa deficiency in the rice endosperm. *Plant Physiol.* **2007**, *144*, 2009–2023. [CrossRef]
44. Ohdan, T.; Francisco, P.B., Jr.; Sawada, T.; Hirose, T.; Terao, T.; Satoh, H.; Nakamura, Y. Expression profiling of genes involved in starch synthesis in sink and source organs of rice. *J. Exp. Bot.* **2005**, *56*, 3229–3244. [CrossRef]
45. Hirose, T.; Terao, T. A comprehensive expression analysis of the starch synthase gene family in rice (*Oryza sativa* L.). *Planta* **2004**, *220*, 9–16. [CrossRef]
46. Pfister, B.; Zeeman, S.C. Formation of starch in plant cells. *Cell Mol. Life Sci.* **2016**, *73*, 2781–2807. [CrossRef]
47. Nakamura, Y. Towards a better understanding of the metabolic system for amylopectin biosynthesis in plants: Rice endosperm as a model tissue. *Plant Cell Physiol.* **2002**, *43*, 718–725. [CrossRef]
48. Zhu, L.; Gu, M.; Meng, X.; Cheung, S.C.K.; Yu, H.; Huang, J.; Sun, Y.; Shi, Y.; Liu, Q. High-amylose rice improves indices of animal health in normal and diabetic rats. *Plant. Biotechnol. J.* **2012**, *10*, 353–362. [CrossRef]
49. Nakamura, Y.; Utsumi, Y.; Sawada, T.; Aihara, S.; Utsumi, C.; Yoshida, M.; Kitamura, S. Characterization of the reactions of starch branching enzymes from rice endosperm. *Plant. Cell Physiol.* **2010**, *51*, 776–794. [CrossRef]
50. Fujita, N.; Toyosawa, Y.; Utsumi, Y.; Higuchi, T.; Hanashiro, I.; Ikegami, A.; Akuzawa, S.; Yoshida, M.; Mori, A.; Inomata, K.; et al. Characterization of pullulanase (PUL)-deficient mutants of rice (*Oryza sativa* L.) and the function of PUL on starch biosynthesis in the developing rice endosperm. *J. Exp. Bot.* **2009**, *60*, 1009–1023. [CrossRef]
51. Hwang, S.-K.; Koper, K.; Okita, T.W. The plastid phosphorylase as a multiple-role player in plant metabolism. *Plant Sci.* **2020**, *290*, 110303. [CrossRef]
52. Van Berkel, J.; Conrads-Strauch, J.; Steup, M. Glucan-phosphorylase forms in cotyledons of Pisum sativum L.: Localization, developmental change, in-vitro translation, and processing. *Planta* **1991**, *185*, 432–439. [CrossRef]

53. Satoh, H.; Shibahara, K.; Tokunaga, T.; Nishi, A.; Tasaki, M.; Hwang, S.-K.; Okita, T.W.; Kaneko, N.; Fujita, N.; Yoshida, M.; et al. Mutation of the plastidial alpha-glucan phosphorylase gene in rice affects the synthesis and structure of starch in the endosperm. *Plant Cell* **2008**, *20*, 1833–1849. [CrossRef]
54. Hwang, S.-K.; Singh, S.; Cakir, B.; Satoh, H.; Okita, T.W. The plastidial starch phosphorylase from rice endosperm: Catalytic properties at low temperature. *Planta* **2016**, *243*, 999–1009. [CrossRef] [PubMed]
55. Lee, S.-K.; Eom, J.-S.; Hwang, S.-K.; Shin, D.; An, G.; Okita, T.W.; Jeon, J.-S. Plastidic phosphoglucomutase and ADP-glucose pyrophosphorylase mutants impair starch synthesis in rice pollen grains and cause male sterility. *J. Exp. Bot.* **2016**, *67*, 5557–5569. [CrossRef]
56. Ritte, G.; Lloyd, J.R.; Eckermann, N.; Rottmann, A.; Kossmann, J.; Steup, M. The starch-related R1 protein is an α-glucan, water dikinase. *Proc. Natl. Acad. Sci. USA* **2002**, *99*, 7166. [CrossRef]
57. Mikkelsen, R.; Baunsgaard, L.; Blennow, A. Functional characterization of alpha-glucan, water dikinase, the starch phosphorylating enzyme. *Biochem. J.* **2004**, *377*, 525–532. [CrossRef]
58. Kötting, O.; Pusch, K.; Tiessen, A.; Geigenberger, P.; Steup, M.; Ritte, G. Identification of a novel enzyme required for starch metabolism in Arabidopsis leaves. The phosphoglucan, Water Dikinase. *Plant Physiol.* **2005**, *137*, 242. [CrossRef]
59. Baunsgaard, L.; Lütken, H.; Mikkelsen, R.; Glaring, M.A.; Pham, T.T.; Blennow, A. A novel isoform of glucan, water dikinase phosphorylates pre-phosphorylated α-glucans and is involved in starch degradation in Arabidopsis. *Plant J.* **2005**, *41*, 595–605. [CrossRef] [PubMed]
60. Blennow, A.; Engelsen, S.B. Helix-breaking news: Fighting crystalline starch energy deposits in the cell. *Trends Plant Sci.* **2010**, *15*, 236–240. [CrossRef] [PubMed]
61. Tagliabracci, V.S.; Roach, P.J. Insights into the mechanism of polysaccharide dephosphorylation by a glucan phosphatase. *Proc. Natl. Acad. Sci. USA* **2010**, *107*, 15312–15313. [CrossRef]
62. Reimann, R.; Ritte, G.; Steup, M.; Appenroth, K.-J. Association of α-amylase and the R1 protein with starch granules precedes the initiation of net starch degradation in turions of *Spirodela polyrhiza*. *Physiol. Plant* **2002**, *114*, 2–12. [CrossRef]
63. Hirose, T.; Aoki, N.; Harada, Y.; Okamura, M.; Hashida, Y.; Ohsugi, R.; Miyao, A.; Hirochika, H.; Terao, T. Disruption of a rice gene for α-glucan water dikinase, *OsGWD1*, leads to hyperaccumulation of starch in leaves but exhibits limited effects on growth. *Front. Plant Sci.* **2013**, *4*, 147. [CrossRef] [PubMed]
64. Chen, Y.; Sun, X.; Zhou, X.; Hebelstrup, K.H.; Blennow, A.; Bao, J. Highly phosphorylated functionalized rice starch produced by transgenic rice expressing the potato *GWD1* gene. *Sci. Rep.* **2017**, *7*, 3339. [CrossRef]
65. Huang, L.-F.; Liu, Y.-K.; Su, S.-C.; Lai, C.-C.; Wu, C.-R.; Chao, T.-J.; Yang, Y.-H. Genetic engineering of transitory starch accumulation by knockdown of *OsSEX4* in rice plants for enhanced bioethanol production. *Biotechnol. Bioeng.* **2020**, *117*, 933–944. [CrossRef]
66. Critchley, J.H.; Zeeman, S.C.; Takaha, T.; Smith, A.M.; Smith, S.M. A critical role for disproportionating enzyme in starch breakdown is revealed by a knock-out mutation in Arabidopsis. *Plant. J.* **2001**, *26*, 89–100. [CrossRef]
67. Blennow, A.; Nielsen, T.H.; Baunsgaard, L.; Mikkelsen, R.; Engelsen, S.B. Starch phosphorylation: A new front line in starch research. *Trends Plant Sci.* **2002**, *7*, 445–450. [CrossRef]
68. Kötting, O.; Santelia, D.; Edner, C.; Eicke, S.; Marthaler, T.; Gentry, M.S.; Comparot-Moss, S.; Chen, J.; Smith, A.M.; Steup, M.; et al. Starch-Excess4 is a laforin-like phosphoglucan phosphatase required for starch degradation in *Arabidopsis thaliana*. *Plant Cell* **2009**, *21*, 334–346. [CrossRef] [PubMed]
69. Santelia, D.; Kötting, O.; Seung, D.; Schubert, M.; Thalmann, M.; Bischof, S.; Meekins, D.A.; Lutz, A.; Patron, N.; Gentry, M.S.; et al. The phosphoglucan phosphatase like sex four2 dephosphorylates starch at the C3-position in *Arabidopsis*. *Plant Cell* **2011**, *23*, 4096–4111. [CrossRef] [PubMed]
70. Zeeman, S.C.; Rees, T.A. Changes in carbohydrate metabolism and assimilate export in starch-excess mutants of *Arabidopsis*. *Plant Cell Environ.* **1999**, *22*, 1445–1453. [CrossRef]
71. Ma, J.; Jiang, Q.-T.; Wei, L.; Yang, Q.; Zhang, X.-W.; Peng, Y.-Y.; Chen, G.-Y.; Wei, Y.-M.; Liu, C.; Zheng, Y.-L. Conserved structure and varied expression reveal key roles of phosphoglucan phosphatase gene *starch excess 4* in barley. *Planta* **2014**, *240*, 1179–1190. [CrossRef] [PubMed]
72. Akdogan, G.; Kubota, J.; Kubo, A.; Takaha, T.; Kitamura, S. Expression and characterization of rice disproportionating enzymes. *J. Appl. Glycosci.* **2011**, *58*, 99–105. [CrossRef]
73. Dong, X.; Zhang, D.; Liu, J.; Liu, Q.Q.; Liu, H.; Tian, L.; Jiang, L.; Qu, L.Q. Plastidial disproportionating enzyme participates in starch synthesis in rice endosperm by transferring maltooligosyl groups from amylose and amylopectin to amylopectin. *Plant Physiol.* **2015**, *169*, 2496. [CrossRef] [PubMed]
74. Chia, T.; Thorneycroft, D.; Chapple, A.; Messerli, G.; Chen, J.; Zeeman, S.C.; Smith, S.M.; Smith, A.M. A cytosolic glucosyltransferase is required for conversion of starch to sucrose in *Arabidopsis* leaves at night. *Plant J.* **2004**, *37*, 853–863. [CrossRef] [PubMed]
75. Seung, D.; Smith, A.M. Starch granule initiation and morphogenesis—Progress in Arabidopsis and cereals. *J. Exp. Bot.* **2019**, *70*, 771–784. [CrossRef]
76. Seung, D. Amylose in starch: Towards an understanding of biosynthesis, structure and function. *New Phytol.* **2020**, *228*, 1490–1504. [CrossRef]

77. Lohmeier-Vogel, E.M.; Kerk, D.; Nimick, M.; Wrobel, S.; Vickerman, L.; Muench, D.G.; Moorhead, G.B.G. Arabidopsis At5g39790 encodes a chloroplast-localized, carbohydrate-binding, coiled-coil domain-containing putative scaffold protein. *BMC Plant Biol.* **2008**, *8*, 120. [CrossRef]
78. Seung, D.; Soyk, S.; Coiro, M.; Maier, B.A.; Eicke, S.; Zeeman, S.C. Protein targeting to starch is required for localising granule-bound starch synthase to starch granules and for normal amylose synthesis in Arabidopsis. *PLoS Biol.* **2015**, *13*, e1002080. [CrossRef]
79. Seung, D.; Echevarría-Poza, A.; Steuernagel, B.; Smith, A.M. Natural polymorphisms in Arabidopsis result in wide variation or loss of the amylose component of starch. *Plant Physiol.* **2020**, *182*, 870. [CrossRef] [PubMed]
80. Seung, D.; Boudet, J.; Monroe, J.; Schreier, T.B.; David, L.C.; Abt, M.; Lu, K.-J.; Zanella, M.; Zeeman, S.C. Homologs of Protein Targeting to Starch control starch granule initiation in Arabidopsis leaves. *Plant Cell* **2017**, *29*, 1657. [CrossRef]
81. Seung, D.; Schreier, T.B.; Bürgy, L.; Eicke, S.; Zeeman, S.C. Two plastidial coiled-coil proteins are essential for normal starch granule initiation in Arabidopsis. *Plant Cell* **2018**, *30*, 1523. [CrossRef]
82. Wang, W.; Wei, X.; Jiao, G.; Chen, W.; Wu, Y.; Sheng, Z.; Hu, S.; Xie, L.; Wang, J.; Tang, S.; et al. *Gbss-Binding Protein*, encoding a CBM48 domain-containing protein, affects rice quality and yield. *J. Integr. Plant Biol.* **2020**, *62*, 948–966. [CrossRef]
83. Peng, C.; Wang, Y.; Liu, F.; Ren, Y.; Zhou, K.; Lv, J.; Zheng, M.; Zhao, S.; Zhang, L.; Wang, C.; et al. *Floury Endosperm6* encodes a CBM48 domain-containing protein involved in compound granule formation and starch synthesis in rice endosperm. *Plant J.* **2014**, *77*, 917–930. [CrossRef] [PubMed]
84. Walsh, C.T.; Garneau-Tsodikova, S.; Gatto Jr, G.J. Protein posttranslational modifications: The chemistry of proteome diversifications. *Angew. Chem. Int. Ed.* **2005**, *44*, 7342–7372. [CrossRef] [PubMed]
85. Sreedhar, A.; Wiese, E.K.; Hitosugi, T. Enzymatic and metabolic regulation of lysine succinylation. *Genes Dis.* **2020**, *7*, 166–171. [CrossRef]
86. Ishimaru, T.; Matsuda, T.; Ohsugi, R.; Yamagishi, T. Morphological development of rice caryopses located at the different positions in a panicle from early to middle stage of grain filling. *Funct Plant Biol.* **2003**, *30*, 1139–1149. [CrossRef] [PubMed]
87. Zhang, Z.; Zhao, H.; Tang, J.; Li, Z.; Li, Z.; Chen, D.; Lin, W. A proteomic study on molecular mechanism of poor grain-filling of rice (*Oryza sativa* L.) inferior spikelets. *PLoS ONE* **2014**, *9*, e89140. [CrossRef]
88. Zhang, Z.; Tang, J.; Du, T.; Zhao, H.; Li, Z.; Li, Z.; Lin, W. Mechanism of developmental stagnancy of rice inferior spikelets at early grain-filling stage as revealed by proteomic analysis. *Plant Mol. Biol. Report.* **2015**, *33*, 1844–1863. [CrossRef]
89. Liao, J.-L.; Zhou, H.-W.; Zhang, H.-Y.; Zhong, P.-A.; Huang, Y.-J. Comparative proteomic analysis of differentially expressed proteins in the early milky stage of rice grains during high temperature stress. *J. Exp. Bot.* **2014**, *65*, 655–671. [CrossRef]
90. Timabud, T.; Yin, X.; Pongdontri, P.; Komatsu, S. Gel-free/label-free proteomic analysis of developing rice grains under heat stress. *J. Proteom.* **2016**, *133*, 1–19. [CrossRef]
91. Li, H.; Chen, Z.; Hu, M.; Wang, Z.; Hua, H.; Yin, C.; Zeng, H. Different effects of night versus day high temperature on rice quality and accumulation profiling of rice grain proteins during grain filling. *Plant Cell Rep.* **2011**, *30*, 1641–1659. [CrossRef]
92. Kaneko, K.; Sasaki, M.; Kuribayashi, N.; Suzuki, H.; Sasuga, Y.; Shiraya, T.; Inomata, T.; Itoh, K.; Baslam, M.; Mitsui, T. Proteomic and glycomic characterization of rice chalky grains produced under moderate and high-temperature conditions in field system. *Rice* **2016**, *9*, 26. [CrossRef]
93. Ishimaru, T.; Horigane, A.K.; Ida, M.; Iwasawa, N.; San-oh, Y.A.; Nakazono, M.; Nishizawa, N.K.; Masumura, T.; Kondo, M.; Yoshida, M. Formation of grain chalkiness and changes in water distribution in developing rice caryopses grown under high-temperature stress. *J. Cereal Sci.* **2009**, *50*, 166–174. [CrossRef]
94. Lemeer, S.; Heck, A.J.R. The phosphoproteomics data explosion. *Curr. Opin. Chem. Biol.* **2009**, *13*, 414–420. [CrossRef] [PubMed]
95. Yang, P. Phosphoproteomics in Cereals. In *Plant Phosphoproteomics: Methods and Protocols*; Schulze, W.X., Ed.; Springer: New York, NY, USA, 2015; pp. 47–57.
96. Reinders, J.; Sickmann, A. State-of-the-art in phosphoproteomics. *Proteomics* **2005**, *5*, 4052–4061. [CrossRef] [PubMed]
97. Sugiyama, N.; Nakagami, H.; Mochida, K.; Daudi, A.; Tomita, M.; Shirasu, K.; Ishihama, Y. Large-scale phosphorylation mapping reveals the extent of tyrosine phosphorylation in Arabidopsis. *Mol. Syst. Biol.* **2008**, *4*, 193. [CrossRef]
98. Adams, J.A. Kinetic and catalytic mechanisms of protein kinases. *Chem. Rev.* **2001**, *101*, 2271–2290. [CrossRef]
99. Fang, Y.; Deng, X.; Lu, X.; Zheng, J.; Jiang, H.; Rao, Y.; Zeng, D.; Hu, J.; Zhang, X.; Xue, D. Differential phosphoproteome study of the response to cadmium stress in rice. *Ecotoxicol. Environ. Saf.* **2019**, *180*, 780–788. [CrossRef]
100. Ye, J.; Zhang, Z.; Long, H.; Zhang, Z.; Hong, Y.; Zhang, X.; You, C.; Liang, W.; Ma, H.; Lu, P. Proteomic and phosphoproteomic analyses reveal extensive phosphorylation of regulatory proteins in developing rice anthers. *Plant. J.* **2015**, *84*, 527–544. [CrossRef]
101. Wang, Y.; Tong, X.; Qiu, J.; Li, Z.; Zhao, J.; Hou, Y.; Tang, L.; Zhang, J. A phosphoproteomic landscape of rice (*Oryza sativa*) tissues. *Physiol. Plant.* **2017**, *160*, 458–475. [CrossRef]
102. Sun, R.; Qin, S.; Zhang, T.; Wang, Z.; Li, H.; Li, Y.; Nie, Y. Comparative phosphoproteomic analysis of blast resistant and susceptible rice cultivars in response to salicylic acid. *BMC Plant Biol.* **2019**, *19*, 454. [CrossRef] [PubMed]
103. Han, C.; Yang, P.; Sakata, K.; Komatsu, S. Quantitative proteomics reveals the role of protein phosphorylation in rice embryos during early stages of germination. *J. Proteome Res.* **2014**, *13*, 1766–1782. [CrossRef] [PubMed]
104. Zhang, M.; Ma, C.-Y.; Lv, D.-W.; Zhen, S.-M.; Li, X.-H.; Yan, Y.-M. Comparative phosphoproteome analysis of the developing grains in bread wheat (*Triticum aestivum* L.) under well-watered and water-deficit conditions. *J. Proteome Res.* **2014**, *13*, 4281–4297. [CrossRef]

105. Vu, L.D.; Zhu, T.; Verstraeten, I.; van de Cotte, B.; The International Wheat Genome Sequencing, C.; Gevaert, K.; De Smet, I. Temperature-induced changes in the wheat phosphoproteome reveal temperature-regulated interconversion of phosphoforms. *J. Exp. Bot.* **2018**, *69*, 4609–4624. [CrossRef] [PubMed]
106. Zhen, S.; Deng, X.; Zhang, M.; Zhu, G.; Lv, D.; Wang, Y.; Zhu, D.; Yan, Y. Comparative phosphoproteomic analysis under high-nitrogen fertilizer reveals central phosphoproteins promoting wheat grain starch and protein synthesis. *Front. Plant. Sci.* **2017**, *8*, 67. [CrossRef]
107. Chen, G.-X.; Zhou, J.-W.; Liu, Y.-L.; Lu, X.-B.; Han, C.-X.; Zhang, W.-Y.; Xu, Y.-H.; Yan, Y.-M. Biosynthesis and regulation of wheat amylose and amylopectin from proteomic and phosphoproteomic characterization of granule-binding proteins. *Sci. Rep.* **2016**, *6*, 33111. [CrossRef] [PubMed]
108. Lv, D.-W.; Zhu, G.-R.; Zhu, D.; Bian, Y.-W.; Liang, X.-N.; Cheng, Z.-W.; Deng, X.; Yan, Y.-M. Proteomic and phosphoproteomic analysis reveals the response and defense mechanism in leaves of diploid wheat *T. monococcum* under salt stress and recovery. *J. Proteom.* **2016**, *143*, 93–105. [CrossRef]
109. Ishikawa, S.; Barrero, J.M.; Takahashi, F.; Nakagami, H.; Peck, S.C.; Gubler, F.; Shinozaki, K.; Umezawa, T. Comparative phosphoproteomic analysis reveals a decay of ABA signaling in barley embryos during after-ripening. *Plant Cell Physiol.* **2019**, *60*, 2758–2768. [CrossRef]
110. Ishikawa, S.; Barrero, J.; Takahashi, F.; Peck, S.; Gubler, F.; Shinozaki, K.; Umezawa, T. Comparative phosphoproteomic analysis of Barley Embryos with different dormancy during imbibition. *Int. J. Mol. Sci.* **2019**, *20*, 451. [CrossRef]
111. Cao, H.; Zhou, Y.; Chang, Y.; Zhang, X.; Li, C.; Ren, D. Comparative phosphoproteomic analysis of developing maize seeds suggests a pivotal role for enolase in promoting starch synthesis. *Plant Sci.* **2019**, *289*, 110243. [CrossRef] [PubMed]
112. Zhao, X.; Bai, X.; Jiang, C.; Li, Z. Phosphoproteomic analysis of two contrasting maize inbred lines provides insights into the mechanism of salt-stress tolerance. *Int. J. Mol. Sci.* **2019**, *20*, 1886. [CrossRef]
113. Hu, X.; Wu, L.; Zhao, F.; Zhang, D.; Li, N.; Zhu, G.; Li, C.; Wang, W. Phosphoproteomic analysis of the response of maize leaves to drought, heat and their combination stress. *Front. Plant Sci.* **2015**, *6*, 298. [CrossRef]
114. Wu, L.; Wang, S.; Wu, J.; Han, Z.; Wang, R.; Wu, L.; Zhang, H.; Chen, Y.; Hu, X. Phosphoproteomic analysis of the resistant and susceptible genotypes of maize infected with sugarcane mosaic virus. *Amino Acids* **2015**, *47*, 483–496. [CrossRef]
115. Lu, T.C.; Meng, L.B.; Yang, C.P.; Liu, G.F.; Liu, G.J.; Ma, W.; Wang, B.C. A shotgun phosphoproteomics analysis of embryos in germinated maize seeds. *Planta* **2008**, *228*, 1029–1041. [CrossRef]
116. Nakagami, H.; Sugiyama, N.; Mochida, K.; Daudi, A.; Yoshida, Y.; Toyoda, T.; Tomita, M.; Ishihama, Y.; Shirasu, K. Large-scale comparative phosphoproteomics identifies conserved phosphorylation sites in plants. *Plant Physiol.* **2010**, *153*, 1161–1174. [CrossRef]
117. Tetlow, I.J.; Wait, R.; Lu, Z.; Akkasaeng, R.; Bowsher, C.G.; Esposito, S.; Kosar-Hashemi, B.; Morell, M.K.; Emes, M.J. Protein phosphorylation in amyloplasts regulates starch branching enzyme activity and protein-protein interactions. *Plant Cell* **2004**, *16*, 694–708. [CrossRef] [PubMed]
118. Liu, F.; Makhmoudova, A.; Lee, E.A.; Wait, R.; Emes, M.J.; Tetlow, I.J. The *amylose extender* mutant of maize conditions novel protein–protein interactions between starch biosynthetic enzymes in amyloplasts. *J. Exp. Bot.* **2009**, *60*, 4423–4440. [CrossRef]
119. Tetlow, I.J.; Beisel, K.G.; Cameron, S.; Makhmoudova, A.; Liu, F.; Bresolin, N.S.; Wait, R.; Morell, M.K.; Emes, M.J. Analysis of protein complexes in wheat amyloplasts reveals functional interactions among starch biosynthetic enzymes. *Plant Physiol.* **2008**, *146*, 1878. [CrossRef] [PubMed]
120. Mehrpouyan, S.; Menon, U.; Tetlow, I.J.; Emes, M.J. Protein phosphorylation regulates maize endosperm starch synthase IIa activity and protein−protein interactions. *Plant J.* **2021**, *105*, 1098–1112. [CrossRef]
121. Liu, F.; Ahmed, Z.; Lee, E.A.; Donner, E.; Liu, Q.; Ahmed, R.; Morell, M.K.; Emes, M.J.; Tetlow, I.J. Allelic variants of the *amylose extender* mutation of maize demonstrate phenotypic variation in starch structure resulting from modified protein–protein interactions. *J. Exp. Bot.* **2012**, *63*, 1167–1183. [CrossRef] [PubMed]
122. Ahmed, Z.; Tetlow, I.J.; Ahmed, R.; Morell, M.K.; Emes, M.J. Protein–protein interactions among enzymes of starch biosynthesis in high-amylose barley genotypes reveal differential roles of heteromeric enzyme complexes in the synthesis of A and B granules. *Plant Sci.* **2015**, *233*, 95–106. [CrossRef]
123. Liu, D.-R.; Huang, W.-X.; Cai, X.-L. Oligomerization of rice granule-bound starch synthase 1 modulates its activity regulation. *Plant Sci.* **2013**, *210*, 141–150. [CrossRef]
124. Wang, S.J.; Liu, L.F.; Chen, C.K.; Chen, L.W. Regulations of granule-bound starch synthase I gene expression in rice leaves by temperature and drought stress. *Biol. Plant.* **2006**, *50*, 537–541. [CrossRef]
125. Crofts, N.; Abe, N.; Oitome, N.F.; Matsushima, R.; Hayashi, M.; Tetlow, I.J.; Emes, M.J.; Nakamura, Y.; Fujita, N. Amylopectin biosynthetic enzymes from developing rice seed form enzymatically active protein complexes. *J. Exp. Bot.* **2015**, *66*, 4469–4482. [CrossRef]
126. Crofts, N.; Iizuka, Y.; Abe, N.; Miura, S.; Kikuchi, K.; Matsushima, R.; Fujita, N. Rice mutants lacking Starch Synthase I or Branching Enzyme IIb activity altered starch biosynthetic protein complexes. *Front. Plant Sci.* **2018**, *9*, 1817. [CrossRef]
127. Chen, Y.; Pang, Y.; Bao, J.S. Expression profiles and protein complexes of starch biosynthetic enzymes from white-core and waxy mutants induced from high amylose *Indica* rice. *Rice Sci.* **2020**, *27*, 152–161. [CrossRef]

128. Hwang, S.-K.; Koper, K.; Satoh, H.; Okita, T.W. Rice endosperm starch phosphorylase (Pho1) assembles with disproportionating enzyme (Dpe1) to form a protein complex that enhances synthesis of malto-oligosaccharides. *J. Biol. Chem.* **2016**, *291*, 19994–20007. [CrossRef]
129. Liu, Z.; Cao, J.; Gao, X.; Zhou, Y.; Wen, L.; Yang, X.; Yao, X.; Ren, J.; Xue, Y. CPLA 1.0: An integrated database of protein lysine acetylation. *Nucleic Acids Res.* **2011**, *39*, D1029–D1034. [CrossRef]
130. Choudhary, C.; Kumar, C.; Gnad, F.; Nielsen, M.L.; Rehman, M.; Walther, T.C.; Olsen, J.V.; Mann, M. Lysine acetylation targets protein complexes and co-regulates major cellular functions. *Science* **2009**, *325*, 834. [CrossRef]
131. Nallamilli, B.R.R.; Edelmann, M.J.; Zhong, X.; Tan, F.; Mujahid, H.; Zhang, J.; Nanduri, B.; Peng, Z. Global analysis of lysine acetylation suggests the involvement of protein acetylation in diverse biological processes in rice (*Oryza sativa*). *PLoS ONE* **2014**, *9*, e89283. [CrossRef]
132. Wang, Q.; Zhang, Y.; Yang, C.; Xiong, H.; Lin, Y.; Yao, J.; Li, H.; Xie, L.; Zhao, W.; Yao, Y.; et al. Acetylation of metabolic enzymes coordinates carbon source utilization and metabolic flux. *Science* **2010**, *327*, 1004–1007. [CrossRef]
133. Glozak, M.A.; Sengupta, N.; Zhang, X.; Seto, E. Acetylation and deacetylation of non-histone proteins. *Gene* **2005**, *363*, 15–23. [CrossRef]
134. Yan, Z.; Shen, Z.; Gao, Z.-F.; Chao, Q.; Qian, C.-R.; Zheng, H.; Wang, B.-C. A comprehensive analysis of the lysine acetylome reveals diverse functions of acetylated proteins during de-etiolation in *Zea mays*. *J. Plant Physiol.* **2020**, *248*, 153158. [CrossRef]
135. Zhen, S.; Deng, X.; Wang, J.; Zhu, G.; Cao, H.; Yuan, L.; Yan, Y. First comprehensive proteome analyses of lysine acetylation and succinylation in seedling leaves of *Brachypodium distachyon* L. *Sci. Rep.* **2016**, *6*, 31576. [CrossRef]
136. Zhang, Z.; Tan, M.; Xie, Z.; Dai, L.; Chen, Y.; Zhao, Y. Identification of lysine succinylation as a new post-translational modification. *Nat. Chem. Biol.* **2011**, *7*, 58–63. [CrossRef]
137. Xie, Z.; Dai, J.; Dai, L.; Tan, M.; Cheng, Z.; Wu, Y.; Boeke, J.D.; Zhao, Y. Lysine succinylation and lysine malonylation in histones. *Mol. Cell. Proteom.: MCP* **2012**, *11*, 100–107. [CrossRef]
138. Xu, H.; Chen, X.; Xu, X.; Shi, R.; Suo, S.; Cheng, K.; Zheng, Z.; Wang, M.; Wang, L.; Zhao, Y.; et al. Lysine acetylation and succinylation in HeLa Cells and their essential roles in response to UV-induced Stress. *Sci. Rep.* **2016**, *6*, 30212. [CrossRef]
139. Wang, C.; Zhang, C.; Li, X.; Shen, J.; Xu, Y.; Shi, H.; Mu, X.; Pan, J.; Zhao, T.; Li, M.; et al. CPT1A-mediated succinylation of S100A10 increases human gastric cancer invasion. *J. Cell Mol. Med.* **2019**, *23*, 293–305. [CrossRef]
140. Kidwai, S.A.; Ansari, A.A.; Salahuddin, A. Effect of succinylation (3-carboxypropionylation) on the conformation and immunological activity of ovalbumin. *Biochem. J.* **1976**, *155*, 171–180. [CrossRef]
141. Kurmi, K.; Hitosugi, S.; Wiese, E.K.; Boakye-Agyeman, F.; Gonsalves, W.I.; Lou, Z.; Karnitz, L.M.; Goetz, M.P.; Hitosugi, T. Carnitine Palmitoyltransferase 1A has a lysine succinyltransferase activity. *Cell Rep.* **2018**, *22*, 1365–1373. [CrossRef]
142. Jin, W.; Wu, F. Proteome-wide identification of lysine succinylation in the proteins of tomato (*Solanum lycopersicum*). *PLoS ONE* **2016**, *11*, e0147586. [CrossRef] [PubMed]
143. Wang, X.; Chen, X.; Li, J.; Zhou, X.; Liu, Y.; Zhong, L.; Tang, Y.; Zheng, H.; Liu, J.; Zhan, R.; et al. Global analysis of lysine succinylation in patchouli plant leaves. *Hortic. Res.* **2019**, *6*, 133. [CrossRef] [PubMed]
144. Qiu, C.; Wang, Y.; Sun, J.H.; Qian, W.J.; Xie, H.; Ding, Y.Q.; Ding, Z.T. A qualitative proteome-wide lysine succinylation profiling of tea revealed its involvement in primary metabolism. *Mol. Biol.* **2020**, *54*, 144–155. [CrossRef]
145. Dai, L.; Peng, C.; Montellier, E.; Lu, Z.; Chen, Y.; Ishii, H.; Debernardi, A.; Buchou, T.; Rousseaux, S.; Jin, F.; et al. Lysine 2-hydroxyisobutyrylation is a widely distributed active histone mark. *Nat. Chem. Biol.* **2014**, *10*, 365–370. [CrossRef]
146. Hirschey, M.D.; Zhao, Y. Metabolic regulation by lysine malonylation, succinylation, and glutarylation. *Mol. Cell. Proteom. MCP* **2015**, *14*, 2308–2315. [CrossRef] [PubMed]
147. Peng, C.; Lu, Z.; Xie, Z.; Cheng, Z.; Chen, Y.; Tan, M.; Luo, H.; Zhang, Y.; He, W.; Yang, K.; et al. The first identification of lysine malonylation substrates and its regulatory enzyme. *Mol. Cell Proteom.* **2011**, *10*, M111.012658. [CrossRef]
148. Huang, H.; Tang, S.; Ji, M.; Tang, Z.; Shimada, M.; Liu, X.; Qi, S.; Locasale, J.W.; Roeder, R.G.; Zhao, Y.; et al. p300-dediated lysine 2-hydroxyisobutyrylation regulates glycolysis. *Mol. Cell* **2018**, *70*, 663–678.e666. [CrossRef]
149. Chen, X.; Li, X.; Li, P.; Chen, X.; Liu, H.; Huang, J.; Luo, C.; Hsiang, T.; Zheng, L. Comprehensive identification of lysine 2-hydroxyisobutyrylated proteins in *Ustilaginoidea virens* reveals the involvement of lysine 2-hydroxyisobutyrylation in fungal virulence. *J. Integr. Plant. Biol.* **2021**, *63*, 409–425. [CrossRef]
150. Du, Y.; Cai, T.; Li, T.; Xue, P.; Zhou, B.; He, X.; Wei, P.; Liu, P.; Yang, F.; Wei, T. Lysine malonylation is elevated in type 2 diabetic mouse models and enriched in metabolic associated proteins. *Mol. Cell. Proteom. MCP* **2015**, *14*, 227–236. [CrossRef]
151. Galván-Peña, S.; Carroll, R.G.; Newman, C.; Hinchy, E.C.; Palsson-McDermott, E.; Robinson, E.K.; Covarrubias, S.; Nadin, A.; James, A.M.; Haneklaus, M.; et al. Malonylation of GAPDH is an inflammatory signal in macrophages. *Nat. Commun.* **2019**, *10*, 338. [CrossRef]
152. Colak, G.; Pougovkina, O.; Dai, L.; Tan, M.; Te Brinke, H.; Huang, H.; Cheng, Z.; Park, J.; Wan, X.; Liu, X.; et al. Proteomic and biochemical studies of lysine malonylation suggest its malonic aciduria-associated regulatory role in mitochondrial function and fatty acid oxidation. *Mol. Cell. Proteom. MCP* **2015**, *14*, 3056–3071. [CrossRef] [PubMed]

International Journal of
Molecular Sciences

MDPI

Review

Posttranslational Modification of Waxy to Genetically Improve Starch Quality in Rice Grain

Tosin Victor Adegoke [1,2], Yifeng Wang [1], Lijuan Chen [1], Huimei Wang [1], Wanning Liu [1], Xingyong Liu [1], Yi-Chen Cheng [1], Xiaohong Tong [1], Jiezheng Ying [1] and Jian Zhang [1,*]

1 State Key Lab of Rice Biology, China National Rice Research Institute, Hangzhou 311400, China; 2019Y90100074@caas.cn (T.V.A.); wangyifeng@caas.cn (Y.W.); chenlijuan0723@gmail.com (L.C.); wanghuimei@caas.cn (H.W.); liuwn123456789@gmail.com (W.L.); liuxinyong1234@gmail.com (X.L.); chengyichen0721@gmail.com (Y.-C.C.); tongxiaohong@caas.cn (X.T.); yingjiezheng@caas.cn (J.Y.)

2 Graduate School of Chinese Academy of Agricultural Sciences, Beijing 100081, China

* Correspondence: zhangjian@caas.cn; Tel.: +86-57163370277

Abstract: The *waxy* (*Wx*) gene, encoding the granule-bound starch synthase (GBSS), is responsible for amylose biosynthesis and plays a crucial role in defining eating and cooking quality. The waxy locus controls both the non-waxy and waxy rice phenotypes. Rice starch can be altered into various forms by either reducing or increasing the amylose content, depending on consumer preference and region. Low-amylose rice is preferred by consumers because of its softness and sticky appearance. A better way of improving crops other than downregulation and overexpression of a gene or genes may be achieved through the posttranslational modification of sites or regulatory enzymes that regulate them because of their significance. The impact of posttranslational GBSSI modifications on extra-long unit chains (ELCs) remains largely unknown. Numerous studies have been reported on different crops, such as wheat, maize, and barley, but the rice starch granule proteome remains largely unknown. There is a need to improve the yield of low-amylose rice by employing posttranslational modification of *Wx*, since the market demand is increasing every day in order to meet the market demand for low-amylose rice in the regional area that prefers low-amylose rice, particularly in China. In this review, we have conducted an in-depth review of waxy rice, starch properties, starch biosynthesis, and posttranslational modification of waxy protein to genetically improve starch quality in rice grains.

Keywords: waxy; amylose; posttranslational modification; GBSSI; rice

Citation: Adegoke, T.V.; Wang, Y.; Chen, L.; Wang, H.; Liu, W.; Liu, X.; Cheng, Y.-C.; Tong, X.; Ying, J.; Zhang, J. Posttranslational Modification of Waxy to Genetically Improve Starch Quality in Rice Grain. *Int. J. Mol. Sci.* **2021**, *22*, 4845. https://doi.org/10.3390/ijms22094845

Academic Editor: Setsuko Komatsu

Received: 25 March 2021
Accepted: 29 April 2021
Published: 3 May 2021

1. Introduction

Rice (*Oryza sativa* L.), one of the most vital crops, is a primary meal for more than half of the world's population and also serves as a source of energy and nutrition for millions of consumers. It is a significant staple food in Asia, West Africa, Latin America, and the Caribbean; the main end-use of rice is human consumption [1]. By 2027, it is expected that total rice consumption will increase by 13% [1]. Its function in food processing is significant, particularly in Asia, including China [2]. About 70% of the expected rise in global rice demand is accounted for by Asian nations, primarily due to growth in population rather than per capita demand [1]. The rice per capita consumption in kg/person/year in 2014–2016 was 77.8 and will increase to 78.9 in 2026 at 0.08% growth increase per annum in Asia and Pacific [1]. As China's population expands, by 2030, China would have to generate 20% more rice to satisfy its domestic needs if the rice per capita demand remains at the current pace [3]. According to Guo et al. [4], China's population will increase to 1.458 billion in 2030 from 1.33 billion in 2010 if moderate growth is maintained. By 2035, it will increase to about 1.46 billion and then decline to 1.38 billion by 2050 [4]. Since China is a regional area that prefers waxy and low-amylose rice, more energy should be channeled on how to improve the yield of such varieties.

The primary carbohydrate in rice is starch, which ranges from 72 to 75% [5], while the protein content of 3805 *Indica* varieties in China ranged from 6.3 to 15.7%, and that of 1518 *Japonica* varieties ranged from 6.0 to 13.6% [6]. The rice quality is mainly determined by the starch content, especially the cooking and eating qualities [7]. Sun et al. [8] also reported that starch properties primarily affect rice-eating quality. Starch is an essential resource for humans and industries and is abundantly present in many varieties of starch-storing crops, e.g., tubers, storage roots, and cereal seeds [9]. Regina et al. [10] also reported that it is commercially isolated from a wide variety of crops, including stem and pith (e.g., sago), roots and tubers (e.g., cassava, sweet potatoes, potatoes, and arrows), and whole grains (e.g., rice, wheat, and sorghum). Furthermore, starch is a cheap, biodegradable, and sustainable industrial raw material that provides sufficient calories for humans and animals [9]. Consumers are changing their eating habits to incorporate rice varieties with good cooking and eating qualities [11]. Cooking and eating properties are closely related to water absorption, increase in volume, and overall firmness of cooked rice. For eating quality and market approval, rice texture is of vital importance [12]; it is a sensory property that influences the stiffness, stickiness, and overall texture of cooked rice [13]. Chen et al. [14] observed that in addition to eating quality and consumer preference, industries have also employed grain starch as an adhesion, sizing, gelling, thickening, and binding agent. Starch obtained from rice has been used in various foodstuffs and consumer items, such as dessert, baking products, and fats, owing to a broad variety of amylose levels [15]. Champagne reported that "the amylose content of milled rice varies from 0.8 to 37% among varieties" [16]. However, the classification of Juliano varies from less than 2% to 33% and consists of the waxy, very low, low, intermediate, and high amylose classes of rice with amylose contents of <2%, 2–12%, 12–20%, 20–25%, and 25–33%, respectively [17]. The amylose content of milled rice can be described as follows: high, >25.0%; intermediate, 20.1–25.0%; low, 12.1–20.0%; very low, 5.1–12.0%; and waxy, 0–5% [18]. Chen et al. [19] also reported that milled rice amylose is typically classified into five classes: (i) high amylose content, (ii) intermediate, (iii) low, (iv) very low, and (v) waxy, with amylose contents of >24%, 20–24%, 10–19%, 3–9%, and 0–2%, respectively. Consumers in China, Indonesia, the Philippines, Lao PDR, Myanmar, the Republic of Korea, Thailand, Japan, and West Malaysia were found to choose sticky or waxy rice over other classes of waxy rice [18], and this result was confirmed by another study among consumers in Lao PDR and Isan, Thailand [11]. In Indonesia, Myanmar, the Philippines, Lao PDR, the Republic of Korea, Brunei, Thailand, and Sarawak in Malaysia, very low amylose rice is preferred by consumers [18]. Low-amylose rice was found to be preferred by consumers in Cambodia, Japan, Thailand, Taiwan, Australia, South Vietnam, northern and southwestern provinces of China, some regions of Lao PDR, and Egypt [11]; in a separate study, the preference for low-amylose rice was recorded for consumers in Taiwan, Turkey, the Republic of Korea, Cambodia, Egypt, France, Japan, Thailand, Portugal, Australia, South Vietnam, some regions of Lao PDR, Argentina, Bulgaria, the northern and southwestern provinces of China, the United States, and the former Soviet Union (Moldova, Georgia, Uzbekistan, Lithuania, Ukraine, Russia, Azerbaijan, Kyrgyzstan, Turkmenistan, Armenia, Estonia, Belarus, Kazakhstan, Tajikistan, and Latvia) [18]. In the Philippines, Iran, Malaysia, Pakistan, several regions in India, Vietnam, Indonesia, Uruguay, and some provinces in China, intermediate amylose rice is preferable [11]. Juliano [18] reported that intermediate amylose is preferable in Nigeria, Hungary, Bhutan, Philippines, Uruguay, Iran, Italy, Greece, Suriname, several regions in India, Vietnam, Pakistan, Indonesia, Malaysia, some provinces in China, Myanmar, Chile, and Venezuela. High-amylose rice is regionally preferred in Senegal, Sri Lanka, Indonesia, Myanmar, and some parts of Uruguay, Colombia, Ghana, and Suriname, in several states of India [11]. In Nigeria, Paraguay, Peru, in several states of India, Ghana, Guatemala, Colombia, Senegal, some parts of Uruguay, Sierra Leone, Venezuela, Suriname, Indonesia, Myanmar, and Sri Lanka, high-amylose rice is desirable/selected [18]. Few countries preferred more than one type of milled rice classification based on the amylose content. Varieties of amylose rice are regionally preferred by different countries (Figure 1).

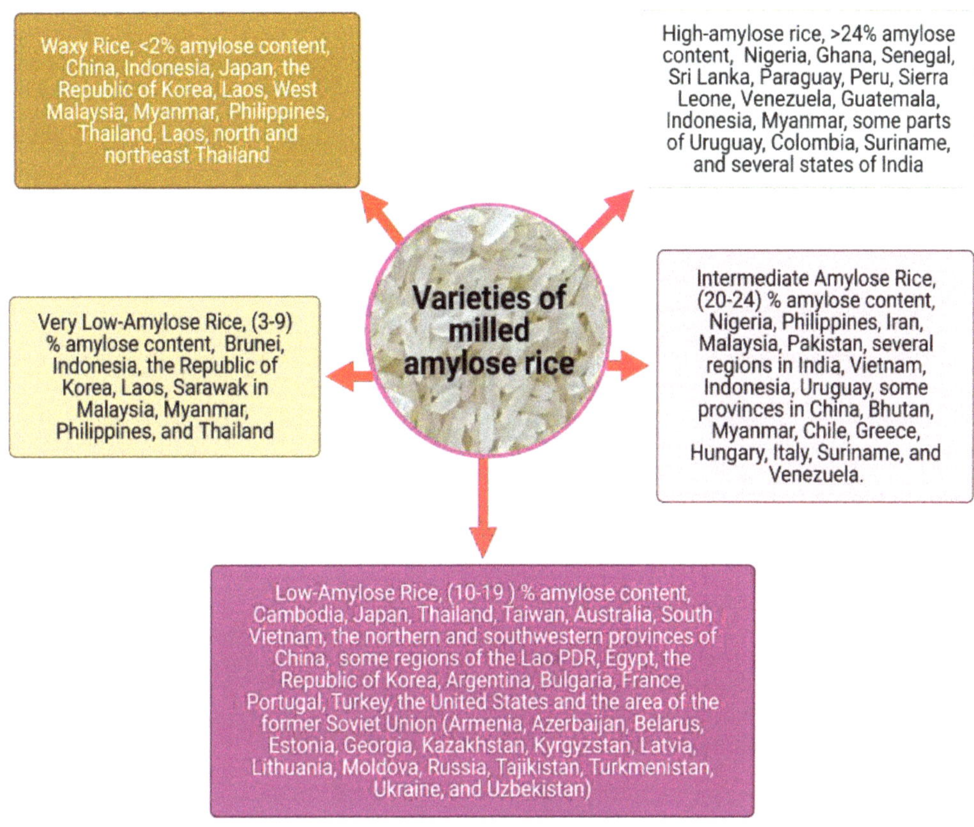

Figure 1. Varieties of amylose rice regionally preferred by different countries. Classification of milled rice based on amylose content and the list of countries that preferred different varieties of amylose rice (created with BioRender.com).

The textural characteristics of cooked rice are mostly distinguished by the content of the starch material, because starch is an essential part of the rice endosperm [20]. The proportions of the short to long chains of waxy rice grains were slightly smaller than those of other grain crop classes [20,21]. The amylose content in the grains is considered an essential factor in the characteristics of cultivated rice. Most efforts to elevate or boost the composition of starch are to increase the ADP-glucose (ADPG) amount required for the biosynthesis of starch [10]. Earlier work included the expression of ADP-glucose pyrophosphorylase-mutated bacteria with decreased allosteric dependency on fructose-6-phosphate activator in potatoes, leading to a rise of >35% in potato tuber starch relative to wild type [22].

Waxy rice is the primary focus of plant breeders in China and is generally used in brewing and conventional Chinese cuisine [23]. The development of amylose is the main determining factor of cooking and eating quality, and the waxy (*Wx*) protein in rice grains is crucial for amylose synthesis [24]. For both gel consistency and amylose content, *waxy* (GBSSI) serves as a vital/critical gene, but it has a minor influence on the temperature of gelatinization [8]. The absence of amylose contributes to comparatively simple starch gelatinization and a lower retrograde tendency, and both results raise the sensitivity of starch to hydrolyzing enzymes with increased digestibility [25]. Granule-bound starch synthase (GBSS) is widely acknowledged to be responsible for amylose synthesis [23]. The GBSS of cereals is subdivided into GBSSI and GBSSII [26]. GBSSI controls the synthesis of amylose in storage tissues (e.g., seed endosperm) [26]. In storage tissues (e.g., seed endosperm), GBSSI is responsible for amylose synthesis [23,27], while GBSSII exists in the green tissues, including the seed pericarp [23,27]. Waxy loci encode granule-bound starch synthase I (GBSSI), which controls the elongation of the endosperm kernel's amylose [25]. Downregulation of GBSSI reduces amylose content and improves the quality of cooking and eating properties [28]. The non-existence or inactivation of waxy (*Wx*) locus-coded GBSSI in plants gives rise to low waxy or waxy rice [29]. Overexpression and downregulation of single or multiple genes have been predominantly crop improvement approaches employed over the past decades [30]. Although some significant advancements have been made, these are comparatively brute force methods, which sometimes contribute to unwanted trade-offs between plant development and stress resilience [30].

In addition to the approaches listed above, a better way of improving crops may be achieved through the posttranslational modification (PTM) of sites or regulatory enzymes that regulate them because of their significance [30] and also because proteins are essential for cell phenotype [31]. Komatsu et al. [32] reported that the ability to discover posttranslational modification (PTM) sites, which is needed to assess the functional effect of protein modification on crop productivity, is a distinct benefit of proteomics over other "omics" techniques. However, genomic editing approaches and transgenic technologies have led to the discovery of novel phenotypes [33]. These methods add new alleles that evoke and enhance stimuli response, resulting in a phenotype improvement; however, the entirety of gene response pathways is still unknown [33]. New strategies must be explored to increase the capacity for plant phenotype improvement [33]. While genomic research can help scientists understand what is potentially conceivable, proteomic research reveals the practical players involved in complex cellular processes [34]. Although functional genomics is effective in elucidating the function of genes, it is necessary to use them in relation to producing a specific phenotype [33]. PTM of proteins is one such avenue. By triggering or repressing protein expression, this mechanism adds another layer of regulation to gene response and allows for fine-tuning of gene response pathways [33]. Graves and Haystead [31] reported that proteomics techniques include PTMs, protein–protein interactions, structural proteomics, functional proteomics, protein mining, and protein expression profiling. Two-dimensional polyacrylamide gel electrophoresis (2D-PAGE) and gel-free-based shotgun procedures have been used extensively in the study of rice grain proteomics [35,36]. Two-dimensional electrophoresis (2-DE) gel is the most commonly employed method for distinguishing proteins in the majority of cited articles [32]. However, in several laboratories, liquid chromatography (LC)-based proteomic analysis is becoming more popular [32]. Both protein separation strategies have their own set of benefits [32]. Protein alteration and degradation can be quickly visualized using a conventional 2-DE technique, while LC-based approaches need a small amount of starting material [32]. Proteomics approaches are great molecular tools that have been extensively employed to explore the molecular basis of many biological mechanisms in plants [37]. The starch granular proteomic study is essential for understanding the starch biosynthesis pathway and its packaging in the amyloplasts of rice in order to improve the quality of grains [38]. Helle et al. [39] also reported that starch granular growth and structural design may be affected by the starch proteome, which leads to a better understanding of starch biosynthesis. Quantitative proteomics has been employed for protein identification and validation of trait-specific markers by coordinating changes in protein levels [40]. To generate mature proteoforms that gradually accumulate in plant cells to form the observed proteome, several proteins undergo PTMs [41,42]. Numerous PTMs have been associated with a wide variety of metabolic roles in recent large-scale proteomic experiments [43].

Posttranslational modification (PTM) processes play vital roles in determining the functional performance of the genome and gene transcription [30]. Virtually any part of protein behavior may be controlled by PTM, including subcellular localization, networks of protein–protein interactions, enzymatic activity, and protein stability [44]. The impact of posttranslational GBSSI modifications on extra-long unit chains (ELCs) remains undetermined, and since serine and tyrosine can be phosphorylated, the importance of posttranslational modifications becomes more noticeable [45]. PTMs have been employed to alter starch quality in various crops to produce different varieties of amylose grains and tubers, depending on consumer preferences and needs. Numerous studies have been reported on various crops, such as wheat, maize, and barley, but the rice starch granule proteome remains largely unknown. However, a large-scale phosphoproteomic study of starch granule binding proteins in cereal crops has not been conducted because of two main setbacks [46]: (i) due to the very low protein content and the high sugar content of the granules and the existence of various compounds that interfere with protein extraction, it is challenging to enrich sufficient proteins from the starch granules for phosphoproteomic analysis, and (ii) there is limited knowledge of phosphorylation modifications in various crops, especially in the very large hexaploid wheat genome, which is ~17 Gb.

In this review, we have conducted an in-depth review of waxy rice, starch properties, starch biosynthesis, and posttranslational modification of waxy protein to genetically improve starch quality in rice grains.

2. Starch Properties

Starch, the primary carbohydrate content in plants, is a crucial natural source of feed, raw materials for industry, and food [9]. It plays a crucial role as the main source of stored carbohydrates for chemical energy in the life cycle of the leaf [47]. To fulfill the continuing energy needs of plant growth (e.g., during the diurnal leaf cycle), transient starch is generated and deteriorates quickly [47]. Conversely, in expectation of potential plant energy demands, for example, germination of seeds or sprouting of the tuber, storage starch accumulates and survives in heterotrophic tissues [47]. It is conserved in sink tissues as an energy source [48]. It exists in the photosynthetic and non-photosynthetic tissue plastids [49]. Starch, the key reserved energy and carbohydrate in the plant, may be classified into two forms, storage starch and transient starch, in relation to their biological role [9]. During the day, transient starch is stored in chloroplasts in photosynthetic tissues [9], and during the night, growth and metabolism are due to transient starch being transferred and depleted to provide nutritious materials and energy [9]. In non-photosynthetic tissues, specialized plastids known as amyloplasts (e.g., storage roots, tubers, and endosperm seeds) store starches that are reserved for a long time [9], and in readiness for propagation, regrowth, or sprouting, they can be remobilized [9].

Starch has two major constituents: α-polyglucan amylose, which is essentially linear, and α-polyglucan amylopectin, which is branched [50,51]. The natural starch composition is uniform with an amylopectin component of 75% and a minor component of amylose of 25%, regardless of the source [10]. The polymers of α-1,4-linked glucan chains of various proportions of α-1,6-linked branch points are both amylose and amylopectin [10]. Although amylose is largely linear with a linkage of about 1% α-1,6, amylopectin is a much bigger and strongly branched molecule with 4–5% α-1,6 connections [10]. Amylopectin has a far more established structure, called the tandem cluster, than glycogen because it consists of tandem-linked clusters (approximately 9–10 nm in length each), where linear α-1,4-glucan chains are regularly branched through α-1,6-glucosidic linkages, as compared to bacterial and animal glycogens which have a more randomly branched structure [51,52].

Starch modification includes an effort to boost or minimize the starch content and adjust the composition and component structure to improve the starch properties by reducing or increasing its content to satisfy consumer preferences and suit particular end uses in industries [10]. Three target points were previously taken to increase starch production in plants: downstream starch biosynthetic enzymes, rate-limiting steps in AGPase-containing starch biosynthesis, and precursor molecules for starch biosynthesis [10]. The physicochemical properties of gelatinized starch are defined primarily by starch properties, and the composition of lipid, protein, and starch granule could also alter the rheological characteristics of cooked starch by reacting with amylose and probably amylopectin [53,54]. The physicochemical characteristics of starch are strongly dependent on its structural characteristics, which affect its activity during production [10]. There has been significant development throughout the last few decades to alter starch amylose content in an essential cereal, either by reducing or increasing the AC compared to the wild-type amylose content [10]. The amylose–amylopectin ratio is the most widely discussed target for starch modification [55]. It is well known that the significant factor affecting the starch physicochemical properties is the amylose-amylopectin ratio [20]. Amylose is capable of forming a solid gel, whereas amylopectin exhibits little gel contraction and great retrograde resistance [20]. GBSSI, starch branching enzymes (SBEs), and soluble starch synthases (SSSs) affect starch synthesis in cereals [56]; GBSSI is involved in amylose synthesis [56,57], while SBE, SS, and DBE affect amylopectin synthesis [58]. Hanashiro et al. [59] reported that GBSSI is responsible not only for the synthesis of amylose but also for the synthesis of amylopectin, particularly for the development of extra-long-chain amylopectin. There is minimal understanding of the starch biosynthesis pathway of biosynthetic starch and the various enzymes involved in this process [9]. Several plants have been improved to develop both high-amylose and high-amylopectin starches via biosynthetic pathways [10]. In raw rice, a higher amylose content is responsible for less sticky and firm rice after cooking [60]. According to recent research, stickiness and amylose content are often negatively associated [61]. Amylose is a straight and long starch molecule that does not gelatinize during cooking. The grain final yield, weight, and grain quality were determined by rice grain filling [56]. Amylopectin is a starch molecule that is strongly branched and is responsible for making rice gelatinous and sticky. When cooked, high-amylopectin rice becomes very sticky, producing high starch content. Typically, short-grain rice has the lowest amylose and amylopectin content (e.g., short grain, Asian-style rice). Stickiness has been shown to improve with a decrease in the amylose content of the whole grain and with a boost in the total volume of amylopectin in leachate, the proportion of small chains of amylopectin, and amylopectin molecular size [12]. Recent research has revealed that amylose content and stiffness are often adversely associated with stickiness; that is, high-amylose rice is firmer and has low stickiness properties, while rice with low amylose content is stickier and softer [13]. The amylopectin-rich *Indica* varieties (waxy) are more resistant to rapid amylase hydrolysis and therefore have a high glycemic index (GI). In the management of diabetes and other diet- or lifestyle-related diseases, in which there is a focus on slowing down digestion of starch and delaying the pace at which glucose breaks down, significant attention has been paid to resistant starch (RS).

3. Waxy Protein

The amylose quality of starch is influenced by many mutated rice genes [20], but the most important gene is the *waxy* gene. Huang et al. [62] reported that *waxy* gene have been prolonged used for the modification of amylose content, and most plant breeders commonly targeted this gene for starch improvement in plants. The starch of waxy mutant contains amylopectin and amylose [20]. Low-amylose rice cultivars displayed higher peak viscosity, and the firmness of the rice flour paste was negatively correlated with the maximum viscosity of the rice starch paste, while the firmness of the paste of the rice flour and the paste of the starch correlated positively with the minimum viscosity of the rice starch paste [21]. Amylose content in rice also defines the transparent properties of the seed; for example, rice with an amylose content of more than 12% is transparent, rice with an amylose content of 8–12% is semi-translucent, and low-amylose (amylose content of less than 8%) is dull or opaque color [62]. The *Wx* gene encoding the enzyme GBSSI primarily regulates the synthesis of amylose in seed production, and the amount of amylose in the grain is closely correlated with the amount of GBSSI in the endosperm seed [63]. The waxy gene is positioned on chromosome 6 and consists of 12 introns and 13 exons [64]. There are two functional alleles in the rice waxy locus, namely (a) Wx^a and (b) Wx^b, which is distinct from a significant discrepancy in the gene quality responsible for *Wx* content in mature seeds [65]. Rice *Wx* protein control is characterized by two functional alleles identified based on *Wx* protein quantity that accumulates in mature seeds [66], including

Wx^a, mainly found in *Indica* rice, and Wx^b, usually found in *Japonica* rice, and is found to prevail in high- and low-amylose rice at the waxy locus, respectively [67]. Zhang et al. [24] reported that numerous allelic variations of *Wx* include Wx^a, Wx^b, Wx^{in}, Wx^{op}, Wx^{mp}, and *Wx*; these variations influenced the geographical variance in amylose content and consumer preferences depending on the region. Waxy proteins, the major regulators of amylose biosynthesis, have been documented [68,69]. Zhang et al. [24] submitted that the Wx^a allele in the grain is responsible for high amylose content, while the Wx^b allele in the grain is responsible for low to moderate amylose content. The allele that regulates the generation of minor/curtailed levels of *Wx* protein is Wx^b, which is primarily expressed in *O. sativa* subsp. *Japonica* [65]. The Wx^a allele generates approximately ten times more *Wx* protein than Wx^b and is commonly dispersed in domestic rice, such as *O. indica sativa*, *O. glaberrim*, and the wild progenitors thereof [65]. Wx^b synthesizes little amylose owing to the mutation on the 5À intron 1 splice site [70]. There are two different pathways by which amylopectin and amylose can be synthesized [71]. Active GBSS is essential for amylose synthesis, while amylopectin is synthesized as a result of a complex pathway involving various isoforms of starch-debranching enzymes (SDBEs), starch branching enzymes (SBEs), and starch synthase [72]. Molecular biologists are now well aware that the absence or inactivation of the *Wx* locus encoding the GBSSI in plants leads to crops with low or no amylose in storage tissues, without apparent total starch content [10]. Park et al. [73] observed that downregulation of GBSSI utilizing a three-prime untranslated region area (3′-UTR) RNA interference (RNAi) structure resulted in a low amylose content (from 5.9 to 9.0%) in transgenic rice lines, in contrast to the wild type with an amylose content of 17.7–18.0%. From this line, starch had no amylose and incredibly short-chain amylopectin, which reduced the number of gel contractions to nearly zero even after many freeze–thaw cycles [10]. Heilersig et al. [74] tested large inverted repeats consisting of the 5′ and 3′ halves of the GBSSI cDNA in potato, and their results showed that the 3IR construct resulted in a notably lower silencing efficacy than the 5IR construct and vice versa. Most waxy and low apparent amylose content (AAC) cultivars verified so far bear this polymorphism, resulting in decreased pre-mRNA splicing efficiency and encouraging alternate splicing at exon 1 cryptic locations, leading to decreased functional enzyme output and triggering the phenotypes of glutinous and low amylose [64]. Maize is the most common crop in which the waxy protein has been modified [10]. It is the most popular crop that is modified to generate waxy corn (protein) [10]. Waxy maize or glutinous maize is a form of cultivated maize categorized by stickiness when cooked because of higher amounts of amylopectin. Waxy kernels from maize plants were first observed in China in 1909. In other maize breeding programs, American breeders used this to tag hidden genes as a genetic marker for a long time because the waxy maize showed several important traits. In 1922 a researcher noted that the endosperm of waxy corn is amylose-free but contains amylopectin only, in contrast to wild maize varieties that contain both. However, the demand for waxy starches has recently increased due to consumer preference for cooking and eating quality and for potato free of amylose starches arising in European countries. AVEBE is an industrialized variety of waxy potato Eliane utilizing conventional mutation breeding methods, while BASF bred Amflora, a genetically modified potato form, by downregulating GBSSI [10]. In rice, a 50-leader/first-intron splicing site G/T polymorphism controls the development of mature GBSS messenger ribonucleic acid (mRNA), which affects amylose content [10]. The existence of G at the splicing position promotes normal splicing, leading to GBSS enhancement and high levels of amylose in *Indica*, while the T present at the junction leads to cryptic splicing, lower GBSS performance, and low amylose levels in *Japonica* [75]. To improve amylose quality, starch is synthesized to generate amylose either by overexpression of sufficient GBSSI [70] or by downregulating the activities of enzymes responsible for the biosynthesis of amylopectin [76,77]. Suppression of SBE and SSIIa in cereals results in higher resistant starch (RS) (e.g., barley and wheat) [78]. The amylose group is correlated with polymorphisms of the waxy protein (*Wx*) [66], which encodes the GBSSI enzyme and controls the synthesis of amylose [79].

4. Other Genes Involved in Starch Modification

Tian et al. [69] reported that in 70 rice varieties, an interaction study of 18 genes involved in gelatinization temperature, starch synthesis, gel consistency, and amylose content revealed that *waxy* (GBSSI) and *ALK* (*SSS2A*) are the major genes responsible for determining the nature of rice cooking and eating quality by influencing the gel consistency, amylose content, and temperature of gelatinization temperature. *ALK* is a major gene that regulates the temperature of grain gelatinization and a minor gene, affects the quality of the gel and amylose content [8]. The genes that influence both gelatinization temperature and gel consistency at the same time are *ISA* and *SBE3* [8]. Two main genes, *SSIIa* and *Wx*, and a variety of minor genes, *SSIV*, *SBEI*, *isoamylase*, and *SBEII3*, influence eating and cooking quality have been identified in rice through the study of various varieties [69,80]. In addition to the major genes, the minor genes included *AGPlar*, *APGiso*, *AGPsma*, *SS1*, *S-II-2*, *SSII-1*, *SSIII-2*, *SSIII-1*, *SSIV-2*, *SSIV-1*, *SBE1*, *SBE3*, *SBE4*, *ISA*, *PUL*, and *GBSSII* [69]. The functions of genes involved in starch modification include amylose synthesis, gel consistency, gel temperature, grain palatability, and amylopectin synthesis (Table 1). Tian et al. [69] stated that on the basis of interaction sites, each of the minor genes that affect eating and cooking quality could be divided into two haplotypes, namely, haplotype I and haplotype II. Haplotypes of individual genes involved in starch synthesis pathways, determined on the basis of interaction sites in 70 rice varieties studied include AGPlar-1633, AGPiso-511, SBE3+3577, *Wx*-1160+111, ISA-1499, ISA-1326, PUL+855, SSII-3+3796, SSI+3216, SSIV-2+437, and SSIII-2-1078 [69], with each of the genes classified as either haplotype I or haplotype II. Starch synthesized from sucrose is regulated by several genes, including *SuS2*, *SuS4*, *SuS3*, *AGPL2*, *AGPL1*, *AGPS*1, *SSSII-3*, *SSSI*, *SSSIII-2*, *SBEI*, *SBEIV*, and *SBEIII* [51,81]. In addition, many minor genes have unique properties; for example, *PUL*, *SSS3A*, *SSS1*, and *AGPlar* affect the amylose level, *AGPiso* affects the quality of gelatinization, and *SSS4B* (*SSSIV-2*) affects the temperature of gelatinization [8]. Correlations between gel consistency, gelatinization temperature, and amylose content are triggered by the combined activity of these relevant genes [8]. The estimation of the haplotype I amylose content is substantially higher than that of haplotype II under the regulation of the *Wx* haplotype for each related gene and vice versa [69]. The measurement of the gelatinization consistency value of haplotype II for each additive gene was substantially higher than that of haplotype I and vice versa, whereas the *Wx* haplotype was controlled [69]. *SSII-3* has a paramount significance for the gelatinization temperature properties of GT, which suggests that *SSII-3* performs a vital role in controlling the gelatinization temperature [69]. There are two types of *SSII-3* alleles: *SSII-3-II* and *SSII-3-I* [69]. Categories with *SSII-3-II* had GT values less than those of *SSII-3-I* [69]. In addition to *SSII-3*, a further study established *ISA*, *SSIV-2*, and *Wx* as minor genes influencing gelatinization temperature additively [69]. Grain palatability decreases with the downregulation of *SBE1* and *SBE3* levels, while overexpression of *SBE1* and *SBE3* increases the palatability of cereal [8]. Collectively, overexpression of *SSS1*, *SSS2A*, *SBE1*, and *SBE3*, especially *SBE3* and *SBE1*, could improve eating quality, but overexpression of GBSSI diminished the eating quality of cereal [8], while downregulation of GBSSI improved cooking and eating quality. Despite the existence of two additional *SSS2* genes, two *SSS3* genes, one *SSS1* gene, and two *SSS4* genes in rice plants, the *SSS2A* action specifies the form of amylopectin formation of rice starch to be either the *Japonica* or *Indica* type by performing a particular function in long B1 chain synthesis by extending the short A and B1 chains [82]. Guo et al. [83] used homozygous mutants to examine the basic function of *TaSSIVb-D* in starch granule synthesis in leaves at various developmental stages and reported that *TaSSIVb-D* mutations decreased the expression of genes and the amount of leaf starch granules.

The transformation of wheat with a changed variant of maize *shrunken 2* gene encodes a modified broad AGP subunit with reduced sensitivity to its adverse allosteric effector, orthophosphate, and more robust connection with small and large subunits, leading to an increased weight of grain per plant up to 38% [84]. Overexpression of *shrunken 2* (encoding large AGPase subunits) and *brittle 2* (encoding small AGPase subunits) at the same time in maize raised the starch content to a higher degree than when both genes were expressed individually [48].

Numerous reports have shown that plant hormone control genes encode enzymes that are essential for starch synthesis enzymes [81]. The production of starch synthesis genes and their enzyme functions are adversely affected by higher levels of abscisic acid (ABA) and ethylene in lower spikelets, resulting in low grain-filling values [85]. Grain filling is an accumulation of starch in the seed endosperm, and starch contributes 80–90% of an unpolished grain's final dry weight in rice [86]. In regulating grain filling, plant hormones, particularly ABA and ethylene, play significant roles [85]. During the seed development stage, higher levels of ethylene are negatively connected to enzyme activity linked to metabolism and contribute to poor grain filling [85,87]. Exogenous ABA supplies decrease the sucrose (SUS) mRNA levels and enzyme activity of the mRNA in different rice grains, whereas the expression of the majority of starch synthetic genes is suppressed by the exogenous supply of ABA by ethephone; these genes are *SUS*, *AGPase*, and *SSS*, and they are depleted by their enzyme activity [85]. The effect of sucrose synthase (SUS) and the manifestation of the gene involved in starch synthesis are improved by a suitable concentration of exogenous abscisic acid, thus improving rice yields [87]. The transportation of sucrose into the grain is reduced by a high concentration of ABA, which reduces the capability of grains to synthesize starch [88], while a suitable ABA concentration boosts the SUS activity [89], increases gene expression related to starch metabolism [90], and improves the yield of grains [87]. The internal starch granule associated proteins in crops[91], are listed in Table 2.

Table 1. Genes involved in starch modification.

Genes	Function	References
Wx, AGPlar, PUL, SSI, ALK (SSII-3), SSIII-2	Amylose synthesis	[69]
Wx, AGPiso, SBE3, ISA, ALK (SSII-3)	Gel consistency	[69]
ALK (SSII-3), Wx, SBE3, ISA, SSIV-2	Gel temperature	[69]
SBE1, SBE3	Grain palatability	[8]
SBE1, SBE3, SBE4, ISA, PUL	Amylopectin synthesis	[92]
BEI, BEIIb, BEIIa, SSIIa, SSIIIa, SSI, PUL, ISA1	Amylopectin synthesis	[93,94]

Table 2. Internal starch granule associated proteins in crops [91].

Crops	Enzymes Identified in Starch Granules	References
Rice	BEIIb (82 kDa), SSIIa (86 kDa), SSI (72 kDa), GBSS (60 kDa)	[95]
Wheat	BEIIa, BEIc (SGP-145, 145 kDa), BEIc (SGP-140, 140 kDa), BEIIb (SGP-2, 92 kDa), SSIIa (SGP-A1, 115 kDa), SSIIa (SGP-D1, 108 kDa), SSIIa (SGP-B1, 100 kDa), SSI (SGP-3, 80 kDa), GBSS (60 kDa)	[96,97]
Barley	BEIc (140 kDa), BEIIb (93 kDa), SSIIa (87 kDa), SSI (71 kDa), GBSS (60 kDa)	[98,99]
Maize	BEIIb (85 kDa), SSIIa (86 kDa), SSI (76 kDa), GBSS (60 kDa)	[100,101]
Potato	R1 (160 kDa), SSII (92 kDa), GBSS (60 kDa)	[102,103]
Pea	R1 (160 kDa), BEI (114 kDa), BEII (100 kDa), SSII (77 kDa), GBSS (60 kDa)	[103,104]

5. Starch Biosynthesis Pathway

Starch biosynthesis from sucrose mainly influences the yield of cereal and rice quality in the developing endosperm [85]. It is a dynamic and highly coordinated mechanism that involves synchronized activities among several enzymes, such as starch debranching enzyme (DBE), starch synthase (SS), starch branching enzyme (SBE), and ADP-glucose pyrophosphorylase (AGPase), to organize activities [9]. According to the Uniprot database (https://www.uniprot.org/keywords/KW-0750 (accessed on the 20 January 2021)), starch (glucans and amylopectins) is synthesized via the ADP-glucose pathway by three main enzymes: (1) starch synthase, (2) ADP-glucose pyrophosphorylase, and (3) starch branching enzyme. To generate amylopectins, the arbitrarily branched glucan molecules are precisely debranched through the debranching enzyme. Both glucose-derived sucrose and glucose-derived fructose are synthesized in the cytoplasm and subsequently transferred to the cytosol for sucrose degradation by the invertase enzyme [8]. Starch is synthesized by multiple isoforms or subunits of five enzyme groups utilizing degraded sucrose, UDP-glucose, and fructose products, which include (1) granule-bound starch synthase (GBSS), (2) soluble starch synthase (SSS), (3) ADP-glucose pyrophosphorylase (AGP), (4) starch debranching enzyme (DBE), and (5) starch branching enzyme (SBE) [105]. Of the five enzymes that synthesize starch, GBSS, SBE, DBE, and SSS contribute to amylopectin structure [106]. Zhu et al. [85] also reported that several primary enzymes are involved in the starch synthesis pathway; these enzymes include SUS, UGPase, AGPase, and SS. The biosynthesis of starch differs quantitatively and qualitatively during the development of storage organs [107]. Most biosynthetic enzyme isoforms could be involved in the early development of starch granules, and others become active later [107]. As a percentage of overall starch in the production of storage organs, the amylose content typically rises, suggesting that it is synthesized late compared to amylopectin content. This may be triggered by the pacing of the synthesis of GBSS, since all the SS enzymes are synthesized faster than the GBSS [108].

The commencement of biosynthesis of starch in storage organs involves sucrose mobilization into glucose-6-phosphate and the importation of G6P into the amyloplast by inorganic phosphate (Pi) exchange; by plastidial phosphoglucomutase (Pi) action, glucose-1-phosphate (G1P) is formed by the transformation of G6P [10]. Geigenberger [109] also stated that the formation of ADP-Glc and inorganic pyrophosphate (PPi) occurs through the transformation of Glc-1-P and ATP, catalyzed by ADP-Glc pyrophosphorylase (AGPase), as the first initiated phase. ADP-Glc acts as a glucosyl donor for various classes of SS that extend the α-1,4-connected glucan chains of the two insoluble amylose and amylopectin starch polymers, from which the granule is developed [109]. By transforming glucose 1-phosphate (Glc-1-P) and ATP to ADP-Glc and inorganic pyrophosphate (PPi) in the amyloplasts, AGPase, the first enzyme that triggers the starch biosynthesis pathway, catalyzes the restricting reaction [9]. Regina et al. [10] also reported that the first step of starch synthesis is ADP glucose (ADPG) production, catalyzed by ADPG pyrophosphorylase (AGPase) via ATP activation of G1P. The synthesis of ADP-glucose occurs both in the cytosol and plastids in the developing endosperm of cereal seeds [105]. For starch synthesis, the ADP-glucose produced inside the cytosol is transferred into the plastid and does not have any other metabolic function [105]. The ADP-glucose importation of ADP-glucose into the plastids is achieved by an ADP antiporter (ADP-glucose transporter)/ADP glucose, which is only located in the endosperm of grass [105]. ATP and glucose-1-phosphate synthesize ADP-glucose in a reaction catalyzed by ADPGPPase, releasing pyrophosphate, synthesized amylose, and amylopectin [107]. Inorganic alkaline pyrophosphatase, which is possibly limited to both non-photosynthetic and photosynthetic tissue plastids, eliminates the pyrophosphate formed by ADPGPPase [107]. Riso 16 mutant barley, the small subunit of cytosolic AGPase inactivation (where plastidial AGPase function is not affected), results in decreased aggregation of ADPG and starch in the endosperm with reduction of storage of protein accumulation and seed size occurring at the same time [10]. 3-Phosphoglyceric acid (3-PGA) acts as an enzyme activator, leading to the catalytic activity of the enzymes, and is blocked by inorganic phosphate (Pi) [9]. Oxidation-mediated development of disulfide bridges between neighboring AGPSSs is also restricted to the action of AGPase, which can contribute to reactivation by decreased thioredoxin (or dithiothreitol in vitro) [110]. Hannah et al. [111] observed that in both the tiny *shrunken 2* (*Sh2*) and the huge *brittle 2* (*Bt2*) gene subunits, mutations in AGPase contribute to a significant decrease in starch content. Starch synthase (SS) can be classified as GBSS, which synthesizes amylose and the extra-long chain fraction of amylopectin, and soluble starch synthase (SSS), which synthesizes amylopectin [112]. Geigenberger [109] reported that in plants, five different types of SS groups are known, namely, GBSS, which is responsible for amylose synthesis, and soluble SSs, which range from I to IV and regulate the synthesis of amylopectin. The synthesis of the α-(1,4) linkage between the non-reducing end of the pre-existing glucan chain and the glucosyl moiety of ADP-glucose is catalyzed by SS, triggering ADP release [107]. Both amylose and amylopectin may be used as substrates for SSs in vitro [107]. Amylose is an essentially linear glucan comprising relatively few branches of α-1,6 and generates up to 20–30% of regular starch, whereas amylopectin is extremely branched [109]. Two groups of SBE (SBEI and SBEII), which vary in relation to the extent of the transferred glucan chains and the specificity of the substrate, add branch points [109]. The branching activities of starch branching enzymes, which belong to the α-amylase family, are controlled by Q enzymes by introducing an arrangement that is branched by cleaving the α-1,4-glucan chain in polyglucans and then re-attaching the cleaved chain through an α-1,6-glucan linkage to the acceptor's chin, thereby forming a branch on the same or another chain [113].

SBE, which hydrolyzes an α-(1,4) linkage within a chain and then catalyzes the creation of an α-(1,6) connection between the decreasing end of the cut glucan chain and another residue of glucose, presumably one from the hydrolyzed chain, creates the α-(1,6) branches in starch polymers [107]. Remarkably, two categories of debranching enzymes are often involved in the synthesis of starch, which split branch points and may have an effect in modifying the branched glucans into a state that can cause crystallization inside the granule [109]. Starch debranching enzymes (DBEs), glucan-modifying enzymes that exist in dual types (PUL and ISA), are also significant in the synthesis of starch [9]. The most significant functional distinction between these ISA types is typically the operation of phytoglycogen and amylopectin by hydrolyzing polyglucan α-1,6 bonds, which play an essential role in altering overly branched chains or eliminating unnecessary amylopectin branches created by branching enzymes to preserve the amylopectin cluster structure [9]. In addition, ISA is likely to have branched amylose chains [9]. PUL naturally cleaves polyglucan α-1,6-linkages in pullulan, while amylopectin has little or no effect on glycogen [114].

Genetic findings demonstrate that various SS, SBE, and debranching enzyme isoforms are likely to play unique roles in establishing complex starch structures [109]. To synthesize the crystalline matrix of starch granules, they must act in close coordination [109]. Interestingly, SSIII and SSIV are involved in the initiation of starch granules [115]. Recent studies have proposed that the plastidial pathway of starch synthesis occurs in all current green algae and higher plants and that the enzymes involved in starch biosynthesis in higher plants underwent a complicated series of modifications during evolution [116]. The main pathway of starch biosynthesis, including sucrose synthesis, sucrose degradation, and starch synthesis in rice [8], is illustrated in Figure 2.

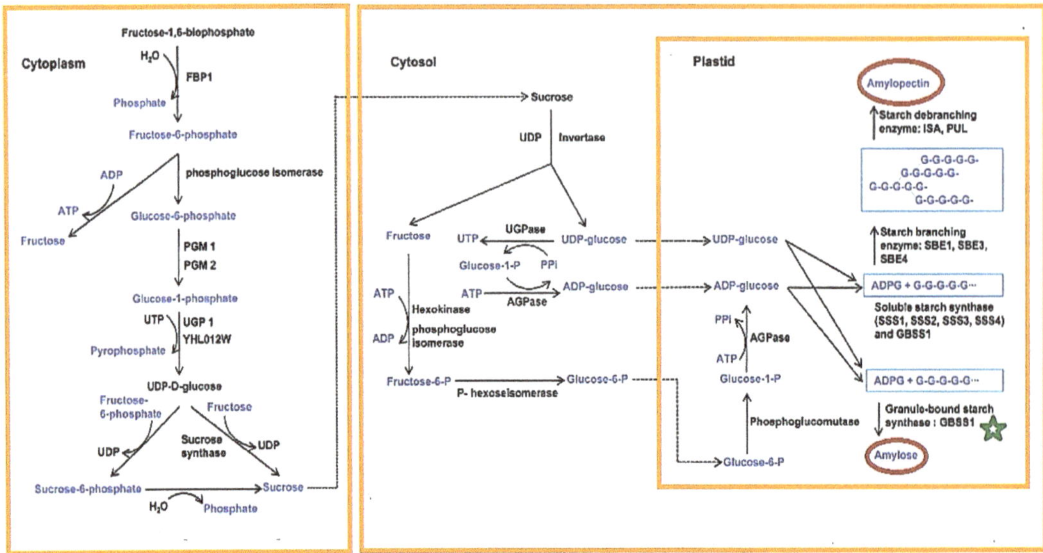

Figure 2. "Main pathway of starch biosynthesis including sucrose synthesis, sucrose degradation and starch synthesis in rice" [8]. "FBP1, fructose-1,6-bisphophatase1; PGM, phosphoglucomutase; AGPase, ADP-glucose pryophosphorylase; PPi, pyrophosphate" [8]. For both gel consistency and amylose content, waxy (GBSSI, marked with a green star) serves as a vital/critical gene, but it has a minor influence on the temperature of gelatinization. The amylose and amylopectin circled in red color are very important in starch modification. Source: Sun et al. [8].

6. Posttranslational Modification (PTM)

PTMs are covalent modifications that alter the principal protein structure in a manner that is sequence-specific, through acylation, phosphorylation, nitration, glycosylation, and ubiquitination, by removal and reversible addition of functional classes [117]. This method is achieved by the induction of a covalent linkage to a new functional group, such as acetyl, phosphate, ubiquitin, or methyl [118]. Proteins are synthesized on ribosomes, generating a nascent polypeptide chain. To generate mature proteoforms that gradually accumulate in plant cells to form the observed proteome, several proteins undergo PTMs [42]. Protein posttranslational modifications greatly enhance functionality, increase proteome diversity, and allow rapid responses, all at reasonably low cell costs [43]. They range from co-translational modifications to position, feature, and signaling enablers and also include stability and degradation markers [42]. PTMs occur on the side chain or on the protein N or C termini and expand the chemical decoration and characteristics of the 20 standard amino acids by changing established functional groups or adding new ones [42]. PTMs play a crucial role in plants via their effects on signaling, enzyme kinetics, protein stability and interaction, and gene expression [43]. After a concise review of the experimental and bioinformatics difficulties of PTM site recognition, position, and occupancy/quantification, a succinct description of the main PTMs and their promising functional effects in plants is presented, with priority on plant metabolism [43]. The center of numerous cellular signaling events is the PTM [119]. Besides the single regulatory PTM, there are other PTMs that act in a coordinated manner [119]. These PTM crosstalks generally function as a modified system for changing cellular responses to the smallest environmental changes [119]. Although PTM crosstalk has been examined in detail in different animals, this aspect is just emerging in plants [119].

Due to the physical–chemical characteristics of reactive amino acids and the cellular atmosphere, including oxygen, pH, metabolites, and reactive oxygen species, or as a result of the activity of particular modifying enzymes, PTMs may occur spontaneously (non-enzymatically) [120]. Significant quantities of proteins and a very sensitive system for their discovery are necessary for identifying PTMs [117]. Kwon et al. [117] also observed that significant quantities of proteins and an extremely sensitive system for their investigation are needed to identify PTMs [117]. Adjacent residues and their exposure to the surface are also important for PTM identification [43]. Over 300 separate forms of PTMs have been reported, and newly discovered types are introduced to the list regularly [121]. Throughout a protein's life cycle, different PTMs exist [42]. PTMs are subdivided into different groups, which include the addition of peptide groups, amino acid modification, the addition of complex molecules, the addition of chemical groups, and cleavage of proteins via a variety of proteolytic mechanisms [30]. The addition of chemical groups can be grouped into acetylation, redox-based modifications, methylation, and phosphorylation; types of polypeptide addition include SUMOylation, ubiquitination, and other ubiquitin-like protein conjugations; the processes by which complex molecules are added include AMPylation, glycosylation, acylation, prenylation, and ADP-ribosylation; and the direct processes by which amino acids are modified are eliminylation and deamidation [30]. Although protein modification is irreversible, the additions of polypeptides, chemical groups, and complex molecules are sometimes reversible [30]. Glycosylation, acetylation, and phosphorylation are primary types of PTMs, of which phosphorylation is one of the most vital, prevalent, and widely employed protein PTMs used in the course of the research by molecular biologists [122]. Vu et al. [119] also stated that in molecular biology research, the most employed PTM is phosphorylation, and its crosstalk with other forms of PTM has been notably observed, particularly in recent studies. The most critical posttranslational modifications detected in plants include phosphorylation (His, Tyr, Ser, Asp, Thr), S-Nitrosylation (Cys) and nitration (Tyr), acetylation (Lys ε-amine, N-terminal α-amine), deamidation (Asn, Gln), lipidation (S-acetylation, N-myristoylation, prenylation; Gly, Cys, Trp, Lys, N terminal), N-Linked glycosylation (Asp) and O-linked glycosylation (Thr, Lys, Trp, Ser), ubiquitination (N terminal, Lys), sumoylation (Lys), carbonylation (Pro, Thr, Arg, Lys), methylation (Lys, Arg, N terminal), glutathionylation (Cys), oxidation (Cys, Met), peptidase (cleavage peptidyl bond), S-Guanylation (Cys), and formylation (Met) [43] (Table 3).

Table 3. The most critical posttranslational modifications detected in plants [43].

Type of PTM (Reversible If Asterisk)	Enzymatic or Spontaneous (Nonenzymatic)	Comment on Subcellular Location and Frequency
Phosphorylation (His, Tyr, Ser, Asp, Thr)	Enzymatic	Phosphorylation of His and Asp have low frequency
S-Nitrosylation (Cys) and nitration* (Tyr)	Spontaneous (RNS), but reversal by thioredoxins is enzymatic for Cys	Throughout the cell
Acetylation (Lys ε-amine, N-terminal α-amine)	Enzymatic	In mitochondria, very little N-terminal acetylation, but high Lys acetylation; Lys acetylation correlates to [acetyl-CoA]
Deamidation (Asn, Gln)	Spontaneous, but isoAsp reversal is enzymatic by isoAsp methyltransferase	Throughout the cell
Lipidation (S-acetylation,N-myristoylation*, prenylation*; Gly, Cys, Trp, Lys, N terminal)	Enzymatic	Not (or not often) within plastids, peroxisomes, mitochondria

Table 3. *Cont.*

Type of PTM (Reversible If Asterisk)	Enzymatic or Spontaneous (Nonenzymatic)	Comment on Subcellular Location and Frequency
N-Linked glycosylation (Asp); O-linked glycosylation (Thr, Lys, Trp, Ser)	Enzymatic	Only proteins passing through the secretory system; O-linked glycosylation in the cell wall
Ubiquitination (N terminal, Lys)	Enzymatic	Not within plastids, peroxisomes, mitochondria
Sumoylation (Lys)	Enzymatic	Not within plastids, peroxisomes, mitochondria
Carbonylation* (Pro, Thr, Arg, Lys)	Spontaneous (ROS)	High amounts in chloroplast and mitochondria
Methylation (Lys, Arg, N terminal)	Enzymatic	Chloroplasts and histones (nucleus); still not fully explored
Glutathionylation (Cys)	Enzymatic	High amounts in chloroplasts
Oxidation (Cys, Met)	Spontaneous (ROS) and enzymatic (by PCOs), but reversal is enzymatic by thioredoxins, Met sulfoxide reductases, and glutaredoxins, except if double oxidized	High amounts in chloroplast and mitochondria
Peptidase* (cleavage peptidyl bond)	Enzymatic	Throughout the cell
S-Guanylation (Cys)	Spontaneous (RNS)	Rare; 8-nitro-cGMP is signaling molecule in guard cells
Formylation (Met)	Spontaneous, but deformylation by peptide deformylase is enzymatic	All chloroplasts and mitochondria-encoded proteins are synthesized with initiating formylated Met

The reversible phosphate group covalently binds to the hydroxyl group of hydroxyl amino acids such as threonine, serine, and tyrosine but rarely covalently binds to hydroxyproline in protein phosphorylation [123]. Tyr, Thr, and Ser account for PTM phosphorylation with percentages of 1–5%, 15–20%, and 75–80%, respectively [124]. Phosphorylation PTM is the most widely used and most significant in providing proteomic dynamics [125]. PTM, particularly phosphorylation of protein, has long been regarded as a crucial regulatory mechanism that regulates transcription factors [126,127]. Its discovery in recent times has propelled decades of potential phosphoproteome studies [30]. Protein phosphorylation has arisen as one of the main PTMs. It plays a key role in DNA replication, cell signaling and differentiation, gene expression, and enzyme activity increases or decreases; it also changes other biological effects, controlling cell proliferation and enlargement, phytohormone biosynthesis and signaling, plant disease resistance, grain-filling, and development of rice seed quality [128].

C3 phosphorylation accounts for 20 to 30% of phosphate monoesters, while about 70 to 80% of phosphate monoesters are attached to the C6 position of the glucosyl unit [129]. In order to catalyze the phosphorylation of starch, two forms of glucan water dikinases have been employed: glucan water dikinase (GWD1) and phosphoglucan water dikinase/glucan water dikinase 3 (PWD/GWD3) [14]. Specifically, GWD1 phosphorylates starch at the C6 position, catalyzing the transition of β-phosphate from ATP to a residue of glucosyl, and then PWD identifies the pre-phosphorylated glucan C6 and ultimately phosphorylates hydroxyl C3 [130]. Glucan water dikinase (GWD1) and phosphoglucan water dikinase/glucan water dikinase 3 (PWD/GWD3) catalyze starch phosphorylation, which occurs naturally during starch metabolism in plants [14]. The only identified naturally occurring PTM of starch is phosphorylation [131]. Phosphorylation often liaises with other forms of PTMs in the cell, most often for protein turnover and protein function attenuation [33]. Nabeshima et al. [132] reported that the incorporation and quantity of phosphate groups differ pertaining to the type of starch, based on the source of the starch, amylose–amylopectin ratio, purity of the starch, microstructural features, conditions during phosphorylation, and various phosphate salts. GWD1 phosphorylation disrupts the starch surface crystalline structure and enhances the hydrolytic action of plastidial β-amylases [133].

The degree of phosphorylation of starch differs greatly because of the sources of starch in the plant [134]. Hebelstrup et al. [131] reported that phosphate levels are higher in tuber starch than in cereal starch. Compared with leaf starch, the content of phosphate detected in starch deposited in seeds is very low [135]. Thorough study on modification of *Wx* has rendered it easier to identify more suitable target locations for editing and enabled the amylose content of crops to be controlled more specifically in a more effective manner [62]. The reduction of phosphorylation at the Ser-34 position affects the activity of GBSSI, synthesizes amylase, and leads to amylose content reduction [136,137]. Phosphate ester removal is important for starch degradation [135]. Liu et al. [50] reported that the substitution of Asp165/Gly165 had no noticeable influence on the activity of GBSSI in vitro; nevertheless, it remarkably caused a reduction in GBSSI binding to the starch granules, which led to amylose content reduction in rice. More knowledge of GBSS posttranslational and transcriptional control may lead to the discovery of new mechanisms that affect amylose content [138]. LC-MS/MS was used to identify the phosphorylation sites in GBSSI [137]. Non-amylolytic GWD glucan phosphorylation is necessary for the turnover of plant starch [135]. Huang et al. [62] reported that in the waxy locus, other sites might be targets for PTM modification or translation (e.g., the stability of protein, starch binding capacity, phosphorylation, ADP-glucose binding, and codons that control the enzymatic function of GBSSI). Zhang et al. [24] located nine phosphorylated PTM sites in rice GBSSI, but only Ser415P had an altered level of phosphorylation, leading to moderate amylose content. One approach for enhancing bakery product quality is through the phosphorylation of rice [139]. Kringel et al. [139] reported that wild-type rice had a lower pasting temperature and higher breakdown rate than phosphorylated rice. Phosphorylation results in a substantial decrease in the paste synthesis and retrogradation of rice flour [139]. Phosphorylated rice flour used for bread baking demonstrated a hardness reduction at all the tested storage temperatures and had an impact on the bulkiness, texture, and color of rice bread, indicating the feasibility of utilizing phosphorylated rice flour in gluten-free bread [139]. Gluten's viscoelastic properties cannot be improved without modifying the fraction of protein, as shown by the use of gluten-free cereals in dough systems [140]. In proteomic studies of gluten proteins, PTM is commonly used to explore the modification of protein [141]. The linearity of amylopectin and amylose molecules can be interrupted by the use of negatively charged phosphate classes [139]. The second cys (Cys529) in rice GBSSI is conserved in plant GBSSIs, and insertions of sequence were likewise found around this cysteine [142]. In the *Poaceae* family, the first cys (Cys337) in GBSSI of rice, rye, corn, sorghum, wheat, and barley is conserved, whereas in non-Poaceae plants such as soybean, pea, *Arabidopsis thaliana*, cassava, kidney stone, sweet potato, and potato, it is substituted with valine [142]. In cultivated rice, the combination of 'GC' (leucine) at SNP4 and 'G' (valine) at SNP3 is necessary for the development of L-type rice starch, which has a higher GT than S-type rice starch [143]. The conversion of L-type starch to S-type starch is caused by changing leucine to phenylalanine or valine to methionine, which leads to low GT [143]. Consumers prefer low-GT rice because of its cooking quality [143]. It is worth noting that in poaceous GBSSIs, disulfide bonds are detected, including in most cereal plants that store large amounts of starch in their seeds [142]. The disulfide bridges cause restricted domain movement, which could help to improve the efficiency of starch biosynthesis [142]. Liu et al. [50] reported that via a posttranslational approach, Wx^{hp} directly affects the *Wx* gene and subsequently influences Haopi amylose content. In addition, by editing the CpG site in the waxy promoter, DNA methylation modification can give rise to rice with preferable eating quality and healthy rice with minimized GI [62]. DNA methylation is linked to amylose and amylopectin synthesis, and altered methylation can affect gene expression [144]. Lysine acetylation has been observed in many primary enzymes in the starch synthesis pathway, including GBSSI, AGPS, SBE3, AGPLar, SBE1, and Pho1 [38]. Amylose chains can depolymerize as a result of acetylation [145]. Out of 247 proteins, 421 malonylated lysine locations were observed and perform a crucial role in multiple essential metabolic processes, including lipid metabolism, central carbon metabolism, photosynthesis, and biosynthesis of starch [146]. Lysine, arginine, and charged amino acids in the lysine flanking position are favored residues [146]. In recent years, lysine succinylation has been recognized as a posttranslational modification (PTM) [147]. Because of the bulkier structural modifications and more important charge variations on the changed lysine residue, succinylation may have a greater practical effect than acetylation [147]. During the development of grains, cereals undergo various PTMs of proteins [148]. Many of the class of enzymes involved in biosynthesis of starch are regulated and coordinated and in a range of ways from gene expression to different posttranslational mechanisms including protein phosphorylation and redox modulation [149]. Meng et al. [147] also observed that the enzymes involved in starch biosynthesis and regulatory pathways, including starch branching enzymes (OsBEI and OsBEIIb), ADP-glucose pyrophosphorylases (OsAGPS2 and OsAGPL2), sucrose synthase (SUS2 and SUS3), starch debranching enzymes (OsPUL), UDP-glucose pyrophosphorylase (UGP), phosphoglycerate mutase (PGM), and starch phosphorylase (OsPHOL), in developing rice seeds undergo succinylation. Surprisingly, heavy succinylation was found on major seed storage proteins, as well as key enzymes involved in central carbon metabolism and starch biosynthetic pathways for the development of rice seed [147]. In order to improve starch quality, most of the starch biosynthesis proteins listed above can be subjected to crotonylation, acetylation, 2-hydroxyisobutyrylation, and malonylation [147]. Multiple (PTM-dependent) approaches carefully control all steps of CO_2 fixation and starch metabolism by balancing the rate of biosynthesis of starch with the availability of carbon and energy in numerous tissues of plants under different environmental conditions [150].

In wheat, Chen et al. [137] detected three GBSSI phosphorylation sites, namely Thr323, Ser34, and Ser450, and three phosphorylation sites in SSIIa, namely Ser776, Tyr358, and Ser228. In both SN119 and ND5181, five starch synthesis enzymes, including SSI, GBSSI, SBEI, SBEII-a, and SSII-a, were phosphorylated, and their phosphorylation sites were conserved [46]. In particular, SSIIa and GBSSI had eight and nine phosphorylated sites, respectively, showing that in starch biosynthesis, many conserved phosphorylated sites play essential roles [46]. In addition, the starch granule-binding proteins were heavily stained with Pro-Q Diamond at five developmental levels [46]. At all five phases, SSI, GBSSI, SBEIIa, and SSIIa were phosphorylated, with phosphorylation levels in ND5181 and SN119 being akin [46]. The characterization of starch granule-binding proteins (SGBPs) by phosphoproteome under water deficit treatment showed reduced phosphorylation levels of major starch synthesis enzymes, notably for GBSSI, SSIII, and SSII-a, that led to total starch reduction [137]. The enzymes involved in starch phosphorylation are attractive candidates for the regulation of flux via degradation of starch [151]. Specifically, SSIII-a site Ser837, GBSSI protein site Ser34, and SSII-a site Tyr358 demonstrated a slight decrease in phosphorylation under water-deficit treatment compared to under well-watered treatment [137]. A higher rate of phosphorylation of starches is characterized by a higher turnover of phosphate groups that exhibit transient esterification [152]. Only in B-type starch granules does the SSI-1 phosphorylation occur [153]. SSI-1 predominantly synthesizes the smallest glucan chain with a DP of approximately 10 glucosylic units or less [154]. The level of phosphorylation at the 36 phosphorylated sites of 19 phosphoproteins was substantially altered [46]. Twenty-two and fourteen phosphorylated sites were discovered in SN119 and ND5181, respectively, out of the thirty-six phosphorylated sites [46]. Glucose-6-phosphate isomerase ser600 was only observed to be phosphorylated in ND5181, while the vacuolar cation or proton exchanger sites Tyr243 and Tyr249 were phosphorylated in SN119 only [46].

Chen et al. [46] observed that the absence of amylose did not influence SGBP expression and the phosphorylation of SGBP is involved in amylopectin biosynthesis.

In maize, a phosphorylation-specific dye study of the granule phosphoproteome showed that starch phosphorylase, GBSS, and BEIIb are phosphorylated as they exist in the granule [91]. Crosslinked maize starch procured by different phosphorylation approaches displayed an elevated percentage reduction in the peak viscosity stretch from 97.1 to 99.6% relative to pulse starches, suggesting that the impact of crosslinks on maize starch pasting properties was more noticeable than that on pulse starches [155]. The probability that starch metabolic enzymes present in granules is controlled by protein–protein interactions and/or modifications was observed [91]. In maize and wheat amyloplasts, the soluble types of starch phosphorylase and BEIIb accept the transition by marking the ATP of a radioactive phosphate group [91]. The crosslinked starch with STMP aqueous solution had a less noticeable effect on the improved crystalline stability than STMP semidry and $POCl_3$ aqueous solution [155]. STMP-aqueous only increased Tc in pulse starches and substantially increased Tp, To, and Tc in maize starch, indicating that crosslinking systems and crosslinking agents have different impacts on different types of starches [155]. Phosphorylation crosslinking considerably improves the pasting temperature, breakdown reduction, gelatinization temperature increase, and thermal stability in various forms, which may permit broader applications in corn, field peas, and faba beans [155].

In Arabidopsis, the low MNF-Pin-39 variety activity indicates that the replacement of Ser-570Tyr is counterproductive to activity and, considering significant GBSS abundance, is attuned with the very low amylose content [156]. The substitution of the amino acid in the GBSS variant does not seem to be harmful to the operation but rather inhibits in vivo amylose synthesis through other techniques; Gn2–3 and TueSB30-3 bear the same GBSS allele, which encodes the substitution of two amino acids (Asn-430Ser and Gly-394Glu) [156]. The lack of amylose means that one or both amino acids are required for in vivo GBSS action [156].

In potato tubers, starch phosphorylation can induce both starch degradation and starch synthesis [157]. Due to the existence of phosphate groups, wild-type potato starch produces a transparent paste, while an increased level of phosphorylation steadily reduces the clarity of the paste [158]. One out of 200–300 glucose units of amylopectin in potato starch is phosphorylated [159]. Phosphate groups may be bound to a glucose residue at the C3 or C6 positions [159]. Potato plants with decreased amounts of R1 protein exhibit much less starch phosphorylation, suggesting that this protein is essential for starch phosphorylation [134]. Phosphorylation carried out at a low substitution degree leads to a significant increase in the swelling power and solubility, whereas the swelling power and solubility decrease steadily with increasing substitution levels [158]. As the degree of phosphorylation increases, viscosity values decrease, while at the lowest degree of substitution of various starch forms apart from maize amylose, the viscosity increase [158]. Therefore, the preparation of phosphorylated starches with very low levels of replacement is suggested to be the best property for paste transparency [158]. Potato tuber reserve starch is another example of a starch phosphorylation feature during starch synthesis [135]. Minor changes in the metabolism of storage starch were recorded in transgenic potato lines with downregulation of *StGWD* [160]. This line of antisense leads to the yield of more, but smaller tubers, and both the viscosity of the starch and the quality of amylose are influenced [135]. In the endosperm amyloplast in barley, *StGWD* overexpression heterologously leads to a 10-fold increase in grain starch-bound phosphate [161]. The development of transformants in which the native catalytic subunit of AGPase (AGPB) is substituted by a modified type of AGPB in which Cys-82 is modified to prevent the formation of an intermolecular bridge is needed for the final genetic proof of posttranslational redox regulation, which is responsible for the inhibition of starch synthesis after tuber detachment [162].

Nabeshima et al. [132] observed the phosphorylation of sago-starch at pH 9.5 and pH 9 at different concentrations of sodium trimetaphosphate (2% and 5%), and they reported that the phosphorylation at pH 9.5, displayed higher cold-paste viscosity and lower hot-paste viscosity than the phosphorylated sago-starches at pH 9 with sodium trypolyphosphate. STMP-semidry reduced the onset gelatinization temperature but greatly elevated the peak and conclusion gelatinization temperatures in all starches [155]. With a rise in the number of phosphate groups introduced in the course of phosphorylation, the values for adhesiveness and firmness declined [132]. The transcriptional expression modifications of many genes that synthesize amylopectin, such as *SBEII-b*, *SSIII*, and *AGPL I*, were substantially decreased at 15 DPA, which is in line with GBSSI, indicating that the impact of drought stress on the biosynthesis of starch occurs at the stage of transcription, translation, and phosphorylation of proteins. Starch crosslinking greatly increases the temperature of gelatinization, whereas the enthalpy is not significantly affected [132]. The To, Tp, and Tc changes in field pear, faba bean, and regular corn showed that field pear had the highest To, Tp, and Tc, followed by faba bean and regular corn [155]. Owing to the restricted enlargement and decreased hydration of starch granules in rice, retrogradation and gelatinization are delayed [132].

7. Conclusions

Significant advancements have been made through the downregulation and overexpression of genes, which are comparatively brute force methods that sometimes contribute to unwanted trade-offs between plant development and stress resilience, and a better way of improving crops may be achieved through the posttranslational modification of sites or regulatory enzymes that regulate them because of their significance. The waxy locus controls both the non-waxy and waxy rice phenotypes. Active granule-bound starch synthase is essential for amylose synthesis, while amylopectin is synthesized as a result of a complex pathway involving various isoforms of starch-debranching enzymes, starch branching enzymes, and starch synthase. An interaction study of 18 genes involved in gelatinization temperature, starch synthesis, gel consistency, and amylose content revealed that waxy and *ALK* are the major genes responsible for determining the nature of rice cooking and eating quality by influencing the gel consistency, amylose content, and temperature of gelatinization. Rice starch can be altered into various forms by either reducing or increasing the amylose content, depending on consumer preference and region. The majority of consumers prefer waxy and low-amylose rice varieties because of their cooking and eating qualities. It is regionally desired and selected for consumption as cooked rice because of its softness and stickiness. To meet the market demand, which is increasing every day, it is important to focus more on yield improvement of *Wx* and low-amylose rice because increases in rice production rely largely on the improvement of yield, which could be achieved by posttranslational modification rather than an extension of the planting field. Therefore, there is a need for government, private, and parastatal organizations to invest more in research to improve the palatability of rice for consumers.

Author Contributions: T.V.A. and J.Z. jointly developed the conceptual structure of the manuscript. T.V.A. wrote the manuscript and prepared all tables and figures. J.Z., Y.W., L.C., H.W., and J.Y. assisted with further modifications of the manuscript. T.V.A. and J.Z. revised the manuscript. L.C., X.L., W.L., Y.-C.C., X.T., and J.Y. provided critical feedback, revised, and approved the manuscript for publication. All authors have read and agreed to the published version of the manuscript.

Funding: This research was funded by the National Natural Science Foundation of China (Grant No. 31871229), the Chinese High-yielding Rice Transgenic Program (Grant No. 2016ZX08001004-001), Central Public-Interest Scientific Institution Basal Research Fund (No. CPSIBRF-CNRRI-202114), and the ASTIP program of CAAS.

Institutional Review Board Statement: Not applicable.

Informed Consent Statement: Not applicable.

Data Availability Statement: Not applicable.

Acknowledgments: This work was supported by the National Natural Science Foundation of China (Grant No. 31871229), the Chinese High-yielding Rice Transgenic Program (Grant No.

2016ZX08001004-001), Central Public-Interest Scientific Institution Basal Research Fund (No. CPSIBRF-CNRRI-202114), and the ASTIP program of CAAS.

Conflicts of Interest: The authors declare no conflict of interest.

References

1. OECD-FAO. OECD-FAO Agricultural Outlook 2013–2022. In *Cereal*; OECD Publishing: Paris, France, 2018; pp. 109–264; ISBN 978-92-64-29721-0.
2. Cao, L.; Chen, X.; Hu, L.; Li, C.; Liu, C.; Lin, W.; Li, Z.; Luo, S.; Khan, M.U.; Azeem, S.; et al. *Agroecological Rice Production in China: Restoring Biological Interactions*; Shiming, L., Ed.; CRC Press: Boca Raton, FL, USA, 2018; pp. 279–299; ISBN 978-92-5-130704-5.
3. Peng, S.; Tang, Q.; Zou, Y.; Peng, S.; Tang, Q.; Zou, Y. Current Status and Challenges of Rice Production in China Current Status and Challenges of Rice Production in China. *Plant Prod. Sci.* **2009**, *12*, 3–10. [CrossRef]
4. Guo, A.; Ding, X.; Zhong, F.; Cheng, Q.; Huang, C. Predicting the future Chinese population using shared socioeconomic pathways, the sixth national population census, and a PDE model. *Sustainability* **2019**, *11*, 3686. [CrossRef]
5. Dipnaik, K.; Kokare, P. Ratio of Amylose and Amylopectin as indicators of glycaemic index and in vitro enzymatic hydrolysis of starches of long, medium and short grain rice. *Int. J. Res. Med Sci.* **2017**, *5*, 4502–4505. [CrossRef]
6. Chen, N.; Luo, Y.-K.; Xie, L.-H.; Zhu, Z.-W.; Duan, B.-W.; Zhang, L.-P. Protein Content and Its Correlation with Other Quality Parameters of Rice in China. *Acta Agron. Sin.* **2006**, *32*, 1193–1196.
7. Han, X.; Hamaker, B.R. Amylopectin Fine Structure and Rice Starch Paste Breakdown. *J. Cereal Sci.* **2001**, *34*, 279–284. [CrossRef]
8. Sun, M.M.; Abdula, S.E.; Lee, H.J.; Cho, Y.C.; Han, L.Z.; Koh, H.J.; Cho, Y.G. Molecular Aspect of Good Eating Quality Formation in Japonica Rice. *PLoS ONE* **2011**, *6*, e18385. [CrossRef] [PubMed]
9. Qu, J.; Xu, S.; Zhang, Z.; Chen, G.; Zhong, Y.; Liu, L.; Zhang, R.; Xue, J.; Guo, D. Evolutionary, structural and expression analysis of core genes involved in starch synthesis. *Sci. Rep.* **2018**, *8*, 1–16. [CrossRef]
10. Regina, A.; Li, Z.; Morell, M.K.; Jobling, S.A. *Genetically Modified Starch: State of Art and Perspectives*; Elsevier: Amsterdam, The Netherlands, 2014; ISBN 9780444537300.
11. Calingacion, M.; Laborte, A.; Nelson, A.; Resurreccion, A.; Concepcion, J.C.; Daygon, V.D.; Mumm, R.; Reinke, R.; Dipti, S.; Bassinello, P.Z.; et al. Diversity of global rice markets and the science required for consumer-targeted rice breeding. *PLoS ONE* **2014**, *9*, e85106. [CrossRef]
12. Li, H.; Fitzgerald, M.A.; Prakash, S.; Nicholson, T.M.; Gilbert, R.G. The molecular structural features controlling stickiness in cooked rice, a major palatability determinant. *Sci. Rep.* **2017**, *7*, 1–12. [CrossRef] [PubMed]
13. Patindol, J.; Gu, X.; Wang, Y.J. Chemometric analysis of cooked rice texture in relation to starch fine structure and leaching characteristics. *Starch/Staerke* **2010**, *62*, 188–197. [CrossRef]
14. Chen, Y.; Sun, X.; Zhou, X.; Hebelstrup, K.H.; Blennow, A.; Bao, J. Highly phosphorylated functionalized rice starch produced by transgenic rice expressing the potato GWD1 gene. *Sci. Rep.* **2017**, *7*, 1–10. [CrossRef]
15. Puchongkavarin, H.; Varavinit, S.; Bergthaller, W. Comparative Study of Pilot Scale Rice Starch Production by an Alkaline and an Enzymatic Process. *Starch/Stärke* **2005**, *57*, 12–14. [CrossRef]
16. Zambrano, A.D.; Bhandari, B.; Ho, B.; Prakash, S. Retrogradation—Digestibility Relationship of Selected Glutinous and Non-Glutinous Fresh and Stale Cooked Rice. *Int. J. Food Prop.* **2016**, *19*, 2608–2622. [CrossRef]
17. Juliano, B.O. The chemical basis of rice end-use quality. In *Proceedings of the Workshop on Chemical Aspects of Rice Grin Quality*; International Rice Research Institute: Los Baños, Laguna, Philippine, 1979; pp. 69–90.
18. Juliano, B.O. *Rice in Human Nutrition*; Internatio; AO Food and Nutrition Series: Rome, Italy, 1993; ISBN 9251031495.
19. Cheng, A.; Ismail, I.; Osman, M.; Hashim, H. Simple and rapid molecular techniques for identification of amylose levels in rice varieties. *Int. J. Mol. Sci.* **2012**, *13*, 6156–6166. [CrossRef] [PubMed]
20. Noda, T.; Nishiba, Y.; Sato, T.; Suda, I. Properties of starches from several low-amylose rice cultivars. *Cereal Chem.* **2003**, *80*, 193–197. [CrossRef]
21. Kuno, M.; Kainuma, K.; Takahashi, S. Characteristics, Physicochemical Starches, Low-amylose Rice. *J. Appl. Glycosci.* **2000**, *47*, 319–326. [CrossRef]
22. Genbiologische, I.; Berlin, F. Approaches to influence starch quantity and starch quality in transgenic plants. *Plant Cell Environ.* **1994**, *17*, 601–613.
23. Zhang, M.Z.; Fang, J.H.; Yan, X.; Liu, J.; Bao, J.S.; Fransson, G.; Andersson, R.; Jansson, C.; Åman, P.; Sun, C. Molecular insights into how a deficiency of amylose affects carbon allocation—Carbohydrate and oil analyses and gene expression profiling in the seeds of a rice waxy mutant. *BMC Plant Biol.* **2012**, *12*, 1–18. [CrossRef]
24. Zhang, C.; Zhu, J.; Chen, S.; Fan, X.; Li, Q.; Lu, Y.; Wang, M.; Yu, H.; Yi, C.; Tang, S.; et al. Wxlv, the Ancestral Allele of Rice Waxy Gene. *Mol. Plant* **2019**, *12*, 1157–1166. [CrossRef] [PubMed]
25. Perin, D.; Murano, E. Starch polysaccharides in the human diet: Effect of the different source and processing on its absorption. *Nat. Prod. Commun.* **2017**, *12*, 837–853. [CrossRef]
26. Vrinten, P.L.; Nakamura, T. Wheat granule-bound starch synthase I and II are encoded by separate genes that are expressed in different tissues. *Plant Physiol.* **2000**, *122*, 255–263. [CrossRef] [PubMed]
27. Li, H.; Liu, Y. Effects of variety and growth location on the chain-length distribution of rice starches. *J. Cereal Sci.* **2019**, *85*, 77–83. [CrossRef]

28. Bao, J. Genes and QTLs for Rice Grain Quality Improvement. *Rice Germplasm Genet. Improv.* **2014**. [CrossRef]
29. Li, Q.F.; Huang, L.C.; Chu, R.; Li, J.; Jiang, M.Y.; Zhang, C.Q.; Fan, X.L.; Yu, H.X.; Gu, M.H.; Liu, Q.Q. Down-Regulation of SSSII-2 Gene Expression Results in Novel Low-Amylose Rice with Soft, Transparent Grains. *J. Agric. Food Chem.* **2018**, *66*, 9750–9760. [CrossRef] [PubMed]
30. Spoel, S.H. Orchestrating the proteome with post-translational modifications. *J. Exp. Bot.* **2018**, *69*, 4499–4503. [CrossRef]
31. Graves, P.R.; Haystead, T.A.J. Molecular Biologist's Guide to Proteomics. *Microbiol. Mol. Biol. Rev.* **2002**, *66*, 39–63. [CrossRef]
32. Komatsu, S.; Mock, H.-P.; Yang, P. and Svensson, B. Application of proteomics for improving crop protection/artificial regulation. *Front. Plant Sci.* **2013**, *4*, 1–3. [CrossRef]
33. Trinidad, J.L.; Pabuayon, I.C.M.; Kohli, A. *Harnessing Protein Posttranslational Modifications for Plant Improvement*; Elsevier Inc.: Amsterdam, The Netherlands, 2020; ISBN 9780128185810.
34. Eldakak, M.; Milad, S.I.M.; Nawar, A.I.; Rohila, J.S. Proteomics: A biotechnology tool for crop improvement. *Front. Plant Sci.* **2013**, *4*, 1–12. [CrossRef]
35. Lin, S.K.; Chang, M.C.; Tsai, Y.G.; Lur, H.S. Proteomic analysis of the expression of proteins related to rice quality during caryopsis development and the effect of high temperature on expression. *Proteomics* **2005**, *5*, 2140–2156. [CrossRef]
36. Lee, J.; Koh, H.J. A label-free quantitative shotgun proteomics analysis of rice grain development. *Proteome Sci.* **2011**, *9*, 1–10. [CrossRef]
37. Zhang, H.; Chen, J.; Shan, S.; Cao, F.; Chen, G.; Zou, Y.; Huang, M.; Abou-Elwafa, S.F. Proteomic profiling reveals differentially expressed proteins associated with amylose accumulation during rice grain filling. *BMC Genomics* **2020**, *21*, 1–11. [CrossRef] [PubMed]
38. Xing, S.; Meng, X.; Zhou, L.; Mujahid, H.; Zhao, C.; Zhang, Y.; Wang, C.; Peng, Z. Proteome Profile of Starch Granules Purified from Rice (*Oryza sativa*) Endosperm. *PLoS ONE* **2016**, *11*, e0168467. [CrossRef] [PubMed]
39. Helle, S.; Bray, F.; Putaux, J.L.; Verbeke, J.; Flament, S.; Rolando, C.; D'Hulst, C.; Szydlowski, N. Intra-sample heterogeneity of potato starch reveals fluctuation of starch-binding proteins according to granule morphology. *Plants* **2019**, *8*, 324. [CrossRef]
40. San-eufrasio, B.; Dar, E.; Guerrero-s, V.M.; Chaturvedi, P.; Jorr, V.; Rey, M. Proteomics Data Analysis for the Identification of Proteins and Derived Proteotypic Peptides of Potential Use as Putative Drought Tolerance Markers for Quercus ilex. *Int. J. Mol. Sci.* **2021**, *22*, 3191. [CrossRef]
41. Smith, L.M.; Kelleher, N.L. Proteoform: A single term describing protein complexity. *Nat. Methods* **2013**, *10*, 186–187. [CrossRef] [PubMed]
42. Millar, A.H.; Heazlewood, J.L.; Giglione, C.; Holdsworth, M.J.; Bachmair, A.; Schulze, W.X. The Scope, Functions, and Dynamics of Posttranslational Protein Modifications. *Annu. Rev. Plant Biol.* **2019**, *70*, 119–151. [CrossRef]
43. Friso, G.; Van Wijk, K.J. Posttranslational protein modifications in plant metabolism. *Plant Physiol.* **2015**, *169*, 1469–1487. [CrossRef] [PubMed]
44. Cruz, E.R.; Nguyen, H.; Nguyen, T.; Wallace, I.S. Functional analysis tools for post-translational modification: A post-translational modification database for analysis of proteins and metabolic pathways. *Plant J.* **2019**, 1–11. [CrossRef] [PubMed]
45. Crofts, N.; Itoh, A.; Abe, M.; Miura, S.; Oitome, N.F.; Bao, J.; Fujita, N. Three Major Nucleotide Polymorphisms in the Waxy Gene Correlated with the Amounts of Extra-long Chains of Amylopectin in Rice Cultivars with S or L-type Amylopectin Nucleotide. *J. Appl. Glycosci.* **2018**, *66*, 37–46. [CrossRef]
46. Chen, G.X.; Zhou, J.W.; Liu, Y.L.; Lu, X.B.; Han, C.X.; Zhang, W.Y.; Xu, Y.H.; Yan, Y.M. Biosynthesis and Regulation of Wheat Amylose and Amylopectin from Proteomic and Phosphoproteomic Characterization of Granule-binding Proteins. *Sci. Rep.* **2016**, *6*, 1–15. [CrossRef] [PubMed]
47. Zhang, X.; Myers, A.M.; James, M.G. Mutations affecting starch synthase III in Arabidopsis alter leaf starch structure and increase the rate of starch synthesis. *Plant Physiol.* **2005**, *138*, 663–674. [CrossRef] [PubMed]
48. Li, N.; Zhang, S.; Zhao, Y. Over-expression of AGPase genes enhances seed weight and starch content in transgenic maize. *Planta* **2011**, *233*, 241–250. [CrossRef] [PubMed]
49. Bahaji, A.; Sánchez-lópez, Á.M.; De Diego, N.; Mu, F.J. Plastidic Phosphoglucose Isomerase Is an Important Determinant of Starch Accumulation in Mesophyll Cells, Growth, Photosynthetic Capacity, and Biosynthesis of Plastidic Cytokinins in Arabidopsis. *PLoS ONE* **2015**, *10*, e0119641. [CrossRef] [PubMed]
50. Liu, L.; Ma, X.; Liu, S.; Zhu, C.; Jiang, L.; Wang, Y.; Shen, Y.; Ren, Y.; Dong, H.; Chen, L.; et al. Identification and characterization of a novel Waxy allele from a Yunnan rice landrace. *Plant Mol. Biol.* **2009**, *71*, 609–626. [CrossRef]
51. Ohdan, T.; Francisco, P.B.; Sawada, T.; Hirose, T.; Terao, T.; Satoh, H.; Nakamura, Y. Expression profiling of genes involved in starch synthesis in sink and source organs of rice. *J. Exp. Bot.* **2005**, *56*, 3229–3244. [CrossRef]
52. Thompson, D.B. On the non-random nature of amylopectin branching. *Carbohydr. Polym.* **2000**, *43*, 223–239. [CrossRef]
53. Ibáñez, A.M.; Wood, D.F.; Yokoyama, W.H.; Park, I.M.; Tinoco, M.A.; Hudson, C.A.; McKenzie, K.S.; Shoemaker, C.F. Viscoelastic properties of waxy and nonwaxy rice flours, their fat and protein-free starch, and the microstructure of their cooked kernels. *J. Agric. Food Chem.* **2007**, *55*, 6761–6771. [CrossRef] [PubMed]
54. Martin, M.; Fitzgerald, M.A.Ã. Proteins in Rice Grains Infuence Cooking Properties! *J. Cereal Sci.* **2002**, *36*, 285–294. [CrossRef]
55. Upadhyaya, C.P.; Bagri, D.S.; Upadhyaya, D.C. Genetically Modified Potato as a Source of Novel Carbohydrates. *Potato Incas World* **2018**, 19–35. [CrossRef]
56. Gupta, P.K.; Varshney, R. *Cereal Genomics II*, 2nd ed. Springer: Heidelberg, Germany, 2013; ISBN 9789400764002.

57. Tetlow, I.J.; Emes, M.J. Starch biosynthesis in the developing endosperms of grasses and cereals. *Agronomy* **2017**, *7*, 81. [CrossRef]
58. Zhu, J.; Yu, W.; Zhang, C.; Zhu, Y.; Xu, J. New insights into amylose and amylopectin biosynthesis in rice endosperm. *Carbohydr. Polym.* **2020**, *230*, 115656. [CrossRef] [PubMed]
59. Hanashiro, I.; Itoh, K.; Kuratomi, Y.; Yamazaki, M.; Igarashi, T.; Matsugasako, J.; Takeda, Y. Granule-Bound Starch Synthase I is Responsible for Biosynthesis of Extra-Long Unit Chains of Amylopectin in Rice. *Plant Cell Physiol.* **2008**, *49*, 925–933. [CrossRef]
60. Ayabe, S.; Kasai, M.; Ohishi, K.; Hatae, K. Textural Properties and Structures of Starches from Indica and Japonica Rice with Similar Amylose Content. *Food Sci. Technol. Res.* **2009**, *15*, 299–306. [CrossRef]
61. Li, H.; Prakash, S.; Nicholson, T.M.; Fitzgerald, M.A.; Gilbert, R.G. The importance of amylose and amylopectin fine structure for textural properties of cooked rice grains. *FOOD Chem.* **2016**, *196*, 702–711. [CrossRef] [PubMed]
62. Huang, L.; Sreenivasulu, N.; Liu, Q. Trends in Plant Science Waxy Editing: Old Meets New Trends in Plant Science. *Trends Plant Sci.* **2020**, *25*, 963–966. [CrossRef] [PubMed]
63. Mikami, I.; Uwatoko, N.; Ikeda, Y.; Yamaguchi, J.; Hirano, H.Y.; Suzuki, Y.; Sano, Y. Allelic diversi W cation at the wx locus in landraces of Asian rice. *Theor. Appl. Genet.* **2008**, *116*, 979–989. [CrossRef]
64. Biselli, C.; Cavalluzzo, D.; Perrini, R.; Gianinetti, A.; Bagnaresi, P.; Urso, S.; Orasen, G.; Desiderio, F.; Lupotto, E.; Cattivelli, L.; et al. Improvement of marker-based predictability of apparent amylose content in japonica rice through GBSSI allele mining. *Rice* **2014**, *7*, 1–18. [CrossRef]
65. Hirano, H.Y.; Eiguchi, M.; Sano, Y. A single base change altered the regulation of the waxy gene at the posttranscriptional level during the domestication of rice. *Mol. Biol. Evol.* **1998**, *15*, 978–987. [CrossRef] [PubMed]
66. Sano, Y. Differential regulation of waxy gene expression in rice endosperm. *Theor. Appl. Genet.* **1984**, *68*, 467–473. [CrossRef]
67. Dobo, M.; Ayres, N.; Walker, G.; Park, W.D. Polymorphism in the GBSS gene affects amylose content in US and European rice germplasm. *J. Cereal Sci.* **2010**, *52*, 450–456. [CrossRef]
68. Wang, Z.; Zheng, F.; Shen, G.; Gao, J.; Snusted, D.P.; Li, M.G.; Zhang, J.L.; Hong, M.M. Post-Transcriptional Regulation of the Waxy Gene. *Plant J.* **1995**, *7*, 613–622. [CrossRef]
69. Tian, Z.; Qian, Q.; Liu, Q.; Yan, M.; Liu, X.; Yan, C.; Liu, G.; Gao, Z.; Tang, S.; Zeng, D.; et al. Allelic diversities in rice starch biosynthesis lead to a diverse array of rice eating and cooking qualities. *Proc. Natl. Acad. Sci. USA* **2009**, *106*, 21760–21765. [CrossRef]
70. Itoh, K.; Ozaki, H.; Okada, K.; Hori, H.; Takeda, Y.; Mitsui, T. Introduction of Wx transgene into rice wx mutants leads to both high- and low-amylose rice. *Plant Cell Physiol.* **2003**, *44*, 473–480. [CrossRef]
71. Sun, Y.; Jiao, G.; Liu, Z.; Zhang, X.; Li, J.; Guo, X.; Du, W.; Du, J.; Francis, F.; Zhao, Y.; et al. Generation of high-amylose rice through CRISPR/Cas9-mediated targeted mutagenesis of starch branching enzymes. *Front. Plant Sci.* **2017**, *8*, 1–15. [CrossRef]
72. Ball, S.G.; Morell, M.K. From Bacterial Glycogen to Starch: Understanding the Biogenesis of the Plant Starch Granule. *Annu. Rev. Plant Biol.* **2003**, *54*, 207–233. [CrossRef]
73. Park, H.-M.; Choi, M.-S.; Chun, A.; Lee, J.-H.; Kim, M.-K.; Kim, Y.-G.; Shin, D.-B.; Lee, J.-Y.; Kim, Y.-H. Variation of Amylose Content Using dsRNAi Vector by Targeting 3′-UTR Region of GBSSI Gene in Rice. *Korean J. Breed. Sci.* **2010**, *42*, 515–524.
74. Heilersig, H.J.B.; Loonen, A.; Bergervoet, M.; Wolters, A.M.A.; Visser, R.G.F. Post-transcriptional gene silencing of GBSSI in potato: Effects of size and sequence of the inverted repeats. *Plant Mol. Biol.* **2006**, *60*, 647–662. [CrossRef] [PubMed]
75. Cai, X.; Wang, Z.; Xing, Y.; Zhang, J.; Hong, M. Aberrant splicing of intron 1 leads to the heterogeneous 5 J UTR and decreased expression of waxy gene in rice cultivars of intermediate amylose content. *Plant J.* **1998**, *14*, 459–465. [CrossRef] [PubMed]
76. Morell, M.K.; Myers, A.M. Towards the rational design of cereal starches. *Curr. Opin. Plant Biol.* **2005**, 204–210. [CrossRef]
77. Regina, A.; Bird, A.; Topping, D.; Bowden, S.; Freeman, J.; Barsby, T.; Kosar-hashemi, B.; Li, Z.; Rahman, S.; Morell, M. High-amylose wheat generated by RNA interference improves indices of large-bowel health in rats. *Proc. Natl. Acad. Sci. USA* **2006**, *103*, 3546–3551. [CrossRef] [PubMed]
78. Zhou, H.; Wang, L.; Liu, G.; Meng, X.; Jing, Y.; Shu, X.; Kong, X.; Sun, J.; Yu, H.; Smith, S.M.; et al. Critical roles of soluble starch synthase SSIIIa and granule-bound starch synthase Waxy in synthesizing resistant starch in rice. *Proc. Natl. Acad. Sci. USA* **2016**, *113*, 12844–12849. [CrossRef]
79. Smith, A.M.; Denyer, K.; Martin, C. The Synthesis of the Starch Granule. *Annu. Rev. Plant Physiol. Plant Mol. Biol.* **1997**, *48*, 67–87. [CrossRef]
80. Kharabian-masouleh, A.; Waters, D.L.E.; Reinke, R.F.; Ward, R.; Henry, R.J. SNP in starch biosynthesis genes associated with nutritional and functional properties of rice. *Sci. Rep.* **2012**, *2*, 1–9. [CrossRef] [PubMed]
81. You, C.; Zhu, H.; Xu, B.; Huang, W.; Wang, S.; Ding, Y.; Liu, Z.; Li, G.; Chen, L.; Ding, C.; et al. Effect of removing superior spikelets on grain filling of inferior spikelets in rice. *Front. Plant Sci.* **2016**, *7*, 1–16. [CrossRef] [PubMed]
82. Nakamura, Y.; Francisco, P.B., Jr.; Hosaka, Y.; Sato, A.; Sawada, T.; Kubo, K.; Fujita, N. Essential amino acids of starch synthase IIa differentiate amylopectin structure and starch quality between japonica and indica rice varieties. *Plant Mol. Biol.* **2005**, *58*, 213–227. [CrossRef] [PubMed]
83. Guo, H.; Liu, Y.; Li, X.; Yan, Z.; Xie, Y.; Xiong, H.; Zhao, L.; Gu, J.; Zhao, S.; Liu, L. Novel mutant alleles of the starch synthesis gene TaSSIVb-D result in the reduction of starch granule number per chloroplast in wheat. *BMC Genomics* **2017**, *18*, 1–10. [CrossRef] [PubMed]
84. Smidansky, E.D.; Clancy, M.; Meyer, F.D.; Lanning, S.P.; Blake, N.K.; Talbert, L.E.; Giroux, M.J. Enhanced ADP-glucose pyrophosphorylase activity in wheat endosperm increases seed yield. *Proc. Natl. Acad. Sci. USA* **2002**, *99*, 1724–1729. [CrossRef]

85. Zhu, G.; Ye, N.; Yang, J.; Peng, X.; Zhang, J. Regulation of expression of starch synthesis genes by ethylene and ABA in relation to the development of rice inferior and superior spikelets. *J. Exp. Bot.* **2011**, *62*, 3907–3916. [CrossRef]
86. Duan, M.; Sun, S.S.M. Profiling the expression of genes controlling rice grain quality. *Plant Mol. Biol.* **2005**, *59*, 165–178. [CrossRef]
87. Yang, J.; Zhang, J.; Wang, Z.; Liu, K.; Wang, P. Post-anthesis development of inferior and superior spikelets in rice in relation to abscisic acid and ethylene. *J. Exp. Bot.* **2006**, *57*, 149–160. [CrossRef]
88. Bhatia, S.; Singh, R. Phytohormone-mediated transformation of sugars to starch in relation to the activities of amylases, sucrose-metabolising enzymes in sorghum grain. *Plant Growth Regul.* **2002**, *55072*, 1–8.
89. Tang, T.; Xie, H.; Wang, Y.; Lu, B.; Liang, J. The effect of sucrose and abscisic acid interaction on sucrose synthase and its relationship to grain filling of rice (*Oryza sativa* L.). *J. Exp. Bot.* **2009**, *60*, 2641–2652. [CrossRef] [PubMed]
90. Akihiro, T.; Umezawa, T.; Ueki, C.; Mohamed, B.; Mizuno, K. Genome wide cDNA-AFLP analysis of genes rapidly induced by combined sucrose and ABA treatment in rice cultured cells. *FEBS Lett.* **2006**, *580*, 5947–5952. [CrossRef]
91. Grimaud, F.; Rogniaux, H.; James, M.G.; Myers, A.M.; Planchot, V. Proteome and phosphoproteome analysis of starch granule-associated proteins from normal maize and mutants affected in starch biosynthesis. *J. Exp. Bot.* **2008**, *59*, 3395–3406. [CrossRef] [PubMed]
92. Fiaz, S.; Ahmad, S.; Ali Noor, M.; Wang, X.; Younas, A.; Riaz, A.; Riaz, A.; Ali, F. Applications of the CRISPR/Cas9 system for rice grain quality improvement: Perspectives and opportunities. *Int. J. Mol. Sci.* **2019**, *20*, 888. [CrossRef] [PubMed]
93. Jeon, J.S.; Ryoo, N.; Hahn, T.R.; Walia, H.; Nakamura, Y. Starch biosynthesis in cereal endosperm. *Plant Physiol. Biochem.* **2010**, *48*, 383–392. [CrossRef]
94. Inukai, T. Differential Regulation of Starch-synthetic Gene Expression in Endosperm Between Indica and Japonica Rice Cultivars. *Rice* **2017**, *10*, 1–10. [CrossRef]
95. Umemoto, T.; Aoki, N. Single-nucleotide polymorphisms in rice starch synthase IIa that alter starch gelatinisation and starch association of the enzyme. *Funct. Plant Biol.* **2005**, *32*, 763–768. [CrossRef]
96. Rahman, S.; Kosar-Hashe, B.; Samuel, M.S.; Hill, A.; Abbott, D.C.; Skerritt, J.H.; Preiss, J.; Appels, R.; Morell, M.K. The Major Proteins of Wheat Endosperm Starch Granules. *Aust. J. Plant Physiol.* **1995**, *22*, 793–803. [CrossRef]
97. Regina, A.; Kosar-Hashemi, B.; Li, Z.; Pedler, A.; Mukai, Y.; Yamamoto, M.; Gale, K.; Sharp, P.J.; Morell, M.K.; Rahman, S. Starch branching enzyme IIb in wheat is expressed at low levels in the endosperm compared to other cereals and encoded at a non-syntenic locus. *Planta* **2005**, *222*, 899–909. [CrossRef]
98. Borén, M.; Larsson, H.; Falk, A.; Jansson, C. The barley starch granule proteome—Internalized granule polypeptides of the mature endosperm. *Plant Sci.* **2004**, *166*, 617–626. [CrossRef]
99. Peng, M.; Gao, M.; Båga, M.; Hucl, P.; Chibbar, R.N. Starch-Branching Enzymes Preferentially Associated with A-Type Starch Granules in Wheat Endosperm 1. *Plant Physiol.* **2020**, *124*, 265–272. [CrossRef]
100. Mu-forster, C.; Huang, R.; Powers, J.R.; Harriman, R.W.; Knight, M.; Singletary, C.W.; Keeling, P.; Wasserman, B.P. Physical Association of Starch Biosynthetic Enz mes with Starch Granules of Maize Endosperm. *Plant Physiol.* **1996**, *111*, 821–829. [CrossRef] [PubMed]
101. Zhang, X.; Colleoni, C.; Ratushna, V.; Sirghie-colleoni, M.; James, M.G.; Myers, A.M. Molecular characterization demonstrates that the Zea mays gene sugary2 codes for the starch synthase isoform SSIIa. *Plant Mol. Biol.* **2004**, *54*, 865–879. [CrossRef]
102. Edwards, A.; Marshall, J.; Sidebottom, C.; Visser, R.G.F.; Smith, A.M.; Martin, C.; Centre, I.; Lane, C.; Nr, N. Biochemical and molecular characterization of a novel starch synthase from potato tubers. *Plant J.* **1995**, *8*, 283–294. [CrossRef]
103. Ritte, G.; Eckermann, N.; Haebel, S.; Lorberth, R.; Steup, M. Compartmentation of the Starch-Related R1 Protein in Higher Plants. *Starch* **2000**, *52*, 145–149. [CrossRef]
104. Denyer, K.; Sidebottom, C.; Hylton, C.M.; Smith, A.M. Soluble isoforms of starch synthase and starch-branching enzyme also occur within starch granules in developing pea embryos. *Plant J.* **1993**, *4*, 191–198. [CrossRef] [PubMed]
105. Comparot-Moss, S.; Denyer, K. The evolution of the starch biosynthetic pathway in cereals and other grasses. *J. Exp. Bot.* **2009**, *60*, 2481–2492. [CrossRef]
106. Nakamura, Y. Towards a Better Understanding of the Metabolic System for Amylopectin Biosynthesis in Plants. *Plant Cell Physiol.* **2002**, *43*, 718–725. [CrossRef] [PubMed]
107. Martin, C.; Smith, A.M. Starch Biosynthesis. *Plant Cell* **1995**, *7*, 971.
108. Dry, I.; Smith, A.; Edwards, A.; Dunn, P.; Martin, C.; Science, P.; Centre, J. Characterization of cDNAs encoding two isoforms of granule-bound starch synthase which show differential expression in developing storage organs of pea and potato. *Plant J.* **1992**, *2*, 193–202. [PubMed]
109. Geigenberger, P. Regulation of starch biosynthesis in response to a fluctuating environment. *Plant Physiol.* **2011**, *155*, 1566–1577. [CrossRef]
110. Cross, J.M.; Clancy, M.; Shaw, J.R.; Greene, T.W.; Schmidt, R.R.; Okita, T.W.; Hannah, L.C. Both subunits of ADP-glucose pyrophosphorylase are regulatory. *Plant Physiol.* **2004**, *135*, 137–144. [CrossRef]
111. Hannaha, L.C.; Girouxb, M.; Boyer, C. Biotechnological modification of carbohydrates for sweet corn and maize improvement. *Sci. Hortic.* **1993**, *55*, 177–197. [CrossRef]
112. Lin, Q.; Huang, B.; Zhang, M.; Zhang, X.; Rivenbark, J.; Lappe, R.L.; James, M.G.; Myers, A.M.; Hennen-bierwagen, T.A. Functional Interactions between Starch Synthase III and Isoamylase-Type Starch-Debranching Enzyme in Maize Endosperm. *Plant Physiol.* **2012**, *158*, 679–692. [CrossRef] [PubMed]

113. Yamanouchi, H.; Agriculture, N.; Nakamura, Y. Organ Specificity of Isoforms of Starch Branching Enzyme (Q-Enzyme) in Rice. *Plant CellPhysiol.* **2015**, *33*, 985–991. [CrossRef]
114. Nakamura, Y.; Umemoto, T.; Ogata, N.; Kuboki, Y.; Yano, M.; Sasaki, T. Starch debranching enzyme (R-enzyme or pullulanase) from developing rice endosperm: Purification, cDNA and chromosomal localization of the gene. *Planta* **1996**, *81*, 209–218. [CrossRef]
115. Szydlowski, N.; Ragel, P.; Raynaud, S.; Lucas, M.M.; Rolda, I.; Ovecka, M.; Bahaji, A.; Jose, F. Starch Granule Initiation in Arabidopsis Requires the Presence of Either Class IV or Class III Starch Synthases. *Plant Cell* **2009**, 1–16. [CrossRef]
116. Deschamps, P.; Colleoni, C.; Nakamura, Y.; Suzuki, E.; Putaux, J.; Bule, A.; Haebel, S.; Ritte, G.; Steup, M.; Falco, L.I.; et al. Metabolic Symbiosis and the Birth of the Plant Kingdom. *Mol. Biol. Evol.* **2008**, *25*, 536–548. [CrossRef]
117. Kwon, S.J.; Choi, E.Y.; Choi, Y.J.; Ahn, J.H.; Park, O.K. Proteomics studies of post-translational modifications in plants. *J. Exp. Bot.* **2006**, *57*, 1547–1551. [CrossRef]
118. Xu, Y.; Wu, W.; Han, Q.; Wang, Y.; Li, C.; Zhang, P.; Xu, H. Post-translational modification control of RNA-binding protein hnRNPK function. *Open Biol.* **2019**, *9*, 1–10. [CrossRef] [PubMed]
119. Vu, L.D.; Gevaert, K.; Smet, I. De Protein Language: Post-Translational Modi fi cations Talking to Each Other. *Trends Plant Sci.* **2018**, *23*, 1068–1080. [CrossRef]
120. Ryšlavá, H.; Doubnerová, V.; Kavan, D.; Vaněk, O. Effect of posttranslational modifications on enzyme function and assembly. *J. Proteomics* **2013**, *92*, 80–109. [CrossRef] [PubMed]
121. Jensen, O.N. Modification-specific proteomics: Characterization of post-translational modifications by mass spectrometry. *Curr. Opin. Chem. Biol.* **2004**, *8*, 33–41. [CrossRef]
122. Hou, Y.; Qiu, J.; Tong, X.; Wei, X.; Nallamilli, B.R.; Wu, W.; Huang, S.; Zhang, J. A comprehensive quantitative phosphoproteome analysis of rice in response to bacterial blight. *BMC Plant Biol.* **2015**, *15*, 163. [CrossRef]
123. Reinders, J.; Sickmann, A. State-of-the-art in phosphoproteomics. *Proteomics* **2005**, *5*, 4052–4061. [CrossRef] [PubMed]
124. Champion, A.; Kreis, M.; Mockaitis, K.; Picaud, A.; Henry, Y. Arabidopsis kinome: After the casting. *Funct. Integr. Genomics* **2004**, *4*, 163–187. [CrossRef]
125. Gelens, L.; Saurin, A.T. Exploring the Function of Dynamic Phosphorylation-Dephosphorylation Cycles. *Dev. Cell* **2018**, *44*, 659–663. [CrossRef]
126. Meng, X.; Xu, J.; He, Y.; Yang, K.-Y.; Mordorski, B.; Liu, Y.; Zhang, S. Phosphorylation of an ERF Transcription Factor by Arabidopsis MPK3_MPK6 Regulates Plant Defense Gene Induction and Fungal Resistance—PubMed. *Plant Cell* **2013**, *25*, 1126–1142. [CrossRef]
127. Yang, W.; Zhang, W.; Wang, X. Post-translational control of ABA signalling: The roles of protein phosphorylation and ubiquitination. *Plant Biotechnol. J.* **2017**, 4–14. [CrossRef]
128. Ajadi, A.A.; Cisse, A.; Ahmad, S.; Yifeng, W.; Yazhou, S.; Shufan, L.; Xixi, L.; Bello, B.K.; Tajo, S.M.; Xiaohong, T.; et al. Protein Phosphorylation and Phosphoproteome: An Overview of Rice. *Rice Sci.* **2020**, *27*, 184–200. [CrossRef]
129. Bay-Smidt, A.M. Starch bound phosphate in potato as studied by a simple method for determination of organic phosphate and (31)P-NMR. *Starch-Stärke* **1994**, *46*, 167–172.
130. Kotting, O.; Pusch, K.; Tiessen, A.; Geigenberger, P.; Steup, M.; Ritte, G. Identification of a Novel Enzyme Required for Starch Metabolism in Arabidopsis Leaves. *Phosphoglucan Water Dikinase.* **2005**, *137*, 242–252. [CrossRef]
131. Hebelstrup, K.H.; Sagnelli, D.; Blennow, A. The future of starch bioengineering: GM microorganisms or GM plants? *Front. Plant Sci.* **2015**, *6*, 1–6. [CrossRef]
132. Nabeshima, E.H.; Bustos, F.M.; Hashimoto, J.M.; El Dash, A.A. Improving functional properties of rice flours through phosphorylation. *Int. J. Food Prop.* **2010**, *13*, 921–930. [CrossRef]
133. Edner, C.; Li, J.; Albrecht, T.; Mahlow, S.; Hejazi, M.; Hussain, H.; Kaplan, F.; Guy, C.; Smith, S.M.; Steup, M.; et al. Glucan, Water Dikinase Activity Stimulates Breakdown of Starch Granules by Plastidial b -Amylases 1 [W][OA]. *Plant Physiol.* **2007**, *145*, 17–28. [CrossRef]
134. Blennow, A.; Engelsen, S.B.; Nielsen, T.H.; Baunsgaard, L.; Mikkelsen, R.; Blennow, A.; Nielsen, T.H. Starch phosphorylation: A new front line in starch research. *TRENDS Plant Sci.* **2002**, *7*, 445–450. [CrossRef]
135. Mahlow, S.; Orzechowski, S.; Fettke, J. Starch phosphorylation: Insights and perspectives. *Cell. Mol. Life Sci.* **2016**. [CrossRef]
136. Tetlow, I.J.; Morell, M.K.; Emes, M.J. Recent developments in understanding the regulation of starch metabolism in higher plants. *J. Exp. Bot.* **2004**, *55*, 2131–2145. [CrossRef]
137. Chen, G.X.; Zhen, S.M.; Liu, Y.L.; Yan, X.; Zhang, M.; Yan, Y.M. In vivo phosphoproteome characterization reveals key starch granule-binding phosphoproteins involved in wheat water-deficit response. *BMC Plant Biol.* **2017**, *17*, 1–13. [CrossRef]
138. Seung, D. Amylose in starch: Towards an understanding of biosynthesis, structure and function. *New Phytol.* **2020**, *228*, 1490–1504. [CrossRef] [PubMed]
139. Kringel, D.H.; Filipini, S.; Salas-mellado, M.D.L.M. Influence of phosphorylated rice flour on the quality of gluten-free bread. *Int. J. Food Sci. Technol.* **2017**, 1–8. [CrossRef]
140. Espinoza-herrera, J.; Mar, L.; Serna-sald, S.O.; Chuck-hern, C. Methods for the Modification and Evaluation of Cereal Proteins for the Substitution of Wheat Gluten in Dough Systems. *Foods* **2021**, *10*, 118. [CrossRef]
141. Wang, D.; Li, F.; Cao, S.; Zhang, K. Genomic and functional genomics analyses of gluten proteins and prospect for simultaneous improvement of end—Use and health—Related traits in wheat. *Theor. Appl. Genet.* **2020**, *133*, 1521–1539. [CrossRef] [PubMed]

142. Momma, M.; Fujimoto, Z. Interdomain Disulfide Bridge in the Rice Granule Bound Starch Synthase I Catalytic Domain as Elucidated by X-ray Structure Analysis Catalytic Domain as Elucidated by X-ray Structure Analysis. *Biosci. Biotechnol. Biochem.* **2014**, *76*, 1591–1595. [CrossRef]
143. Waters, D.L.E.; Henry, R.J.; Reinke, R.F.; Fitzgerald, M.A. Gelatinization temperature of rice explained by polymorphisms in starch synthase. *Plant Biotechnol. J.* **2006**, 115–122. [CrossRef]
144. Oh, J.; Jeong, H.; Kang, Y.; Lee, G.; Kim, Y.; Kim, Y.; Kim, K. DNA Methylation Biosynthesis in Rice for Transcriptome Changes during Starch. *Int. J. Agric. Biol.* **2018**, *20*, 149–156. [CrossRef]
145. Berski, W.; Ptaszek, A.; Ptaszek, P.; Ziobro, R.; Kowalski, G.; Grzesik, M.; Achremowicz, B. Pasting and rheological properties of oat starch and its derivatives. *Carbohydr. Polym.* **2011**, *83*, 665–671. [CrossRef]
146. Mujahid, H.; Meng, X.; Xing, S.; Peng, X.; Wang, C.; Peng, Z. Malonylome analysis in developing rice (Oryza sativa) seeds suggesting that protein lysine malonylation is well-conserved and overlaps with acetylation and succinylation substantially. *J. Proteomics* **2017**, *6*, 88–98. [CrossRef]
147. Meng, X.; Mujahid, H.; Zhang, Y.; Peng, X.; Redon, E.D.; Wang, C.; Meng, X.; Mujahid, H.; Zhang, Y.; Peng, X.; et al. Comprehensive Analysis of the Lysine Succinylome and Protein Co-modifications in Developing Rice Authors Comprehensive Analysis of the Lysine Succinylome and Protein Co-modifications in Developing Rice Seeds. *Mol. Cell. Proteom.* **2019**, 2359–2372. [CrossRef] [PubMed]
148. Guo, G.; Lv, D.; Yan, X.; Subburaj, S.; Ge, P.; Li, X.; Hu, Y. Proteome characterization of developing grains in bread wheat cultivars (Triticum aestivum L.). *BMC Plant Biol.* **2012**, *12*, 1–24. [CrossRef]
149. Tetlow, I.J.; Bertoft, E. A review of starch biosynthesis in relation to the building block-backbone model. *Int. J. Mol. Sci.* **2020**, *21*, 7011. [CrossRef]
150. Grabsztunowicz, M.; Koskela, M.M.; Mulo, P. Post-translational Modifications in Regulation of Chloroplast Function: Recent Advances. *Front. Plant Sci.* **2017**, *8*, 1–12. [CrossRef]
151. Skeffington, A.W.; Graf, A.; Duxbury, Z.; Gruissem, W.; Smith, A.M. Glucan, water dikinase exerts little control over starch degradation in arabidopsis leaves at night. *Plant Physiol.* **2014**, *165*, 866–879. [CrossRef]
152. Ritte, G.; Scharf, A.; Eckermann, N.; Haebel, S.; Steup, M. Phosphorylation of Transitory Starch Is Increased during Degradation 1 [w]. *Plant Physiol.* **2004**, *135*, 2068–2077. [CrossRef] [PubMed]
153. Cao, H.; Yan, X.; Chen, G.; Zhou, J.; Li, X.; Ma, W.; Yan, Y. ScienceDirect Comparative proteome analysis of A- and B-type starch granule-associated proteins in bread wheat (*Triticum aestivum* L.) and Aegilops crassa. *J. Proteomics* **2014**, *112*. [CrossRef]
154. Fujita, N.; Yoshida, M.; Asakura, N.; Ohdan, T.; Miyao, A. Function and Characterization of Starch Synthase I Using Mutants in Rice. *Plant Physiol.* **2006**, *140*, 1070–1084. [CrossRef]
155. Dong, H.; Vasanthan, T. Effect of phosphorylation techniques on structural, thermal, and pasting properties of pulse starches in comparison with corn starch. *Food Hydrocoll.* **2020**, 106078. [CrossRef]
156. Seung, D.; Echevarría-poza, A.; Steuernagel, B.; Smith, A.M. Natural Polymorphisms in Arabidopsis Result in Wide Variation or Loss of the Amylose Component. *Plant Physiol.* **2020**, *182*, 870–881. [CrossRef] [PubMed]
157. Xu, X.; Dees, D.; Dechesne, A.; Huang, X.F.; Visser, R.G.F.; Trindade, L.M. Starch phosphorylation plays an important role in starch biosynthesis. *Carbohydr. Polym.* **2017**, *157*, 1628–1637. [CrossRef]
158. Sitohy, M.Z.; El-saadany, S.S.; Labib, S.M.; Rama-, M.F. Physicochemical Properties of Different Types of Starch Phosphate Monoesters. *Starch/Stärke* **2000**, *4*, 101–105. [CrossRef]
159. Nazarian-Firouzabadi, F.; Visser, R.G.F. Potato starch synthases: Functions and relationships. *Biochem. Biophys. Rep.* **2017**, *10*, 7–16. [CrossRef]
160. Kozlov, S.S.; Blennow, A.; Krivandin, A.V.; Yuryev, V.P. Structural and thermodynamic properties of starches extracted from GBSS and GWD suppressed potato lines. *Int. J. Biol. Macromol.* **2007**, *40*, 449–460. [CrossRef] [PubMed]
161. Carciofi, M.; Shaik, S.S.; Jensen, S.L.; Blennow, A.; Svensson, J.T.; Vincze, É.; Hebelstrup, K.H. Hyperphosphorylation of cereal starch. *J. Cereal Sci.* **2011**, *54*, 339–346. [CrossRef]
162. Tiessen, A.; Hendriks, J.H.M.; Stitt, M.; Branscheid, A.; Gibon, Y.; Farré, E.M.; Geigenberger, P. Starch synthesis in potato tubers is regulated by post-translational redox modification of ADP-glucose pyrophosphorylase: A novel regulatory mechanism linking starch synthesis to the sucrose supply. *Plant Cell* **2002**, *14*, 2191–2213. [CrossRef] [PubMed]

International Journal of
Molecular Sciences

MDPI

Article

Comparative Phosphoproteomic Analysis Reveals the Response of Starch Metabolism to High-Temperature Stress in Rice Endosperm

Yuehan Pang, Yaqi Hu and Jinsong Bao *

Institute of Nuclear Agricultural Sciences, College of Agriculture and Biotechnology, Zhejiang University, Zijingang Campus, Hangzhou 310058, China; pyh_tt@163.com (Y.P.); 11916013@zju.edu.cn (Y.H.)
* Correspondence: jsbao@zju.edu.cn

Abstract: High-temperature stress severely affects rice grain quality. While extensive research has been conducted at the physiological, transcriptional, and protein levels, it is still unknown how protein phosphorylation regulates seed development in high-temperature environments. Here, we explore the impact of high-temperature stress on the phosphoproteome of developing grains from two indica rice varieties, 9311 and Guangluai4 (GLA4), with different starch qualities. A total of 9994 phosphosites from 3216 phosphoproteins were identified in all endosperm samples. We identified several consensus phosphorylation motifs ([sP], [LxRxxs], [Rxxs], [tP]) induced by high-temperature treatment and revealed a core set of protein kinases, splicing factors, and regulatory factors in response to high-temperature stress, especially those involved in starch metabolism. A detailed phosphorylation scenario in the regulation of starch biosynthesis (AGPase, GBSSI, SSIIa, SSIIIa, BEI, BEIIb, ISA1, PUL, PHO1, PTST) in rice endosperm was proposed. Furthermore, the dynamic changes in phosphorylated enzymes related to starch synthesis (SSIIIa-Ser94, BEI-Ser562, BEI-Ser620, BEI-Ser821, BEIIb-Ser685, BEIIb-Ser715) were confirmed by Western blot analysis, which revealed that phosphorylation might play specific roles in amylopectin biosynthesis in response to high-temperature stress. The link between phosphorylation-mediated regulation and starch metabolism will provide new insights into the mechanism underlying grain quality development in response to high-temperature stress.

Keywords: rice endosperm; phosphorylation; high temperature; sucrose and starch metabolism; starch biosynthesis

Citation: Pang, Y.; Hu, Y.; Bao, J. Comparative Phosphoproteomic Analysis Reveals the Response of Starch Metabolism to High-Temperature Stress in Rice Endosperm. *Int. J. Mol. Sci.* **2021**, *22*, 10546. https://doi.org/10.3390/ijms221910546

Academic Editors: Sixue Chen and Setsuko Komatsu

Received: 25 August 2021
Accepted: 27 September 2021
Published: 29 September 2021

Publisher's Note: MDPI stays neutral with regard to jurisdictional claims in published maps and institutional affiliations.

1. Introduction

Rice is an agriculturally major cereal crop worldwide. The yield and quality of rice are often severely affected by heat stress, and this phenomenon is further aggravated with the intensified global warming [1]. During endosperm development, exposure to high-temperature environment results in an accelerated filling rate of rice grains, which eventually leads to poor grain quality and a severe reduction in yield [2,3]. The impact of high-temperature stress on physiological metabolism has been intensively investigated [2,4]. Furthermore, transcriptome and proteome profiles in rice endosperm have been used to explore differences under high-temperature stress at gene [5,6] and protein [7–9] expression levels. However, a notable paucity exists focusing specifically on phosphorylation-mediated regulation under heat stress during the grain-filling stage.

It is now well accepted that phosphorylation plays a vital role in the regulation of many intracellular processes during plant growth and development. The possible role of protein phosphorylation in the formation of a protein complex participating in starch synthesis in wheat endosperm was first proposed by Tetlow et al. [10,11]. Later studies conducted in maize have also revealed a vital regulation role of phosphorylation in forming a starch-synthesizing protein complex [12,13]. Further examinations of individual enzymes basically came from maize endosperm [13,14]. In particular, Makhmoudova et al. identified the phosphorylation status of SBEIIb at three sites (Ser 286, Ser297, and Ser649) by Ca^{2+}-dependent protein kinase [15]. With the development of mass-spectrometry-based techniques [16], several sites of phosphorylation that may regulate starch biosynthesis were discovered in cereal endosperm [17,18], despite the real functions of those phosphosites being still unclear [19]. Qiu et al. identified some phosphorylated proteins in rice pistils and seeds and focused on the differentially phosphorylated proteins at early seed development [20]. Developing rice seeds undergo active cell division and differentiation at the early stage, and 6 to 20 days after flowering (DAF) is the critical period for grain filling and starch accumulation [21]. Hence, there has been little systematic discussion at the phosphorylation level from rice endosperm as compared to the other main crops, especially at the critical periods of starch accumulation. Although several phosphosites involved in starch synthesis were identified in our previous research [22], an in-depth investigation under abiotic stress is necessary to gain a more comprehensive understanding of the regulation pathway.

Here, a label-free quantitative phosphoproteomic analysis was applied to examine the heat-induced phosphorylation change in two rice varieties with different starch qualities. As a result, a series of phosphorylation motifs, enzymes, protein kinases, splicing factors, and other potential regulators involved in seed development were revealed, particularly the phosphoproteins involved in starch metabolism. The properties of phosphorylation sites involved in starch synthesis and their change trends in response to heat stress were comprehensively explored through sequence alignment and site conservation analysis. Western blot with site-specific phosphopeptide antibodies was used to verify and explore the dynamic change in phosphorylation related to starch synthesis under high-temperature stress. Taken together, an in-depth understanding of phosphorylation-mediated regulation in rice endosperm under heat stress will shed new light on thermal signaling transduction and functional phosphosites related to starch metabolism.

2. Results

We sought to assess dynamic changes in protein phosphorylation in the developing rice seed under high-temperature stress. The control and high-temperature treatment groups were designed for two indica rice varieties, 9311 and Guangluai4 (GLA4). In the treatment group, heat stress was performed on the fifth DAF, which prevented the potential effects of high temperature on pollination and seed setting at the early developmental stage. In addition, the 6, 10, and 14 DAF were designed as sampling time points, corresponding to the 1, 5, and 9 days of high-temperature treatment, respectively (Figures 1a and S1). The average temperature of the treatment group (HT; 30–38 °C; Figure S1) was 10 °C higher than that of the control group (CT; 20–28 °C; Figure S1), while other parameters were kept the same between two artificial climatic chambers. Immature endosperm samples were analyzed by LC-MS/MS and Western blot, and the corresponding mature seeds were harvested for the determination of starch thermal properties.

2.1. *Dynamic Changes in Rice Grain Appearance and Thermal Properties under High-Temperature Treatment*

Grain chalkiness is an indicator of abnormally developed endosperm and the most sensitive trait in response to heat stress [2,3]. It is evident that the chalkiness degree gradually increased with increasing days of high-temperature treatment (Figure 1b). After 5 days of treatment, noticeable chalkiness changes were observed in both varieties, demonstrating the effectiveness of high-temperature treatment (Figure 1b). Besides, shrunken grains were found after 9 days of treatment, especially in the GLA4-H9 group (Figure 1b).

The thermal properties of the flour samples were determined by differential scanning calorimetry (DSC) (Figure 1c and Table S1). All GLA4 samples had significantly higher gelatinization temperatures than the 9311 samples (Figure 1c), indicating a dramatic difference between the two varieties in starch physicochemical properties, which is in good agreement with our previous results [22]. One-day treatment only appeared to affect the thermal properties of 9311 plants (Table S1). After 5 days of exposure to high-temperature stress, a significantly higher gelatinization temperature was observed in both varieties, and this phenomenon was more noticeable after 9 days of treatment (Table S1).

Overall, results of grain chalkiness and thermal properties indicated that high-temperature treatment for 5 days is sufficient to cause irreversible damage to rice grain quality. Accordingly, samples collected at 10 DAF with 5 days of treatment were the optimal treated group for subsequent phosphoproteomic analysis.

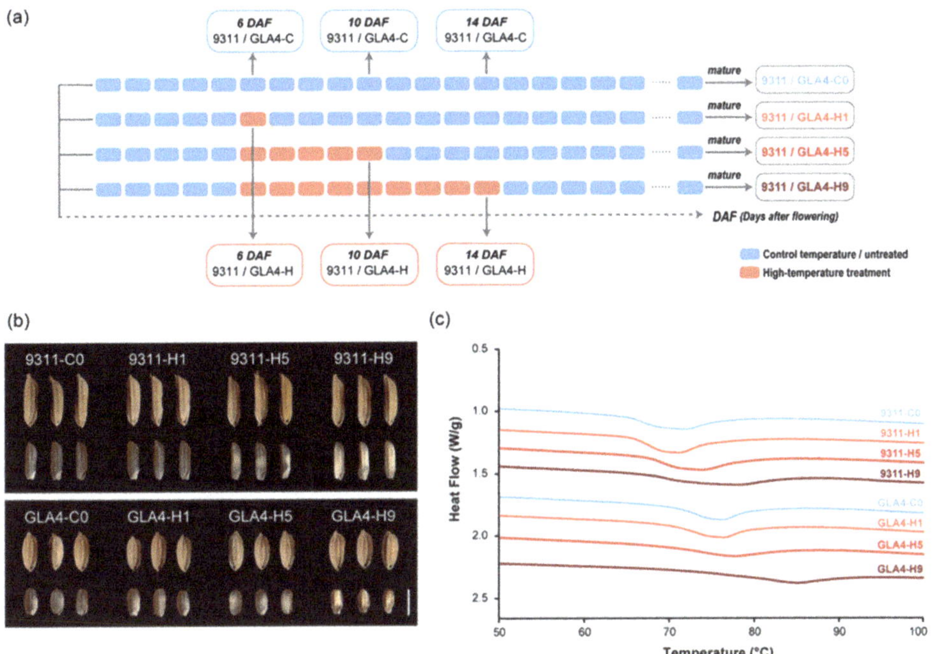

Figure 1. Effect of high-temperature stress on the rice grain. (**a**) Schematic diagram of the experimental design and sampling time points during the grain-filling period with high-temperature treatment. (**b**) Effect of three high-temperature treatments on grain morphologies of 9311 and GLA4. Scale bar = 5 mm. (**c**) Impact of three high-temperature treatments on DSC thermograms of 9311 and GLA4.

2.2. Phosphoproteins Identified in Rice Endosperm

To elucidate how heat stress influences rice endosperm at the phosphorylation level, a label-free analysis was performed to quantify phosphoproteome from two indica rice cultivars under normal and high-temperature conditions. As depicted in Figure S2a, samples of 9311-C, 9311-H, GLA4-C, and GLA4-H were collected, pretreated, lysed, digested, and enriched, and then analyzed by LC-MS/MS. To validate the accuracy of MS data, we confirmed the mass error of all identified peptides and found the distribution satisfied an expected error control (Figure S2b). Meanwhile, the distribution of peptide length was checked to ensure that sample preparation reached standard conditions (Figure S2c).

In all endosperm samples, our workflow led to the identification of 9994 phosphosites located on 3216 proteins, of which 7604 phosphosites were quantifiable (Figure 2a and Table S2). The majority of the phosphosites quantified were identified as serine (89.2%), followed by threonine (10.5%) and tyrosine (0.3%) residues (Figure 2b). Among the 5801 peptides quantified, singly phosphorylated peptides were dominant (80.2%), and around 3.7% peptides carried three or more phosphorylation modifications (Figure 2b). The number of phosphorylated sites in a single protein also varied considerably, with the range from 1 to 55 residues (Figure 2c). Over half phosphoproteins possessed two or more sites, indicating the functional importance of proteins with multiple phosphosites in regulatory networks. Distribution of the phosphorylation sites in specific protein regions suggested that N- and C-terminal regions are preferentially phosphorylated (Figure 2d), which were consistent with data obtained in Arabidopsis anthers [23]. A comparative analysis between our dataset and the japonica dataset (P3DB database) [24] and the previously published phosphoproteome of japonica rice endosperm [20] was performed (Figure 2e). Over 1000 phosphorylated proteins were common to all three datasets, and 1207 phosphoproteins were newly identified in this study. Moreover, we discovered 7365 novel phosphosites involved in 1875 phosphoproteins compared with our previous research (Figure 2f) [22].

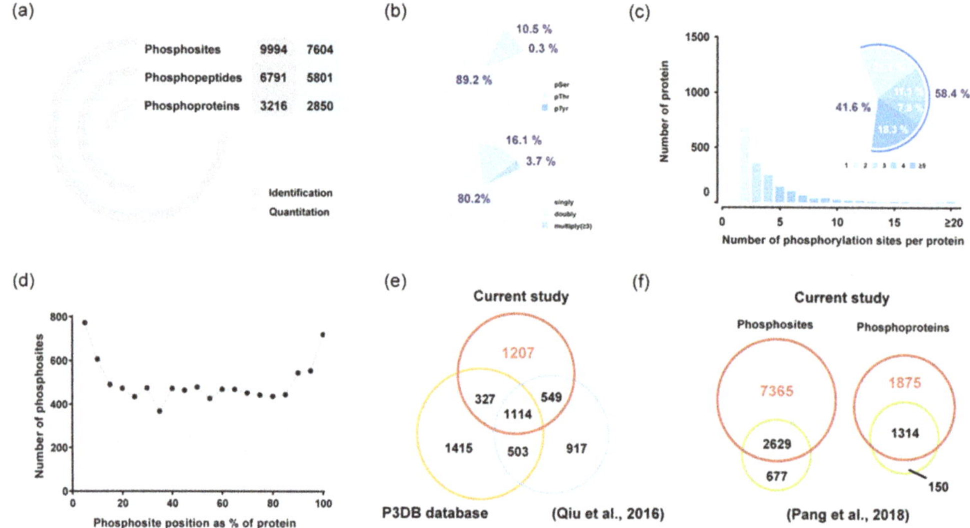

Figure 2. Characteristics of the phosphoproteome of rice endosperm at the critical grain-filling stage. (**a**) Summary of phosphoproteome analysis in rice endosperm. (**b**) Distribution of the number of phosphates and phospho-amino acid residues for all quantifiable phosphopeptides. (**c**) Frequency distribution of phosphoproteins according to the number of phosphosites identified. (**d**) Positional distribution of the identified phosphosites in protein sequences. (**e**) Overlap of the identified phosphoproteins in our study with phosphoproteins in the japonica datasets—the P3DB database [24] and the previously published phosphoproteome of japonica rice endosperm [20]. (**f**) Comparative analysis of phosphosites and phosphoproteins between the current phosphoproteome and our previous research [22].

2.3. A Temperature-Regulated Rice Endosperm Phosphoproteome

To detect possible changes in the phosphoproteome attributable to heat stress, we then performed label-free quantification analysis on all quantifiable phosphosites within our dataset. Only 2680 common phosphosites were quantifiable for all sample groups due to reversible phosphorylation induced by high temperature (Figure 3a). More phosphosites, phosphopeptides, and phosphoproteins were identified in 9311-H and GLA4-H groups (Figure 3b), suggesting that exposure to heat stress may increase the phosphorylation events in rice endosperm. Principal component analysis (PCA) showed that three repeats of each sample clustered together, and four groups were clearly separated (Figure 3c). Pearson's correlation coefficients were also generated, suggesting good reproducibility and consistency between replicates (Figure S2d).

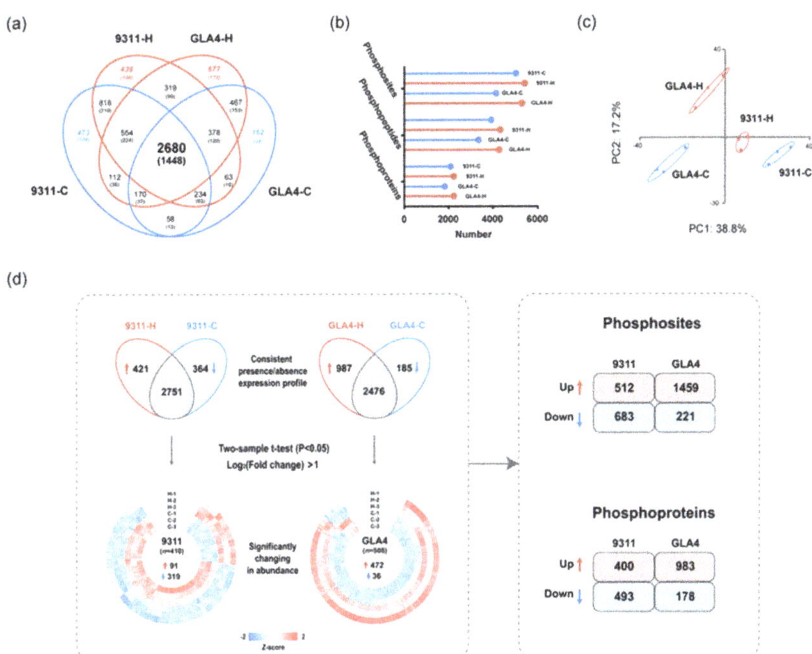

Figure 3. A temperature regulated rice endosperm phosphoproteome. (**a**) Venn diagram depicting the comparison of phosphosites (phosphoproteins) from four sample groups. (**b**) Number of phosphosites, phosphopeptides, and phosphoproteins detected in 9311-C, 9311-H, GLA4-C, and GLA4-H. (**c**) Principal component analysis (PCA) based on phosphorylation intensity across all four sample groups with three biological repetitions. (**d**) Differentially expression profiles of phosphosites (phosphoproteins) in 9311 and GLA4 under high-temperature stress. The expression profiles of selected phosphosites ($p < 0.05$, $\log_2$ (fold change) >1) were normalized using the Z-score and presented in a heatmap. In each variety, phosphosites (phosphoproteins) with a consistent presence/absence expression pattern and significantly regulated from the statistical test were combined for subsequent comparative analysis.

To detect possible changes in the phosphoproteome attributable to heat stress, we then performed label-free quantification (LFQ) analysis on all phosphosites identified within our dataset (Figure 3d). Where LFQ values were missing, the data were filtered to identify those phosphosites with a consistent presence/absence expression pattern. These analyses yielded 421 phosphosites that were only present in 9311-H and 364 that were only present in 9311-C (Figure 3d and Table S3). Similarly, 987 differentially abundant phosphosites were present in GLA4-H and 185 phosphosites that only occurred in GLA4-C (Figure 3d and Table S4). Beyond that, a total of 410 and 508 significantly changed phosphorylation sites ($p < 0.05$, $\log_2$ (fold-change) >1) were screened out in 9311 and GLA4, respectively, where LFQ data was available in both conditions (Figure 3d and Tables S5 and S6). For subsequent comparative analysis, phosphorylation sites that were uniquely identified in either condition and significantly regulated from the statistical test were combined and divided into four groups (9311-Up, 9311-Down, GLA4-Up, and GLA4-Down; Figure 3d). The number of significantly down-regulated phosphosites was far greater than up-regulated phosphosites in the 9311 variety. However, the opposite trend was observed in GLA4 plants. Comparing four sets of differential phosphorylated sites, we found only 132 phosphosites were commonly up-regulated in both rice varieties and 24 were commonly down-regulated (Figure S3a). In addition, there were 74 phosphosites showing completely opposite regulatory trends in two cultivars induced by high temperature (Figure S3a). When all significantly changed phosphosites corresponded to the specific protein, comparison results become even more complicated. There was a compelling phenomenon that 39 phosphoproteins of 9311 and 43 of GLA4 displayed a combination of up- and down-regulated phosphosites (Figure S3b). It is possible, therefore, that the status of these phosphosites was directly controlled by associated kinases and phosphatases.

2.4. Regulation of Phosphorylation Motifs and Kinases

The in vivo phosphorylation status induced by heat stress is often inseparable from protein kinase activity, which is usually regulated by upstream kinases or autophosphorylation. Up to now, few studies have examined the association between high-temperature response and protein phosphorylation involved in the signaling pathway [25,26]. It is now well established from a variety of studies that candidate substrates for the specific kinase are identified based on motif analysis [27]. A detailed investigation with a focus on potential phosphorylation motifs was, therefore, first conducted. We retrieved 8 and 15 over-represented motifs from the Ser-/Thr-containing differential phosphopeptides in 9311 and GLA4, respectively (Figure 4a).

(a)

9311

Common

GLA4

(b)

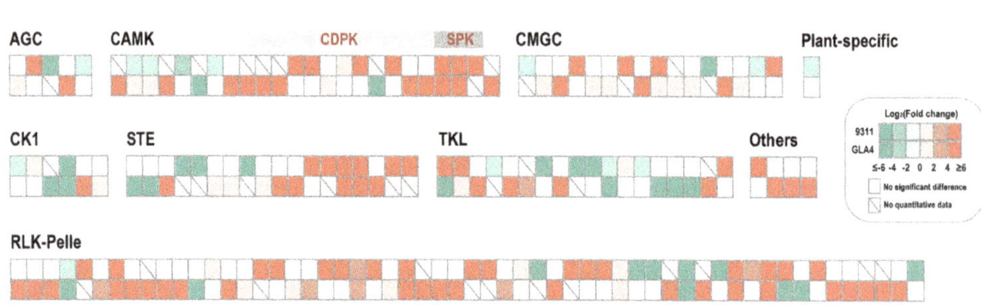

Figure 4. Motif enrichment analysis of differential phosphopeptides in 9311 and GLA4 (**a**) and phosphorylated kinases with significant differences in rice endosperm under high-temperature stress (**b**).

Both 9311 and GLA4 shared a number of consensus motifs ([sP], [LxRxxs], [Rxxs], [tP]; Figure 4a). Motifs presented are the results induced by multiple kinases, which were activated by high-temperature stress. Proline-directed motifs, such as [sP] and [tP], were recognized by kinases CDK, RLK, RLCK, MPK, SnRK2, CDPK, and SLK [27]. Note that [RxxS], which could be recognized by CDPKs and SnRKs, was also the 14-3-3 binding motif [28,29]. [LxRxxs] was known to be targeted by CDPKs [29]. The acidic motif [SDxE] was unique to 9311 and known to be targeted by CDPK, RLK, and AGC [27]. Notably, it is plausible that CDPKs were the key kinases in response to high-temperature stress. because all consensus motifs and [SDxE] were potential substrates for CDPKs. Apart from common motifs, the phosphosites from GLA4 samples yielded more enriched motifs than 9311, indicating a more complicated kinase system in response to heat stress in GLA4 plants (Figure 4a). In support of this, we found the number of up-regulated phosphosites in GLA4 kinases is considerably greater than that in 9311 kinases (Figure 3d).

In total, 192 protein kinases with 691 phosphorylated sites were identified in our phosphoproteomic dataset (Table S7), including RLKs (87, 45.3%), TKL (22, 11.5%), CMGC (21, 10.9%), and CAMK (20, 10.4%) [30]. Further enrichment analysis indicated that the family of TKL, CMGC, CAMK, STE, and CK1 was over-represented (Figure S4a). Under heat stress, only 47 phosphosites of kinases showed significant up-regulation in 9311, whereas the up-regulated sites in GLA4 were up to 77 (Table S7). Of the 148 phosphosites that were significantly regulated (Figure 4b), only 16 phosphosites showed the same regulatory trend in both varieties, while 4 phosphosites showed opposite regulatory trends (Figure S4b). In this sense, it could reasonably explain the vast difference in the phosphorylation regulation pattern between the two varieties (Figure 3d).

2.5. Functions for Differentially Phosphorylated (DP) Proteins

GO enrichment was applied to analyze the DP phosphoproteins to obtain an overview of the phosphorylation events during grain development. As expected, the up-regulated phosphoproteins in both 9311 and GLA4 were highly enriched in terms of heat response, such as heat acclimation, response to heat [26], and response to temperature stimulus. Among proteins involved in the heat response, phosphorylation levels of 12 heat shock proteins (HSPs) increased significantly under heat stresses. Of these, five phosphorylated sites were found common to both varieties (Tables S3–S7).

In the biological process, the most interesting aspect was the metabolic process in which abundant phosphoproteins in GLA4-Up and 9311-Down were enriched (Figure S5a and Table S8). Besides, the up-regulated functional phosphoproteins of 9311 were enriched in glucan polysaccharide and transduction phosphorylation, while the down-regulated ones were over-represented in negative regulation. From the molecular function perspective, ATPase activity, kinase activity, phosphotransferase activity, and binding for GTP and nucleic acid binding were mainly enriched, indicating the importance of kinases, phosphatases, and transcription factors in the phosphorylation regulatory network (Figure S5b and Table S9). Figure S6 provides the summary statistics for the phosphorylated TFs. Overall, 140 phosphorylated TFs that were divided into 35 families were identified [31]. The largest fraction was identified as the C3H family (21, 15%), followed by bZIP (18, 12.9%) and Trihelix (11, 7.9%) (Figure S6). Meanwhile, these phosphorylated TF families were highly enriched when using the total TFs identified in the indica rice database as a reference.

KEGG database annotation was then applied to predict the potential metabolic pathways. In the 9311 group, DP phosphoproteins were mainly over-represented in the pathway of spliceosome, transcription, starch and sucrose metabolism, and aminoacyl-tRNA biosynthesis (Figure S5c). In the GLA4 group, three genetic information pathways (ribosome biogenesis in eukaryotes, mismatch repair, and nucleotide excision repair) and the spliceosome pathway were enriched (Figure S5c).

2.6. Phosphoproteins Identified in Starch Metabolism

There is no doubt that sucrose and starch metabolism was the most noteworthy pathway with a large number of phosphoproteins involved (Table S10). A systematic and detailed investigation was then conducted of the specific proteins involved in sucrose and starch metabolism (Figure 5). The critical functions of phosphoproteins involved can be listed as follows: sucrose hydrolysis (SUS, INV, SPS, FK, HK, PGI, PGM, and UGPase), starch synthesis (AGPase, GBSSI, SSIIa, SSIIIa, BEI, BEIIb, ISA1, PUL, Pho1, and PTST), starch hydrolysis (BAM), and protein transport (SUT1, BT1, and GPT). To assess how high-temperature stress affects the crucial pathway, the significantly differential phosphosites in 9311 and GLA4 were displayed in the heatmap of specific proteins (Figure 5). From the perspective of sucrose hydrolysis, almost all enzymes that provide G1P for starch synthesis possessed phosphorylation sites (Figure 5). Reversible phosphorylation events were flexibly regulated by high-temperature stress in starch biosynthesis, whereas only one enzyme possessed phosphorylation modification in the starch degradation pathway (Figure 5).

In starch biosynthesis, the key rate-limiting enzyme AGPase, including AGPL1, AGPL2, AGPL3, AGPS1, and AGPS2, was phosphorylated among all sample groups (Figure 5 and Table S10). Interestingly, heat stress down-regulates the phosphosites of AGPL2 located in the N-terminal and up-regulates the sites in C-terminal, even though those sites in 9311 and GLA4 were found at different positions (Figure 5). In GLA4, most phosphosites of AGPS2 were down-regulated, probably owing to the regulation of protein abundance of AGPase under high-temperature stress [7].

Our study identified a large number of phosphorylation sites in GBSSI and found that 6 phosphosites were located at the glycosyltransferase 5 domain (Figure 6a). However, no valid phosphorylation intensity value was detected in 9311 due to a relatively lower phosphorylation level caused by the low abundance of GBSSI protein (Figure S7 and S8). In GLA4 groups, it is obvious that exposure to heat stress resulted in increasing GBSSI phosphorylation intensity at S123, S169, and S553. Phosphorylation events were prevalent in SSIIa and SSIIIa with more than 20 phosphosites but absent in SSI as well as SSIV (Figures 5 and S9). In GLA4 groups, heat stress triggered the increasing phosphorylation intensity of SSIIIa-T98. Likewise, the phosphorylation intensity of S915 and S1058 was greatly enhanced in the 9311 heat-stressed group. SSIIIa-S915 was conserved among all plants' SSIII and located at the CBM53 domain (Figure S9), which has been shown to be necessary for enzyme activity and affinities toward various glucans. Interestingly, a single peptide of SSIIa could only be detected under normal conditions in 9311, although the phosphosite could not be accurately localized (either S260 or S261; Figure S9).

Three BE isozymes possessed the same CBM48, GH13, and Aamy_C domains (Figure 6b). However, the phosphorylation events among BEs were somewhat discordant, with 10, 2, and 11 phosphosites involved in BEI, BEIIa, and BEIIb (Figure 6b and Table S10), respectively. There was an intriguing correlation among OsBEI, OsBEIIa, and OsBEIIb in that five serine residues of the three isozymes were phosphorylated at the same position (Figure 6b). More concretely, the S562, S611, and S749 of OsBEI corresponded to the S562, S685, and S808 of OsBEIIb, respectively, and the OsBEIIb-S323/324 corresponded to OsBEIIa-S466/S467 (Figure 6b). In the 9311 group, heat stress triggered phosphorylation at S11 and T580 of BEIIb and exerted suppressive effects on BEIIa-S467 (Figure 5). Only one phosphosite (BEIIb-S173) in GLA4 was up-regulated when exposed to heat stress (Figure 5).

Among 11 phosphosites identified in PUL, phosphoserine 221 in 9311, within the PULN2 domain of PUL, was 4.9-fold up-regulated at high temperature. Another phenomenon we observed is that S795 and S405 appeared only in 9311-C and GLA4-C, respectively (Figure 5). We found a widespread occurrence of Pho1 phosphorylation events in rice endosperm, and three residues were phosphorylated in the rice L80 region (Figure S10 and Table S10). Here, phosphoserine 376 in 9311 disappeared as the temperature increased (Figure 5). Curiously, this phosphosite in GLA4 could only be detected under high-temperature conditions. Besides, another three phosphosites (S741, S932, T947) were significantly up-regulated in GLA4 (Figure 5).

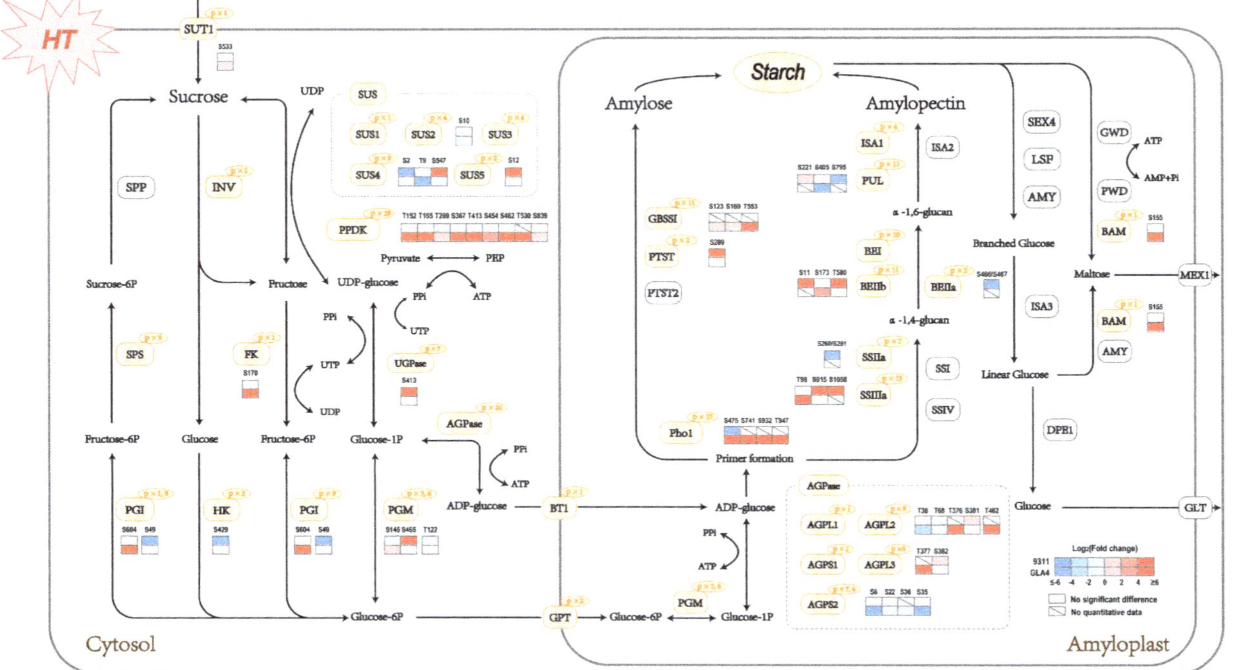

Figure 5. Sucrose and starch pathway at the phosphorylation levels in rice endosperm under high-temperature stress. Orange and gray shadings represent phosphorylated and non-phosphorylated proteins, respectively. The number of phosphosites is shown at the top right of rounded boxes. Sucrose hydrolysis: SUS, sucrose synthase; INV, invertase; SPP, sucrose-phosphate phosphatase; SPS, sucrose-phosphate synthase; FK, fructokinase; HK, hexokinase; PGI, glucose-6-phosphate isomerase; PGM, phosphoglucomutase; UGPase, UDP-glucose pyrophosphorylase. Starch synthesis: AGPase, ADP-glucose pyrophosphorylase; GBSSI, granule-bound starch synthase I; PTST, protein targeting to starch; SSI, SSIIa, SSIIIa, and SSIV, starch synthase I, IIa, IIIa, and IV, respectively; BEI, BEIIa, and BEIIb, starch branching enzyme I, IIa, and IIb, respectively; ISA1, ISA2, isoamylase isoform 1 and 2, respectively; PUL, pullulanase; Pho1, plastidial phosphorylase. Starch hydrolysis: AMY, α-amylase; BAM, β-amylase; LSF, like sex four; SEX4, starch excess 4; ISA3, isoamylase isoform 3; GWD, glucan water dikinase; PWD, phosphoglucan water dikinase; DPE1, disproportionating enzyme 1. Protein transport: SUT1, sucrose transporter; BT1, small-solute transporter; GPT, phosphometabolite transporter; MEX1, maltose transporter; GLT, glucose transporter.

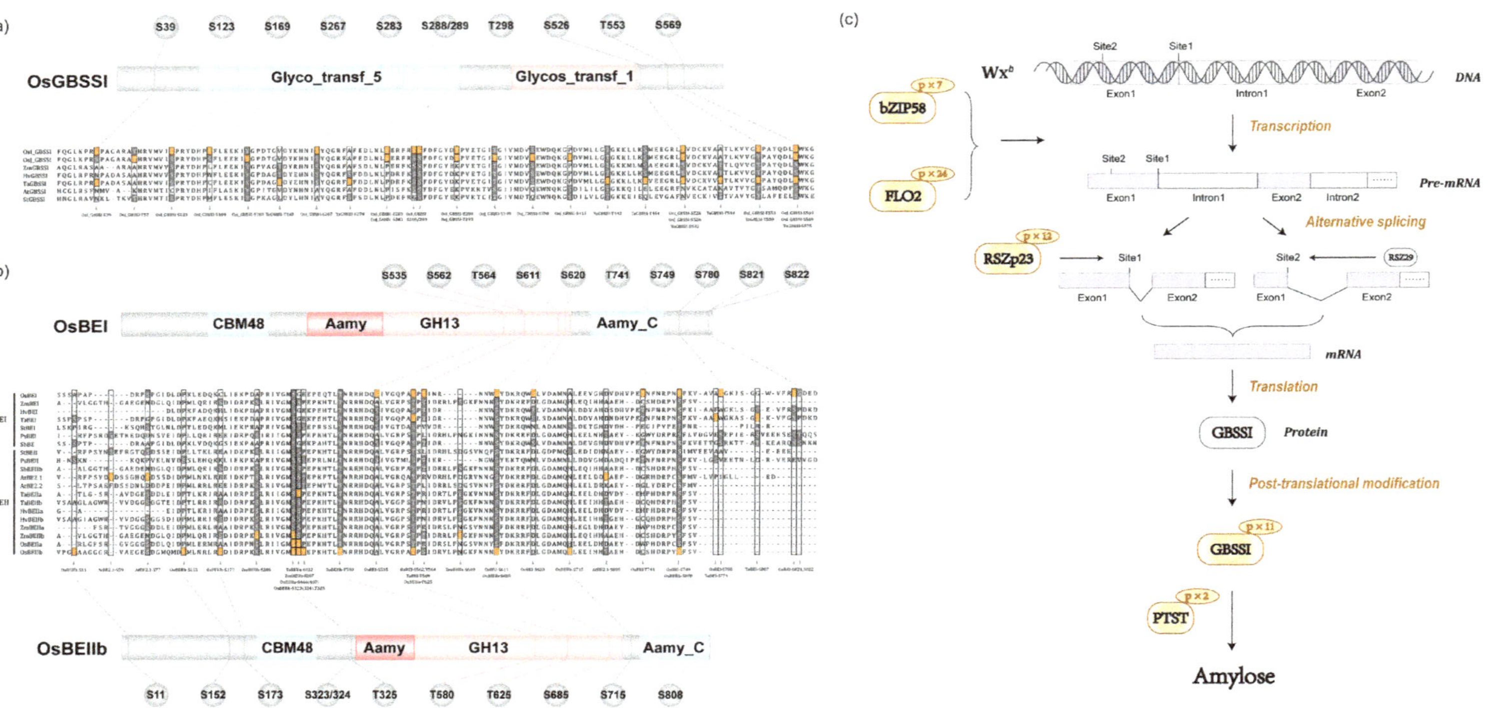

Figure 6. Domain structure and amino acid sequence alignments of GBSSI (**a**) and BEs (**b**). Residues in yellow indicate the phosphorylation site. Non-phosphorylated residues are shown in dark gray. (**c**) Potential effect of phosphorylation regulation on amylose biosynthesis by regulatory factors (OsbZIP58 and FLO2), RSZp23, GBSSI, and PTST.

Apart from the enzymes mentioned above, phosphorylated regulatory factors related to starch metabolism are given in Figure S11a. Among all seven phosphosites of OsbZIP58, three sites (S46, S183, S277) of GLA4 samples were up-regulated, while no significantly regulated phosphosites were observed in 9311. There were 24 residues phosphorylated at FLO2. Specifically, 9311 witnessed different degrees of significant down-regulation in nine phosphosites, while for GLA4, only one phosphosite increased considerably. Phosphoserine 277 is located within the bZIP_1 domain, whose function is to mediate sequence-specific DNA binding properties and the leucine zipper.

2.7. Dynamic Change in Phosphorylation Status Related to Starch Synthesis

To further validate and explore the dynamic change in the phosphorylation status involved in starch synthesis under high-temperature stress, site-specific phosphopeptide antibodies (BEI-Ser562, BEI-Ser620, BEI-Ser821, BEIIb-Ser685, BEIIb-Ser715, SSIIIa-Ser94) were prepared, which were identified in a previous study [22] and the present study (Table S11). The mass spectrum of synthetic peptides ensures the quality of blocking peptides (Figure S12). Only when specific phosphopeptides were incubated with antibodies did the specific band disappear (Figure S13), confirming the high efficiency and specificity of the phosphor-antibodies. The phosphorylation status of SSIIIa (SSIIIa-S94), BEI (BEI-Ser562, BEI-Ser620, BEI-Ser821), and BEIIb (BEIIb-Ser685, BEIIb-Ser715) was further confirmed by Western blot analysis (Figure 7a–i).

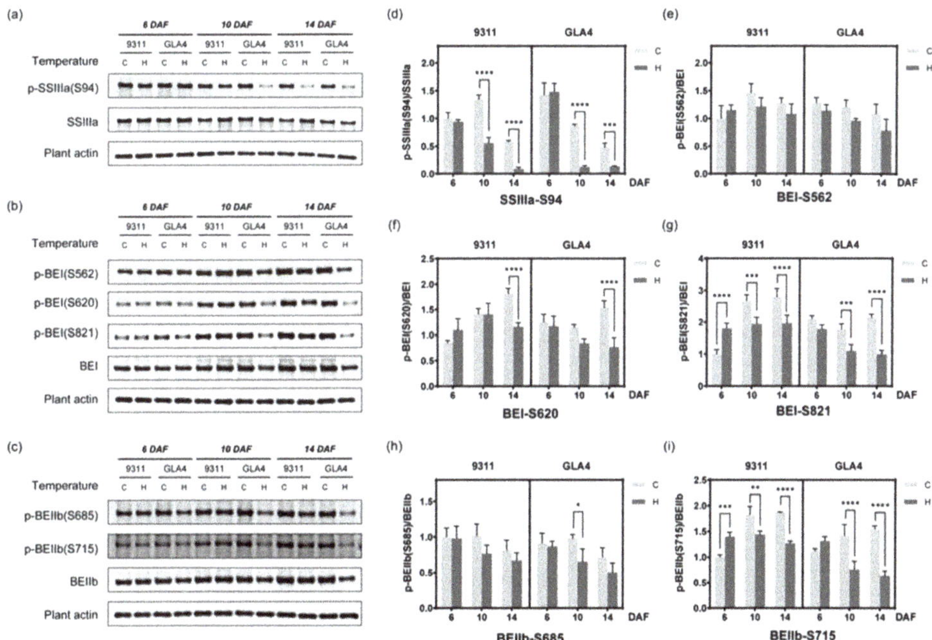

Figure 7. Western blot assay of some phosphorylated proteins. (**a–c**) Western blot assay of the phosphorylation of SSIIIa-S94, BEI-S562, BEI-S620, BEI-S821, BEIIb-S685, and BEIIb-S715. The best blot of three independent experiments is shown here. The results of the independent experiments followed a similar trend in expression. Uncropped gels are shown in Figure S19. (**d–i**) Evaluation of the relative phosphorylation intensity of starch-synthesis-related enzymes at the three grain-filling stages under high-temperature stress. * $p < 0.05$, ** $p < 0.01$, *** $p < 0.001$, and **** $p < 0.0001$.

Starch synthase IIIa (SSIIIa) has the second-highest starch synthase (SS) activity and plays a critical role in forming long B chains, most notably B2 and B3 chains [32,33]. Considering the phosphorylation event of SSIIIa-S94 was identified multiple times in rice endosperm [20,22], we speculated that S94 is a crucial site for protein function and prepared a site-specific phosphopeptide antibody for Western blot (Table S11). To eliminate the impact of protein abundance on the phosphorylation level, the relative phosphorylation intensity was used to evaluate the regulation of the modification level of the phosphosite. As shown in Figure 7d, the relative phosphorylation intensity of SSIIIa-S94 decreased with the growth period, and high-temperature treatment exacerbated the inhibitory tendency. In the meantime, SSIIIa protein expression witnessed significantly and relatively mild rises in 9311 and GLA4 at 10 DAF, respectively (Figures 7a and S8), which was inversely associated with the phosphorylation status of S94. Results from immunoblotting analysis at 10 DAF were consistent with the trend tested by the mass spectrum except for the GLA4-C group (Figure S14). We attributed this slight inconsistency to the different characteristics of the normalization methods. In other words, high temperature severely inhibits the phosphorylation level of SSIIIa-S94, and the increase in protein expression does not appear to compensate for the phosphorylation reduction.

Branching enzymes (BEs) including BEI, BEIIa, and BEIIb, are fundamental to form a distinct fine structure of amylopectin [34]. Given the repeated identification in the phosphoproteome [22] and high probabilities of these phosphosites, it is worth to determine the phosphorylation intensity of BEI-S562, BEI-S620, BEI-S821, BEIIb-S685, and BEIIb-S715 (Figure 7b,c). Large proportion of missing values and poor repeatability of LC/MS-MS data in these specific phosphosites made the trend difficult to evaluate (Figure S14), and statistics from immunoblotting analysis may compensate for the deficiency. A significant decline in both varieties was observed in the relative phosphorylation intensity of BEI-S620, BEI-S821, and BEIIb-S715 at 14 DAF (Figure 7f,g,i). Besides, evident alterations in the phosphorylation status of BEI-S821 and BEIIb-S715 already appeared at 6 DAF (Figure 7g,i). On the contrary, BEI-S562 and BEIIb-S685 changed their phosphorylation status based on the BE protein expression with temperature stimulus (Figure 7e,h).

To gain a better understanding of the phosphorylation-mediated regulation mode of starch-synthesis-related enzymes under abiotic stress, we counted the number of significant differences for each of the phosphosites mentioned above and analyzed their relationship with site conservation (Table S12). A meaningful outcome may be that phosphorylation levels of conserved phosphosites (BEI-S562 and BEIIb-S685) were not significantly affected by heat stress, while significant phosphorylation changes were observed in non-conserved phosphosites (SSIIIa-S94, BEI-S620, BEI-S821, and BEIIb-S715) at different periods of grain development (Table S12).

3. Discussion

Generally, japonica rice is used as the primary material for phosphoproteomic investigations [20,24]. However, research focus on indica rice seems even more necessary because it is the most widely cultivated rice in Asia, with over 70% of rice production worldwide. As shown in Figure 2e, 1114 phosphoproteins are common to all three datasets, whose phosphorylation status is ubiquitous across different rice varieties or tissues (Figure 2e). Notably, 1207 phosphoproteins were newly identified in this study, substantially filling the missing information in the rice phosphorylation database. In particular, a large number of phosphoproteins related to starch metabolism, including abundant enzymes, transcription factors, and kinases, were newly identified. Compared with our previous research [22], the current phosphoproteome increased 73.7% identification (7365 phosphosites; Figure 2f), which likely benefited from the improved experimental technique and stress treatments. The current research covered 89.8% phosphoproteins detected previously (Figure 2f), further confirming the reliability of the experiment. Taken together, our indica rice phosphoproteome, as the most extensive set of identified phosphosites in rice endosperm, greatly enriches the plant post-translational modification information.

Phosphorylation-mediated regulation associated with high-temperature stress has been explored in rice leaves [35]. Researchers have found the dephosphorylation of ribulose bisphosphate carboxylase (RuBisCo) and the phosphorylation of ATP synthase subunit-β under heat stress [35]. However, the potential phosphorylation-mediated regulation remains to be disclosed owing to the limited data. A recent large-scale phosphoproteomic study of wheat leaf and spikelet revealed temperature-induced interconversion of neighboring phosphorylation residues [26]. So far, the phosphorylation-mediated potential in cereal endosperm under heat stress remains unexplored. This study is the first large-scale investigation to focus specifically on the phosphorylation status of rice endosperm under high-temperature stress.

3.1. An Essential Role of CDPKs against Heat Stress

CDPKs belong to the CAMK kinase family and sense changes in the cytoplasmic Ca^{2+} concentration in response to abiotic stress and translate these perceived signals into subsequent downstream signaling events to trigger response mechanisms/pathways [36]. Among all differentially regulated kinases (Table S7), 9 CDPKs with 15 residues occurred most in all kinase subfamily (Figures 4b and S15), and 14 phosphosites tended to be up-regulated in a high-temperature environment (Figure 4b), suggesting that CDPKs is likely a critical factor activated by high temperature. Similarly, this inference was supported by the result of phosphorylation motif analysis presented above that consensus motifs ([sP], [LxRxxs], [Rxxs], [tP]) induced by heat stress are potential substrates for CDPKs (Figure 4a). In other words, evidence from two aspects may provide novel insights for subsequent studies of CDPKs in response to heat stress.

In the literature, so far, kinase–substrate networks induced by high-temperature have not been investigated in detail [26]. Here, we focused on a potential regulatory network mediated by SPK, a kind of CDPK specifically expressed in immature endosperm. SPK showed increasing the phosphorylation intensity of three sites induced by heat stress, and two up-regulated phosphosites (S303, S317) were shared by both vaterites (Figures 4b and S11b and Table S7). Based on this finding, we boldly speculated that the kinase activity of SPK is subsequently activated to deal with the possible deficiency in starch accumulation induced by heat stress. Using the approach of in vitro phosphorylation, Asano et al. found that a serine residue at the N-terminal region of sucrose synthase is a target of SPK [37]. However, it is still unclear which sucrose synthase isoform(s) is phosphorylated actually in vivo. In this study, four possible target phosphosites were screened from developing seeds for the first time: SUS2-S10, SUS3-S15, SUS4-S11, and SUS5-S12 (Figure S11b). Under high-temperature treatment, only SUS5-S12 in 9311 was significantly up-regulated as we expected, but the phosphorylation intensity of SUS2-S10 was down-regulated in GLA4 (Figure 5). In addition, no significant differences emerged in the phosphorylation intensity of SUS3-S15 and SUS4-S11. Two plausible reasons could explain those findings. One possibility is that SPK alone does not induce these phosphosites. Indeed, the phosphorylation status of the specific sites is likely a consequence of multiple protein kinases and phosphatases. Another potential explanation is that the protein expression of SUSs is severely inhibited by heat stress [7] so that the up-regulated phosphorylation is not sufficient to compensate for the loss of protein expression.

3.2. RNA Splicing Is a Critical Pathway in Response to Heat Stress

RNA splicing, a form of RNA processing, removes introns and joins exons together to make the pre-mRNA transform into a mature messenger RNA (mRNA). Recent observations have suggested that post-transcriptional regulation, especially alternative splicing (AS), appears to function in plant responses to environmental stress [38]. In our phosphoproteomic dataset, the spliceosome pathway was greatly over-represented in both varieties by KEGG pathway predication (Figure S5c). Phosphorylation events (430 phosphosites corresponding to 80 proteins) occurred within almost all spliceosome complexes (Figure S16), indicating that RNA splicing might be a critical pathway response to high-temperature stress in rice endosperm.

Ser/Arg (SR)-rich proteins are a group of RNA-binding proteins that finely regulate alternative splicing by interacting with pre-mRNA sequences and splicing factors during spliceosome assembly [39]. Our work found a substantial number of phosphorylation sites on rice SR proteins (RSZp21a, RSZp21b, and RSZ23; Figure S17). In rice, a prior study noted the importance of SR protein in constitutive and alternative splicing of Pre-mRNA, and RSZp23 enhanced the splicing of the *Wx*^{b} gene at the proximal sites [40]. It is worth mentioning that the RS domain of Arabidopsis RSZp22, homologous to rice RSZp23, regulates its shuttling between nucleoplasm and nucleolus through its level of phosphorylation [41,42]. Therefore, in the present study, we focused on the 12 phosphorylated serine sites of RSZp23 and found that nine phosphosites were located in the sequence of RSYSRSP at the RS domain (Figure S17). We considered whether there was a plausible mechanism by which the phosphorylation of the RS domain in RSZp23 might influence the splicing efficiency of *Wx*^{b}. The indica variety 9311 carrying *Wx*^{b} observed a down-regulated phosphorylation trend in RSZp23 induced by heat stress (Figure S18). In contrast, GLA4 exhibited up-regulation trends (Figure S18). Under high-temperature treatment (10 DAF), the GBSSI protein in 9311 slightly increased, whereas that of GLA4 declined (Figures S7 and S8). This inconsistency may be due to post-transcriptional regulation of RSZp23, which is likely a major factor affecting the alternative splicing of *Wx*.

Previous studies have revealed the association between the splicing efficiency of *Waxy* and temperature in that high temperature caused a significant decrease in the number of mature mRNAs [43]. Overall, the results presented in this study provide valuable insights into the hypothesis that adjustment of the RSZp23 phosphorylation pattern induced by high temperature affects the splicing efficiency of *Waxy* and eventually influences amylose biosynthesis. In this sense, the potentially essential differences of SR proteins between 9311 and GLA4 at the phosphorylation level may provide novel ideas for the improvement of starch quality.

3.3. *Phosphorylation Regulates Amylose Biosynthesis*

Amylose content is one of the key factors that strongly influence rice grain quality. Regretfully, to date, few studies have been able to carry out any systematic research of amylose biosynthesis mediated by phosphorylation, particularly under abiotic stress. The potential roles of phosphorylation in regulating amylose biosynthesis were addressed in three dimensions (Figure 6c).

First, the phosphorylation of regulatory factors and splicing factors may affect the expression of GBSSI (Figure 6c). In rice endosperm, OsbZIP58 bonds directly to the promoters of multiple rice starch biosynthetic genes in vivo, including AGPL3, Wx, SSIIa, SBEI, BEIIb, and ISA2, which affects the accumulation of starch during grain development [44]. A mutant analysis reported that FLO2 alters the expression of various genes involved in sucrose and starch metabolism by mediating protein–protein interactions [45]. Although there is currently no evidence of the biological functions of phosphorylated regulatory factors, a large number of phosphosites detected in the present study still provide a possible connection between the gene expression of GBSSI and the phosphorylation status of those regulatory factors. In addition, as mentioned above, the phosphorylation status of RSZp23 may affect the splicing efficiency of *Wx*^{b}, modulating GBSSI expression at the post-transcriptional level.

Second, the most crucial point to note is that 11 phosphosites of GBSSI were identified in GLA4 (Figure 6a). In rice endosperm, GBSSI is a key enzyme specifically responsible for elongating amylose polymers and serves as the only protein known to be required for amylose synthesis [46]. In recent years, Pro-Q Diamond dye and LC-MS/MS have been employed to determine the phosphosites of GBSSI in rice and wheat endosperm [47,48]. More specifically, Zhang et al. found at least nine phosphosites in rice GBSSI and inferred that the S415P substitution may modulate its enzyme activity regulated by phosphorylation levels [49]. As far as we know, our study is the most extensive-scale study of the GBSSI phosphorylation event. Sequence alignment analyses indicated that the six residues are phosphorylated at the same position that corresponds to japonica rice and wheat (Figure 6a) [48,49]. Remarkably, all significantly up-regulated phosphosites (S123, S169, and S553) are conserved among plant GBSSI, suggesting their functional importance. Despite these promising results, further data collection is required to determine precisely how heat affects the phosphorylation level of GBSSI and subsequently influences amylose synthesis.

Third, protein targeting starch (PTST) was phosphorylated in rice endosperm (Figure 6c). PTST is originally a carbohydrate-binding scaffold protein that interacts with GBSSI as a pre-requisite for subsequent starch granule binding in Arabidopsis [50]. Zhong et al. carried out a series of experiments and confirmed the vital importance of PTST for amylose biosynthesis in maize endosperm [51]. Our study revealed two novel phosphosites (S213 and S289) in indica rice endosperm. The phosphorylation of S289 was uniquely detected at 9311-H. Similarly, the intensity of GLA4-H was 3.71-fold in comparison with GLA4-C, although this difference did not reach statistical significance (Figure 5). It is thus likely that kinases targeting this phosphosite might be activated by high temperature and eventually affects the localization of GBSSI.

3.4. Phosphorylation Regulates Amylopectin Biosynthesis

Amylopectin is the major component of starch and serves as a key substance in rice grains. Several works on maize and wheat endosperm have emphasized the vital importance of protein phosphorylation in the formation of a starch-synthesizing protein complex [10,11]. Dephosphorylation of BEs and SSs is associated with reduced activity and protein complex formation [10–12,52]. In rice endosperm, although several investigations observed the importance of high-molecular-weight complexes for maintaining the activities of corresponding key enzymes [53], the phosphorylation-dependent formation of the complexes remains unknown [54]. In this sense, sequence alignment analysis was carried out to obtain the information about the similarities and differences between rice and other plants.

SS, covering four isoforms, SSI, SSII, SSIII, and SSIV, has been attributed to catalyze the chain elongation reaction of α-1,4-glucosidic linkages in amylopectin synthesis [55]. Analysis of cereal endosperm appears to show that each isoform of SS performs a specific role in amylopectin synthesis [10]. Consistent with the review by Croft et al., sequence alignment analysis revealed that approximately all SSIIa phosphosites identified are not conserved among various plant species [19]. It has been demonstrated that wheat starch synthesis enzymes SSI and SSIIa are phosphorylated [11,48] and the protein complex including SSII is phosphorylation dependent [10,11]. However, relevant evidence in rice SSIIa is still lacking. Nakamura et al. concluded that SSIIa activity is a determinant affecting thermal gelatinization properties [46]. Gelatinization temperature witnessed a clear trend of significantly increasing in 9311 under heat stress (Table S1). Interestingly, the phosphorylation intensity of a single peptide of SSIIa was only detected in 9311-C (Figure 5), suggesting that enzyme activity and protein complex formation are likely to be determined by the phosphorylation status rather than the protein abundance.

BEs are responsible for catalyzing the formation of α-1,6-glucosidic linkages of amylopectin in rice endosperm [55]. The phosphorylation sites of BEs with different conservations may display variations in protein structure, binding location, and kinase specificity [19]. To probe such possibility in plant BEs, we, thus, performed sequence alignment analyses of BEs from eight plants (rice, maize, barley, wheat, potato, sorghum, pea, and Arabidopsis) and focused on novel identified phosphorylation sites (Figure 6b). As is apparent, some phosphosites were conserved among all plant BEs, while others were species specific (Figure 6b). In particular, some residues of OsBEIIb (S323/S324, S625), OsBEIIa (S466/467), and OsBEI (S562) were phosphorylated at the same alignment positions when compared with maize (Figure 6b) [15,56–58] and wheat (Figure 6b) BEs [59,60]. Moreover, a close link was observed in three BE isoforms in that they possessed overlapping phosphosites (Figure 6b). In the present case, plant BEs possess the most conserved phosphorylation pattern among all starch-synthesis-related enzymes, suggesting a high probability of the conserved regulation mechanism in cereal endosperm.

Western blot analysis of five residues of rice BEs (BEI-S562, BEI-S620, BEI-S821, BEIIb-S685, and BEIIb-S715) further verified that the degree of phosphorylation change is possibly significantly associated with corresponding site conservation (Figure 7e–i and Table S12). In other words, high-temperature treatment tends to have a greater effect on non-conserved phosphosites. This phenomenon can also be confirmed in phosphoproteomic analysis in that heat stress triggered phosphorylation at S11 and S173 of BEIIb but exerted suppressive effects on BEIIa-S467. Only the conserved phosphosite BEIIb-T580 located at GH13 was an exception, showing an up-regulation trend under high-temperature stress in 9311 groups. Overall, our findings still support the hypothesis that BEI-S562 and BEIIb-S685, as conserved sites in the functional domain, reflect steady-state phosphorylation levels during heat stress to maintain the relative stability of essential functions. As non-conserved sites that may play specific roles, BEI-S620, BEI-S821, and BEIIb-S715 might exhibit more flexible phosphorylation regulation patterns when exposed to an adverse environment. These findings contribute to a better understanding of phosphorylation-mediated regulation under abiotic stress and provide a solid basis for subsequent preparation of corresponding mutants to verify the specific function in the near future.

DBE consists of isoamylase (ISA) and Pullulanase (PUL) with catalytic function for hydrolyzing α-1,6-glucosic linkages [55]. Recent experiments have shown that the function of PUL appeared to be positively regulated by high temperature [61]. However, a postulated and legitimately interpreted mechanism for this phenomenon remains absent [61]. Our present study aims to fill this gap at the phosphorylation level. As shown in Figure 5, the phosphorylation intensity of S221 was significantly enhanced. It is thus most likely that the phosphosite (PUL-S221) within the functional domain is regulated by potential kinases and subsequently affects the enzymatic activity of the domain.

Plastidial phosphorylase (Pho1), a temperature-dependent enzyme, is considered crucial not only during the maturation of amylopectin but also in the initiation process of starch synthesis [19]. Satoh et al. examined the effects of low temperature on the *pho1* mutant and found an essential role of Pho1 during its initial stages of α-glucan biogenesis, especially under conditions of low temperature [62]. Hence, they speculated that one or more other factors are involved to replace Pho1 in a high-temperature environment. Most phosphosites are conserved among monocotyledonous plants, and five phosphosites (S341, S932, S939, S944, T974) explicitly localized at the phosphorylase domain are highly conserved among all plants (Figure S10). In particular, S341 was identified as an active site pocket of glycogen phosphorylase and similar proteins (cd04300; InterPro), suggesting a link may exist between phosphorylation and functional activity. An early report has demonstrated that the serine residue in the L80 region from the sweet potato root is phosphorylated [63], and we found three phosphosites in the same region, although their alignment positions were not completely consistent (Figure S10). Under high-temperature stress, the phosphorylation intensity of T376 exhibited a reverse trend between two varieties (Figure 5). These phosphosites identified in rice Pho1 with significant differences under high-temperature stress are expected to become a new breakthrough point for functional activity research.

4. Materials and Methods

4.1. Plant Materials and Experimental Design

Two indica rice varieties (*Oryza sativa* L.) 9311 and Guangluai4 (GLA4) were used to examine the effect of high temperature on the developing rice endosperm with different starch qualities. The detailed parameters of starch qualities are provided by Pang et al. [22]. All plant materials were planted at the experimental farm of Zhejiang University under normal rice cultivation conditions until heading. Before the flowering stage, all rice plants were transferred into artificial climatic chambers at a suitable temperature (20–28 °C) with a 14/10 h photoperiod (day/night). Each panicle, on the day of rice flowering, was tagged to facilitate collecting samples at defined developmental stages. Five days later, rice plants of the treatment group were moved into a high-temperature chamber (HT; 30–38 °C;

Figure S1) and then exposed to high temperature for 1, 5, and 9 days. In the meantime, other plants were still cultivated at a suitable temperature as controls (CT; 20–28 °C; Figure S1). All temperature treatments were completed under other consistent growth conditions, and the relevant parameters were as follows: the illumination time was 14 h/day (5:00–19:00), the illumination intensity during the light phase was 10,000–12,000 Lux, the relative humidity was 70–80%, the maximum temperature difference was 8 °C, and the average wind speed was 0.5 m/s. During grain development, rice panicles were handpicked from fresh plants at different periods (6, 10, and 14 DAF) in three separate biological replicates and immediately frozen in liquid nitrogen (9311/GLA4-C, 9311/GLA4-H; Figure 1a). For phosphorylation studies, developing seeds collected at 10 DAF were first determined by LC-MS/MS and then all samples in three periods were analyzed by Western blot using site-specific phosphopeptide antibodies. In addition, the remaining seeds were harvested at maturity for chalkiness degree and starch quality analysis (9311/GLA4-C0, 9311/GLA4-H1, 9311/GLA4-H3, 9311/GLA4-H5; Figure 1a).

4.2. Determination of Starch Quality

Under flowing nitrogen conditions, differential scanning calorimetry (DSC) measurements were conducted using a TA instrument Q20 (TA Instruments, New Castle, DE, USA) at a heating rate of 10 °C /min according to the method described by Bao et al. [64]. The onset temperature (To), peak temperature (Tp), conclusion temperature (Tc), gelatinization enthalpy (ΔH) and width at half peak height ($\Delta T_{1/2}$) were obtained from the DSC thermogram. Three replicates were performed per sample. Duncan's multiple-range test of ANOVA was performed in SPSS (IBM SPSS Statistic 20). Statistical significance was defined at the level of $p < 0.05$.

4.3. Protein Preparation, Digestion, and Phosphopeptide Enrichment

For the extraction of endosperm proteins, other tissues (husk, pericarp, and embryo) were removed from immature rice grains [53]. The procedures were quickly carried out on ice. Rice endosperm was then homogenized by grinding in the presence of liquid nitrogen prior to protein extraction. In brief, rice endosperm was extracted with extraction buffer (4% SDS, 1mM DTT, 100 mM Tris–HCl, pH 7.6) supplemented with EDTA-free protease and phosphatase inhibitor cocktails (Sigma-Aldrich, St. Louis, MO, USA). Sonication was performed using 10 rounds of 10 s sonication and 3 s off-sonication. Following boiling for 10 min, protein samples were centrifuged at 12,000× *g* for 40 min at 4 °C. Supernatants were gathered and kept in a freezer at −80 °C for subsequent phosphoproteomic pretreatment. The protein amount was estimated by BCA assay (Pierce BCA Protein assay kit, Thermo Fisher Scientific, Waltham, MA, USA).

For FASP digestion, samples were treated as previously described [65]. SDS, DTT, and other low-molecular-weight components were removed using UA buffer (8M Urea, 150 mM Tris-HCl, pH 8.0) by repeated ultrafiltration (10 kDa, Satorious, Gottingen, Germany). Then protein mixtures were alkylated with 100 mM iodoacetamide (IAA) for 30 min in darkness. After repeated washing with UA buffer and 25 mM NH_4HCO_3, the protein suspensions were digested with trypsin (Promega, Madison, WI, USA) overnight at 37 °C. The enzymatic peptides were desalted on a Sep-Pak C18 cartridge (Waters, Milford, MA, USA) and subjected to TiO_2-based phosphopeptide enrichment, as previously described by Pang et al. [22]. The enriched eluates of each sample were concentrated by vacuum evaporation and reconstituted in 0.1% formic acid for MS analysis.

4.4. LC-MS/MS and Data Analysis

Liquid chromatography–tandem mass spectrometry (LC-MS/MS) analysis was performed using an Easy-nLC System coupled with a Q-Exactive Plus mass spectrometer (Thermo Fisher Scientific, Waltham, MA, USA). The mobile phases consisted of 0.1% formic acid (A) and 0.1% formic acid in 84% *v/v* acetonitrile (B). The column was equilibrated with 95% solution A. A volume of 6 µL of phosphopeptide solution was loaded onto Acclaim PepMap100 (100 μm × 2 cm, nanoViper C18, 3 μm, 100 Å; Thermo Fisher Scientific, Waltham, MA, USA), and separated by an EASY column (10 cm, ID75 μm, 3 μm, C18-A2; Thermo Fisher Scientific, Waltham, MA, USA) at a flow rate of 300 nL/min. Over a period of 0–2 min, the concentration of solution B rose linearly from 5% to 7%; from 2 to 162 min, it increased from 7% to 25%; from 162 to 225 min, it increased from 25% to 40%; from 225 to 230 min, it increased from 40% to 100%; and from 230 to 240 min, it was maintained at 100%.

The Q-Exactive Plus mass spectrometer was operated in positive ion mode over 240 min. Full-scan mass spectra were acquired over a mass range of 300–1800 m/z. The resolution of first-order mass spectrometry was 70,000, the AGC target was 1×10^6, and the first-order maximum IT was 50 ms. For subsequent MS2 analysis, only the top 10 precursors were selected. HCD-MS2 spectra were acquired with 1 microscan at a resolution of 17,500, and the AGC target was 1×10^5. The MS2 scan range was set from 200 to 2000 m/z, the maximum IT was 110 ms, and the isolation window was 2.0 m/z. Dynamic exclusion was employed with an exclusion duration of 60 s. Three biological replicates were performed independently for each group (Figure S2a).

Raw mass spectrometric data were processed with MaxQuant software (version 1.5.5.1) and compared with the indica rice protein sequence database (*Oryza sativa* subsp. *indica*-ASM465v1). Trypsin/P was specified as the enzyme, and two missed cleavages were allowed. The precursor mass tolerance was set to 20 ppm for the first search (used for mass re-calibration) and to 4.5 ppm for the main search. The MS/MS mass tolerance was set to 20 ppm. Carbamidomethylation of cysteine residues was selected as a fixed modification, while protein N-terminal acetylation, methionine oxidation, and phosphorylation on serine/threonine/tyrosine were allowed as variable modifications. False discovery rate (FDR) thresholds for protein, peptide, and modification sites were specified at 1%. A minimum peptide length of seven amino acids was required. The mass spectrometry proteomics data have been deposited to the ProteomeXchange Consortium (http://proteomecentral.proteomexchange.org, accessed on 2 July 2021) via the iProX partner repository [66] with the dataset identifier PXD027052 and the subject ID IPX0003230000. Groups N1, N2, L1, and L2 in the PXD027052 project correspond to groups 9311-C, 9311-H, GLA4-C, and GLA4-H in this study, respectively.

4.5. Statistical and Bioinformatic Analyses

Quantification of the modified peptides was performed using the label-free quantification algorithm. In general, phosphosites that exhibited valid values in one condition (at least 2 of 3 replicates) and none in the other indicate a massive change in phosphorylation levels. We, therefore, opted to select those phosphosites that feature a consistent presence/absence profile in 9311 and GLA4, respectively. On the rest of the dataset, we performed Student's t-test ($p < 0.05$, $\log_2$ (fold-change) > 1) on phosphosites with at least two valid values in any condition. For a more comprehensive understanding of phosphosites showing significant differences in expression, phosphorylation sites that were uniquely identified in either condition and significantly regulated from the statistical test were combined and divided into four groups (9311-Up, 9311-Down, GLA4-Up, and GLA4-Down; Figure 3d).

A total of 13 amino acid (AA) sequences centered by the phosphosite were extracted, and the enriched phosphorylation motifs induced by high-temperature stress were predicted using the MoMo tool (http://meme-suite.org/tools/Momo, accessed on 9 January 2020) with the motif-x algorithm. The conserved motif patterns were then redrawn and visualized by TBtools [67]. Heatmaps for the relative abundances, Venn diagrams, and upset plots for various lists were also produced using TBtools. GO enrichment analysis was conducted using the AgriGO website (http://bioinfo.cau.edu.cn/agriGO/, accessed on 6 January 2020). Cytoscape (http://www.cytoscape.org, accessed on 6 January 2020) was used to generate network visuals. KEGG pathway annotation was performed by using KEGG Automatic Annotation Server (KAAS) software (http://www.genome.jp/kegg/kaas, accessed on 18 December 2019). A p-value of <0.05 was used as the threshold of significant enrichment. Pfam (www.sanger.ac.uk/Software/Pfam/, accessed on 12 February 2020) and InterPro (https://www.ebi.ac.uk/interpro/, accessed on 12 February 2020) were used to identify functional domains. Protein sequences were aligned using T-coffee. The figures were annotated with Adobe Illustrator (Adobe Systems, San Jose, CA, USA).

4.6. Preparation of Primary Antibodies

Abundantly conventional and site-specific phosphopeptide antibodies were used for this experiment.

Conventional antibodies: Anti-rice GBSSI, SSIIIa, BEI, and BEIIb antibodies) were kindly gifted by Prof. Naoko Fujita (Akita Prefectural University, Akita, Japan). Mouse monoclonal antibodies against plant actin were obtained from Sigma-Aldrich (Sigma-Aldrich, St. Louis, MO, USA).

Site-specific phosphopeptide antibodies: Modification peptides of interest from phosphoproteome were selected and are listed in Table S11. The synthetic phosphorylated peptides were coupled to keyhole limpet hemocyanin (KLH) before immunization of rabbits. The antibodies were purified from rabbit polyclonal antiserum by affinity purification via sequential chromatography on phosphopeptide and non-phospho-peptide affinity columns (Affinity Biosciences, Changzhou, China). The efficiency of antibody production was monitored using ELISA.

4.7. Western Blotting

Protein expression of starch-synthesis-related enzymes as well as the site-specific phosphorylation intensity were examined by Western blot assay [53]. Total proteins were separated by SDS-PAGE after loading buffer was added. Immediately, separated proteins were transferred to a PVDF membrane (Merck Millipore, Billerica, MA, USA) using a Trans-Blot Cell system (Bio-Rad Laboratories, Hercules, CA, USA). The membranes proceeded directly to the blocking step and then were exposed to appropriate antibodies. Chemiluminescence signals were developed with an ECL kit (SuperSignal West pico; Thermo Fisher Scientific, Waltham, MA, USA) and detected by a ChemiDoc imaging system (Bio-Rad Laboratories, Hercules, CA, USA). Correspondingly, analysis of bands was performed with Image Lab™ software (Bio-Rad Laboratories). For normalization, within the same membrane, plant actin was used as a loading control. For evaluating the relative phosphorylation intensity, phosphorylated phosphosites were corrected to the levels of the corresponding total protein. Statistical analysis was performed using ANOVA with Tukey's post-test in GraphPad Prism (GraphPad Software, San Diego, CA, USA).

5. Conclusions

In conclusion, this study set out to provide the first systematic investigation of the phosphoproteome induced by high-temperature stress in rice endosperm. Comparative analysis of the temperature-induced phosphorylation status revealed some interesting similarities and differences between two indica rice varieties (9311 and GLA4). On the one hand, both 9311 and GLA4 shared several consensus motifs ([sP], [LxRxxs], [Rxxs], [tP]) and were highly enriched in terms of heat response (GO) and the spliceosome pathway (KEGG). On the other hand, a dramatic difference was observed in the phosphorylation status of kinases induced by heat stress, which could reasonably explain the different phosphorylation regulatory patterns of 9311 and GLA4. More importantly, we detailed the most comprehensive starch metabolism pathway at the phosphorylation level in rice endosperm, including starch-synthesis-related enzymes (AGPase, GBSSI, SSIIa, SSIIIa, BEI, BEIIb, ISA1, PUL, PHO1, PTST), transcription factors (OsbZIP58 and FLO2), SR protein (RSZp23), and CDPK kinase (SPK). Western blot with site-specific phosphor-antibodies was used to verify and explore the dynamic change of the phosphorylation status of SSIIIa (SSIIIa-S94), BEI (BEI-Ser562, BEI-Ser620, BEI-Ser821), and BEIIb (BEIIb-Ser685, BEIIb-Ser715), which might play specific roles in amylopectin biosynthesis in response to high-temperature stress. An exciting phenomenon was discovered in that conserved sites tend to reflect steady-state phosphorylation levels. However, non-conserved sites exhibit more flexible phosphorylation regulation patterns when exposed to high-temperature stress. These findings provide valuable insights into the role of phosphorylation response to high-temperature stress, adding unprecedented depth and breadth to the cereal research community.

Supplementary Materials: The following are available online at https://www.mdpi.com/article/10.3390/ijms221910546/s1.

Author Contributions: J.B. conceived and designed the experiments. Y.P. and Y.H. performed the experiments. Y.P. analyzed the data. Y.P. and J.B. wrote the paper. All authors have read and agreed to the published version of the manuscript.

Funding: This work was financially supported by the National Key Research and Development Program of China (grant no. 2016YFD0400104) and the Natural Science Foundation of China (grant no. 31871531) and Natural Science Foundation of Zhejiang Province (LZ21C130003).

Data Availability Statement: The mass spectrometry proteomics data have been deposited to the ProteomeXchange Consortium (http://proteomecentral.proteomexchange.org, accessed on 2 July 2021) via the iProX partner repository [66] with the dataset identifier PXD027052 and the subject ID IPX0003230000 (https://www.iprox.cn/page/project.html?id=IPX0003230000, accessed on 2 July 2021).

Acknowledgments: Authors sincerely thank Naoko Fujita for providing some antisera used in this study. Technical supports from Shanghai Applied Protein Technology, Co. with mass spectrometry are highly acknowledged.

Conflicts of Interest: The authors declare no conflict of interest.

Abbreviations

AGPase	ADP-glucose pyrophosphorylase
AMY	α-amylase
BAM	β-amylase
BE	Starch branching enzyme
DAF	Days after flowering
DBE	Starch debranching enzyme
DSC	Differential scanning calorimetry
FK	Fructokinase
GBSS	Granule-bound starch synthase
ΔH	Enthalpy of gelatinization
HK	Hexokinase
INV	Invertase
ISA	Isoamylase
PGI	Glucose-6-phosphate isomerase
PGM	Phosphoglucomutase
PHO1	Plastidial phosphorylase
PTM	Post-translational modification
PUL	Pullulanase
SPP	Sucrose-phosphate phosphatase
SPS	Sucrose-phosphate synthase
SS	Starch synthase
SUS	Sucrose synthase
SUT	Sucrose transporter
Tc	Conclusion temperature
To	Onset temperature
Tp	Peak temperature
UDPG	UDP-glucose
UGPase	UDP-glucose pyrophosphorylase

References

1. Jagadish, S.V.K.; Murty, M.V.R.; Quick, W.P. Rice responses to rising temperatures-challenges, perspectives and future directions. *Plant Cell Environ.* **2015**, *38*, 1686–1698. [CrossRef]
2. Jiang, H.; Dian, W.; Wu, P. Effect of high temperature on fine structure of amylopectin in rice endosperm by reducing the activity of the starch branching enzyme. *Phytochemistry* **2003**, *63*, 53–59. [CrossRef]
3. Lin, C.-J.; Li, C.-Y.; Lin, S.-K.; Yang, F.-H.; Huang, J.-J.; Liu, Y.-H.; Lur, H.-S. Influence of high temperature during grain filling on the accumulation of storage proteins and grain quality in rice (*Oryza sativa* L.). *J. Agric. Food Chem.* **2010**, *58*, 10545–10552. [CrossRef]
4. Tang, R.-S.; Zheng, J.-C.; Jin, Z.-Q.; Zhang, D.-D.; Huang, Y.-H.; Chen, L.-G. Possible correlation between high temperature-induced floret sterility and endogenous levels of IAA, GAs and ABA in rice (*Oryza sativa* L.). *Plant Growth Regul.* **2008**, *54*, 37–43. [CrossRef]
5. Yamakawa, H.; Hirose, T.; Kuroda, M.; Yamaguchi, T. Comprehensive expression profiling of rice grain filling-related genes under high temperature using DNA microarray. *Plant Physiol.* **2007**, *144*, 258–277. [CrossRef] [PubMed]
6. Liao, J.-L.; Zhou, H.-W.; Peng, Q.; Zhong, P.-A.; Zhang, H.-Y.; He, C.; Huang, Y.-J. Transcriptome changes in rice (*Oryza sativa* L.) in response to high night temperature stress at the early milky stage. *BMC Genom.* **2015**, *16*, 18. [CrossRef]
7. Liao, J.-L.; Zhou, H.-W.; Zhang, H.-Y.; Zhong, P.-A.; Huang, Y.-J. Comparative proteomic analysis of differentially expressed proteins in the early milky stage of rice grains during high temperature stress. *J. Exp. Bot.* **2014**, *65*, 655–671. [CrossRef] [PubMed]
8. Timabud, T.; Yin, X.; Pongdontri, P.; Komatsu, S. Gel-free/label-free proteomic analysis of developing rice grains under heat stress. *J. Proteom.* **2016**, *133*, 1–19. [CrossRef]
9. Zhang, H.-Y.; Lei, G.; Zhou, H.-W.; He, C.; Liao, J.-L.; Huang, Y.-J. Quantitative iTRAQ-based proteomic analysis of rice grains to assess high night temperature stress. *Proteomics* **2017**, *17*, 1600365. [CrossRef] [PubMed]
10. Tetlow, I.J.; Wait, R.; Lu, Z.; Akkasaeng, R.; Bowsher, C.G.; Esposito, S.; Kosar-Hashemi, B.; Morell, M.K.; Emes, M.J. Protein phosphorylation in amyloplasts regulates starch branching enzyme activity and protein–protein interactions. *Plant Cell* **2004**, *16*, 694–708. [CrossRef]
11. Tetlow, I.J.; Beisel, K.G.; Cameron, S.; Makhmoudova, A.; Liu, F.; Bresolin, N.S.; Wait, R.; Morell, M.K.; Emes, M.J. Analysis of protein complexes in wheat amyloplasts reveals functional interactions among starch biosynthetic enzymes. *Plant Physiol.* **2008**, *146*, 1878–1891. [CrossRef] [PubMed]
12. Hennen-Bierwagen, T.A.; Liu, F.; Marsh, R.S.; Kim, S.; Gan, Q.; Tetlow, I.J.; Emes, M.J.; James, M.G.; Myers, A.M. Starch biosynthetic enzymes from developing maize endosperm associate in multisubunit complexes. *Plant Physiol.* **2008**, *146*, 1892–1908. [CrossRef] [PubMed]
13. Hennen-Bierwagen, T.A.; Lin, Q.; Grimaud, F.; Planchot, V.; Keeling, P.L.; James, M.G.; Myers, A.M. Proteins from multiple metabolic pathways associate with starch biosynthetic enzymes in high molecular weight complexes: A model for regulation of carbon allocation in maize amyloplasts. *Plant Physiol.* **2009**, *149*, 1541–1559. [CrossRef] [PubMed]
14. Liu, F.; Ahmed, Z.; Lee, E.A.; Donner, E.; Liu, Q.; Ahmed, R.; Morell, M.K.; Emes, M.J.; Tetlow, I.J. Allelic variants of the amylose extender mutation of maize demonstrate phenotypic variation in starch structure resulting from modified protein-protein interactions. *J. Exp. Bot.* **2012**, *63*, 1167–1183. [CrossRef] [PubMed]
15. Makhmoudova, A.; Williams, D.; Brewer, D.; Massey, S.; Patterson, J.; Silva, A.; Vassall, K.A.; Liu, F.; Subedi, S.; Harauz, G.; et al. Identification of multiple phosphorylation sites on maize endosperm starch branching enzyme IIb, a key enzyme in amylopectin biosynthesis. *J. Biol. Chem.* **2014**, *289*, 9233–9246. [CrossRef]
16. Silva-Sanchez, C.; Li, H.; Chen, S. Recent advances and challenges in plant phosphoproteomics. *Proteomics* **2015**, *15*, 1127–1141. [CrossRef]
17. Meyer, L.J.; Gao, J.; Xu, D.; Thelen, J.J. Phosphoproteomic analysis of seed maturation in Arabidopsis, rapeseed, and soybean. *Plant Physiol.* **2012**, *159*, 517–528. [CrossRef]
18. Walley, J.W.; Shen, Z.; Sartor, R.; Wu, K.J.; Osborn, J.; Smith, L.G.; Briggs, S.P. Reconstruction of protein networks from an atlas of maize seed proteotypes. *Proc. Natl. Acad. Sci. USA* **2013**, *110*, E4808–E4817. [CrossRef]
19. Crofts, N.; Nakamura, Y.; Fujita, N. Critical and speculative review of the roles of multi-protein complexes in starch biosynthesis in cereals. *Plant Sci.* **2017**, *262*, 1–8. [CrossRef]
20. Qiu, J.; Hou, Y.; Tong, X.; Wang, Y.; Lin, H.; Liu, Q.; Zhang, W.; Li, Z.; Nallamilli, B.R.; Zhang, J. Quantitative phosphoproteomic analysis of early seed development in rice (*Oryza sativa* L.). *Plant Mol. Biol.* **2016**, *90*, 249–265. [CrossRef]
21. Xu, S.B.; Li, T.; Deng, Z.Y.; Chong, K.; Xue, Y.; Wang, T. Dynamic proteomic analysis reveals a switch between central carbon metabolism and alcoholic fermentation in rice filling grains. *Plant Physiol.* **2008**, *148*, 908–925. [CrossRef]
22. Pang, Y.; Zhou, X.; Chen, Y.; Bao, J.S. Comparative phosphoproteomic analysis of the developing seeds in two indica rice (*Oryza sativa* L.) cultivars with different starch quality. *J. Agric. Food Chem.* **2018**, *66*, 3030–3037. [CrossRef]
23. Ye, J.; Zhang, Z.; You, C.; Zhang, X.; Lu, J.; Ma, H. Abundant protein phosphorylation potentially regulates Arabidopsis anther development. *J. Exp. Bot.* **2016**, *67*, 4993–5008. [CrossRef] [PubMed]
24. Yao, Q.; Ge, H.; Wu, S.; Zhang, N.; Chen, W.; Xu, C.; Gao, J.; Thelen, J.J.; Xu, D. P^3DB 3.0: From plant phosphorylation sites to protein networks. *Nucl. Acids Res.* **2014**, *42*, D1206–D1213. [CrossRef]
25. Yu, J.; Han, J.; Kim, Y.-J.; Song, M.; Yang, Z.; He, Y.; Fu, R.; Luo, Z.; Hu, J.; Liang, W.; et al. Two rice receptor-like kinases maintain male fertility under changing temperatures. *Proc. Natl. Acad. Sci. USA* **2017**, *114*, 12327–12332. [CrossRef]

26. Vu, L.D.; Zhu, T.; Verstraeten, I.; van de Cotte, B.; The International Wheat Genome Sequencing Consortium; Gevaert, K.; De Smet, I. Temperature-induced changes in the wheat phosphoproteome reveal temperature-regulated interconversion of phosphoforms. *J. Exp. Bot.* **2018**, *69*, 4609–4624. [CrossRef]
27. Van Wijk, K.J.; Friso, G.; Walther, D.; Schulze, W.X. Meta-analysis of *Arabidopsis thaliana* phospho-proteomics data reveals compartmentalization of phosphorylation motifs. *Plant Cell* **2014**, *26*, 2367–2389. [CrossRef]
28. Zhang, S.-H.; Kobayashi, R.; Graves, P.R.; Piwnica-Worms, H.; Tonks, N.K. Serine phosphorylation-dependent association of the band 4.1-related protein-tyrosine phosphatase PTPH1 with 14-3-3 protein. *J. Biol. Chem.* **1997**, *272*, 27281–27287. [CrossRef] [PubMed]
29. Ku, N.-O.; Liao, J.; Omary, M.B. Phosphorylation of human keratin 18 serine 33 regulates binding to 14-3-3 proteins. *EMBO J.* **1998**, *17*, 1892–1906. [CrossRef] [PubMed]
30. Zheng, Y.; Jiao, C.; Sun, H.; Rosli, H.G.; Pombo, M.A.; Zhang, P.; Banf, M.; Dai, X.; Martin, G.B.; Giovannoni, J.J.; et al. ITAK: A program for genome-wide prediction and classification of plant transcription factors, transcriptional regulators, and protein kinases. *Mol. Plant* **2016**, *9*, 1667–1670. [CrossRef] [PubMed]
31. Jin, J.; Tian, F.; Yang, D.-C.; Meng, Y.-Q.; Kong, L.; Luo, J.; Gao, G. PlantTFDB 4.0: Toward a central hub for transcription factors and regulatory interactions in plants. *Nucl. Acids Res.* **2017**, *45*, D1040–D1045. [CrossRef]
32. Fujita, N.; Yoshida, M.; Kondo, T.; Saito, K.; Utsumi, Y.; Tokunaga, T.; Nishi, A.; Satoh, H.; Park, J.-H.; Jane, J.-L.; et al. Characterization of SSIIIa-deficient mutants of rice: The function of SSIIIa and pleiotropic effects by SSIIIa deficiency in the rice endosperm. *Plant Physiol.* **2007**, *144*, 2009–2023. [CrossRef]
33. Nakamura, Y.; Aihara, S.; Crofts, N.; Sawada, T.; Fujita, N. In vitro studies of enzymatic properties of starch synthases and interactions between starch synthase I and starch branching enzymes from rice. *Plant Sci.* **2014**, *224*, 1–8. [CrossRef] [PubMed]
34. Nakamura, Y.; Utsumi, Y.; Sawada, T.; Aihara, S.; Utsumi, C.; Yoshida, M.; Kitamura, S. Characterization of the reactions of starch branching enzymes from rice endosperm. *Plant Cell Physiol.* **2010**, *51*, 776–794. [CrossRef] [PubMed]
35. Chen, X.; Zhang, W.; Zhang, B.; Zhou, J.; Wang, Y.; Yang, Q.; Ke, Y.; He, H. Phosphoproteins regulated by heat stress in rice leaves. *Proteome Sci.* **2011**, *9*, 37. [CrossRef] [PubMed]
36. Dodd, A.N.; Kudla, J.; Sanders, D. The language of calcium signaling. *Annu. Rev. Plant Biol.* **2010**, *61*, 593–620. [CrossRef]
37. Asano, T.; Kunieda, N.; Omura, Y.; Ibe, H.; Kawasaki, T.; Takano, M.; Sato, M.; Furuhashi, H.; Mujin, T.; Takaiwa, F.; et al. Rice SPK, a calmodulin-like domain protein kinase, is required for storage product accumulation during seed development: Phosphorylation of sucrose synthase is a possible factor. *Plant Cell* **2002**, *14*, 619–628. [CrossRef]
38. Dong, C.; He, F.; Berkowitz, O.; Liu, J.; Cao, P.; Tang, M.; Shi, H.; Wang, W.; Li, Q.; Shen, Z.; et al. Alternative splicing plays a critical role in maintaining mineral nutrient homeostasis in rice (*Oryza sativa*). *Plant Cell* **2018**, *30*, 2267–2285. [CrossRef]
39. Reddy, A.S.N.; Marquez, Y.; Kalyna, M.; Barta, A. Complexity of the alternative splicing landscape in plants. *Plant Cell* **2013**, *25*, 3657–3683. [CrossRef]
40. Isshiki, M.; Tsumoto, A.; Shimamoto, K. The serine/arginine-rich protein family in rice plays important roles in constitutive and alternative splicing of pre-mRNA. *Plant J.* **2005**, *18*, 146–158. [CrossRef]
41. Tillemans, V.; Leponce, I.; Rausin, G.; Dispa, L.; Motte, P. Insights into nuclear organization in plants as revealed by the dynamic distribution of *Arabidopsis* SR splicing factors. *Plant Cell* **2006**, *18*, 3218–3234. [CrossRef]
42. Rausin, G.; Tillemans, V.; Stankovic, N.; Hanikenne, M.; Motte, P. Dynamic nucleocytoplasmic shuttling of an Arabidopsis SR splicing factor: Role of the RNA-binding domains. *Plant Physiol.* **2010**, *153*, 273–284. [CrossRef]
43. Larkin, P.D.; Park, W.D. Transcript accumulation and utilization of alternate and non-consensus splice sites in rice granule-bound starch synthase are temperature-sensitive and controlled by a single-nucleotide polymorphism. *Plant Mol. Biol.* **1999**, *40*, 719–727. [CrossRef]
44. Wang, J.-C.; Xu, H.; Zhu, Y.; Liu, Q.-Q.; Cai, X.-L. OsbZIP58, a basic leucine zipper transcription factor, regulates starch biosynthesis in rice endosperm. *J. Exp. Bot.* **2013**, *64*, 3453–3466. [CrossRef]
45. She, K.-C.; Kusano, H.; Koizumi, K.; Yamakawa, H.; Hakata, M.; Imamura, T.; Fukuda, M.; Naito, N.; Tsurumaki, Y.; Yaeshima, M.; et al. A novel factor *FLOURY ENDOSPERM2* is involved in regulation of rice grain size and starch quality. *Plant Cell* **2010**, *22*, 3280–3294. [CrossRef]
46. Nakamura, Y. Rice starch biotechnology: Rice endosperm as a model of cereal endosperms. *Starch-Stärke* **2018**, *70*, 1600375. [CrossRef]
47. Zhang, Z.; Zhao, H.; Tang, J.; Li, Z.; Li, Z.; Chen, D.; Lin, W. A proteomic study on molecular mechanism of poor grain-filling of rice (*Oryza sativa* L.) inferior spikelets. *PLoS ONE* **2014**, *9*, e89140. [CrossRef]
48. Chen, G.-X.; Zhou, J.-W.; Liu, Y.-L.; Lu, X.-B.; Han, C.-X.; Zhang, W.-Y.; Xu, Y.-H.; Yan, Y.-M. Biosynthesis and regulation of wheat amylose and amylopectin from proteomic and phosphoproteomic characterization of granule-binding proteins. *Sci. Rep.* **2016**, *6*, 33111. [CrossRef]
49. Zhang, C.; Zhu, J.; Chen, S.; Fan, X.; Li, Q.; Lu, Y.; Wang, M.; Yu, H.; Yi, C.; Tang, S.; et al. Wx^{lv}, the ancestral allele of rice *Waxy* gene. *Mol. Plant* **2019**, *12*, 1157–1166. [CrossRef]
50. Seung, D.; Soyk, S.; Coiro, M.; Maier, B.A.; Eicke, S.; Zeeman, S.C. PROTEIN TARGETING TO STARCH is required for localising GRANULE-BOUND STARCH SYNTHASE to starch granules and for normal amylose synthesis in Arabidopsis. *PLoS Biol.* **2015**, *13*, e1002080. [CrossRef]

51. Zhong, Y.; Blennow, A.; Kofoed-Enevoldsen, O.; Jiang, D.; Hebelstrup, K.H. Protein Targeting to Starch 1 is essential for starchy endosperm development in barley. *J. Exp. Bot.* **2019**, *70*, 485–496. [CrossRef]
52. Lin, Q.; Huang, B.; Zhang, M.; Zhang, X.; Rivenbark, J.; Lappe, R.L.; James, M.G.; Myers, A.M.; Hennen-Bierwagen, T.A. Functional interactions between starch synthase III and isoamylase-type starch-debranching enzyme in maize endosperm. *Plant Physiol.* **2012**, *158*, 679–692. [CrossRef]
53. Crofts, N.; Abe, N.; Oitome, N.F.; Matsushima, R.; Hayashi, M.; Tetlow, I.J.; Emes, M.J.; Nakamura, Y.; Fujita, N. Amylopectin biosynthetic enzymes from developing rice seed form enzymatically active protein complexes. *J. Exp. Bot.* **2015**, *66*, 4469–4482. [CrossRef]
54. Tappiban, P.; Ying, Y.; Xu, F.; Bao, J.S. Proteomics and post-translational modifications of starch biosynthesis-related proteins in developing seeds of rice. *Int. J. Mol. Sci.* **2021**, *22*, 5901. [CrossRef]
55. Nakamura, Y. Towards a better understanding of the metabolic system for amylopectin biosynthesis in plants: Rice endosperm as a model tissue. *Plant Cell Physiol.* **2002**, *43*, 718–725. [CrossRef]
56. Grimaud, F.; Rogniaux, H.; James, M.G.; Myers, A.M.; Planchot, V. Proteome and phosphoproteome analysis of starch granule-associated proteins from normal maize and mutants affected in starch biosynthesis. *J. Exp. Bot.* **2008**, *59*, 3395–3406. [CrossRef]
57. Chao, Q.; Gao, Z.; Wang, Y.; Li, Z.; Huang, X.; Wang, Y.; Mei, Y.; Zhao, B.; Li, L.; Jiang, Y.; et al. The proteome and phosphoproteome of maize pollen uncovers fertility candidate proteins. *Plant Mol. Biol.* **2016**, *91*, 287–304. [CrossRef]
58. Stokes, M.P.; Farnsworth, C.L.; Moritz, A.; Silva, J.C.; Jia, X.; Lee, K.A.; Guo, A.; Polakiewicz, R.D.; Comb, M.J. PTMScan direct: Identification and quantification of peptides from critical signaling proteins by immunoaffinity enrichment coupled with LC-MS/MS. *Mol. Cell Proteom.* **2012**, *11*, 187–201. [CrossRef]
59. Chen, G.-X.; Zhen, S.-M.; Liu, Y.-L.; Yan, X.; Zhang, M.; Yan, Y.-M. In vivo phosphoproteome characterization reveals key starch granule-binding phosphoproteins involved in wheat water-deficit response. *BMC Plant Biol.* **2017**, *17*, 168. [CrossRef]
60. Willems, P.; Horne, A.; Van Parys, T.; Goormachtig, S.; De Smet, I.; Botzki, A.; Van Breusegem, F.; Gevaert, K. The Plant PTM Viewer, a central resource for exploring plant protein modifications. *Plant J.* **2019**, *99*, 752–762. [CrossRef]
61. Li, H.; Chen, Z.; Hu, M.; Wang, Z.; Hua, H.; Yin, C.; Zeng, H. Different effects of night versus day high temperature on rice quality and accumulation profiling of rice grain proteins during grain filling. *Plant Cell Rep.* **2011**, *30*, 1641–1659. [CrossRef] [PubMed]
62. Satoh, H.; Shibahara, K.; Tokunaga, T.; Nishi, A.; Tasaki, M.; Hwang, S.-K.; Okita, T.W.; Kaneko, N.; Fujita, N.; Yoshida, M.; et al. Mutation of the plastidial α-glucan phosphorylase gene in rice affects the synthesis and structure of starch in the endosperm. *Plant Cell* **2008**, *20*, 1833–1849. [CrossRef]
63. Young, G.-H.; Chen, H.-M.; Lin, C.-T.; Tseng, K.-C.; Wu, J.-S.; Juang, R.-H. Site-specific phosphorylation of L-form starch phosphorylase by the protein kinase activity from sweet potato roots. *Planta* **2006**, *223*, 468–478. [CrossRef]
64. Bao, J.; Sun, M.; Corke, H. Analysis of genotypic diversity in starch thermal and retrogradation properties in nonwaxy rice. *Carbohydr. Polym.* **2007**, *67*, 174–181. [CrossRef]
65. Wiśniewski, J.R.; Zougman, A.; Nagaraj, N.; Mann, M. Universal sample preparation method for proteome analysis. *Nat. Methods* **2009**, *6*, 359–362. [CrossRef]
66. Ma, J.; Chen, T.; Wu, S.; Yang, C.; Bai, M.; Shu, K.; Li, K.; Zhang, G.; Jin, Z.; He, F.; et al. IProX: An integrated proteome resource. *Nucl. Acids Res.* **2019**, *47*, D1211–D1217. [CrossRef]
67. Chen, C.; Chen, H.; Zhang, Y.; Thomas, H.R.; Frank, M.H.; He, Y.; Xia, R. TBtools: An integrative toolkit developed for interactive analyses of big biological data. *Mol. Plant* **2020**, *13*, 1194–1202. [CrossRef]

International Journal of
Molecular Sciences

MDPI

Article

Characterization of the Heat-Stable Proteome during Seed Germination in Arabidopsis with Special Focus on LEA Proteins

Orarat Ginsawaeng [1], Michal Gorka [1,2], Alexander Erban [1], Carolin Heise [1,3], Franziska Brueckner [1], Rainer Hoefgen [1], Joachim Kopka [1], Aleksandra Skirycz [1,4], Dirk K. Hincha [1,†] and Ellen Zuther [1,*]

1 Max Planck Institute of Molecular Plant Physiology, Am Muehlenberg 1, 14476 Potsdam, Germany; ginsawaeng@mpimp-golm.mpg.de (O.G.); michal.gorka@celonpharma.com (M.G.); Erban@mpimp-golm.mpg.de (A.E.); carolin.heise@uni-rostock.de (C.H.); brueckner@mpimp-golm.mpg.de (F.B.); hoefgen@mpimp-golm.mpg.de (R.H.); kopka@mpimp-golm.mpg.de (J.K.); skirycz@mpimp-golm.mpg.de (A.S.); hincha@mpimp-golm.mpg.de (D.K.H.)
2 Celon Pharma S.A., Marymoncka 15, Kazun Nowy, 05-152 Czosnow, Poland
3 Institute of Life Science, University of Rostock, Albert-Einstein-Str. 3, 18059 Rostock, Germany
4 Boyce Thompson Institute, Cornell University, Ithaca, NY 14853, USA
* Correspondence: zuther@mpimp-golm.mpg.de; Tel.: +49-331-5678251
† We dedicate this paper to the memory of Dirk K. Hincha.

Citation: Ginsawaeng, O.; Gorka, M.; Erban, A.; Heise, C.; Brueckner, F.; Hoefgen, R.; Kopka, J.; Skirycz, A.; Hincha, D.K..; Zuther, E. Characterization of the Heat-Stable Proteome during Seed Germination in Arabidopsis with Special Focus on LEA Proteins. *Int. J. Mol. Sci.* **2021**, *22*, 8172. https://doi.org/10.3390/ijms22158172

Academic Editors: Sixue Chen

Received: 8 June 2021
Accepted: 28 July 2021
Published: 29 July 2021

Abstract: During seed germination, desiccation tolerance is lost in the radicle with progressing radicle protrusion and seedling establishment. This process is accompanied by comprehensive changes in the metabolome and proteome. Germination of Arabidopsis seeds was investigated over 72 h with special focus on the heat-stable proteome including late embryogenesis abundant (LEA) proteins together with changes in primary metabolites. Six metabolites in dry seeds known to be important for seed longevity decreased during germination and seedling establishment, while all other metabolites increased simultaneously with activation of growth and development. Thermo-stable proteins were associated with a multitude of biological processes. In the heat-stable proteome, a relatively similar proportion of fully ordered and fully intrinsically disordered proteins (IDP) was discovered. Highly disordered proteins were found to be associated with functional categories development, protein, RNA and stress. As expected, the majority of LEA proteins decreased during germination and seedling establishment. However, four germination-specific dehydrins were identified, not present in dry seeds. A network analysis of proteins, metabolites and amino acids generated during the course of germination revealed a highly connected LEA protein network.

Keywords: late embryogenesis abundant; Arabidopsis; seed germination; metabolomics; heat-stable proteome; dehydrins; LEA transcripts; intrinsically disordered proteins

1. Introduction

Seeds are the desiccation-tolerant life stage of most plant species. However, seedlings lose their desiccation tolerance during the early stages of germination [1–3]. In the vegetative stage, with the exception of a small number of resurrection plants, the vast majority of species are desiccation-sensitive (see [4] for a recent review).

During germination, the quiescent embryo resumes growth and development, resulting in seedling establishment under favorable environmental conditions. The process of germination is initiated upon imbibition and has been divided into three phases [5,6]. Phase I consists of a rapid uptake of water which allows the initiation of metabolic activity, such as protein synthesis from stored mRNAs. During Phase II, further water uptake is limited. Phenotypically, testa and endosperm rupture are observed, followed by radicle protrusion, which marks the end of this phase. Gene transcription from newly synthesized mRNAs and protein synthesis results in further activation of metabolism. In Phase III, also referred to as the post-germination phase, water uptake increases again and seedling development is observed. This includes cell division, elongation of the radicle and mobilization of stored reserves to fuel seedling development before plants become photoautotrophic. After emergence of the radicle, seedling establishment progresses by emergence of hypocotyl and cotyledons and a new growth stage begins with the full opening of the cotyledons [7].

This complex developmental process has been investigated on metabolic [8–11], transcriptomic [10,12–17] and proteomic [18–26] levels. In all cases, distinct regulatory patterns have been observed for different metabolites, transcripts and proteins during germination and seedling development in different species. Among the metabolites that are typically found in high concentrations in dry seeds are sucrose and raffinose-family oligosaccharides (RFO) such as raffinose, stachyose and verbascose (see [27] for review). These sugars are mainly responsible for the formation of a cytosolic glass phase during maturation drying, which protects the cells from damage as water content declines [28–30]. Obviously, this process needs to be reversed during germination to allow full metabolic activity.

In addition to seed storage proteins, late embryogenesis abundant (LEA) proteins are highly abundant in dry seeds [31]. They were first discovered in cotton seeds during the late stages of seed development [32–34]. Besides seeds, they are also found in vegetative plant organs, mainly in response to dehydration stress. For example, it has been shown that LEA proteins can protect enzymes and membranes during freezing and drying [35–38]. LEA proteins are also found in other organisms including algae, anhydrobiotic invertebrates and bacteria [35,37,39]. In plants, LEA proteins have been linked to the desiccation tolerance of seeds as their accumulation coincides with the development of desiccation tolerance in orthodox seeds and their accumulation is decreasing in germinating seeds when they become desiccation sensitive [18,40,41]. In Arabidopsis, the transcripts encoding several LEA proteins have also been detected in dry, mature seeds [42]. However, the role of specific LEA proteins during seed germination is not known yet. A detailed temporal proteome analysis during seed germination and seedling establishment will help shed light on their importance during these processes.

Most LEA proteins contain a high fraction of hydrophilic amino acid residues and have been computationally and experimentally classified as intrinsically disordered proteins (IDPs) [38,42–45]. This lack of folding in solution and the low fraction of hydrophobic amino acids make most LEA proteins heat-stable [35–38]. This property has been used, on the one hand, for the purification of recombinant LEA proteins (see, e.g., [45,46] for reviews), and on the other hand, to enrich LEA proteins from seed extracts of *Medicago truncatula* in a heat-stable proteome [41,47]. In *Arabidopsis thaliana*, a total of 51 LEA proteins across nine different groups based on Pfam family domains have been identified, of which 34 were classified as seed-expressed based on the presence of the corresponding mRNAs in dry seeds [42].

Here, we explored the heat-stable proteome in dry and germinating seeds and young seedlings of Arabidopsis with a special focus on low abundant proteins. The assignment of heat-stable proteins to functional metabolic categories was investigated. LEA proteins and the abundance of LEA transcripts was analyzed using proteomics and qRT-PCR approaches, respectively. In addition, we investigated the dynamics of the seed metabolome for comparison with previous publications to improve our understanding of the complex metabolic changes taking place during germination and seedling establishment.

2. Results

To investigate the early stages of germination in Arabidopsis, seeds were imbibed and then subsequently collected at different time points. We collected dry seeds (DS) and germinating seeds and seedlings (Supplemental Figure S1). All treatments, except DS, also experienced a stratification period, which might have an additional impact on the results. After imbibition, we observed testa rupture (24 HAI; hours after imbibition), elongation of radicles (36 HAI), greening of cotyledons (48 HAI), hypocotyl and cotyledon emergence (60 HAI) and fully opened cotyledons (72 HAI).

2.1. Dynamics of Seed Metabolites and Amino Acids after Imbibition

As metabolic levels during seed germination of Arabidopsis were previously reported [8,10,11], we firstly characterized the molecular status during imbibition on the metabolic level for potential comparisons and additional evaluation of the complex changes.

Small hydrophilic metabolites were analyzed using gas chromatography-mass spectrometry analysis. After excluding known contaminants and potential contaminants identified by hierarchal cluster analysis, the data set contained 100 metabolites (see Supplemental Table S1 for the complete data set). More than half of the metabolites detected in all time points increased in abundance during germination and seedling establishment. Six metabolites (A148003, A225007, erythronic acid, mannitol, uracil and uric acid) were only found in DS.

Principal component analysis (PCA) showed a clear separation between the metabolite composition of DS and all other time points along principal component 1 (PC1), which explained 61.8% of the total variance in the data set (Figure 1A). The different time points after imbibition were mainly separated by PC2, explaining 17.7% of the total variance, indicating that imbibition had the most drastic effect on metabolite composition, compared to further germination. Data obtained from some data points clustered more closely together, suggesting that the metabolite composition was similar in testa ruptured and radicle protruding seeds (24 and 36 HAI) and also before and after cotyledon opening in seedlings (60 and 72 HAI).

To resolve the different kinetic patterns of the metabolites (Supplemental Figure S2), we subjected the data of those that were detected at all time points (70% of the data set) to k-means clustering (Figure 2A). We observed six clearly resolved kinetic patterns, four of which showed an overall increase in metabolite content over 72 HAI. Cluster 1 and cluster 4 represented metabolites with a gradual increase over time from DS to 72 HAI with a higher increase in cluster 4. Similarly, metabolites in cluster 5 were highly increased from 24 HAI onwards. Cluster 6 was characterized by a higher abundance of the respective metabolites already in DS and more gradual changes over 72 HAI. The largest cluster 2 represented metabolites with only small fluctuations, especially at early time points after imbibition. The smallest cluster 3, contained metabolites with a high abundance in dry seeds and a steep decrease after 24 HAI or 36 HAI.

Metabolites of the six clusters were assigned to nine metabolite classes (Figure 2B). The most diverse clusters were cluster 1 (acids, unknowns (MSTs; mass spectral tags), N-compounds, phosphates, polyols and sugars) and cluster 2 (acids, MSTs, N-compounds, phenylpropanoids, sugar conjugates and sugars). Sugars was the only class found in all clusters, with the highest proportion in cluster 5 (40%). Acids were detected in almost every cluster except for cluster 4. MSTs, found in all clusters except for cluster 3, was the class with the highest number of members in the clustered data set (42.9%). Phenylpropanoids were exclusively observed in cluster 2.

The abundance of 20 amino acids in DS, germinating seeds and seedlings was measured using reverse High-Performance Liquid Chromatography (HPLC) (Supplemental Table S2). A PCA plot of amino acid concentrations (Figure 1B) showed a similar pattern to the PCA of metabolites. A separation of DS samples from other samples collected 24–72 HAI could be observed by PC1, which explained 72.6% of the total variance. Mainly PC2, explaining 15.9% of the total variance, separated samples collected after seed germination. The changes in amino acids during seed germination clustered into four groups according to their kinetic patterns (Figure 3). It revealed that most amino acids (75%) increased in abundance over time from DS to seedlings (Figure 3A). Amino acids with increased abundance clustered into cluster 2 and also cluster 3 with the largest number of members (55%) and a more pronounced increase (Figure 3B). The abundance of amino acids in cluster 1 slightly fluctuated during DS-to-seedling transition. Lastly, the abundance of γ-aminobutyric acid (gABA) as the only member of cluster 4 decreased drastically 24 HAI and recovered 48 HAI.

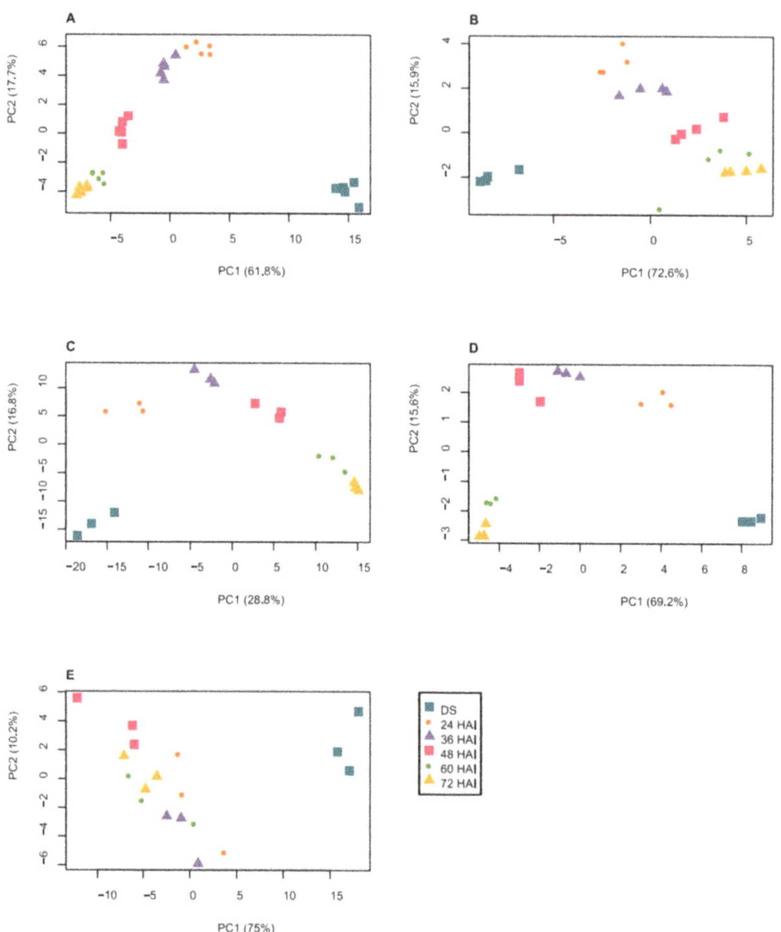

Figure 1. PCA of (**A**) metabolites, (**B**) amino acid composition, (**C**) heat-stable proteome, (**D**) LEA proteins and (**E**) expression of genes encoding LEA proteins of dry seeds (DS), germinating seeds and seedlings at different hours after imbibition (HAI). Data are log_2-transformed, median-normalized mass spectral intensities (n = 5) (**A**), concentration of amino acid (nmol/gDW) after log_2 transformation and median normalization (n = 4) (**B**), mass spectral label-free quantification (LFQ) intensities after log_2 transformation and median normalization (n = 3) (**C**,**D**) or DW-normalized absolute transcript copy number in log_2 with median normalization (n = 3) (**E**). Missing values were replaced with 0 prior to the analysis. Samples at different HAI are as indicated by the legend.

2.2. Changes in the Heat-Stable Proteome during Germination and Seedling Development

The strong domination of seed storage proteins in dry seeds [48] makes it difficult to detect lower abundant proteins, e.g., LEA proteins. Thus, the heat-stable proteome was characterized for the detection of lower abundant proteins, especially chaperones, which may be important for seed germination and seedling establishment. We used a heat treatment to remove seed storage proteins and other heat-sensitive proteins before analyzing the composition of the remaining heat-stable proteome in seeds before and after imbibition by Liquid Chromatography-Mass Spectrometry (LC-MS/MS) analysis of proteolytic peptides. This proteomic approach revealed the presence of 1545 different proteins. After removing known contaminants, proteins represented by only one unique peptide and proteins with missing values in at least two of the three replicates in all time points, 898 proteins remained in the data set (see Supplemental Table S3 for the complete data set). Not all proteins of the heat-stable proteome were detectable at every time point—72.6%, 45.3%, 31.3%, 22.3%, 28.4% and 24.3% of the total number of proteins were missing from the samples collected at DS, 24, 36, 48, 60 and 72 HAI, respectively.

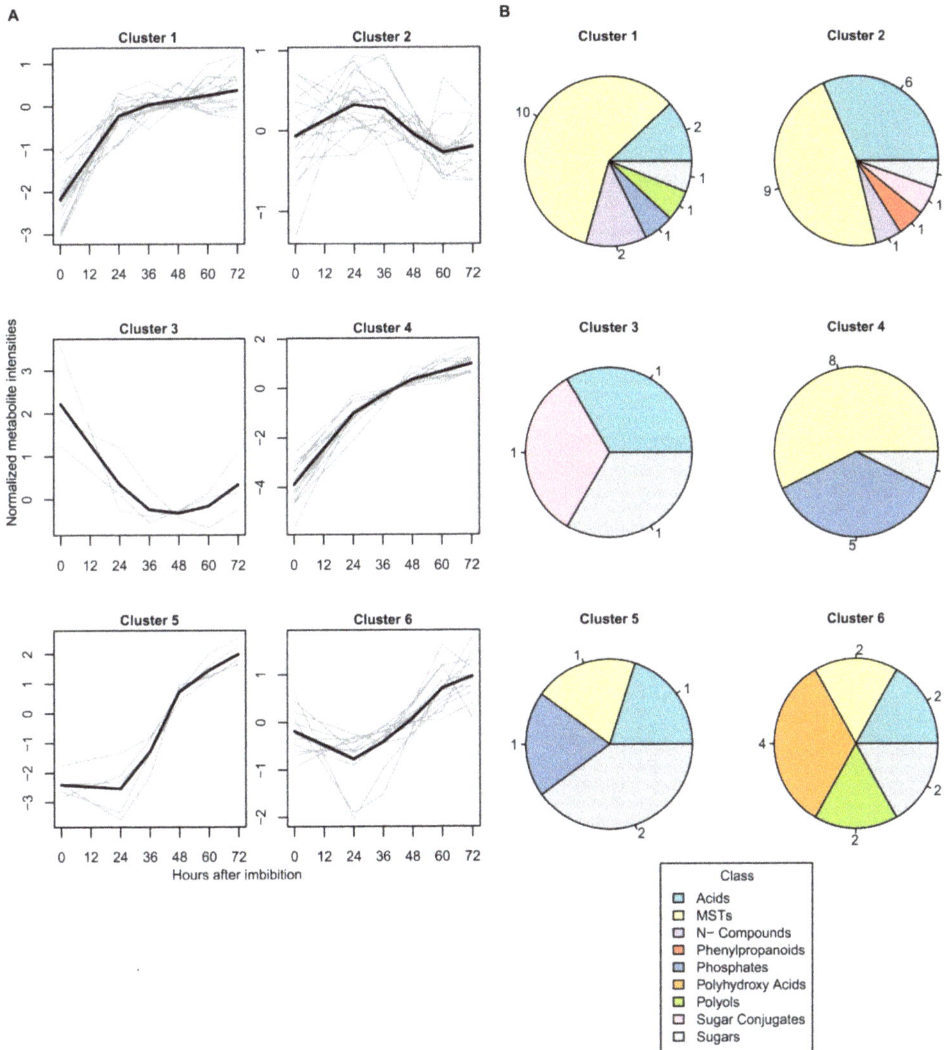

Figure 2. Patterns of metabolite abundance as determined by k-means clustering (clusters 1–6) (see Supplemental Figure S2) (**A**). Gray lines represent individual metabolites while black lines show the mean of each cluster. (**B**) Distribution of metabolite classes in each cluster. Amino acids were considered separately. Each chemical class is represented by a unique color as shown (MSTs = mass-spectral tags). Numbers around the pie charts represent the numbers of class members.

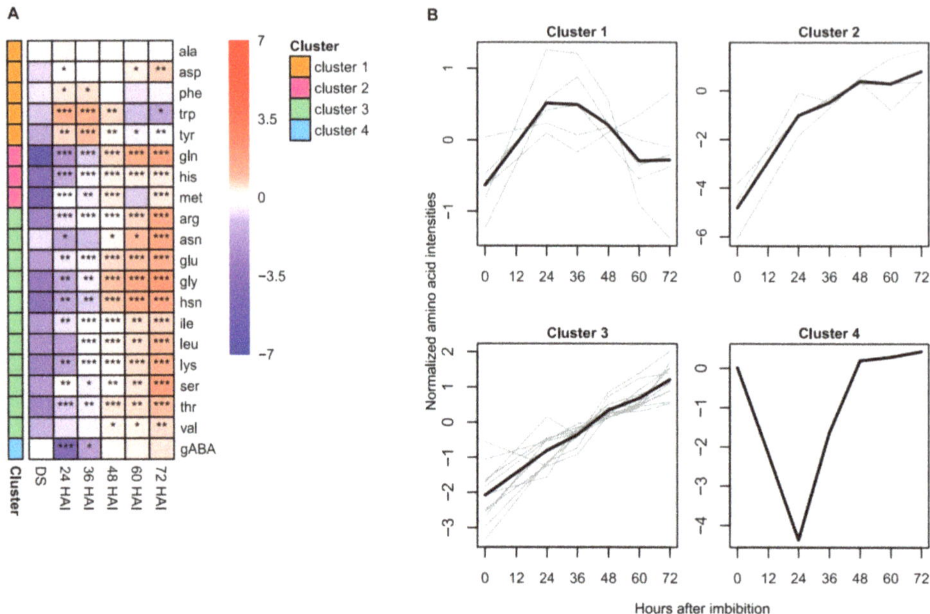

Figure 3. Amino acid levels in dry seeds (DS), germinating seeds and seedlings and their kinetic patterns. (**A**) Abundance of amino acids in DS, germinating seeds and seedlings collected at the indicated time points after imbibition. The color gradient represents mean concentration of amino acid (nmol/gDW) after log_2 transformation and median normalization ($n = 4$). Levels of significance comparing the amino acids in germination seeds and seedlings to DS (unpaired two-sided *t*-test) are indicated by asterisks (* $p < 0.05$; ** $p < 0.01$; *** $p < 0.001$). Amino acids are listed alphabetically within their clusters (cluster 1–4) as determined by k-means clustering. Each cluster is indicated by a unique color. (**B**) Patterns of amino acid abundance as determined by k-means clustering. Gray lines represent individual metabolites while black lines show the mean of each cluster. gABA–γ-aminobutyric acid.

PCA revealed a similar separation of the heat-stable proteins from different time points after imbibition (Figure 1C) as observed for amino acids (Figure 1B). PC1 separated DS from germinating seeds and seedlings, explaining 28.8% of the total variance in the proteomic data set, while PC2 (16.8% of the total variance) separated samples at different time points after imbibition. Data obtained from seedlings 60 and 72 HAI clustered closely together, suggesting that protein composition remained similar in seedlings once cotyledons had emerged.

The heatmap of 898 heat-stable proteins (Supplemental Figure S3) revealed that most of the proteins had one directional change in abundance, i.e., either increased or decreased during germination and seedling establishment. Some clusters showed also a transient increase for 24 to 48 HAI. The number of detected proteins in each time point is shown in a Venn diagram using proteins from DS, 24 HAI, 36 together with 48 HAI and 60 together with 72 HAI data sets (Figure 4A, Supplemental Table S4). The total number of detected proteins is increasing over time. The two time points with the least number of common proteins were DS and 60 and 72 HAI (193 proteins). On the contrary, the highest common protein number was found when overlapping 36 and 48 HAI proteins and 60 and 72 HAI proteins (613 proteins). After excluding eight proteins, which were only present in either one of the combined time points (marked in yellow in Supplemental Table S4a), we were left with 182 common heat-stable proteins that were found throughout germination and early seedling development. Among them, CpHsc70-1 (AT4G24280) and LEA20 (AT2G40170) had the highest increase and decrease in abundance in 72 HAI seedlings compared to DS, respectively (Supplemental Table S3).

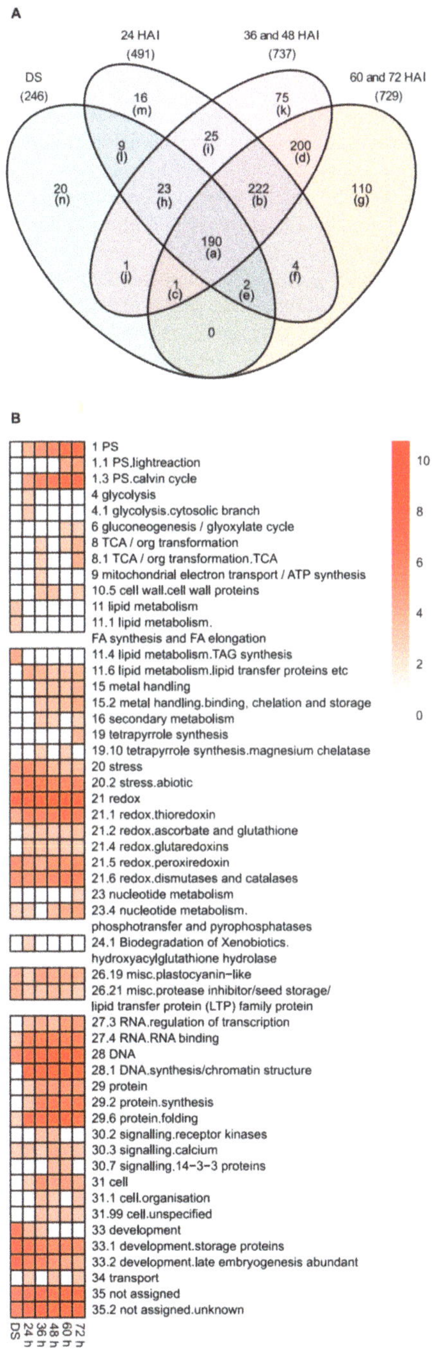

Figure 4. Heat-stable proteins in dry seeds (DS), germinating seeds and seedlings. (**A**) Venn diagram showing common and unique heat-stable proteins at the indicated time points after imbibition. (**B**) Functional enrichment of MapMan terms in the heat-stable proteome of DS, germinating seeds and seedlings. The white/red color gradient represents z-scores calculated from p-values ($p < 0.01$).

To explore the assignment of the 898 heat-stable proteins to main metabolic pathways and other processes we performed a functional enrichment analysis (Figure 4B, Supplementary Table S5) as statistical measure of whether a particular functional group (bin) of proteins contains more up- or down-regulated proteins than expected if all detected proteins would be equally distributed among all bins. We observed significant enrichment of 23 MapMan bins in DS, germinating seeds and seedlings. Nine bins were enriched in all observed time points; stress (20) with sub-bin abiotic stress (20.2), redox (21) with sub-bins thioredoxin (21.1), peroxiredoxin (21.5), dismutases and catalases (21.6), misc-related proteins including plastocyanin-like (26.19) and inhibitor/seed storage/lipid transfer protein (LTP) family protein (26.21), RNA binding (27.4), DNA (28), protein folding (29.6), signaling calcium (30.3), development-related proteins including storage protein (33.1) and late embryogenesis abundant proteins (33.2) and not assigned (35 and 35.2). Among these, redox, DNA and not assigned bins were highly enriched throughout the observation window. RNA binding (27.4) and protein folding (29.6), on the other hand, were highly enriched after 24 HAI.

Proteins from some bins and sub-bins were absent in DS but found in germinating seeds and young seedlings. These bins were PS (1) with sub-bin Calvin cycle (1.3), lipid transfer proteins (11.6), ascorbate and glutathione (21.2) and glutaredoxins (21.4), regulation of transcription (27.3), DNA synthesis/chromatin structure (28.1), protein (29) and its sub-bin synthesis (29.2) and cell (31). Enrichment of PS and Calvin cycle proteins clearly increased with HAI where it reached its peaks in young seedlings, coinciding with the transition from heterotrophic seeds to autotrophic seedlings. Most bins were enriched at more than one observed time point. However, proteins belonging to lipid metabolisms, (11) including FA synthesis and elongation (11.1) and TAG synthesis (11.4), were enriched only in DS while the enrichment of tetrapyrrole synthesis (19) and nucleotide (23) proteins was specific to seedlings 72 HAI.

Additionally, kinetic patterns within the heat-stable proteome were investigated for the 182 proteins, which were common among DS and all samples after imbibition using k-means clustering. Clustering revealed seven patterns of changes in protein abundance (Supplemental Figure S4A) which contained proteins assigned to 21 bins (Supplementary Figure S4B). The five bins with the highest proportion were: not assigned (17.6%), proteins (synthesis/post-translational modification/degradation/folding) (12.6%), redox (11.0%), development (9.90%) and stress (8.24%).

Among the clusters with a larger number of members (clusters 2, 3, 4 and 5), no single functional bin dominated. Excluding cluster 6 with only four members, the most common bins found across all clusters were stress (six clusters) and proteins (five clusters). Some bins were shared among clusters with the same protein abundance pattern. Heat-stable proteins that belonged to the PS, proteins and stress bins were found in all clusters with an overall increase in protein abundance after imbibition (clusters 1, 2 and 4). On the other hand, development was the common bin among clusters with a decrease in abundance (clusters 3, 6 and 7).

2.3. *The Heat-Stable Proteome Exhibited an Equal Distribution of Intrinsically Ordered and Disordered Proteins*

Since it is well-established that most LEA proteins are intrinsically disordered proteins [38] and that these proteins are heat-stable [49], we analyzed these sequences within the heat-stable proteome to predict disorder content in each protein using the MFDp2 online tool [50] (Table 1, Supplemental Table S3). Of the 898 heat-stable proteins, 701 amino acid sequences were retrieved from the UniProtKB/Swiss-Prot database (Supplemental Table S3). Various degrees of disorder content were found in the heat-stable proteins, from 0% disorder (12%) to 100% (10.8%). Hence, an almost equal fractioning of fully ordered and fully intrinsically disordered proteins (IDP) was observed. Proteins with a predicted disorder content between 0 and 10% comprised the biggest group in our data set (20.5%) followed by the group with a predicted disorder content of 10–20% (15.8%). The protein number tended to be smaller with increasing degree of disorder. In fact, proteins with at least 50% disorder content comprised only 21.8% of the whole data set, clearly indicating that disorder is not a prerequisite for heat stability. Likewise, small and large proteins were found in all disorder categories. The mean size of proteins was similar among most categories. The smallest and largest means were found in 70–80% and 90–100% disorder predicted protein, respectively (199.1 vs. 461.4 residues).

Table 1. Heat-stable proteins with different predicted disorder content. A total of 701 protein sequences were retrieved from UniProtKB/Swiss-Prot. The prediction was performed by MFDp2. The table shows protein counts and other information (Min = minimum, Max = maximum, Mean of length of residues in each group of predicted disorder content).

Predicted Disorder Content (%)	Counts	Min	Max	Mean
			(Residues)	
0	84	80	843	255.6
0–10	144	56	990	342.3
10–20	111	82	974	310.5
20–30	97	69	976	348.9
30–40	62	87	987	341
40–50	50	72	953	309.8
50–60	27	98	724	294.2
60–70	18	93	772	303.4
70–80	9	129	367	199.1
80–90	7	111	577	277
90–100	16	131	891	461.4
100	76	62	956	284.9

To investigate if the grade of disorder is related to functional categories, the protein abundance for proteins with 0% and 100% disorder content was exemplarily assigned to MapMan bins using TAIR10 annotations [51] during the investigated germination time-course (Figure 5, Supplemental Table S6). While there was no distinct pattern found among fully folded proteins in terms of protein abundance among the bins identified, we could see some pattern among the fully disordered proteins. For example, there was an increasing abundance within 72 HAI for fully disordered proteins in the following bins: cell, protein, signaling and transport. The majority of these proteins was undetected in DS and early stages of germination, and therefore, a significance test to protein abundance in DS was often impossible due to missing values at this condition. Significant decrease in abundance of 100% disordered proteins within 72 HAI was found in the bins development and hormone metabolism.

2.4. LEA Protein Dynamics during Seed Germination

From the 898 heat-stable proteins in the data set, 29 LEA proteins in eight of the nine LEA Pfam groups [42] were identified (Supplemental Table S7). Only members of the Pfam LEA_3 group were not detected. The Pfam group with the largest number of members was LEA_4 (37.9%) followed by dehydrin (20.7%) and seed maturation protein (SMP) (10.3%). Out of 29 LEA proteins in the heat-stable proteome, 17 sequences were retrieved from the UniProtKB/Swiss-Prot database. The length of these LEA proteins ranged between 97 and 479 amino acids. Most of them (76.5%) were predicted by MFDp2 to have 100% disorder content (Supplemental Table S7). Only four of them were predicted with a smaller fraction of disorder content: LEA21 with 16.4%, LEA22 with 12.1%, LEA31 with 24% and LEA32 with 48.8%.

To observe whether the high proportion of fully disordered LEA proteins is only found in the heat-stable proteome, we investigated sequences of LEA proteins which were reported to be seed-expressed [42] but were not detected in our data set Supplemental Table S8). We were able to retrieve 10 of these from the UniProtKB/Swiss-Prot database and found that 30% of them were predicted to be 100% disordered. The missing LEA protein with the lowest disorder content was LEA50 with 0%, followed by LEA1 (3.3%) and LEA27 (13.9%).

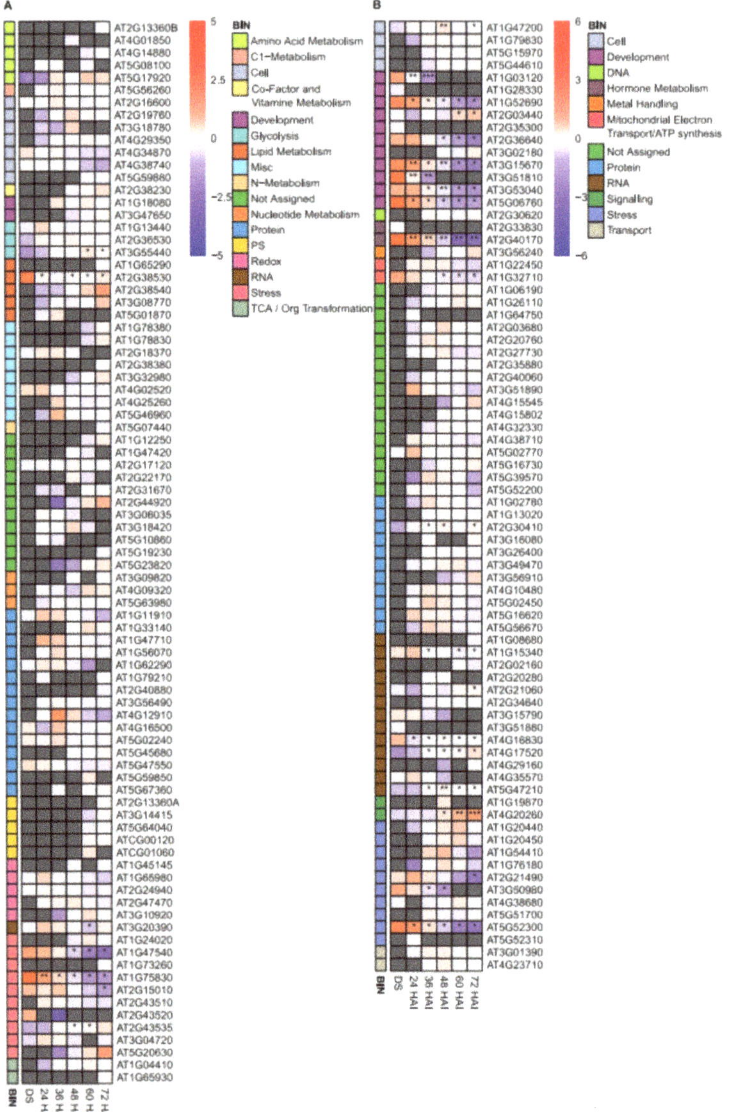

Figure 5. Abundance of proteins with (**A**) 0% and (**B**) 100% disorder content in the heat-stable proteome in dry seeds (DS), germinating seeds and seedlings. The color gradient represents the $\log_2$-median transformed means of mass spectral LFQ intensities (n = 3). Gray boxes represent missing values at the indicated time point. Levels of significance comparing protein abundance in germinating seeds and seedlings to DS (unpaired two-sided t-test) are indicated by asterisks (* $p < 0.05$; ** $p < 0.01$; *** $p < 0.001$). Proteins are assigned to MapMan bins, each bin is indicated by a unique color as shown. The transcript AT2G13360 mapped to two bins and was therefore marked with A and B. For additional information, see Supplemental Table S6.

A PCA of the LEA protein data set (Figure 1D) showed a clear separation between samples from different time points. Similar to the heat-stable proteome data set, PC1 explained 69.2% of the total variance and separated DS from all time points after imbibition, while PC2 (15.6% of the total variance) separated germinating seeds and young seedling samples.

Most of the detected LEA proteins (79.3%) were found in DS (Figure 6A) and most of these showed a significant reduction in abundance after imbibition that became more pronounced with time. Some of these LEA proteins (34.5% and 37.9%) were no longer detecTable 60 and 72 HAI. Noticeably, three LEA proteins (LEA18 from LEA_1 and LEA21 and LEA22 from the AtM Pfam groups) were specific to DS. On the other hand, six LEA proteins (LEA4, LEA5, LEA8 and LEA10 from the dehydrin, LEA12 from the LEA_4 and LEA26 from the LEA_2 Pfam groups) became only detectable in germinating seeds from 24 or 36 HAI onwards. Thus, they were referred to as germination-specific. LEA12, however, was already undetectable again 60 HAI, indicating a very specific function during early germination.

To observe patterns of changes in LEA protein abundance, k-means clustering was applied to 13 LEA proteins found across all time points (Figure 6B). Although members of all clusters were reduced after imbibition, three clusters of LEA proteins from different Pfam groups with different kinetics were identified (Figure 6C). LEA_4 proteins were the most commonly found in the cluster data set with 69.2%. Proteins in cluster 1 showed a constant gradual decrease. Proteins in the largest cluster 2 (53.8% of cluster data set) showed a steady abundance up to 24 HAI, followed by a gradual decrease except for LEA16. Proteins in cluster 3 largely decreased in abundance up to 60 HAI with little further change over the last 12 h. LEA20 was the LEA protein with the largest reduction in abundance between DS and 72 HAI among all clustered proteins.

2.5. Transcript Abundance of Genes Encoding LEA Proteins

While we detected 29 LEA proteins in the heat-stable proteome, transcripts encoding 34 LEA proteins were previously detected by qRT-PCR in dry Arabidopsis seeds [42]. Here, we investigated the transcript levels of these LEA genes in DS and after imbibition along with genes encoding LEA proteins in the heat-stable proteome that were not among those 34. Therefore, we investigated in total 41 LEA genes. The abundance of LEA transcripts was expressed in absolute copy number of transcripts normalized to dry weight (DW) of samples used for RNA extraction (copy/gDW). Copy numbers of LEA transcripts were highest in DS, with a maximum average of 3.38×10^{13} copy/gDW for LEA20, and lowest in seedlings 72 HAI, with a minimum average of 9.47×10^{6} copy/gDW for LEA50 (see Supplemental Table S9 for the complete data set). We detected transcripts for LEA1, LEA6, LEA9, LEA13, LEA27, LEA33, LEA38, LEA41, LEA45, LEA49, LEA50 and LEA51 (for LEA27, LEA38 and LEA41 with higher abundance at 36 and 60 DAI) even though no corresponding protein was detected (Figure 7A).

A PCA of the LEA transcript data (Figure 1E) showed a separation between DS and all time points after imbibition by PC1, which explained 75% of the total variance. However, unlike for the LEA protein data (Figure 1D), there was no clear separation of the time points after imbibition by PC2, suggesting that the kinetics of LEA transcript appearance were less defined than those of the proteins.

K-means clustering showed three different transcript abundance patterns of the observed LEA transcripts (Figure 7B). Due to presence of missing values, LEA5, LEA12 and LEA45 were excluded from k-means clustering. All LEA Pfam groups were represented across the three clusters (Figure 7C). Cluster 1 comprised LEA transcripts that increased in abundance after imbibition. The increase was gradual up to 36 HAI, and then transcript levels generally remained constant. The transcript of LEA38 was the only one in cluster 1, which did not encode germination-specific LEA proteins as described above. Cluster 2 contained LEA transcripts with similar abundance over time and only small fluctuations. Cluster 3 contained transcripts with reduced abundance after imbibition with the lowest abundance at 48 HAI. This cluster represented the majority of the observed LEA transcripts

(68.4% of cluster data set) and was dominated by LEA_4 transcripts (38.5%). Cluster 3 was the most diverse cluster consisting of LEA transcripts from 7 Pfam groups, only missing members from LEA_2 and LEA_3 groups, which were both detected in cluster 1 and cluster 2 (Figure 7C). Cluster 3 was the only cluster with LEA transcripts from AtM, LEA_1, LEA_5 and SMP Pfam groups.

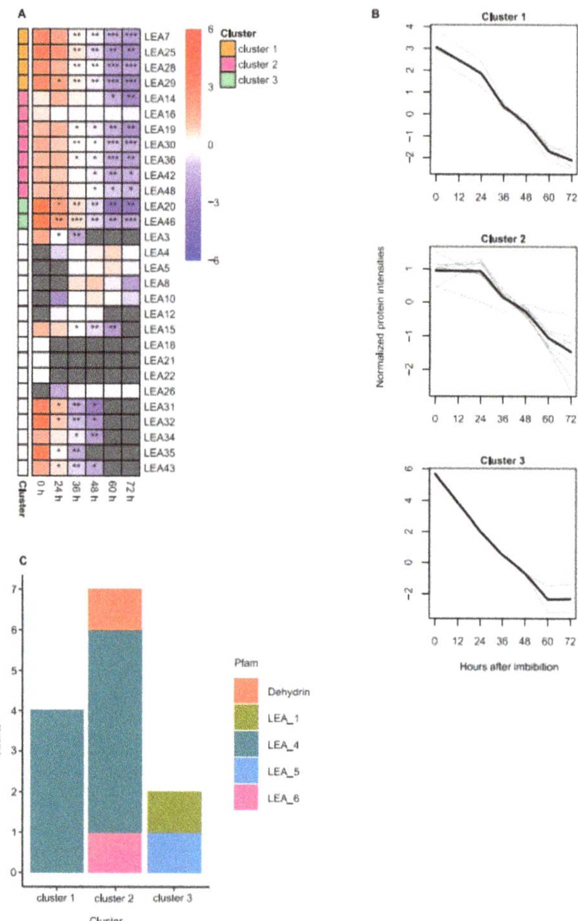

Figure 6. LEA protein abundance in dry seeds (DS), germinating seeds and seedlings and their kinetic patterns and Pfam distributions. (**A**) LEA protein abundance in DS and at different time points after imbibition. The color gradient represents the $\log_2$-median transformed means of mass spectral LFQ intensities (n = 3). Gray boxes represent missing values at the indicated time point. Levels of significance comparing LEA protein abundance in germinating seeds and seedlings to DS (unpaired two-sided t-test) are indicated by asterisks (* $p < 0.05$; ** $p < 0.01$; *** $p < 0.001$). LEA proteins are listed according to their LEA ID and corresponding clusters (cluster 1–3) as determined by k-means clustering. Each cluster is represented by a unique color. LEA proteins excluded from k-means clustering are depicted in white. (**B**) Patterns of LEA protein abundance as determined by k-means clustering. Gray lines represent individual proteins while black lines show the mean of each cluster. (**C**) Counts of LEA proteins in different Pfam groups in each cluster. Each Pfam group is represented by a unique color.

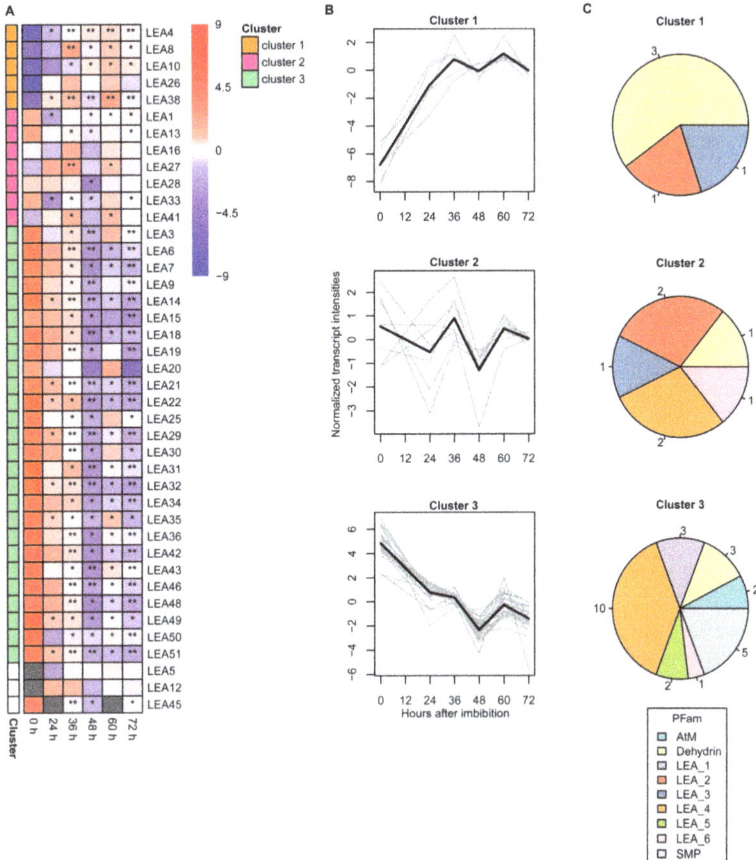

Figure 7. Transcript levels of genes encoding LEA proteins in dry seeds (DS), germinating seeds and seedlings after imbibition. (**A**) LEA transcripts at different time points after imbibition. The color gradient represents means of DW-normalized absolute transcript copy number in $\log_2$ with median normalization ($n = 3$). Gray boxes represent missing values at the indicated time point. Level of significance comparing LEA transcript abundance in germinating seeds and seedlings to DS (unpaired two-sided t-test) are indicated by asterisks (* $p < 0.05$; ** $p < 0.01$; *** $p < 0.001$). LEA transcripts are listed according to their LEA ID and corresponding clusters as determined by k-means clustering. Each cluster is represented by a unique color. LEA transcripts excluded from k-means clustering are depicted in white. (**B**) Patterns of LEA transcript abundance as determined by k-means clustering. Gray lines represent individual transcripts while black lines show the mean of each cluster. (**C**) Distribution of Pfam of LEA transcripts in each cluster. Each Pfam is represented by a unique color. Numbers in pie charts represent the numbers of Pfam members.

Since we determined both LEA protein and transcript abundances, we were able to investigate the correlation between mRNA and protein levels. We only investigated the correlations between 23 LEA proteins and their transcripts when detected in at least four of the six observed time points. Significant correlations were found in nine of these twenty-three cases, which came from four different Pfam groups of LEA_1, LEA_4, dehydrin and SMP (Table 2). The highest correlation between the abundance of a protein and the corresponding transcript was found for LEA32 from SMP Pfam group. Of the nine significantly correlated pairs of proteins and transcripts, only the germination-specific dehydrins LEA4 and LEA10 were upregulated, while the other seven were downregulated during germination. A delay in protein accumulation following the transcript accumulation was observed for some LEA proteins. For example, for LEA10, a delay of at least 24 h has been found between the increase in transcript and protein detection.

Table 2. Pearson correlation between the abundance of LEA proteins and their corresponding transcripts. LEA proteins with missing means mass spectral LFQ intensities in more than two observed time points were excluded from analysis. Levels of significance of the correlations are indicated by asterisks (** $p < 0.01$; *** $p < 0.001$).

LEA ID	Pfam (Hundertmark and Hincha, 2008)	Pearson Correlation	p value
LEA4	dehydrin	0.703 **	0.010
LEA5	dehydrin	−0.091	0.839
LEA7	LEA_4	0.669 **	0.009
LEA8	dehydrin	0.052	0.873
LEA10	dehydrin	0.789 **	0.005
LEA14	dehydrin	0.212	0.577
LEA15	LEA_6	0.581	0.061
LEA16	LEA_6	−0.169	0.664
LEA19	LEA_4	0.461	0.131
LEA20	LEA_5	0.250	0.507
LEA25	LEA_4	0.409	0.191
LEA26	LEA_2	0.050	0.873
LEA28	LEA_4	0.070	0.839
LEA29	LEA_4	0.695 **	0.006
LEA30	LEA_4	0.297	0.393
LEA31	SMP	0.839 **	0.005
LEA32	SMP	0.934 ***	0.0003
LEA34	dehydrin	0.813 **	0.006
LEA36	LEA_4	0.127	0.715
LEA42	LEA_4	0.346	0.310
LEA43	LEA_4	0.777 **	0.009
LEA46	LEA_1	0.824 ***	0.0004
LEA48	LEA_4	0.243	0.507

2.6. Protein–Metabolite Network Analyses Reveal a LEA Cluster with High Connectivity

As the majority of LEA proteins common for all time points showed a fast decrease during seed germination and seedling establishment, we were interested in other proteins and metabolites following the same pattern to identify highly correlated protein or protein–metabolite networks associated with seed germination. With this approach, we wanted to integrate omics data for an interpretable network for the discovery of molecular interactions previously not discovered for heat-stable proteins under these conditions. A protein–metabolite network analysis was performed, including all heat stable proteins (182) together with all metabolites (70) and amino acids (20) common for all six time points. For reduction in edges, a cut off ($r \geq 0.99$ and a p-value < 0.05) was chosen (Supplemental Table S10). The identified network included 91 proteins (including 11 LEA proteins), 42 metabolites and 13 amino acids with a varying range of correlations and a high connectivity level (Supplemental Figure S5). LEA proteins (colored in purple) had multiple relationships to each other and revealed positive connections building a small sub-cluster with LEA7, LEA19, LEA25, LEA28, LEA29, LEA30, LEA36, LEA42 and LEA48. Among them, LEA7, LEA25, LEA28, LEA29 and LEA30 were identified as network hubs with a high level of inter-connectivity represented by eight to eleven edges. These LEA proteins were closely correlated with several heat-stable proteins as the seed storage proteins SESA1, a bi-functional inhibitor/lipid-transfer protein/seed storage 2S albumin superfamily protein (AT4G30880), the ABA-inducible dehydration-responsive protein RD29B/LTI65, the eukaryotic elongation factor 5a (ELF5a) and a protein of unknown function (AT1G16850). The large subunit of Rubisco (RBCL) displayed several negative correlations with LEA proteins (LEA25, LEA29, LEA42) and other members in this sub-network. Only three metabolites were involved in this cluster, namely phosphoric acid, 1,5-anhydro-sorbitol and an unknown analyte (A167004).

This large LEA cluster was connected to another large cluster including mainly metabolites of the central metabolism, most of the amino acids, sugar phosphates, *myo*-inositol-phosphate and phosphoglyceric acid via a positive correlation to LEA20. LEA20 together with LEA46 was more closely correlated with several other heat-stable proteins and metabolites. For LEA20, correlations were found with the seed storage protein SESA4, the Low-Molecular-Weight Cysteine-Rich 67 (LCR67) involved in pathogen response and an unknown protein (AT3G12960). LEA46 showed also positive connections to LCR67, a cytochrome c oxidase (AT1G32710) and a serine protease inhibitor (AT2G38870), possibly involved in biotic stress response. LEA46 showed in contrast to the other LEA proteins a high number of negative correlations with metabolites and amino acids, including mannose-6-phosphate, myo-inositol-phosphate, isoleucine, glutamine and threonine.

3. Discussion

Seed germination is a complex process, which is accompanied by comprehensive changes and a concerted action of molecular players represented by gene expression, protein abundance and metabolite levels. Dry seeds with a quiescent metabolic state undergo a transition to proliferative metabolic state for plant propagation [22]. Molecular omics changes during seed germination in Arabidopsis were studied before but not much attention was directed to variations of the heat-stable proteome including intrinsically disordered proteins and especially changes in LEA proteins and their correlation network with metabolic changes.

3.1. Degradation of Storage Compounds to Central Metabolites Contributes to the Transition from Heterotrophic to Autotrophic Growth

Fifty-six known and forty-three unknown metabolites were identified over a time frame of 72 HAI, which is comparable to other publications on Arabidopsis [8,11], wheat [16] and tomato seeds [17]. Following a previous classification [11], metabolite profiles showed major shifts between three developmental stages: (1) the dry state, (2) initial stages of germination (24 HAI and 36 HAI) and (3) greening and fully opened cotyledons (48 HAI to 72 HAI). Similar to our findings, a PCA clearly distinguished metabolite profiles of wheat embryo and endosperm after 24 h of germination from the profiles at 0 h [16]. Six metabolites, namely fumaric acid, galactinol, sucrose, raffinose, 2,5-dihydroxybenzaldehyde and 3,4-dihydroxybenzoic acid, revealed high levels in DS and were decreased during seed germination and seedling establishment, underlining their role as storage reserves and protective substances for dry seeds [30]. Sucrose or raffinose family oligosaccharides (RFOs) are the most abundant non-reducing sugars in dry mature seeds [36] and serve as storage reserves during germination, resulting in their depletion for energy production and heterotrophic growth before the transition to autotrophic growth [10]. A continuous decrease in sucrose and galactinol as precursor of RFO during seed to seedling transition in Arabidopsis was reported previously [10,11]. Sucrose was additionally suggested as a signaling metabolite during seed germination as sucrose levels showed correlations with 129 transcripts including high numbers of sucrose-responsive genes [10].

Galactinol was described as marker for seed longevity in Brassicaceae and tomato [52]. An overexpression of galactinol synthase of *Cicer arietinum* in Arabidopsis led to higher seed longevity by reducing reactive oxygen substance (ROS) accumulation [53]. On the other hand, galactinol synthase 1 was reported as negative regulator of seed germination [54]. For fumaric acid, an increase during seed maturation was shown, especially during desiccation, resulting in high levels in dry seeds [8].

The two metabolites 2,5-dihydroxybenzaldehyde and 3,4-dihydroxybenzoic acid, which were present in DS, 24 HAI and 36 HAI, have been, to our knowledge, not described in seed germination studies before. 3,4-dihydroxybenzoic acid (protocatechuic acid) was widely detected in cereals and legumes and other species [55,56]. It affects growth of tobacco callus and organs, with small concentrations inducing root growth while high concentrations inhibited shoot and root growth [57]. Furthermore, positive correlations between antioxidant enzyme activity and 3,4-dihydroxybenzoic acid were found in Miscanthus [58].

Most of the other identified metabolites, sugars and sugar phosphates, amino acids and most organic acids were increased with proceeding germination following the activation of biosynthetic processes for growth and development. They were found previously within 24 h germination or till opening of the cotyledons [8,11].

Consistent with other studies, intermediates of the TCA cycle were rather reduced during germination whereas metabolites involved in glycolysis, such as sucrose-derived sugar-phosphates, were increased, supporting the finding that the main energy source for seed development is glycolysis [8,11,16].

A release of amino acids anticipating radicle protrusion and seedling growth was described to start at 24 HAI [22]. Canonical Correlation Analysis (CCA) of metabolite changes during 6 h of seed germination in tomato revealed close correlations of characteristic combinatorial changes in a larger set of metabolites but not individual metabolites with germination parameters. This larger set included metabolites of central metabolic pathways as sucrose, glucose, fructose, metabolites of the TCA cycle as fumarate and succinate and in addition glycerol-3-phosphate, the sugar alcohols *myo*-inositol and galactinol and several amino acids [17]. Altogether, the metabolic status of the here-described experiment was similar to previously published studies, which allows us to compare our data on the heat-stable proteome with previous publications.

3.2. Thermal Stability of Proteins Is Related to Biological Processes

Whereas metabolic shifts during seed germination were previously investigated, less is known about the heat-stable proteome including low abundant proteins at different stages of seed germination. A total of 898 heat-stable proteins were identified in seeds over 72 h of germination in our study, which was the equivalent of around 30% of 2967 proteins with high melting temperature (T_m) or no melting point (nonmelters) in Arabidopsis [59]. Shifts in protein abundance of the heat-stable proteome during seed germination over time differed from that of metabolites. Proteome samples from all time points were clearly separated from each other (except 60 HAI and 72 HAI), whereas metabolite profiles of 24 HAI and 36 HAI were closely clustered.

Heat-stable proteins were enriched in the following bins at all time points: stress, redox, RNA binding, DNA, protein folding and development-related proteins including storage and late embryogenesis abundant proteins. The functional bin photosynthesis with the sub-bin Calvin cycle was highly enriched from 24 HAI. Previously, heat-stable proteins were described to be enriched for ribosomal, RNA-binding and protein biosynthesis processes [60], and functional enrichment analysis of heat-stable Arabidopsis proteins by thermal protein profiling (TPP) pointed to bins related to protein folding, carbon fixation and the proteasome [61], thus revealing an overlap with our findings. Interestingly, an enrichment of proteins involved in carbon fixation was observed including RuBisCo subunits [61], as we also discovered in our data set.

For human cell lines, it was suggested that differences in thermal stability might reflect differences in the activity of biological processes [59]. It was also hypothesized that heat-induced cell damage is mainly caused by the denaturation of a relatively small set of functionally essential "hub" proteins [60]. Following our functional enrichment analysis, this hypothesis can be confirmed as heat-stable proteins were enriched in biological processes important for ongoing seed germination and greening of the cotyledons as photosynthesis, Calvin cycle, abiotic stress and redox reactions.

A comparison of the here-discovered heat-stable proteome of 898 proteins of the Arabidopsis accession Col-0 with 256 proteins previously identified with two-dimensional gel-electrophoresis in the accession Landsberg erecta (Ler) showed an overlap of 149 proteins, revealing a heat-stable percentage of 58% in the Ler data set [22]. The overlap included a large amount of heat-stable proteins involved in biological processes, e.g., several seed storage proteins as cruciferins (CRA1, AT5G44120; CRU2, AT1G03880; CRU3, AT4G28520) belonging to the most abundant cruciferins in DS and serving as nitrogen and amino acid source for seedling development. Additionally, abundant 12S cruciferins were hypothesized to be predominant ROS scavengers in seeds being highly exposed to oxidative stress [22].

In the early germination phases, the de novo synthesis of key enzymes of the glyoxylate cycle, such as isocitrate lyase (AT3G21720), aconitase 3 (AT2G05710) and malate synthase (AT5G03860), was reported [22], pointing to lipid remobilization. All of them were also included in the heat-stable proteome of Col-0. Malate dehydrogenase (AT1G04410), converting malate into oxaloacetate for amino acid synthesis pathways or conversion to phosphoenolpyruvate by phosphoenolpyruvate carboxykinase 1 (AT4G37870), was also found in both studies.

Furthermore, heat-stable proteins common in both studies are involved in antioxidant defense and detoxification: iron/manganese superoxide dismutase (AT3G56350), 1-cysteine peroxiredoxin (AT1G48130), catalase 2 (AT4G35090), thioredoxin-dependent peroxidase 1 (AT1G65980), ferritin 2 (AT3G11050) and monodehydroascorbate reductase 6 (AT1G63940). Generation of ROS takes place during water uptake, which might have negative effects on proteins and other cell components [62]. In dry pea seeds, superoxide dismutase, catalase, ascorbate peroxidase, dehydroascorbate reductase and glutathione reductase were detected as being involved in radical scavenging during imbibition [62].

The restart of the cellular metabolism during germination involves elements of the cytoskeleton, including actin and tubulin, and consequently, tubulin 3 (AT5G62700), actin 2 (AT3G18780) and actin 7 (AT5G09810), which were identified in both data sets and are heat stable. Finally, proteins involved in stress response as heat shock proteins (HSP70, AT3G12580; HSC70.5, AT5G09590; HSP60.2, AT2G33210) [22] overlapped between the studies, underlying the thermal stability of proteins involved in important biological processes during germination. The classification of heat-stable proteins into main metabolic processes and defense responses and the enrichment in functional bins representing biological processes point to mobilization of storage proteins and lipids, an activated heterotrophic and later autotrophic metabolism and several stress response mechanisms during seed germination.

3.3. *An Equal Distribution of Fully Ordered and Fully Intrinsically Disordered Proteins Characterized a Part of the Heat-Stable Proteome of Germinating Seeds*

The loose structure of intrinsically disordered proteins (IDP), e.g., LEA proteins, contributes to their heat-stability when globular proteins aggregate [63]. Often, IDPs are enriched after heating, e.g., when comparing heat-stable proteins with the whole proteome of soybean [64]. An increased content of IDPs was also reported in the heat-stable protein fraction of imbibed radicles of *Medicago truncatula* [41]. In a study compiling an atlas of the thermal stability of 48,000 proteins across 13 species ranging from archaea to humans, nonmelters were most strongly enriched in disordered regions [59].

Nevertheless, a surprisingly high percentage of low disorder content was found for the heat-stable proteome in our study, independent of protein length. In contrast, a study estimating the relative content of intrinsic protein disorder in 96 plant proteomes reported an inverse relationship between the degree of intrinsic protein disorder and protein length [65]. A high intrinsic disorder does not seem to be the main requirement for the heat stability of proteins. A more complex relationship between disorder and thermal protein stability may exist [59]. Recently, features as molecular weight, hydrophobicity, charged versus polar (CvP) bias and protein helix and sheet composition were shown to be highly correlated with thermal stability of Arabidopsis proteins. In addition, cytoplasmic protein concentration, interactions of small molecules and cellular localization were suggested to have an effect [59,61], whereas protein abundance was controversially discussed as either not a good predictor of thermal stability or important for stability [59]. Thermal stability of proteins can be additionally changed by phosphorylation [66] or the redox state [67].

While no functional category was favored in proteins with 0% disorder content, 100% disordered proteins of the heat-stable proteome during seed germination and seedling establishment were mainly assigned to the functional MapMan bins development, DNA, protein, RNA and stress. The disorder-related functions of plant proteins were previously related to stress tolerance, transcription regulation, cell cycle regulation, molecular chaperones and developmental regulation [64]. In addition, IDPs were mainly involved in the GO terms regulation of nucleus activities, regulation of metabolic processes, response to biotic and abiotic stimuli and signaling, depicting an overlap with our results even that functional bins were differently classified [65]. When investigating proteins from 13 species, highly unstructured proteins were enriched in nuclear and phosphorylated proteins [59]. On the other hand, nonmelters with low disorder were enriched in transmembrane-domain-containing proteins, proteins with extracellular domains, glycosylated proteins, secreted proteins with signal peptides and proteins containing disulfide bonds [59].

3.4. Dehydrins Were Identified as Seed Germination-Specific LEA Proteins

Within the heat-stable proteome 29 LEA, proteins were detected with the vast majority of them having 100% disorder. An analysis of the heat-stable proteome of imbibed radicles or seeds of *Medicago truncatula* aiming to investigate proteins linked to desiccation tolerance identified 11 to 16 LEA proteins [41,47].

The ABA-induced accumulation of LEA proteins in parallel to the accumulation of non-reducing sugars during seed maturation provide resistance to desiccation [36,68]. Seed rehydration imposes severe stress, including leakage of solutes and temporary membrane and organelle damage from free radicals. Consequently, synthesis of proteins and compounds which limit and repair cell damage is prevalently needed [6]. With proceeding water uptake during seed germination and the loss of desiccation tolerance, the abundance of most LEA proteins declines sharply [24,68].

The 10 LEA proteins previously described as seed-expressed in Arabidopsis [42] were not detected in the heat-stable proteome of germinating seeds (LEA1, 6, 27, 33, 37, 38, 41, 49, 50, 51), even if 100% disordered (LEA6, 33, 51). Most LEA proteins were detected in DS, including three LEA, specific to DS and decreased during germination. LEA proteins were decreasing and partially remaining during germination in Arabidopsis [18] and pea [69] with a protective function postulated for the remaining proteins.

Only a small number of LEA proteins appeared at later time points. LEA12 was only present from 24 HAI to 48 HAI and was previously described as bud specific [42]. Five other LEA proteins were only present from 36 HAI and were defined as germination specific (LEA4-synonym COR47, LEA5-ERD10/LTI45/LTI29, LEA8-HIRD11, LEA10-ERD14, LEA26), most of them belonging to the class of dehydrins and predicted to be 100% disordered. Three of the five germination-specific LEA proteins were previously described as stress induced [42], which suggests the possibility of contributing to stress tolerance during the development of seedlings.

The protein abundance of LEA4 and LEA10 was highly correlated with the related transcript levels, whereas for the other germination-specific LEAs (LEA5, 8, 26), transcripts were present earlier compared to proteins, suggesting the possibility of a fast de-novo synthesis during seed germination.

Dehydrins contain highly conserved stretches of 7-17 residues that are repetitively scattered in their sequences, the K-, S- and Y-rich segments, and accumulate during seed maturation and in response to abiotic stress as cold, dehydration, osmotic stress or ABA [70]. Some of them are phosphorylated in response to stress and are then capable of calcium binding (e.g., ERD14). Dehydrins are able to bind to phospholipids and thereby to modulate membrane properties [39]. In most studies, their abundance is reduced during seed germination (for a review, see [68]). Nevertheless, these proteins were also shown to be involved in seed germination and development [71]. LEA5/ERD10 is important for completion of seed development as *erd10* mutants revealed abnormal shape and reduced germination [71]. Results on the expression of *ERD10* in DS are controversial, from no expression (our results and [72]) to detection in DS [71].

LEA4/COR47, LEA5/ERD10 and LEA10/ERD14 were previously identified as members of protein complexes. Homodimeric and heterodimeric interactions were verified for LEA4/COR47 and LEA5/ERD10 in the cytosol of tobacco cells by bimolecular fluorescence complementation, but also heterodimeric associations between LEA51/RAB18 or PIP2B, and these dehydrins took place in the cytosol [73,74]. For LEA4/COR47, the transcription factor WRKY63, involved in seedling establishment and growth, was identified as an upstream regulator, pointing to a possible role in seedling development [75].

LEA8/HIRD11 has RNA binding activity, containing an RNA G-quadruplex structure (RG4) in its 3′UTR, which folds with guanine-rich sequences [76]. These complexes play a role in post-transcriptional regulation of gene expression, plant development and growth modulation [76]. An *HIRD11* mutant showed reduced root growth and retarded shoot growth.

LEA26, a germination-specific LEA beside LEA12 not belonging to the class of dehydrins, was shown to function as a protector (chaperon) of cellular components from stress, as this protein prevented the inactivation of the enzyme lactate dehydrogenase (LDH) during freezing and stabilized membrane structures [77].

All of these germination-specific, stress-related dehydrins in addition to their specific functions might protect germinating seeds from environmental stresses after the reduction in osmoprotective substances such as raffinose and other protecting LEA proteins. As an example, they were able to provide cold stress resistance to highly watered seed [68].

3.5. LEA Proteins Build a Highly Correlated Cluster in the Seed Germination Specific Common Protein–Metabolite Network

The construction of a protein–metabolite network helped to develop a deeper understanding of their pattern during seed germination, even though only heat-stable proteins and metabolites were included common for all investigated time points. The discovery of a tight LEA protein cluster among all common proteins during germination is mainly based on a common decrease in abundance of all LEA proteins at the same time during imbibition, but is still mentionable, as only very little proteins follow the same pattern. For that reason, the following discussion of this network focuses mainly on the roles of the specific LEA proteins in DS and very early germination rather than for seedling establishment. Several LEA proteins were identified as network hubs, including LEA7, LEA25, LEA28, LEA29 and LEA30, all belonging to the Pfam group LEA_4.

As a high number of metabolites in the network were unknown, not many conclusions could be drawn regarding common protein–metabolite accumulation patterns that might reflect interactions. Nevertheless, the LEA proteins of the LEA cluster were separated from almost all other metabolites. All of these LEA proteins play important roles for seed protection in the dry state also represented by connections to seed storage proteins and a seed storage 2S albumin superfamily protein. LEA7 in leaves had a protective function on enzymes during freezing and drying [78] and its structural transitions upon drying were modulated by the presence of membranes [79]. A connection of several LEA proteins to the ABA-inducible dehydration-responsive protein RD29B/LTI65 might be related to the induction of LEA proteins by ABA [80,81]. In *nced2569* (9-cis epoxycarotenoid dioxygenase), a quadruple mutant with ABA deficiency, expression of genes encoding LEA28, LEA29, LEA48 and LEA42 was strongly downregulated together with 1-Cys peroxiredoxin1 (Per1), all involved in our LEA cluster [81]. The negative connectivity to RBCL was also interesting, which might point to a protective role of LEA proteins as chaperones for this protein before the induction of photosynthesis during development.

Only LEA20 and LEA46 showed connectivity to a cluster, including metabolites and amino acids of the central metabolism and were only weakly connected to the LEA cluster. LEA20 from Pfam group LEA_5, also named EM6, is induced by ABA and salt [42] and mutant analysis revealed that LEA20 is required for normal seed development in Arabidopsis [82]. LEA20 has a role in water binding and loss during embryo maturation, as mature seed from these mutants lose more water than wild-type seeds during air drying and show an accelerated acquisition of desiccation tolerance [83].

The hub LEA46 (Pfam group LEA_1) with high connectivity might have a protective function, e.g., for enzymes as cytochrome-c-oxidase involved in respiration. The overexpression of LEA46 conferred drought tolerance to severe stress [84]. This might be due to the chaperone-like activity after conformational transition from disorder to alpha-helix folding [85]. Negative correlation of abundance patterns of LEA46 with that of isoleucine, glutamine and threonine might point to a mechanism switching from protection of enzymes during the dry state to securing biosynthesis of crucial metabolites during the developmental stages. Threonine serves as precursor of isoleucine, important for root development, as isoleucine deficiency impairs root development in Arabidopsis [86]. Furthermore, all three amino acids might serve as precursor for glutathione, important as antioxidants and for detoxification [23].

A demonstration of the functional relevance of the identified putative interactions will be necessary in the future to fully understand the importance of the putative protein–metabolite network during seed germination.

4. Materials and Methods

4.1. Plant Materials and Experimental Conditions

Arabidopsis thaliana (accession Col-0) plants were grown on soil in a climate chamber with 20 °C/60% relative humidity (RH) during the day and 6 °C/70% RH during the night in a 14 h/10 h light/dark cycle with of 180 µmol m^{-2} s^{-1} light intensity. After a week, plants were moved to a phytotron with an 8 h light period at a light intensity of 200 µmol m^{-2} s^{-1}, 20 °C, 60% RH and 16 °C during the 16 h night at 75% RH for another week before pricking. After pricking, plants were kept at these conditions for two additional weeks to gain more leaf biomass before being transferred to long-day conditions in a greenhouse (16 h light at 21 °C, 8 h night at 19 °C and 50% RH during day and night) until seed harvest. Plants were bagged after siliques matured to minimize seed loss. After bagging, plants were left to dry for two weeks in the greenhouse. Seeds were collected in the bags and stored in the dark at 15 °C/15% RH for three weeks before they were used for experiments.

Samples for metabolomics, proteomics and transcript analysis were taken from dry seeds (DS), germinating seeds and young seedlings (24, 36, 48, 60 and 72 HAI). Seeds (except for DS) (75 mg of DS per replicate, 3–5 replicates per time point) were sterilized with 70% ethanol for 2 min, followed by 9% bleach solution containing 0.02% Triton X-100 for 17 min and five washings with sterile water. Seeds were then sown on blue germinating paper (grade 190, 300 g/m^2, Sartorius, Göttingen, Germany) on agar plates containing half-strength MS medium without sucrose. Seeds on plates were stratified at 4 °C in the dark for 24 h before transfer to a growth chamber (22 °C, 16 h light period). Germinating seeds and seedlings were collected 24, 36, 48, 60 and 72 HAI by flash freezing with liquid nitrogen in a mortar. After homogenizing dry seed samples using mixer mill MM200 (Retsch, Haan, Germany) and germinating seeds and seedlings with mortar and pestle, fresh weight (FW) of the samples was determined before extractions for further analysis. Dry weight (DW) of samples was determined from 60 mg FW of plant material after drying overnight at 4 °C in a freeze dryer.

4.2. Metabolite Profiling

Aliquots of small polar metabolites were prepared from five biological replicates of each time point using 60 mg of ground fresh material. The extraction used water/methanol: chloroform with lipid partitioning as previously described [87]. $^{13}C_6$-sorbitol (Sigma-Aldrich, Taufkirchen, Germany) was added during the extraction and used as internal standard. The analysis, using 160 µL from the upper polar phase, dried overnight in a vacuum concentrator, was carried out using gas chromatography after derivatization with 40 µL methoxymation reagent in pyridine and 80 µL silylation-mix for tri-methyl-silylation in split 1/30 and splitless mode with 1 µL injection volume coupled to electron impact ionization-time of flight-mass spectrometry (GC/EI-TOF-MS) [87]. ChromaTOF software (LECO Instrumente GmbH, Mönchengladbach, Germany) was used to process the acquired chromatograms. Identification of metabolites was conducted using TagFinder [88], the NIST08 software (http://chemdata.nist.gov/dokuwiki/doku.php?id=start (accessed on 28 January 2020)) and the mass spectral and retention time index reference collection of the Golm Metabolome Database [89–92]. Statistical analysis was based on mass spectral intensities were normalized to dry weight and $^{13}C_6$-sorbitol.

The mass spectrometry metabolomics data have been deposited into the MetaboLights database [93] with the dataset identifier MTBLS2980.

4.3. Amino Acid Analysis

Dried 160 µL of the upper polar phase extracted for metabolite profiling were used for the analysis. Four biological replicates per time point were resuspended in 65 µL of 0.1 M HCl and centrifuged at 14,000 rpm at 4 °C for 15 min. The measurement of amino acids was carried out using reverse HPLC as previously described [94] with small adjustments in LC gradient. Amino acid content was calculated based on the peak area of the mass fragments normalized to DW of the sample.

4.4. Proteomic Analysis of Heat-Stable Proteins

Extraction of total soluble protein and extraction of the heat-stable proteome (95 °C, 10 min) were performed as described [41] from three biological replicates from each time point. Protein concentration was determined using Bradford reagent (B6916, Sigma-Aldrich, Taufkirchen, Germany)/(Bio-Rad Laboratories, Hercules, CA, USA) and bovine serum albumin as a standard. Then, 100 µg of protein per sample was mixed with 100 µL of 8 M urea in 10 mM Tris-HCl, pH 8.0. Samples were loaded on filter columns (Microcon-30 kDa Centrifugal Filter Unit with Ultracel-30 membrane). Columns were washed with 8 M urea, 10 mM Tris-HCl, pH 8.0. Next, proteins were reduced using 10 mM DTT in 8 M urea, 10 mM Tris-HCl, pH 8.0 and alkylated using 27 mM iodoacetamide in the same buffer. Samples were mixed at 600 rpm for 1 min and then incubated without mixing for an additional 5 min. Subsequently, columns were washed using 8 M urea, 10 mM Tris-HCl, pH 8.0 and centrifuged. Proteins in the columns were digested for 14 h using mass spec grade Trypsin/Lys-C mix (Promega, Madison, WI, USA). The resulting peptides were desalted on C18 SepPack columns (Teknokroma, Barcelona, Spain) and eluted with 800 µL 60% acetonitrile, 0.1% trifluoroacetic acid, dried in the speed vacuum concentrator and stored at −80 °C. Measurements were performed on a Q Exactive HF high-resolution mass spectrometer (Thermo Scientific, Waltham, MA, USA) coupled to an ACQUITY UPLC M-Class System (Waters, Milford, MA, USA). Peptide samples were loaded onto an Acclaim PepMap RSLC reversed-phase column (75 µm inner diameter, 25 cm length, 2 µm bead size (Thermo Scientific, Waltham, MA, USA) at a flow rate of 0.4 µL min^{-1} in 3% (*v*/*v*) acetonitrile, 0.1% (*v*/*v*) formic acid. Peptides were eluted by an acetonitrile gradient from 3% to 80% (*v/v*) over 120 min at a flow rate of 0.5 µL/min. Peptide ions were detected in a full scan from mass-to-charge ratio 300 to 1600 at resolution of 60,000. MS/MS scans were performed for the top ten MS scans with the highest MS signal (ddMS2 resolution of 15,000, AGC target 3e5, isolation width mass-to-charge ratio 1.4 m/z, relative collision energy 30). Peptides for which MS/MS spectra had been recorded were excluded from further MS/MS scans for 20 s.

Raw data were processed using MaxQuant software [95] and the *A. thaliana* TAIR10 annotations (Arabidopsis TAIR database Version 10, The Arabidopsis Information Resource, www.Arabidopsis.org (accessed on 27 February 2019)) in combination with the search engine Andromeda [96]. The settings for MaxQuant analysis were set as follows: trypsin and lysine selected as digesting enzyme, two missed cleavages allowed, fixed modification was set to carbamidomethylation (cysteine) and oxidation of methionine was set as variable modification. Spectra were also searched against a decoy database of the *A. thaliana* proteome and results were filtered to obtain a FDR below 1% on the protein level. The "label-free quantification" and "match between runs" options were selected. A minimum peptide length of six amino acids was used. The quantification was performed for proteins with a minimum of one unique and one razor peptide. Known contaminants, such as keratins, and proteins, which were identified with only one unique peptide, were removed from further analysis. Statistical analysis was based on mass spectral label-free quantification (LFQ) intensities.

The mass spectrometry proteomics data have been deposited into the ProteomeXchange Consortium via the PRIDE [97] partner repository with the dataset identifier PXD027546.

4.5. *qRT-PCR Analysis*

Total RNA was extracted from three biological replicates of each time point. The extraction was performed as described [98], with minor modifications. During the first step of RNA extraction, six artificial RNAs from the ArrayControl RNA Spikes kit (Ambion, Austin, TX, USA) were added to the samples as internal standard to allow calculation of the absolute number of transcript copies [99,100]. The number of copies of the different RNA spikes per extract were as follows: Spike 1 (6.08×10^9), Spike 2 (1.52×10^9), Spike 3 (4.56×10^8), Spike 4 (1.14×10^8), Spike 5 (2.94×10^7) and Spike 8 (7.60×10^5). RNA concentration and quality were assessed using a Nanodrop One UV/VIS spectrophotometer (Thermo Scientific, Waltham, MA, USA). Genomic DNA in the RNA samples were removed using RapidOut DNA removal kit (Thermo Scientific, Waltham, MA, USA). Absence of DNA was confirmed by qPCR using primers designed for an intron sequence of *rbcS* (Supplemental Table S11) [101]. All qPCR reactions used 2× SYBR Green Master Mix Reagent (Applied Biosystems, San Francisco, CA, USA) in optical 384-well plates using 7900HT Fast Real-Time PCR System (Thermo Scientific, Waltham, MA, USA). cDNA was synthesized using PrimeScript RT reagent kit with the use of Oligo dT primers (Takara Bio, Kusatsu, Japan). To check the quality of cDNA, qPCR was performed using primers designed for 5′ and 3′ ends of *GAPDH* (Supplemental Table S11) [99]. All samples were found with good quality, as $|Ct5'–Ct3'|$ of *GAPDH* was between 0–1.

To observe the dynamics of LEA transcripts at different HAI, we used primers specific for 41 LEA genes and the six RNA spikes. This included 34 LEA genes that were previously identified as expressed in dry seeds [42] and eight additional genes encoding LEA proteins identified by our proteomics analysis. Sequences of all primers targeting LEA transcripts except LEA35 and RNA spikes have been published previously [42,100] and are listed in Supplemental Table S11. The absolute copy number of each LEA transcript was calculated based on the linear correlation of the CT values and $\log_{10}$ copy number of RNA spikes in all samples. Extrapolation based on the linear correlation was carried out for transcripts with a copy number of more than 6.08×10^9. Absolute copy numbers were normalized to sample DW prior to statistical analysis.

4.6. *Statistical and Other Analyses*

All statistical analyses were performed with RStudio version 1.2.5033 [102]. Prior to performing Principal Component Analysis (PCA) and unpaired two-sided *t*-test, data imputation was performed if a metabolite was detected in three out of five and a protein or transcript was detected in two out of three replicates per time point. The imputation was performed using half the minimum value of either DW-normalized mass-spectral intensities (for metabolites), LFQ intensity (for heat-stable proteins) or DW-normalized absolute copy number (for LEA transcripts). Analytes with more than 40% missing value at a time point were not considered for further evaluation. All data were $\log_2$-transformed and median-normalized for statistical analyses.

In addition, for the heat-stable protein data set, prediction of disorder content of the proteins was carried out using the MFDp2 online tool [50]. The protein sequences of 701 proteins from our data set were available on the UniProtKB/Swiss-Prot database [103].

Unique and common proteins in each time point were shown in a Venn diagram from "VennDiagram" package version 1.6.20. Moreover, functional enrichment of heat-stable proteins was performed according to bins assigned by MapMan using TAIR10 annotations [51]. Enrichment of MapMan bins in the heat-stable proteome was analyzed based on LFQ intensities, which were added with 1 after imputation, thus allowing log_{10} transformation of proteins with previous intensities of 0 without affecting the intensities of other proteins. CorTo v.1.0.3 (http://www.usadellab.org/cms/index.php?page=corto (accessed on 6 January 2021) was used to analyze the enrichment by conducting Fisher's exact test followed by the Benjamini and Hochberg correction with of p value of 0.01 as a threshold. Z-scores were calculated from p values using the inverse normal cumulative distribution function.

PCA was performed using the "pcaMethods" package version 1.74.0 [104]. For PCA, missing values were replaced with 0 and mean values were Pareto-scaled and mean-centered using the probabilistic method. Abundance of analytes from DS to 72 HAI were shown in heat maps which illustrated the log_2-median transformed mean of each analyte at each observed time point. Only the missing values in the heat map showing the abundance of all heat-stable proteins were replaced with 0 to allow clustering using Euclidean clustering on the heat map, otherwise, the missing values remained. Clustering for other heat maps was performed based on the k-means clustering approach by excluding metabolites, proteins or transcripts with missing values from k-means clustering. Clusters were further characterized based on classes of metabolites, protein bins as assigned by MapMan [51] and Pfam groups of LEA proteins [42]. Unpaired two-sided t-testwas performed to test for significant differences between abundance at DS and subsequent time points. Pearson correlation tests were performed between the abundance of LEA proteins and their corresponding transcripts for those with protein and transcript detection from at least four of the six investigated time points. All p value were adjusted using the Benjamini–Hochberg procedure [105] for false discovery rate correction.

4.7. Network Analysis of Common Metabolites, Amino Acids, Heat-Stable Proteins

Mean of log_2-transformed and median-normalized data of metabolites, amino acids and heat-stable proteins which could be detected in DS and samples from all observed time points after imbibition were included in a network analysis using Cytoscape version 3.7.2 [106] with the threshold cut-off of Benjamini–Hochberg adjusted $p < 0.05$ and Pearson correlation $r \geq |0.99|$, as described previously [107]. Network statistics of the undirected protein–metabolite interaction network revealed a power law with $R^2 = 0.782$, indicating that the network is scale-free.

Supplementary Materials: The following are available online at https://www.mdpi.com/article/10.3390/ijms22158172/s1. Figure S1: Seeds and seedlings collected after imbibition. Figure S2: Relative metabolite abundance in dry seeds (DS), germinating seeds and seedlings and their kinetic patterns and class distributions. Figure S3: Relative abundance of heat-stable proteins in dry seeds (DS), germinating seeds and seedlings collected at the indicated time points after imbibition. Figure S4: Kinetic patterns of 182 heat-stable proteins common in dry seeds (DS), germinating seeds and seedlings at indicated time points after imbibition. Figure S5: Network analysis of heat-stable proteins with metabolites and amino acids common in dry seeds (DS), germinating seeds and seedlings. Table S1: Mass-spectral intensities of metabolites normalized to the internal standard $^{13}C_6$-sorbitol and dry weigh in dry seeds (DS), germinating seeds and seedlings at indicated time points after imbibition. Table S2: Amino acid content normalized to dry weight (nmol/gDW) in dry seeds (DS), germinating seeds and seedlings at indicated time points after imbibition. Table S3: LFQ intensities, annotation and disorder content of heat-stable proteins in dry seeds, germinating seeds and seedlings at indicated time points after imbibition. Table S4: Proteins in different section of the Venn diagram (Figure 4A) with their TAIR10 annotation. Table S5: Functional enrichment of MapMan bins of heat-stable proteins found in dry seeds, germinating seeds and seedlings at indicated time points after imbibition. Table S6: LFQ intensities, annotation and disorder content of fully folded (0% disorder) and fully disordered (100% disorder) heat-stable proteins in dry seeds, germinating seeds and seedlings at indicated time points after imbibition (see Figure 5). Table S7: LFQ intensities and disorder content of LEA proteins found in heat-stable proteome in dry seeds, germinating seeds and seedlings at indicated time points after imbibition. Table S8: Disorder content of seed-expressed LEA proteins not found in the heat-stable proteome of dry seeds, germinating seeds and seedlings. Table S9: Copy number of LEA transcript normalized to dry weight (copy/gDW) in dry seeds, germinating seeds and seedlings at the indicated time points after imbibition. Table S10: Pearson correlations and Benjamini–Hochberg adjusted p value of analytes included in the network. Table S11: List of primers used for RNA quality check and qRT-PCR.

Author Contributions: Conceptualization, D.K.H., E.Z.; methodology, O.G.; M.G.; A.E., R.H., J.K., A.S.; investigation, O.G., M.G., A.E., C.H., F.B.; data curation, O.G., M.G., A.E., C.H., F.B., R.H., J.K., A.S., E.Z.; writing—original draft preparation, O.G., E.Z.; writing—review and editing, O.G., E.Z., with contribution of all authors; visualization, O.G.; supervision, D.K.H., E.Z.; funding acquisition, D.K.H., O.G. All authors have read and agreed to the published version of the manuscript.

Funding: This research was funded by the Max-Planck-Society. O.G. was funded by a Ph.D. scholarship of the Royal Thai Government.

Institutional Review Board Statement: Not applicable.

Informed Consent Statement: Not applicable.

Data Availability Statement: All data are available in the Supplementary Materials. In addition, the proteomics data are available at ProteomeXchange via the accession number PXD027546. The metabolomics data are available at MetaboLights via MTBLS2980.

Acknowledgments: We thank Jessica Alpers and Ines Fehrle (Max Planck Institute of Molecular Plant Physiology) for technical support, Virginie Mengin for advice on RNA spikes and Stephanie Schaarschmidt (Max Planck Institute of Molecular Plant Physiology) for support and advice with R programming.

Conflicts of Interest: The authors declare no conflict of interest. The funders had no role in the design of the study; in the collection, analyses or interpretation of data; in the writing of the manuscript, or in the decision to publish the results.

References

1. Maia, J.; Dekkers, B.J.W.; Provart, N.J.; Ligterink, W.; Hilhorst, H.W.M. The Re-Establishment of Desiccation Tolerance in Germinated Arabidopsis thaliana Seeds and Its Associated Transcriptome. *PLoS ONE* **2011**, *6*, e29123. [CrossRef] [PubMed]
2. Daws, M.; Bolton, S.; Burslem, D.F.R.P.; Garwood, N.C.; Mullins, C.E. Loss of desiccation tolerance during germinationin neo-tropical pioneer seeds: Implications forseed mortality and germination characteristics. *Seed Sci. Res.* **2007**, *17*, 273–281. [CrossRef]
3. Buitink, J.; Vu, B.L.; Satour, P.; Leprince, O. The re-establishment of desiccation tolerance in germinated radicles ofMedicago truncatulaGaertn. seeds. *Seed Sci. Res.* **2003**, *13*, 273–286. [CrossRef]
4. Oliver, M.J.; Farrant, J.M.; Hilhorst, H.W.; Mundree, S.; Williams, B.; Bewley, J.D. Desiccation Tolerance: Avoiding Cellular Damage During Drying and Rehydration. *Annu. Rev. Plant Biol.* **2020**, *71*, 435–460. [CrossRef] [PubMed]
5. Bewley, J.D. Seed germination and dormancy. *Plant Cell* **1997**, *9*, 1055–1066. [CrossRef]
6. Nonogaki, H.; Bassel, G.; Bewley, D. Germination—Still a mystery. *Plant Sci.* **2010**, *179*, 574–581. [CrossRef]
7. Boyes, D.C.; Zayed, A.M.; Ascenzi, R.; McCaskill, A.J.; Hoffman, N.E.; Davis, K.R.; Gorlach, J. Growth stage-based phenotypic analysis of Arabidopsis: A model for high throughput functional genomics in plants. *Plant Cell* **2001**, *13*, 1499–1510. [CrossRef]
8. Fait, A.; Angelovici, R.; Less, H.; Ohad, I.; Urbanczyk-Wochniak, E.; Fernie, A.R.; Galili, G. Arabidopsis Seed Development and Germination Is Associated with Temporally Distinct Metabolic Switches. *Plant Physiol.* **2006**, *142*, 839–854. [CrossRef]
9. Sreenivasulu, N.; Usadel, B.; Winter, A.; Radchuk, V.; Scholz, U.; Stein, N.; Weschke, W.; Strickert, M.; Close, T.J.; Stitt, M.; et al. Barley Grain Maturation and Germination: Metabolic Pathway and Regulatory Network Commonalities and Differences Highlighted by New MapMan/PageMan Profiling Tools. *Plant Physiol.* **2008**, *146*, 1738–1758. [CrossRef]
10. Allen, E.; Moing, A.; Ebbels, T.M.; Maucourt, M.; Tomos, A.D.; Rolin, D.; Hooks, M.A. Correlation Network Analysis reveals a sequential reorganization of metabolic and transcriptional states during germination and gene-metabolite relationships in developing seedlings of Arabidopsis. *BMC Syst. Biol.* **2010**, *4*, 62. [CrossRef]
11. Silva, A.T.; Ligterink, W.; Hilhorst, H.W.M. Metabolite profiling and associated gene expression reveal two metabolic shifts during the seed-to-seedling transition in Arabidopsis thaliana. *Plant Mol. Biol.* **2017**, *95*, 481–496. [CrossRef]
12. Nakabayashi, K.; Okamoto, M.; Koshiba, T.; Kamiya, Y.; Nambara, E. Genome-wide profiling of stored mRNA in Arabidopsis thaliana seed germination: Epigenetic and genetic regulation of transcription in seed. *Plant J.* **2005**, *41*, 697–709. [CrossRef]

13. Silva, A.T.; Ribone, P.A.; Chan, R.L.; Ligterink, W.; Hilhorst, H.W. A Predictive Coexpression Network Identifies Novel Genes Controlling the Seed-to-Seedling Phase Transition in Arabidopsis thaliana. *Plant Physiol.* **2016**, *170*, 2218–2231. [CrossRef] [PubMed]
14. Bai, B.; Peviani, A.; Van Der Horst, S.; Gamm, M.; Snel, B.; Bentsink, L.; Hanson, J. Extensive translational regulation during seed germination revealed by polysomal profiling. *New Phytol.* **2016**, *214*, 233–244. [CrossRef]
15. Bai, B.; Van Der Horst, S.; Cordewener, J.H.; America, T.A.H.P.; Hanson, J.; Bentsink, L. Seed-Stored mRNAs that Are Specifically Associated to Monosomes Are Translationally Regulated during Germination. *Plant Physiol.* **2020**, *182*, 378–392. [CrossRef] [PubMed]
16. Han, C.; Zhen, S.; Zhu, G.; Bian, Y.; Yan, Y. Comparative metabolome analysis of wheat embryo and endosperm reveals the dynamic changes of metabolites during seed germination. *Plant Physiol. Biochem.* **2017**, *115*, 320–327. [CrossRef]
17. Kazmi, R.H.; Willems, L.A.J.; Joosen, R.V.L.; Khan, N.; Ligterink, W.; Hilhorst, H.W.M. Metabolomic analysis of tomato seed germination. *Metabolomics* **2017**, *13*, 145. [CrossRef]
18. Gallardo, K.; Job, C.; Groot, S.P.; Puype, M.; Demol, H.; Vandekerckhove, J.; Job, D. Proteomic Analysis of Arabidopsis Seed Germination and Priming. *Plant Physiol.* **2001**, *126*, 835–848. [CrossRef]
19. Rajjou, L.; Gallardo, K.; Debeaujon, I.; Vandekerckhove, J.; Job, C.; Job, D. The Effect of α-Amanitin on the Arabidopsis Seed Proteome Highlights the Distinct Roles of Stored and Neosynthesized mRNAs during Germination. *Plant Physiol.* **2004**, *134*, 1598–1613. [CrossRef] [PubMed]
20. Fu, Q.; Wang, B.-C.; Jin, X.; Li, H.-B.; Han, P.; Wei, K.-H.; Zhang, X.-M.; Zhu, Y.-X. Proteomic Analysis and Extensive Protein Identification from Dry, Germinating Arabidopsis Seeds and Young Seedlings. *BMB Rep.* **2005**, *38*, 650–660. [CrossRef]
21. Yang, P.; Li, X.; Wang, X.; Chen, H.; Chen, F.; Shen, S. Proteomic analysis of rice (*Oryza sativa*) seeds during germination. *Proteomics* **2007**, *7*, 3358–3368. [CrossRef]
22. Galland, M.; Huguet, R.; Arc, E.; Cueff, G.; Job, D.; Rajjou, L. Dynamic Proteomics Emphasizes the Importance of Selective mRNA Translation and Protein Turnover during Arabidopsis Seed Germination. *Mol. Cell. Proteom.* **2014**, *13*, 252–268. [CrossRef]
23. Gu, J.; Chao, H.; Gan, L.; Guo, L.; Zhang, K.; Li, Y.; Wang, H.; Raboanatahiry, N.; Li, M. Proteomic Dissection of Seed Germination and Seedling Establishment in Brassica napus. *Front. Plant Sci.* **2016**, *7*, 1482. [CrossRef] [PubMed]
24. Kretzschmar, F.K.; Doner, N.M.; Krawczyk, H.E.; Scholz, P.; Schmitt, K.; Valerius, O.; Braus, G.H.; Mullen, R.T.; Ischebeck, T. Identification of Low-Abundance Lipid Droplet Proteins in Seeds and Seedlings. *Plant Physiol.* **2020**, *182*, 1326–1345. [CrossRef]
25. Sano, N.; Permana, H.; Kumada, R.; Shinozaki, Y.; Tanabata, T.; Yamada, T.; Hirasawa, T.; Kanekatsu, M. Proteomic Analysis of Embryonic Proteins Synthesized from Long-Lived mRNAs During Germination of Rice Seeds. *Plant Cell Physiol.* **2012**, *53*, 687–698. [CrossRef] [PubMed]
26. Sano, N.; Takebayashi, Y.; To, A.; Mhiri, C.; Rajjou, L.; Nakagami, H.; Kanekatsu, M. Shotgun Proteomic Analysis Highlights the Roles of Long-Lived mRNAs and De Novo Transcribed mRNAs in Rice Seeds upon Imbibition. *Plant Cell Physiol.* **2019**, *60*, 2584–2596. [CrossRef] [PubMed]
27. Obendorf, R.L. Oligosaccharides and galactosyl cyclitols in seed desiccation tolerance. *Seed Sci. Res.* **1997**, *7*, 63–74. [CrossRef]
28. Ballesteros, D.; Walters, C. Detailed characterization of mechanical properties and molecular mobility within dry seed glasses: Relevance to the physiology of dry biological systems. *Plant J.* **2011**, *68*, 607–619. [CrossRef]
29. Ballesteros, D.; Walters, C. Solid-State Biology and Seed Longevity: A Mechanical Analysis of Glasses in Pea and Soybean Embryonic Axes. *Front. Plant Sci.* **2019**, *10*, 920. [CrossRef]
30. Buitink, J.; Leprince, O. Glass formation in plant anhydrobiotes: Survival in the dry state. *Cryobiology* **2004**, *48*, 215–228. [CrossRef]
31. Roberts, J.K.; DeSimone, N.A.; Lingle, W.L.; Dure, L., 3rd. Cellular concentrations and uniformity of cell-type accumulation of two LEA proteins in cotton embryos. *Plant Cell* **1993**, *5*, 769–780. [CrossRef]
32. Dure, L.; Chlan, C. Developmental biochemistry of cottonseed embryogenesis and germination: Xii. Purification and properties of principal storage proteins. *Plant Physiol.* **1981**, *68*, 180–186. [CrossRef]
33. Dure, L.; Galau, G.A. Developmental biochemistry of cottonseed embryogenesis and germination: Xiii. Regulation of biosynthesis of principal storage proteins. *Plant Physiol.* **1981**, *68*, 187–194. [CrossRef] [PubMed]
34. Dure, L., 3rd; Greenway, S.C.; Galau, G.A. Developmental biochemistry of cottonseed embryogenesis and germination: Changing messenger ribonucleic acid populations as shown by in vitro and in vivo protein synthesis. *Biochemistry* **1981**, *20*, 4162–4168. [CrossRef] [PubMed]
35. Tunnacliffe, A.; Wise, M.J. The continuing conundrum of the LEA proteins. *Naturwissenschaften* **2007**, *94*, 791–812. [CrossRef] [PubMed]
36. Leprince, O.; Pellizzaro, A.; Berriri, S.; Buitink, J. Late seed maturation: Drying without dying. *J. Exp. Bot.* **2016**, *68*, 827–841. [CrossRef]
37. Hand, S.C.; Menze, M.A.; Toner, M.; Boswell, L.; Moore, D. LEA proteins during water stress: Not just for plants anymore. *Annu. Rev. Physiol.* **2011**, *73*, 115–134. [CrossRef]
38. Hincha, D.K.; Thalhammer, A. LEA proteins: IDPs with versatile functions in cellular dehydration tolerance. *Biochem. Soc. Trans.* **2012**, *40*, 1000–1003. [CrossRef]
39. Tunnacliffe, A.; Hincha, D.; Leprince, O.; Macherel, D. LEA proteins: Versatility of forms and function. In *Dormancy and Resistance in Harsh Environments*; Lubzens, E., Cerda, J., Clark, M., Eds.; Springer: Berlin, Germany, 2010; pp. 91–108.

40. Blackman, S.A.; Wettlaufer, S.H.; Obendorf, R.L.; Leopold, A.C. Maturation Proteins Associated with Desiccation Tolerance in Soybean. *Plant Physiol.* **1991**, *96*, 868–874. [CrossRef] [PubMed]
41. Boudet, J.; Buitink, J.; Hoekstra, F.A.; Rogniaux, H.; Larre, C.; Satour, P.; Leprince, O. Comparative Analysis of the Heat Stable Proteome of Radicles of Medicago truncatula Seeds during Germination Identifies Late Embryogenesis Abundant Proteins Associated with Desiccation Tolerance. *Plant Physiol.* **2006**, *140*, 1418–1436. [CrossRef]
42. Hundertmark, M.; Hincha, D.K. LEA (Late Embryogenesis Abundant) proteins and their encoding genes in Arabidopsis thaliana. *BMC Genom.* **2008**, *9*, 118. [CrossRef]
43. Knox-Brown, P.; Rindfleisch, T.; Günther, A.; Balow, K.; Bremer, A.; Walther, D.; Miettinen, M.S.; Hincha, D.K.; Thalhammer, A. Similar Yet Different–Structural and Functional Diversity among Arabidopsis thaliana LEA_4 Proteins. *Int. J. Mol. Sci.* **2020**, *21*, 2794. [CrossRef] [PubMed]
44. Sun, X.; Rikkerink, E.; Jones, W.T.; Uversky, V.N. Multifarious Roles of Intrinsic Disorder in Proteins Illustrate Its Broad Impact on Plant Biology. *Plant Cell* **2013**, *25*, 38–55. [CrossRef] [PubMed]
45. Covarrubias, A.A.; Romero-Pérez, P.S.; Cuevas-Velazquez, C.L.; Rendón-Luna, D.F. The functional diversity of structural disorder in plant proteins. *Arch. Biochem. Biophys.* **2020**, *680*, 108229. [CrossRef] [PubMed]
46. Graether, S.P.; Boddington, K.F. Disorder and function: A review of the dehydrin protein family. *Front. Plant Sci.* **2014**, *5*, 576. [CrossRef]
47. Chatelain, E.; Hundertmark, M.; Leprince, O.; Le Gall, S.; Satour, P.; Deligny-Penninck, S.; Rogniaux, H.; Buitink, J. Temporal profiling of the heat-stable proteome during late maturation of Medicago truncatula seeds identifies a restricted subset of late embryogenesis abundant proteins associated with longevity. *Plant Cell Environ.* **2012**, *35*, 1440–1455. [CrossRef] [PubMed]
48. Higashi, Y.; Hirai, M.; Fujiwara, T.; Naito, S.; Noji, M.; Saito, K. Proteomic and transcriptomic analysis of Arabidopsis seeds: Molecular evidence for successive processing of seed proteins and its implication in the stress response to sulfur nutrition. *Plant J.* **2006**, *48*, 557–571. [CrossRef]
49. Receveur-Brechot, V.; Bourhis, J.-M.; Uversky, V.N.; Canard, B.; Longhi, S. Assessing protein disorder and induced folding. *Proteins: Struct. Funct. Bioinform.* **2005**, *62*, 24–45. [CrossRef] [PubMed]
50. Mizianty, M.J.; Peng, Z.; Kurgan, L. Mfdp2: Accurate predictor of disorder in proteins by fusion of disorder probabilities, content and profiles. *Intrinsically Disord Proteins* **2013**, *1*, e24428. [CrossRef]
51. Thimm, O.; Bläsing, O.; Gibon, Y.; Nagel, A.; Meyer, S.; Krüger, P.; Selbig, J.; Müller, L.A.; Rhee, S.Y.; Stitt, M. mapman: A user-driven tool to display genomics data sets onto diagrams of metabolic pathways and other biological processes. *Plant J.* **2004**, *37*, 914–939. [CrossRef]
52. Vidigal, D.D.S.; Willems, L.; Van Arkel, J.; Dekkers, B.J.; Hilhorst, H.W.; Bentsink, L. Galactinol as marker for seed longevity. *Plant Sci.* **2016**, *246*, 112–118. [CrossRef]
53. Salvi, P.; Saxena, S.C.; Petla, B.P.; Kamble, N.U.; Kaur, H.; Verma, P.; Rao, V.; Ghosh, S.; Majee, M. Differentially expressed galactinol synthase(s) in chickpea are implicated in seed vigor and longevity by limiting the age induced ROS accumulation. *Sci. Rep.* **2016**, *6*, 35088. [CrossRef]
54. Jang, J.-H.; Shang, Y.; Kang, H.K.; Kim, S.Y.; Kim, B.H.; Nam, K.H. Arabidopsis galactinol synthases 1 (AtGOLS1) negatively regulates seed germination. *Plant Sci.* **2018**, *267*, 94–101. [CrossRef] [PubMed]
55. Gennari, L.; Felletti, M.; Blasa, M.; Angelino, D.; Celeghini, C.; Corallini, A.; Ninfali, P. Total extract of Beta Vulgaris var. Cicla seeds versus its purified phenolic components: Antioxidant activities and antiproliferative effects against colon cancer cells. *Phytochem. Anal.* **2011**, *22*, 272–279. [CrossRef] [PubMed]
56. Xu, H.-Y.; Zheng, H.-C.; Zhang, H.-W.; Zhang, J.-Y.; Ma, C.-M. Comparison of Antioxidant Constituents of Agriophyllum squarrosum Seed with Conventional Crop Seeds. *J. Food Sci.* **2018**, *83*, 1823–1831. [CrossRef] [PubMed]
57. Mucciarelli, M.; Gallino, M.; Maffei, M.; Scannerini, S. Effects of 3,4-dihydroxybenzoic acid on tobacco (*Nicotiana tabacum* L.) culturedin vitro. Growth regulation in callus and organ cultures. *Plant Biosyst. Int. J. Deal. All Asp. Plant Biol.* **2000**, *134*, 185–192. [CrossRef]
58. Ghimire, B.K.; Hwang, M.H.; Sacks, E.J.; Yu, C.Y.; Kim, S.H.; Chung, I.M. Screening of Allelochemicals in *Miscanthus sacchariflorus* Extracts and Assessment of Their Effects on Germination and Seedling Growth of Common Weeds. *Plants* **2020**, *9*, 1313. [CrossRef]
59. Jarzab, A.; Kurzawa, N.; Hopf, T.; Moerch, M.; Zecha, J.; Leijten, N.; Bian, Y.; Musiol, E.; Maschberger, M.; Stoehr, G.; et al. Meltome atlas—thermal proteome stability across the tree of life. *Nat. Methods* **2020**, *17*, 495–503. [CrossRef] [PubMed]
60. Leuenberger, P.; Ganscha, S.; Kahraman, A.; Cappelletti, V.; Boersema, P.J.; von Mering, C.; Claassen, M.; Picotti, P. Cell-wide analysis of protein thermal unfolding reveals determinants of thermostability. *Science* **2017**, *355*, eaai7825. [CrossRef]
61. Volkening, J.D.; Stecker, K.E.; Sussman, M.R. Proteome-wide Analysis of Protein Thermal Stability in the Model Higher Plant Arabidopsis thaliana. *Mol. Cell. Proteom.* **2019**, *18*, 308–319. [CrossRef] [PubMed]
62. Wojtyla, Ł.; Garnczarska, M.; Zalewski, T.; Bednarski, W.; Ratajczak, L.; Jurga, S. A comparative study of water distribution, free radical production and activation of antioxidative metabolism in germinating pea seeds. *J. Plant Physiol.* **2006**, *163*, 1207–1220. [CrossRef] [PubMed]
63. Tsvetkov, P.; Myers, N.; Moscovitz, O.; Sharon, M.; Prilusky, J.; Shaul, Y. Thermo-resistant intrinsically disordered proteins are efficient 20S proteasome substrates. *Mol. BioSyst.* **2011**, *8*, 368–373. [CrossRef] [PubMed]
64. Liu, Y.; Wu, J.; Sun, N.; Tu, C.; Shi, X.; Cheng, H.; Liu, S.; Li, S.; Wang, Y.; Zheng, Y.; et al. Intrinsically Disordered Proteins as Important Players during Desiccation Stress of Soybean Radicles. *J. Proteome Res.* **2017**, *16*, 2393–2409. [CrossRef] [PubMed]

65. Zamora-Briseño, J.A.; Pereira-Santana, A.; Reyes-Hernández, S.J.; Cerqueda-García, D.; Castaño, E.; Rodríguez-Zapata, L.C. Towards an understanding of the role of intrinsic protein disorder on plant adaptation to environmental challenges. *Cell Stress Chaperon* **2021**, *26*, 141–150. [CrossRef] [PubMed]
66. Savitski, M.M.; Reinhard, F.B.M.; Franken, H.; Werner, T.; Savitski, M.F.; Eberhard, D.; Molina, D.M.; Jafari, R.; Dovega, R.B.; Klaeger, S.; et al. Tracking cancer drugs in living cells by thermal profiling of the proteome. *Science* **2014**, *346*, 1255784. [CrossRef] [PubMed]
67. Sun, W.; Dai, L.; Yu, H.; Puspita, B.; Zhao, T.; Li, F.; Tan, J.; Lim, Y.T.; Chen, M.W.; Sobota, R.; et al. Monitoring structural modulation of redox-sensitive proteins in cells with MS-CETSA. *Redox Biol.* **2019**, *24*, 101168. [CrossRef]
68. Azarkovich, M.I. Dehydrins in Orthodox and Recalcitrant Seeds. *Russ. J. Plant Physiol.* **2020**, *67*, 221–230. [CrossRef]
69. DeRocher, A.E.; Vierling, E. Developmental control of small heat shock protein expression during pea seed maturation. *Plant J.* **1994**, *5*, 93–102. [CrossRef]
70. Hundertmark, M.; Buitink, J.; Leprince, O.; Hincha, D.K. The reduction of seed-specific dehydrins reduces seed longevity in Arabidopsis thaliana. *Seed Sci. Res.* **2011**, *21*, 165–173. [CrossRef]
71. Kim, S.Y.; Nam, K.H. Physiological roles of ERD10 in abiotic stresses and seed germination of Arabidopsis. *Plant Cell Rep.* **2010**, *29*, 203–209. [CrossRef] [PubMed]
72. Kiyosue, T.; Yamaguchi-Shinozaki, K.; Shinozaki, K. Characterization of Two cDNAs (ERD10 and ERD14) Corresponding to Genes That Respond Rapidly to Dehydration Stress in Arabidopsis thaliana. *Plant Cell Physiol.* **1994**, *35*, 225–231. [CrossRef] [PubMed]
73. Sánchez, I.E.H.; Maruri-López, I.; Graether, S.P.; Jiménez-Bremont, J.F. In vivo evidence for homo- and heterodimeric interactions of Arabidopsis thaliana dehydrins AtCOR47, AtERD10, and AtRAB18. *Sci. Rep.* **2017**, *7*, 17036. [CrossRef]
74. Hernández-Sánchez, I.E.; Maruri-López, I.; Molphe-Balch, E.P.; Becerra-Flora, A.; Jaimes-Miranda, F.; Jiménez-Bremont, J.F. Evidence for in vivo interactions between dehydrins and the aquaporin AtPIP2B. *Biochem. Biophys. Res. Commun.* **2019**, *510*, 545–550. [CrossRef] [PubMed]
75. Rushton, D.L.; Tripathi, P.; Rabara, R.; Lin, J.; Ringler, P.; Boken, A.K.; Langum, T.J.; Smidt, L.; Boomsma, D.D.; Emme, N.J.; et al. WRKY transcription factors: Key components in abscisic acid signalling. *Plant Biotechnol. J.* **2011**, *10*, 2–11. [CrossRef]
76. Yang, X.; Cheema, J.; Zhang, Y.; Deng, H.; Duncan, S.; Umar, M.I.; Zhao, J.; Liu, Q.; Cao, X.; Kwok, C.K.; et al. RNA G-quadruplex structures exist and function in vivo in plants. *Genome Biol.* **2020**, *21*, 226. [CrossRef]
77. Dang, N.X.; Popova, A.; Hundertmark, M.; Hincha, D.K. Functional characterization of selected LEA proteins from Arabidopsis thaliana in yeast and in vitro. *Planta* **2014**, *240*, 325–336. [CrossRef] [PubMed]
78. Popova, A.; Rausch, S.; Hundertmark, M.; Gibon, Y.; Hincha, D.K. The intrinsically disordered protein LEA7 from Arabidopsis thaliana protects the isolated enzyme lactate dehydrogenase and enzymes in a soluble leaf proteome during freezing and drying. *Biochim. Biophys. Acta BBA Proteins Proteom.* **2015**, *1854*, 1517–1525. [CrossRef] [PubMed]
79. Popova, A.; Hundertmark, M.; Seckler, R.; Hincha, D.K. Structural transitions in the intrinsically disordered plant dehydration stress protein LEA7 upon drying are modulated by the presence of membranes. *Biochim. Biophys. Acta BBA Biomembr.* **2011**, *1808*, 1879–1887. [CrossRef]
80. Zamora-Briseño, J.A.; de Jiménez, E.S. A LEA 4 protein up-regulated by ANA is involved in drought response in maize roots. *Mol. Biol. Rep.* **2016**, *43*, 221–228. [CrossRef]
81. Chauffour, F.; Bailly, M.; Perreau, F.; Cueff, G.; Suzuki, H.; Collet, B.; Frey, A.; Clément, G.; Soubigou-Taconnat, L.; Balliau, T.; et al. Multi-omics Analysis Reveals Sequential Roles for ABA during Seed Maturation. *Plant Physiol.* **2019**, *180*, 1198–1218. [CrossRef]
82. Manfre, A.J.; Lanni, L.M.; Marcotte, W.R. The Arabidopsis Group 1 Late Embryogenesis Abundant Protein ATEM6 Is Required for Normal Seed Development. *Plant Physiol.* **2006**, *140*, 140–149. [CrossRef]
83. Manfre, A.J.; LaHatte, G.A.; Climer, C.R.; Marcotte, W.R. Seed Dehydration and the Establishment of Desiccation Tolerance During Seed Maturation is Altered in the Arabidopsis thaliana Mutant atem6-1. *Plant Cell Physiol.* **2008**, *50*, 243–253. [CrossRef] [PubMed]
84. Olvera-Carrillo, Y.; Campos, F.; Reyes, J.L.; Garciarrubio, A.; Covarrubias, A.A. Functional Analysis of the Group 4 Late Embryogenesis Abundant Proteins Reveals Their Relevance in the Adaptive Response during Water Deficit in Arabidopsis. *Plant Physiol.* **2010**, *154*, 373–390. [CrossRef]
85. Cuevas-Velazquez, C.; Saab-Rincón, G.; Reyes, J.L.; Covarrubias, A.A. The Unstructured N-terminal Region of Arabidopsis Group 4 Late Embryogenesis Abundant (LEA) Proteins Is Required for Folding and for Chaperone-like Activity under Water Deficit. *J. Biol. Chem.* **2016**, *291*, 10893–10903. [CrossRef] [PubMed]
86. Yu, H.; Zhang, F.; Wang, G.; Liu, Y.; Liu, D. Partial deficiency of isoleucine impairs root development and alters transcript levels of the genes involved in branched-chain amino acid and glucosinolate metabolism in Arabidopsis. *J. Exp. Bot.* **2012**, *64*, 599–612. [CrossRef]
87. Erban, A.; Martinez-Seidel, F.; Rajarathinam, Y.; Dethloff, F.; Orf, I.; Fehrle, I.; Alpers, J.; Beine-Golovchuk, O.; Kopka, J. Multiplexed profiling and data processing methods to identify temperature-regulated primary metabolites using gas chromatography coupled to mass spectrometry. *Methods Mol. Biol.* **2020**, *2156*, 203–239. [PubMed]
88. Luedemann, A.; Strassburg, K.; Erban, A.; Kopka, J. TagFinder for the quantitative analysis of gas chromatography—Mass spectrometry (GC-MS)-based metabolite profiling experiments. *Bioinformatics* **2008**, *24*, 732–737. [CrossRef]

89. Hummel, J.; Strehmel, N.; Selbig, J.; Walther, D.; Kopka, J. Decision tree supported substructure prediction of metabolites from GC-MS profiles. *Metabolomics* **2010**, *6*, 322–333. [CrossRef]
90. Kopka, J.; Schauer, N.; Krueger, S.; Birkemeyer, C.; Usadel, B.; Bergmuller, E.; Dormann, P.; Weckwerth, W.; Gibon, Y.; Stitt, M.; et al. GMD@CSB.DB: The Golm Metabolome Database. *Bioinformatics* **2004**, *21*, 1635–1638. [CrossRef] [PubMed]
91. Allwood, J.W.; Erban, A.; De Koning, S.; Dunn, W.; Luedemann, A.; Lommen, A.; Kay, L.; Löscher, R.; Kopka, J.; Goodacre, R. Inter-laboratory reproducibility of fast gas chromatography–electron impact–time of flight mass spectrometry (GC–EI–TOF/MS) based plant metabolomics. *Metabolomics* **2009**, *5*, 479–496. [CrossRef]
92. Strehmel, N.; Hummel, J.; Erban, A.; Strassburg, K.; Kopka, J. Retention index thresholds for compound matching in GC–MS metabolite profiling. *J. Chromatogr. B* **2008**, *871*, 182–190. [CrossRef]
93. Haug, K.; Cochrane, K.; Nainala, V.C.; Williams, M.; Chang, J.; Jayaseelan, K.V.; O'Donovan, C. MetaboLights: A resource evolving in response to the needs of its scientific community. *Nucleic Acids Res.* **2019**, *48*, D440–D444. [CrossRef] [PubMed]
94. Watanabe, M.; Tohge, T.; Balazadeh, S.; Erban, A.; Giavalisco, P.; Kopka, J.; Mueller-Roeber, B.; Fernie, A.R.; Hoefgen, R. Comprehensive Metabolomics Studies of Plant Developmental Senescence. *Adv. Struct. Saf. Stud.* **2018**, *1744*, 339–358. [CrossRef]
95. Cox, J.; Mann, M. MaxQuant enables high peptide identification rates, individualized p.p.b.-range mass accuracies and proteome-wide protein quantification. *Nat. Biotechnol.* **2008**, *26*, 1367–1372. [CrossRef] [PubMed]
96. Cox, J.; Neuhauser, N.; Michalski, A.; Scheltema, R.; Olsen, J.V.; Mann, M. Andromeda: A Peptide Search Engine Integrated into the MaxQuant Environment. *J. Proteome Res.* **2011**, *10*, 1794–1805. [CrossRef]
97. Perez-Riverol, Y.; Csordas, A.; Bai, J.; Bernal-Llinares, M.; Hewapathirana, S.; Kundu, D.J.; Inuganti, A.; Griss, J.; Mayer, G.; Eisenacher, M.; et al. The PRIDE database and related tools and resources in 2019: Improving support for quantification data. *Nucleic Acids Res.* **2019**, *47*, D442–D450. [CrossRef]
98. Vicient, C.M.; Delseny, M. Isolation of Total RNA fromArabidopsis thalianaSeeds. *Anal. Biochem.* **1999**, *268*, 412–413. [CrossRef]
99. Piques, M.; Schulze, W.X.; Höhne, M.; Usadel, B.; Gibon, Y.; Rohwer, J.; Stitt, M. Ribosome and transcript copy numbers, polysome occupancy and enzyme dynamics in Arabidopsis. *Mol. Syst. Biol.* **2009**, *5*, 314. [CrossRef]
100. Flis, A.; Fernández, A.P.; Zielinski, T.; Mengin, V.; Sulpice, R.; Stratford, K.; Hume, A.; Pokhilko, A.; Southern, M.M.; Seaton, D.D.; et al. Defining the robust behaviour of the plant clock gene circuit with absolute RNA timeseries and open infrastructure. *Open Biol.* **2015**, *5*, 5. [CrossRef]
101. Czechowski, T.; Stitt, M.; Altmann, T.; Udvardi, M.K.; Scheible, W.-R. Genome-Wide Identification and Testing of Superior Reference Genes for Transcript Normalization in Arabidopsis. *Plant Physiol.* **2005**, *139*, 5–17. [CrossRef]
102. RStudio Team. *RStudio: Integrated Development for R*; R Studio, Inc.: Boston, MA, USA, 2019.
103. UniProt, Consortium. UniProt: A worldwide hub of protein knowledge. *Nucleic Acids Res.* **2019**, *47*, D506–D515. [CrossRef]
104. Stacklies, W.; Redestig, H.; Scholz, M.; Walther, D.; Selbig, J. pcaMethods a bioconductor package providing PCA methods for incomplete data. *Bioinformatics* **2007**, *23*, 1164–1167. [CrossRef] [PubMed]
105. Benjamini, Y.; Hochberg, Y. Controlling the false discovery rate: A practical and powerful approach to multiple testing. *J. R. Stat. Soc. Ser. B* **1995**, *57*, 289–300. [CrossRef]
106. Shannon, P.; Markiel, A.; Ozier, O.; Baliga, N.S.; Wang, J.T.; Ramage, D.; Amin, N.; Schwikowski, B.; Ideker, T. Cytoscape: A Software Environment for Integrated Models of Biomolecular Interaction Networks. *Genome Res.* **2003**, *13*, 2498–2504. [CrossRef] [PubMed]
107. Contreras-López, O.; Moyano, T.C.; Soto, D.; Gutiérrez, R.A. Step-by-Step Construction of Gene Co-expression Networks from High-Throughput Arabidopsis RNA Sequencing Data. *Methods Mol. Biol.* **2018**, *1761*, 275–301. [CrossRef] [PubMed]

International Journal of *Molecular Sciences*

MDPI

Article

DIA-Based Quantitative Proteomics Reveals the Protein Regulatory Networks of Floral Thermogenesis in *Nelumbo nucifera*

Yueyang Sun [1,†], Yu Zou [1,†], Jing Jin [2], Hao Chen [3], Zhiying Liu [1], Qinru Zi [1], Zeyang Xiong [1], Ying Wang [1], Qian Li [1], Jing Peng [3] and Yi Ding [1,*]

1 State Key Laboratory of Hybrid Rice, Department of Genetics, College of Life Sciences, Wuhan University, Wuhan 430072, China; yueyangsun@whu.edu.cn (Y.S.); zouxiaoyu@whu.edu.cn (Y.Z.); zhiying-liu@whu.edu.cn (Z.L.); qinruzi@whu.edu.cn (Q.Z.); zeyangxiong@whu.edu.cn (Z.X.); yingwang@whu.edu.cn (Y.W.); liqian0074@whu.edu.cn (Q.L.)

2 Department of Biotechnology, College of Life Sciences, Guizhou University, Guiyang 550025, China; jinjing1130@whu.edu.cn

3 Institute of Vegetables, Wuhan Academy of Agricultural Sciences, Wuhan 430065, China; haochen2013@whu.edu.cn (H.C.); pengjing67@163.com (J.P.)

* Correspondence: yiding@whu.edu.cn; Tel./Fax: +86-27-6875-4319

† These authors contributed equally to this work.

Abstract: The sacred lotus (*Nelumbo nucifera*) can maintain a stable floral chamber temperature between 30 and 35 °C when blooming despite fluctuations in ambient temperatures between about 8 and 45 °C, but the regulatory mechanism of floral thermogenesis remains unclear. Here, we obtained comprehensive protein profiles from receptacle tissue at five developmental stages using data-independent acquisition (DIA)-based quantitative proteomics technology to reveal the molecular basis of floral thermogenesis of *N. nucifera*. A total of 6913 proteins were identified and quantified, of which 3513 differentially abundant proteins (DAPs) were screened. Among them, 640 highly abundant proteins during the thermogenic stages were mainly involved in carbon metabolism processes such as the tricarboxylic acid (TCA) cycle. Citrate synthase was identified as the most connected protein in the protein-protein interaction (PPI) network. Next, the content of alternative oxidase (AOX) and plant uncoupling protein (pUCP) in different tissues indicated that AOX was specifically abundant in the receptacles. Subsequently, a protein module highly related to the thermogenic phenotype was identified by the weighted gene co-expression network analysis (WGCNA). In summary, the regulation mechanism of floral thermogenesis in *N. nucifera* involves complex regulatory networks, including TCA cycle metabolism, starch and sucrose metabolism, fatty acid degradation, and ubiquinone synthesis, etc.

Keywords: floral thermogenesis; *Nelumbo nucifera;* DIA-based quantitative proteomics; time series analysis; protein-protein interaction network; WGCNA

Citation: Sun, Y.; Zou, Y.; Jin, J.; Chen, H.; Liu, Z.; Zi, Q.; Xiong, Z.; Wang, Y.; Li, Q.; Peng, J.; et al. DIA-Based Quantitative Proteomics Reveals the Protein Regulatory Networks of Floral Thermogenesis in *Nelumbo nucifera*. *Int. J. Mol. Sci.* **2021**, *22*, 8251. https://doi.org/10.3390/ijms22158251

Academic Editors: Sixue Chen and Setsuko Komatsu

Received: 8 July 2021
Accepted: 29 July 2021
Published: 31 July 2021

1. Introduction

Body heat production is generally considered to occur in homeothermic animals, which can adapt to changes in environmental temperature through thermoregulation [1]. However, similar thermogenesis has also been found and described in the floral organs of some flowering plants, including Araceae [2–5], Nelumbonaceae [6], Nymphaeaceae [7], Magnoliaceae [8], Cycadaceae [9], and so on. The appropriate temperature generated by thermogenesis contributes to the prevention of the flower organs from suffering low temperatures [3], promoting the success of double fertilization [10], facilitating pollen germination and pollen tube growth [11], and attracting insect pollinators by emitting odors and providing heat rewards [12,13].

With the deepening of research on thermogenesis in plants, the molecular mechanism of floral thermogenesis has been gradually explored. There are two energy-dissipating systems thought to be related to the thermogenesis of plants [14,15]. One is the alternative pathway of plant mitochondria mediated by alternative oxidases (AOXs) and the other involves plant uncoupling proteins (pUCPs). In the alternative pathway, electrons that bypass cytochrome c reductase (complex III) and cytochrome c oxidase (COX, complex IV) are directly transferred from ubiquinone to AOX and then transferred to oxygen through AOX. Therefore, most of the energy is released as heat with no energy conservation for the electron flux through the alternative pathway [16]. AOX protein is a nuclear-encoded mitochondrial inner membrane protein that functions as the terminal oxidase of the alternative pathway [17]. In thermogenic plants, the AOX activity is related to thermogenesis, which can characterize the capacity of the alternative pathway and increases in thermogenic tissue when thermogenesis occurs [18,19]. For example, the increased AOX activity was detected during the thermogenic stages in *Nelumbo nucifera*, while it returned to the initial level when thermogenesis ended [20].

Uncoupling proteins (UCPs) form a subfamily of mitochondrial carrier protein, which are located on the inner membrane of mitochondria, and can dissipate the $\Delta\mu H^+$ generated by the respiratory chain so that protons can bypass the ATP synthase and be transported from the membrane space to the mitochondrial matrix [21–23]. This process leads to the uncoupling of respiration and phosphorylation, and electrochemistry energy is released as heat [23]. It was found that pUCPs might participate in thermogenesis in the thermogenic plant, such as skunk cabbage (*Symplocarpus renifolius*). SrUCPA, the main pUCPs of *S. renifolius*, had a high abundance in spadix mitochondria, meanwhile SrUCPA and SrAOX were specifically co-expressed in the thermogenic tissue or stage, which indicated that they might play a role in the thermogenesis of skunk cabbage [24].

N. nucifera, an important aquatic vegetable crop that blooms in early summer, is also a thermogenic plant that has been described in some detail [25]. The relationship between growth development and the thermogenic period of *N. nucifera* flowers can be divided into three distinct physiological phases: Pre-thermogenic, thermogenic, and post-thermogenic phases. Thermogenesis mainly occurs in the receptacle tissue with increased activity at the thermogenic phase and cessation at the post-thermogenic phase [20]. *N. nucifera* is the only thermogenic plant to date that the electron flux flowing to AOX (AOX flux) and COX (COX flux) has been quantified *in vivo* using the stable oxygen isotope technique [20,26]. Measurement of the electron flux through the AOX pathway can accurately determine AOX activity [27]. The AOX flux increased significantly during the thermogenic stages and could even account for 93% of the total respiratory flux in the hottest receptacles, while the COX flux did not change significantly [20]. There was a significantly positive correlation between the thermogenesis levels in thermogenic receptacles and both total respiratory flux and AOX flux, but there was almost no correlation with COX flux [20]. Therefore, AOX was considered to play a role in the regulation of the floral thermogenesis of *N. nucifera*. Unlike *Sauromatum guttatum* [28] and *Arum italicum* [19], the increase of AOX protein in *N. nucifera* receptacles was synchronized with the increase in thermogenic activity [20]. Amino acid multiple sequence alignment and phylogenetic analysis showed that no specific sequence of AOX protein related to thermogenic activity was found in *N. nucifera* [27]. Similarly, no specific primary structures of AOXs and pUCPs related to thermogenesis were found in *S. renifolius* [29]. Hence, there are probably additional factors involved in the regulation of floral thermogenesis. Recently, with the development of high-throughput sequencing technology, several transcriptome-level studies have been performed to reveal the molecular basis of floral thermogenesis. In *S. renifolius*, the utilization of SuperSAGE sequencing revealed gene expression profiles related to the maintenance and termination of thermogenesis [30]. In *Arum concinnatum*, 1266 genes highly correlated with the temperature trend of the flower organs were identified by transcriptome analysis [31]. Then, integrated analysis of small RNA and transcriptome sequencing also revealed the role of miRNAs (microRNAs) in the regulation of the floral thermogenesis of *N. nucifera* [32].

However, so far, there have been no proteomic studies on plant thermogenesis that have been reported.

Data-independent acquisition (DIA)-based quantitative proteomics technology is an emerging technology for large-scale protein identification and quantification in recent years. It has the advantages of deep proteome coverage, high quantitative reproducibility, and accuracy, and has been applied in biomarker research, clinical research, basic research, and other fields [33–35]. In the present study, we provided comprehensive protein profiles from receptacle tissue using DIA-based quantitative proteomics technology to further reveal the molecular basis of floral thermogenesis in *N. nucifera*. The protein regulatory networks related to thermogenesis were identified based on the correlation between protein abundance and the thermogenesis pattern (Figure S1). These global protein profiles during floral thermogenesis in *N. nucifera* have provided new insights into plant thermogenesis and lays a solid foundation for in-depth study on the regulatory mechanisms of floral thermogenesis.

2. Results

2.1. An Overview of the Quantitative Proteomics Analysis

The experimental strategy of data-independent acquisition mass spectrometry (DIA-MS) identification technology in the present study is shown in Figure S2. DIA-MS analysis requires the construction of a data-dependent acquisition (DDA)-based spectral library before protein quantification. As a result, 44976 peptides corresponding to 9461 proteins were identified in the spectral library (Table S1). The DIA data with the spectra library were analyzed by Spectronaut Pulsar 11.0 (Biognosys AG, Switzerland), and the results showed that 6913 proteins were further quantified (Table S2). After being mapped to multiple public databases, the 6913 identified proteins were annotated (Table S3). Statistics of the proteins and peptides identified for each sample by DIA-MS are shown in Figure 1A.

To assess the quality of the proteomic data, quality control (QC) analysis was performed. The median of intra-group coefficient of variation (CV) distribution for each sample was less than 10% (Figure 1B). Principle component analysis (PCA) showed that the samples from five different stages could be well separated, and samples from the same stage could be clustered well (Figure 1C). Pearson correlation coefficient of the protein levels between every two samples showed that three replicates in the same stages had a high correlation (Figure S3). All of these analyses indicated the good repeatability and reliable quality of the proteomic data in this study.

Proteins with a fold change (FC) ≥ 2 and q-value < 0.05 were defined as significantly differentially abundant proteins (DAPs). In total, 3513 DAPs of ten comparison groups were identified in the present study (Figure 1D and Table S4). The DAPs increased with the development of flowers when using stage 1 (S1) as a control. Moreover, the DAPs between adjacent stages (stage 2 (S2) vs. S1, stage 3 (S3) vs. S2, stage 4 (S4) vs. S3, stage 5 (S5) vs. S4) showed that S3 vs. S2 and S5 vs. S4 had more DAPs, which might be related to the larger developmental changes of *N. nucifera* flowers in these two comparative stages (from S2 to S3, the flowers went from buds to blooms, and from S4 to S5, the petals withered gradually). When using stage 3 as a control, the number of DAPs of the comparison groups between different thermogenic stages (S3 vs. S2 and S3 vs. S4) was less than the comparison groups between thermogenic and non-thermogenic stages (S3 vs. S1 and S3 vs. S5).

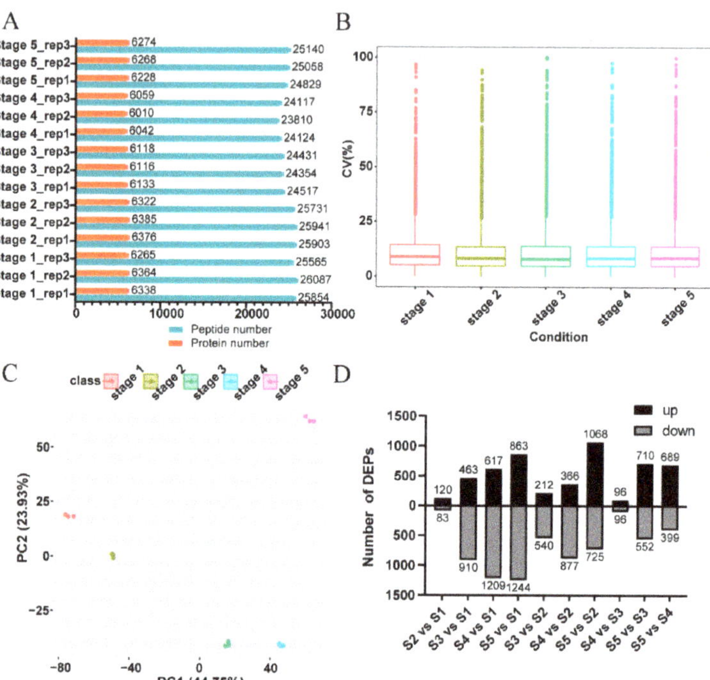

Figure 1. An overview of the quantitative proteomics analysis: (**A**) Statistics of the proteins and peptides identified in each sample. Each stage includes three replications. The 'rep' represents 'replication' on the y-axis. (**B**) Distribution of intra-group coefficient of variation (CV) for each sample. (**C**) Principal component analysis (PCA) of the quantified proteins for each sample. (**D**) Statistics of differentially abundant proteins (DAPs) in each comparison group. S1, S2, S3, S4, and S5 represent stage 1, stage 2, stage 3, stage 4, and stage 5, respectively.

2.2. Time-Series Analysis of DAPs

To reveal the time dynamics of the abundance of all DAPs, we performed the time-series cluster analysis based on five continuous developmental stages. As a result, all DAPs were classified into six clusters with different expression patterns (Figure 2 and Table S5). Among the six clusters, the proteins in cluster 3 (C3) which had a high abundance during thermogenesis increased at stage 2, reached a peak at stage 3 or stage 4, and decreased gradually at stage 5, while the proteins in cluster 5 (C5) had a low abundance in the thermogenic stages (Figure S4). The expression patterns of the proteins in C3 were basically consistent with the thermogenesis pattern while that of C5 were opposite. A total of 640 and 429 proteins were respectively classified into C3 and C5 (Table S5). Considering that the expression patterns of the proteins in C3 and C5 were related to the thermogenesis pattern in the thermogenic receptacles during the five stages, we speculated that these two protein clusters were related to the thermogenic process, which occurred significantly from stage 2 to stage 4 (Figure S1).

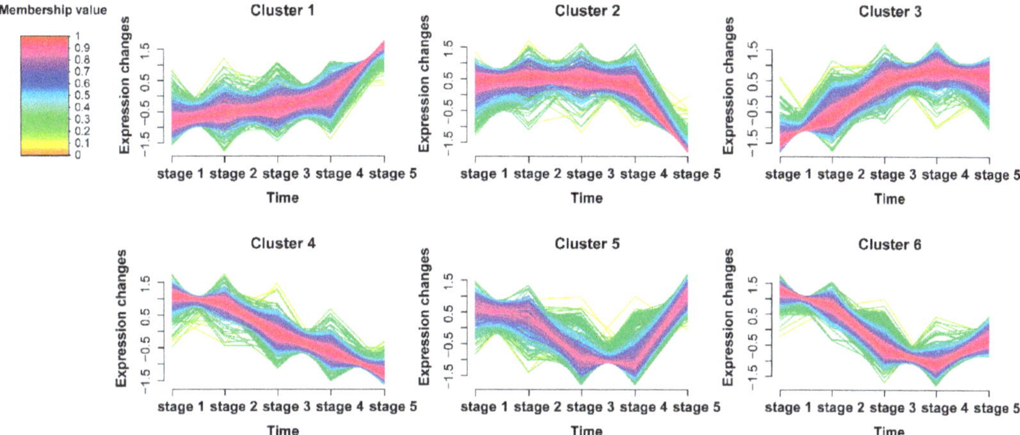

Figure 2. Time series analysis of DAPs: Time series analysis identified six distinct temporal patterns of proteins abundance using fuzzy c-means algorithm. The x-axis represents five developmental stages, while the y-axis represents normalized abundance changes of DAPs. The membership value represents the strength of the relationship between a protein and a corresponding cluster.

2.3. Functional Annotation Analysis of the Protein Clusters Related to the Thermogenic Process

To further understand the function of the proteins related to thermogenesis in C3 and C5, the gene ontology (GO), Kyoto encyclopedia of genes and genomes (KEGG) pathway enrichment analysis, and Eukaryotic orthologous groups (KOG) annotation were performed (Table S6). For the annotation of C3, the main significantly enriched GO terms of the biological process (BP) included tricarboxylic acid cycle (TCA cycle), aerobic respiration, and oxidation-reduction process (Figure S5A, Table S6). Oxidoreductase activity, catalytic activity, and pyruvate dehydrogenase activity were the main enriched GO terms of molecular function (MF). Moreover, the main representative GO terms of the cellular component (CC) were the TCA cycle enzyme complex, mitochondrion, and the dihydrolipoyl dehydrogenase complex. KEGG pathway enrichment results showed that citrate cycle, biosynthesis of secondary metabolites, carbon metabolism, and so on, were the most significantly enriched (Figure 3A, Table S6). KOG annotation showed that many proteins were concentrated in energy production and conversion, carbohydrate transport and metabolism, and post-translational modification (Figure S6A). For the annotation of C5, photosynthesis, signal transducer activity, and photosystem were the most enriched GO terms (Figure S5B). Photosynthesis, plant hormone signal transduction, and MAPK signaling pathway were the top-three enriched KEGG pathways (Figure 3B, Table S6). KOG annotation showed that signal transduction mechanisms covered the most proteins except for only the general function prediction (Figure S6B).

AOX proteins that had been proposed to play an important role in the thermogenic process of *N. nucifera* [20,27] were identified in cluster 3, including NNU_06050-RA, NNU_06051-RA, and NNU_10976-RA (in the present study, NNU_06050-RA represents the corresponding protein ID of gene *NNU_06050*, as are other protein IDs). Photosynthesis, the most enriched pathway of cluster 5, was reported to be related to the function of the receptacles after the end of thermogenesis [36]. This indicated that the DAPs in cluster 3 were highly abundant during the thermogenic stages and might be closely related to floral thermogenesis in *N. nucifera*.

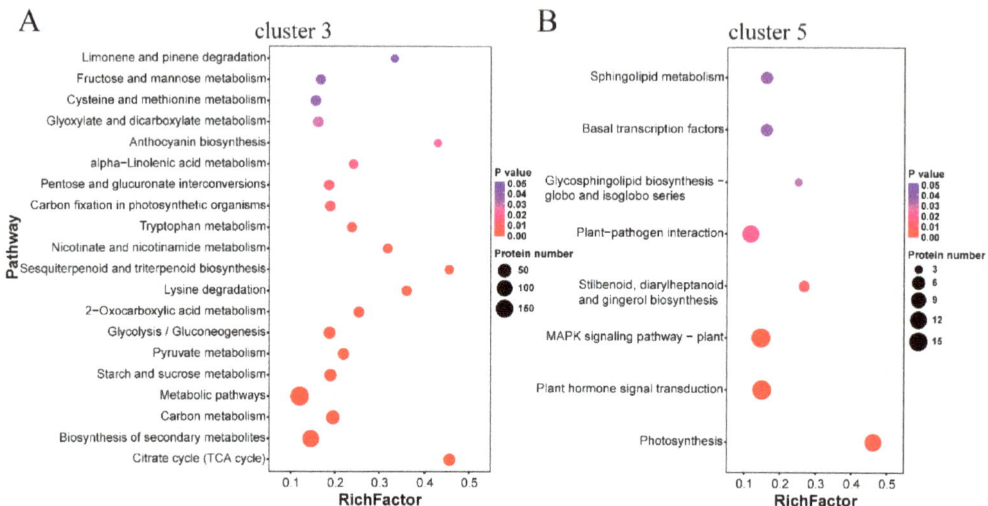

Figure 3. Kyoto encyclopedia of genes and genomes (KEGG) pathway enrichment analysis and Eukaryotic orthologous groups (KOG) annotation: (**A**,**B**) The main enriched KEGG pathways of cluster 3 and 5. The x-axis indicates the enrichment factor (RichFactor) which represents the number of DAPs annotated to each pathway divided by the number of all identified proteins annotated to the same pathway. A greater RichFactor indicates greater enrichment. The y-axis represents significantly enriched pathways with increasing significance from top to bottom.

2.4. Validation of the Proteomic Data by qRT-PCR and Western Blot

To verify the expression patterns of the proteins identified by DIA-MS technology, nine randomly selected DAPs and three interested proteins were verified by qRT-PCR, which reflected their expression at the transcriptional level. It has been proposed that the protein AOX or pUCP participates in the thermogenesis of some thermogenic plants [20,24]. Our results indicated that AOX1a (NNU_06050-RA) and pUCPs (NNU_03736-RA, NNU_13276-RA) showed the same expression pattern between the relative expression level in qRT–PCR, fold changes in RNA–seq data (accessed from Zou et al. [32]), and the proteomic data (Figure 4A–D). Similarly, the expression patterns of the transcripts of seven DAPs related to the TCA cycle, carbon metabolism, and oxidative phosphorylation in cluster 3 were basically consistent with that of the RNA–seq data and proteomic data (Figure 4E–M). Although they all had almost the same expression pattern between the proteomic level, RNA-seq level, and qRT-PCR level, the proteomic abundance peaks of the above ten proteins basically lagged behind the corresponding transcripts expression peaks. However, the transcript of a protein (probably inactive leucine-rich repeat receptor-like protein kinase IMK2, NNU_06853-RA) in cluster 5 had a consistent expression change with RNA-seq data but not with proteomic data. Our qRT-PCR and RNA-seq results both showed that *NNU_06853* had the lowest expression at stage 5 while the proteomic data showed that NNU_06853-RA had the lowest abundance at stage 4, and its abundance went up in stage 5 (Figure 4L,N). This different expression pattern between protein level and transcript level might be caused by the post-transcriptional regulation of the transcripts. ATP synthase subunit delta, chloroplastic, NNU_09824-RA (ATPD), a protein in cluster 1 related to photosynthesis, had a high abundance at stage 5, and its corresponding transcript also had the same expression pattern at the qRT-PCR and RNA-seq levels (Figure 4L,O).

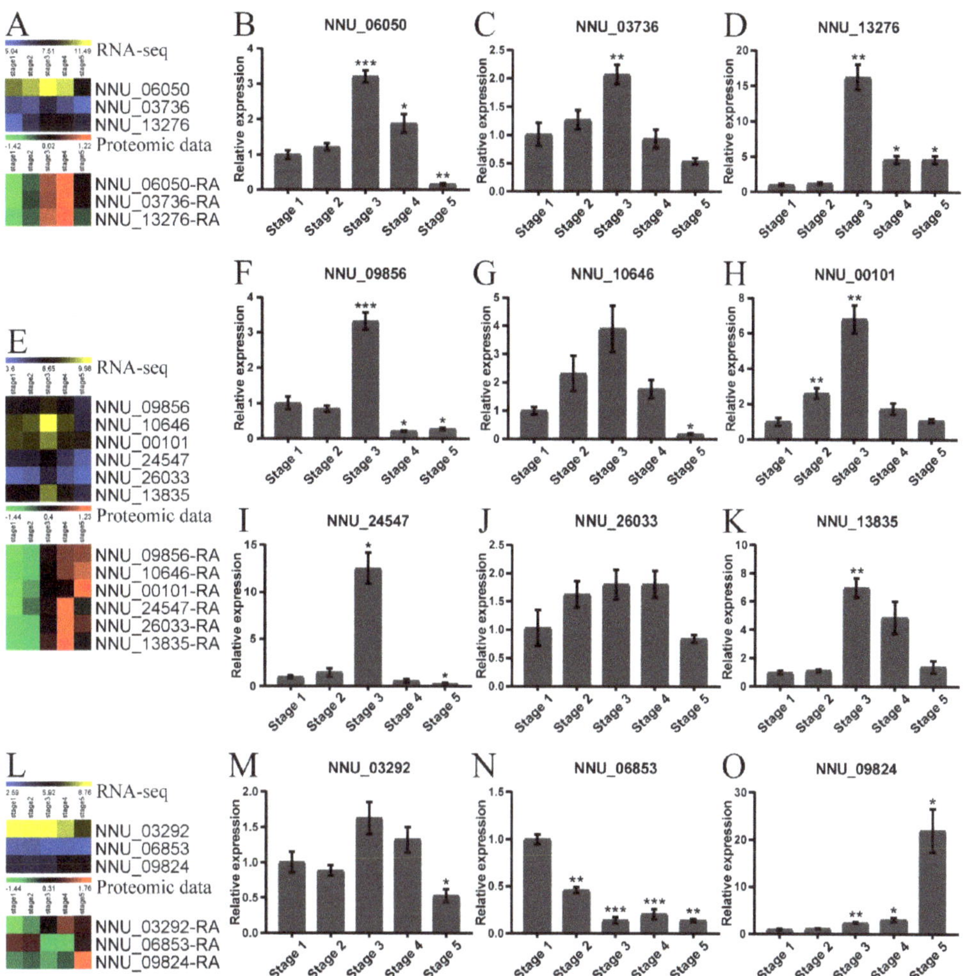

Figure 4. Validation of 12 DAPs by qRT-PCR: (**A–D**) Heatmap and relative transcript expression of AOX1a (alternative oxidase 1a, NNU_06050-RA) and pUCPs (plant uncoupling proteins, NNU_03736-RA and NNU_13276-RA). (**E–K**) Heatmap and relative transcript expression of NAD-ME1 (NAD-dependent malic enzyme 1, mitochondrial, NNU_09856-RA), IDH3 (Isocitrate dehydrogenase (NAD) regulatory subunit 3, mitochondrial, NNU_10646-RA), CYTC (Cytochrome c, NNU_00101-RA), LPD1 (Dihydrolipoyl dehydrogenase 1, mitochondrial, NNU_24547-RA), HXK1 (Hexokinase-1, NNU_26033-RA) and GAL1 (Galactokinase, NNU_13835-RA). (**L–O**) Heatmap and relative transcript expression of SDH2-2 (Succinate dehydrogenase (ubiquinone) iron-sulfur subunit 2, mitochondrial, NNU_03292-RA), probably inactive leucine-rich repeat receptor-like protein kinase IMK2 (NNU_06853-RA) and ATPD (ATP synthase subunit delta, chloroplastic, NNU_09824-RA, NNU_09824-RA). The heatmaps were used to visualize the transcript levels in the RNA-seq data and the protein levels in the proteomic data. The error bars indicate the standard deviation of three replicates. Asterisks indicate a significant difference as determined by one-way ANOVA using stage 1 as a control (* $p < 0.05$, ** $p < 0.01$, *** $p < 0.001$).

The expression patterns of three proteins, including AOX1a, pUCP1 (NNU_03736-RA), and catalase (CAT, NNU_23981-RA), were further validated by our Western blot analysis (Figure 5A–C). The expression patterns of the three proteins in the Western blot experiments (Figure 5A) were highly consistent with the proteomic data (Figure 5B). When using stage 1 as a control, the abundance of AOX increased significantly in the other stages, while only pUCP had a significant increase in stage 4, and there was no significant difference in the abundance of CAT (Figure 5C). Coincidentally, the significant analysis of the Western blot results was consistent with the DAPs analysis of the proteomic data. Overall, all the above results showed that our proteomic data were reliable.

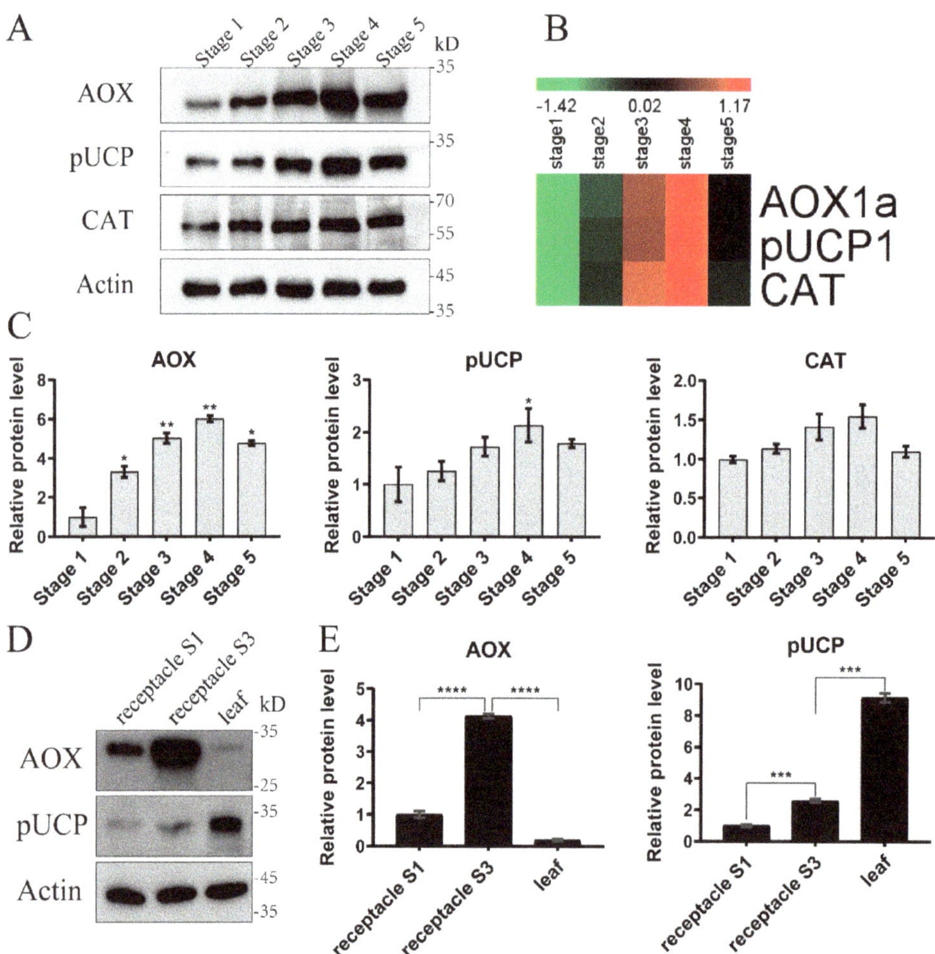

Figure 5. Western blot assays: (**A**) The expression patterns of AOX, pUCP, and catalase (CAT) validated by Western blot. (**B**) Quantitative proteomic data of AOX, pUCP, and CAT visualized by heatmap. (**C**) Grayscale value analysis of the results in Figure 5A using Image J. (**D**) The abundance of AOX and pUCP in different tissues of *N. nucifera*. (**E**) Grayscale value analysis of the results in Figure 5D. S1 and S3 represent stage 1 and stage 3, respectively. The error bars indicate the standard deviation of three replicates. Asterisks indicate a significant difference as determined by one-way ANOVA using stage 1 as a control in Figure 5C and 'receptacle S3' as a control in Figure 5E (* $p < 0.05$, ** $p < 0.01$, *** $p < 0.001$, **** $p < 0.0001$).

2.5. Abundance of the Proteins AOX and pUCP between Thermogenic and non-Thermogenic Tissues

To further understand the abundance of AOX and pUCP in different tissues, we detected the content of these two proteins between thermogenic receptacles and non-thermogenic leaves of *N. nucifera*. Total protein was extracted and analyzed by Western blot using anti-AOX and anti-UCP antibodies (Figure 5D). The expression patterns of

AOX and pUCP were opposite when compared between thermogenic receptacles and non-thermogenic leaves. AOX was extremely abundant in the thermogenic receptacles during thermogenic stage 3 and could be detected in non-thermogenic stage 1. However, comparing to thermogenic receptacles, the AOX protein in non-thermogenic leaves had a much weaker band which was almost undetectable. In contrast, pUCP had a very low content in thermogenic receptacles and a high content in non-thermogenic leaves. Gray value analysis showed significant differences between the thermogenic stage and the non-thermogenic stage or tissue (Figure 5E). AOX was abundant in the thermogenic stage of receptacles while pUCP was abundant in non-thermogenic leaves.

2.6. Protein-Protein Interaction (PPI) Network and Weighted Gene Co-Expression Network Analysis (WGCNA)

To reveal protein regulatory networks, a PPI network analysis of the proteins in cluster 3 was performed using the STRING PPI database. Protein-protein relationships in the top 200 confidence intervals were used to construct the PPI network (Figure 6). Several highly connected proteins were identified. For example, CIT (citrate synthase, mitochondrial, NNU_11668-RA), a protein involved in the TCA cycle, had the most edges and represented the top hub protein (highly connected protein) in cluster 3. Moreover, succinate dehydrogenase (ubiquinone) iron-sulfur subunit 2, mitochondrial, NNU_03292-RA (SDH2-2), dihydrolipoyl dehydrogenase 1, mitochondrial, NNU_24547-RA (LPD1), and isocitrate dehydrogenase (NADP) isoform X1, NNU_21884-RA (IDH) were the other hub proteins with no less than 10 edges.

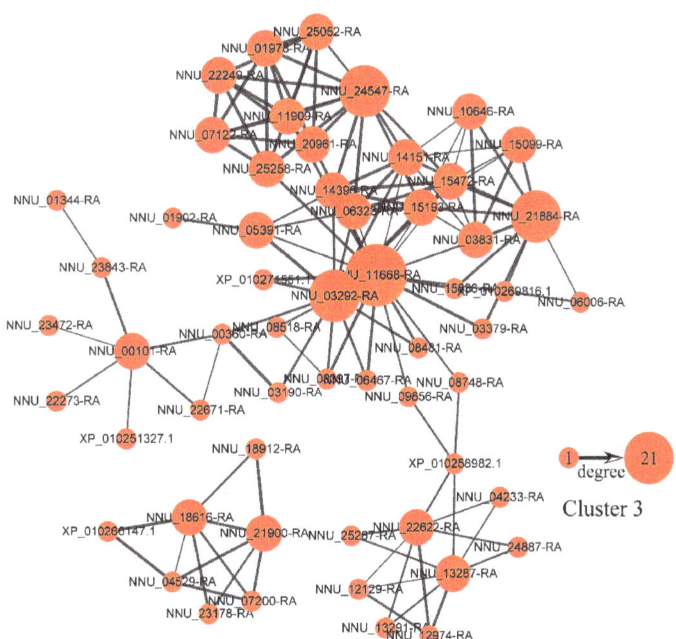

Figure 6. Protein-protein interaction (PPI) network of the proteins in cluster 3: The edge numbers of the proteins range from 1 to 21.

WGCNA [37,38] was performed to reveal protein modules related to the thermogenic phenotype and developmental stages. As a result, ten modules with different expression patterns were identified (Figure 7A,B, Figure S7, and Table S7). Based on the criteria (Pearson $r \geq 0.8$ or $r \leq -0.8$, p-value ≤ 0.05), we found that the red module of 78 proteins was highly correlated with the thermogenic phenotype (Figure 7C). The co-expressed proteins network of the red module was visualized, of which several hub proteins were identified (Figure 7D). For example, the top hub protein was glucose-1-phosphate adenylyltransferase small subunit, chloroplastic/amyloplastic (APS1, NNU_20629-RA) involved in the process of starch biosynthesis. Then, KEGG function enrichment analysis showed that metabolic processes and fatty acid degradation were the main enriched pathways (Figure 7E, Table S8). We also found that ubiquinone and other terpenoid-quinone biosynthesis were enriched, of which ubiquinone was an essential component in the mitochondrial electron transport chain.

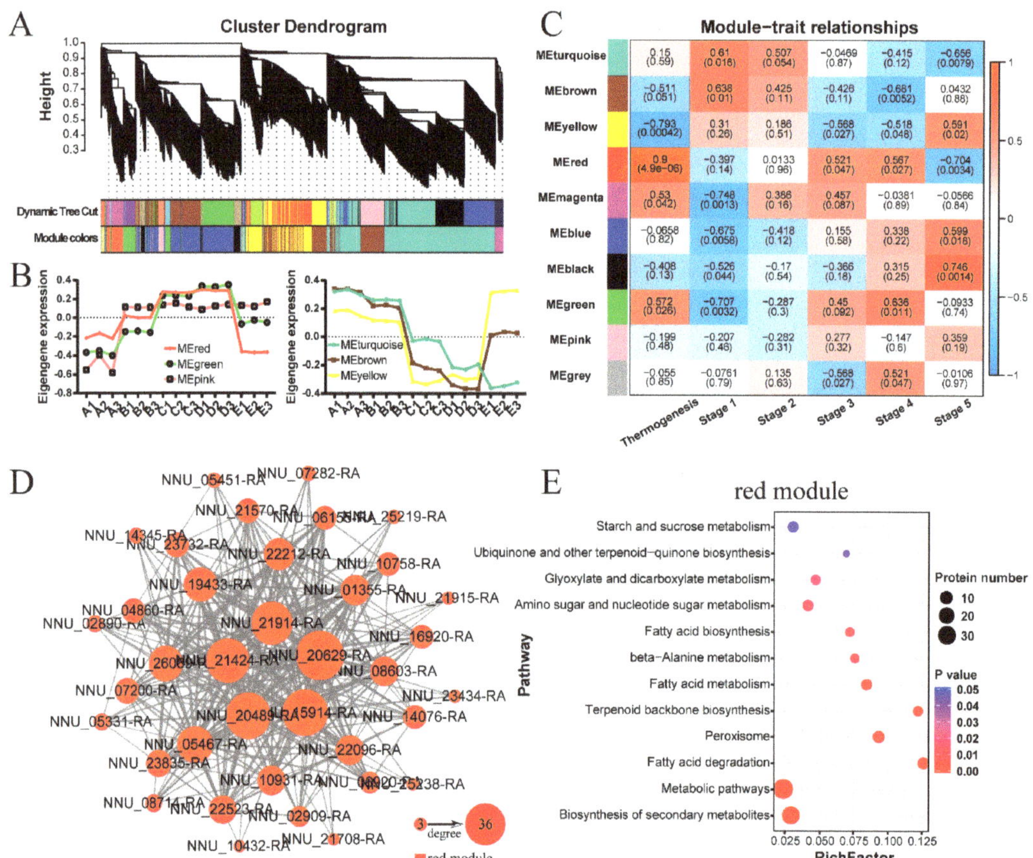

Figure 7. Weighted gene co-expression network analysis: (**A**) Hierarchical clustering dendrogram shows co-expression modules that are color-coded. 'Dynamic Tree Cut' represents initial modules. Module colors represent the final modules. Each branch in the hierarchical tree or each vertical line in color bars represents one protein. (**B**) Temporal expression patterns of different modules. Eigengene is the first principal component of a given module, which can be considered a representative of the protein profiles in a module. A, B, C, D, and E represent stage 1, stage 2, stage 3, stage 4, and stage 5, respectively. Each stage includes three replications. (**C**) Module trait correlation plot. Each row represents one module. Each column represents one trait attribute. The blue color represents negative correlation and the red color represents positive correlation. (**D**) The co-expressed proteins network of the red module with edge weight $\geq$ 0.52. The edge numbers of the proteins range from 3 to 36. (**E**) KEGG enrichment analysis of the proteins in the red module.

3. Discussion

3.1. The Low-Correlation Omics Data Showed Similar Functions That Related to Thermogenesis

In the present study, we identified 640 DAPs in cluster 3 with high abundance and 429 DAPs in cluster 5 with low abundance during thermogenesis of *N. nucifera*. A previous study from our laboratory has identified 3054 thermogenesis-related genes in *N. nucifera* at the transcript level [32], of which 1795 differentially expressed genes were highly expressed and 1259 DEGs were low-expressed genes during thermogenesis. As far as these DEGs/DAPs related to thermogenesis were concerned, only 155 DEGs/DAPs with high abundance and 57 DEGs/DAPs with low abundance during thermogenesis were co-expressed at both transcription and protein levels (Figure S8). Therefore, the transcriptome data and proteome data were less correlated. However, the functional annotation analysis at the two omics levels suggested that the low-correlation omics data showed similar functions that related to thermogenesis. The DEGs/DAPs highly expressed during thermogenesis at the two omics levels were enriched in the KEGG pathways related to the tricarboxylic acid cycle, carbon metabolism, 2-carboxylic acid metabolism, glycolysis/gluconeogenesis, lysine degradation, and other metabolic processes. Moreover, in enriched GO terms of cellular component, mitochondrion was one of the most enriched terms. These results indicated that thermogenesis was related to respiratory metabolism and mitochondrial function, which was consistent with the study of gene expression profiles in *S. renifolius* [30]. Previously studies have suggested that carbohydrates, especially starch, were the respiratory substrates for floral thermogenesis in *N. nucifera* [20,25]. KOG annotation of cluster 3 showed that these proteins with high abundance during thermogenesis were concentrated in energy production and conversion as well as carbohydrate transport and metabolism, and starch and sucrose metabolism was one of the most significantly enriched KEGG pathways (Figure 3). Moreover, APS1, a protein involved in the synthesis of starch, was identified as a hub protein of the co-expressed proteins network (Figure 7D). These results indicated that the enhancement of cell respiration metabolism and the consumption of carbohydrate substrates help the flowers of *N. nucifera* to generate enough heat during anthesis. Similarly, in the present study, the DEGs/DAPs with low abundance during thermogenesis were significantly enriched in the photosynthetic metabolic pathway. The function of the receptacles during the thermogenic periods was mainly to generate heat. After thermogenesis was over, the color of the receptacles gradually changed from yellow to green with its function changing to photosynthesis [36].

3.2. The Protein Expression Patterns of AOX and pUCP between Thermogenic and non-Thermogenic Tissues were Completely Different

Although the COX flux did not change significantly in the receptacles of *N. nucifera* during the five stages, it implied that pUCP might not play a role in the thermogenesis of *N. nucifera* [20]. However, there was no direct evidence showing the abundance of pUCP in receptacles tissues. Grant et al. [20] tried to detect pUCP in the receptacles of *N. nucifera* by immunoblotting, but no band was detected. In the present study, the abundance of pUCP was determined by Western blot assays in the receptacles at the five stages (Figure 5). The abundance change of pUCP was similar to AOX, but there was almost no significant change in comparison with AOX (Figure 5C). The insignificant change in pUCP abundance was consistent with the insignificant change in COX flux. Then, in the comparison between non-thermogenic and thermogenic tissues, pUCP accumulated in non-thermogenic leaves, while AOX accumulated in thermogenic receptacles (Figure 5D,E). Therefore, the expression patterns of AOX and pUCP were opposite between non-thermogenic and thermogenic tissues. Coincidentally, the same situation was detected in the spadices of thermogenic *Symplocarpus renifolius* and non-thermogenic *Lysichiton camtschatcensis*, which are closely related to skunk cabbages in morphology and phylogeny [29]. AOX accumulated in thermogenic *S. renifolius* while pUCP accumulated in non-thermogenic *L. camtschatcensis*. These results suggested the important role of AOX, rather than pUCP, in the thermogenesis of these thermogenic tissues. Taken together, we have provided more direct evidence that pUCP is unlikely to play a role in thermogenic receptacles. In addition, AOX accumulated abundantly in the receptacles and could even be detected in non-thermogenic stage 1 (Figure 5D), which indicated that AOX protein was specifically abundant in thermogenic tissue.

3.3. *TCA Cycle Metabolism, Fatty Acids Degradation, and Ubiquinone Biosynthesis Participate in Regulating Floral Thermogenesis of N. nucifera*

TCA cycle intermediate metabolites can affect the synthesis of AOX protein [39]. Research has reported that an increase in the content of citrate (1–10 mM) caused a dramatic increase in *AOX1* mRNA, AOX protein, and AOX respiratory capacity in tobacco suspension cells [17]. Citrate could induce a stable increase in *AOX1* mRNA, even at a condition close to physiological concentration (0.1 mM) or at a high concentration of 20 mM that affected cell viability [40]. In *Arabidopsis*, perturbations in cellular concentrations of citrate also increased *AOX1a* transcript and had a major impact on nucleus-encoded transcript abundance [41]. In the present study, the TCA cycle was the most significantly enriched KEGG pathway, and several hub proteins identified by the PPI network were key enzymes of the TCA cycle (Figure 6). The top hub protein was citrate synthase, which catalyzed the irreversible condensation reaction of oxaloacetate and acetyl-CoA in the TCA cycle to form citrate [42]. The reaction catalyzed by citric synthase is the initial and key step of the TCA cycle, which can affect the metabolic rate of the entire cycle [43,44]. Our results suggested that the high abundance of citrate synthase during thermogenesis might promote the increase of citrate content. Increased citrate could be a physiological signal between mitochondrial metabolic state and nuclear gene expression, which then affects the expression of the *AOX* gene [45]. Although the mechanism that mediates this physiological signal remains to be discovered [46], it is clear that changes in the TCA cycle metabolism during thermogenesis affect the expression of the nuclear-encoding *AOX* gene, thereby contributing to thermogenesis.

Our WGCNA analysis obtained a protein module related to the thermogenic phenotype. KEGG enrichment analysis of the proteins in this module showed that fatty acid degradation was one of the most enriched pathways in this module (Figure 7E). Since the increase in free fatty acid concentration inhibits AOX activity [47], and *N. nucifera* depends on the AOX-mediated alternative pathway for thermogenesis [20,26], the metabolic activity of fatty acid degradation occurring in the receptacles may facilitate the maintenance of normal activity of AOX, thereby contributing to heat production.

The redox state of ubiquinone and the absolute amount of ubiquinone could affect AOX activity [48,49]. It has been suggested that a low ubiquinone level might limit AOX activity [49]. In the present study, our KEGG enrichment analysis of the proteins in the red module also showed that ubiquinone biosynthesis was significantly enriched (Figure 7E). Thus, the biosynthesis metabolism of ubiquinone can also affect AOX activity, thereby regulating floral thermogenesis.

3.4. *Putative Model for the Regulatory Networks Involved in the Regulation Mechanism of Floral Thermogenesis in N. nucifera*

To summarize, based on the above results and discussion, we proposed a model that demonstrated the regulatory networks involved in the regulation mechanism of floral thermogenesis in *N. nucifera* (Figure 8). During the thermogenic stages of development, mitochondria first respond to ambient temperature changes, and thus the functional status of mitochondria that includes TCA cycle metabolism is altered. Some TCA intermediate metabolites are considered as physiological signals to induce the expression of nuclear-encoded *AOX* genes [45,46]. The signals of mitochondria transmitted to the nucleus to regulate the expression of nuclear-encoded mitochondrial proteins are called mitochondrial retrograde regulation (MRR) [50]. The induction of the *AOX* gene increases the content of AOX protein, which is beneficial to thermogenesis. In addition, succinate as a TCA intermediate has been shown to stimulate the activity of AOX in *N. nucifera* [27]. At the same time, the enhancement of the metabolic activity of fatty acid degradation and ubiquinone synthesis also contributes to the increase of AOX activity, thereby facilitating thermogenesis. In fatty acid degradation, 3-hydroxyacyl-CoA dehydrogenase (HAD, NNU_21424-RA), which was identified as a hub protein (Figure 7D), participates in the β-oxidation of fatty acids that is the main pathway of fatty acid degradation. Then, the increase in the metabolic activity of starch and sucrose provides essential respiratory substrates for the maintenance of thermogenesis. In summary, the regulation mechanism of floral thermogenesis in *N. nucifera* involves complex regulatory networks. In addition to the above mechanisms, changes in ROS homeostasis may also act as retrograde signals to affect the expression of *AOX* genes in *N. nucifera*. This requires further research to explore more comprehensive regulatory networks.

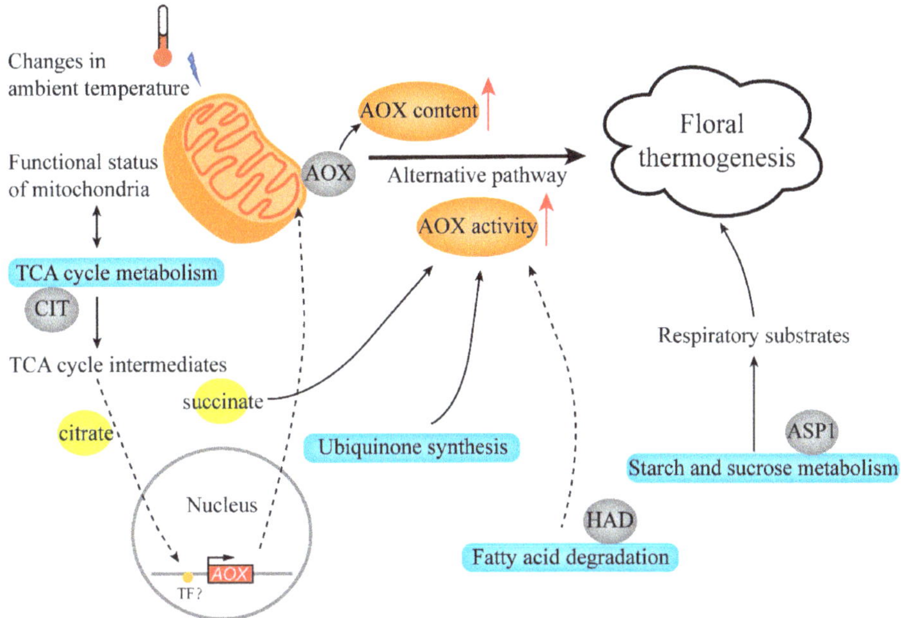

Figure 8. Putative model for the regulatory networks involved in the regulation mechanism of floral thermogenesis in *N. nucifera*. TCA, tricarboxylic acid; CIT, citrate synthase; HAD, 3-hydroxyacyl-CoA dehydrogenase; ASP1, glucose-1-phosphate adenylyltransferase small subunit.

4. Materials and Methods

4.1. Plant Materials

A cultivar 'E Zilian 1′ of *N. nucifera*, which was provided by the Institute of Vegetables, Wuhan Academy of Agricultural Sciences, Wuhan, Hubei, China, was used for DIA-based quantitative proteomic analysis. The cultivar had a new plant variety right (CNA20130464.0) granted by the Ministry of Agriculture of the People's Republic of China. The materials were preserved and cultivated in Wuhan National Germplasm Repository for Aquatic Vegetables (30°12′ N, 111°20′ E), with the preservation number V11A0692. As previously described, the development of *N. nucifera* flowers was divided into five sequential periods: firstly, the flowers were still small green buds at stage 1 and the flower buds became larger with pink tips and were close to blooming at stage 2; then, at stage 3, the full pink petals were half-open, showing mature pistils and immature stamens closely appressed to the receptacles; next, at stage 4, the petals were open horizontally and the stamens with mature pollen were spread out; finally, the petals and stamens withered at stage 5 (Table S9) [20,32]. The temperature of each receptacle was measured with a needle thermocouple and a RDXL4SD digital thermometer (OMEGA, USA) (Table S10). The receptacles of these five periods were collected, with three biological replicates in each period, immediately frozen in liquid nitrogen, and stored at −80 °C.

4.2. Protein Preparation and Digestion

Protein preparation was performed as follows: (1) Weigh approximately 0.4 g of each sample into a 2 mL centrifuge tube. Add 5% polyvinylpolypyrrolidone (PVPP) powder, a 5 mm magnetic bead, 1 mL lysis buffer (7 M Urea, 2 M Thiourea, 0.2% SDS, 20 mM Tris-Cl, pH 8.0), Phenylmethylsulfonyl fluoride (PMSF) with a final concentration of 1 mM and 2 mM Ethylenediaminetetraacetic acid (EDTA) to each tube, then vortex and incubate for 5 min, and add dithiothreitol (DTT) with a final concentration of 10 mM. (2) Shake the above mixture with a tissue grinder for 2 min (power = 50 Hz, time = 120 s) and centrifuge at 25,000 g for 20 min at 4 °C. (3) Transfer the supernatant into new centrifuge tubes and add DTT at a final concentration of 10 mM, followed by a water bath at 56 °C for 1 h. (4) After returning to room temperature, add iodoacetamide at a final concentration of 55 mM in the darkroom and incubate for 45 min. (5) Later, add four times the sample volume of cold acetone and keep at −20 °C for 2 h, followed by centrifugation at 25,000 g for 20 min at 4 °C. Repeat this step two to three times until the supernatant is colorless. (6) After the supernatant is carefully discarded, add a 5 mm magnetic bead and an appropriate amount of lysis buffer without SDS to the precipitate of each tube. (7) Shake the precipitate mixed with the lysis buffer using a tissue grinder (power = 50 Hz, time = 120 s) and centrifuge at 25,000 g for 20 min at 4 °C to obtain the supernatant, which is used for quantification, and then store at −80 °C in a refrigerator. Protein concentration was measured using the Bradford assay with bovine serum albumin as the standard [51]. SDS-PAGE and Coomassie brilliant blue R-250 staining assay were taken to evaluate the quality of the samples.

For protein digestion, the protein solution was diluted four-fold to reduce the concentration of urea before digestion. Each protein sample (100 μg) was digested for 4 h with trypsin at 37 °C (trypsin: protein = 1:40 (v/v), Hualishi Tech. Ltd., Beijing, China). Then, trypsin was added once more at the above ratio, and the hydrolysis was continued at 37 °C for 8 h. The enzymatically hydrolyzed peptides were desalted by Strata X column and dried under vacuum.

4.3. High pH Reverse Phase Fractionation

All samples were mixed equally (10 μg/sample) and diluted with 2 mL mobile phase A (5% acetonitrile (CAN) pH 9.8) and injected into Shimadzu LC-20AB liquid system (Shimadzu, Kyoto, Japan). A 5 um, 4.6 × 250 mm Gemini C18 column was used for liquid phase separation of the samples. The peptides were eluted at a 1 mL/min flow rate with a set of gradients: 5% buffer B (95% ACN, pH 9.8) for 10 min, 5% to 35% buffer B for 40 min, 35% to 95% buffer B for 1 min, buffer B maintained for 3 min and 5% buffer B equilibrated for 10 min. To collect the almost equivalent amounts of peptides from each pool, the elution peak was monitored at a wavelength of 214 nm and the components were collected every minute. Components were combined into a total of 10 fractions, which were then freeze-dried.

4.4. High-Performance Liquid Chromatography (HPLC)

The dried peptide samples were reconstituted with mobile phase A (2% ACN, 0.1% formic acid (FA)), centrifuged at 20,000 g for 10 min, and the supernatant was taken for injection. Separation was carried out with an UltiMate 3000 UHPLC system (Thermo Scientific, Waltham, MA, USA). The samples first entered the trap column to be enriched and desalted, and then were connected in series with a self-packing C18 column (150 μm inner diameter, 1.8 μm column material particle size, 25 cm column length), and were separated by the following effective gradient at a flow rate of 500 nL/min: 5% buffer B (98% ACN, 0.1% FA) for 0–5 min; 5% to 35% buffer B for 5–160 min; 35% to 80% buffer B for 160–170 min; 80% buffer B for 170–175 min; 5% buffer B for 176–180 min. The end of the nanoliter liquid phase separation was directly connected to the mass spectrometer.

4.5. Data-Dependent Acquisition Mass Spectrometry

The peptides separated by the liquid phase were ionized by the nanoESI source and then entered the tandem mass spectrometer Q-Exactive HF (Thermo Fisher Scientific, San Jose, CA, USA) for DDA mode detection. The main parameters were set as follows: Ion source voltage: 1.6 kV; MS scan range: 350–1500 m/z; MS resolution: 60,000; Maximal injection time (MIT): 100 ms; MS/MS collision type: Higher-energy collisional dissociation (HCD); Normalized collision energy (NCE): 28; MS/MS resolution: 30,000; MIT: 100 ms; Dynamic exclusion duration: 30 s. The start m/z for MS/MS was fixed to 100. Precursor for MS/MS scan satisfied: Charge range 2+ to 7+, top 20 precursors with intensity over 10,000. The automatic gain control (AGC) target was set to 3e6 for MS and 1e5 for MS/MS.

4.6. Data-Independent Acquisition Mass Spectrometry

DIA-MS was performed under the same LC system conditions as described above. The mass spectrometer was run under DIA mode and automatically switched between MS and MS/MS modes. The full scan was performed between 350–1500 m/z at 120,000 resolution, and the AGC target for the MS scan was set to 3e6 and the MIT was 50 ms. The full scan range (350–1500 m/z) was equally divided to 40 continuous windows for MS/MS scan with parameters set as follows: MS/MS collision type: HCD; MIT: Auto mode; MS/MS resolution: 30,000; Dynamic exclusion duration: 30 s; Stepped collision energy: 22.5, 25, 27.5; AGC target: 1e5.

4.7. Data Analysis

The raw data of DDA-MS were analyzed in the MaxQuant environment v.1.5.3.30 [52], employing the Andromeda search engine [53]. The MS/MS spectra were searched against the *N. nucifera* genome database LOTUS-DB (26641 entries, downloaded on 15 December 2018, http://lotus-db.wbgcas.cn) [54]. The search parameters were as follows: (1) Digestion enzyme: Trypsin with a maximum of two missed cleavages; (2) Minimal peptide length: 7; (3) Fixed modifications: Carbamidomethyl (C); and (4) Variable modifications: Oxidation (M) and Acetyl (Protein N-term). A false discovery rate (FDR) of 0.01 was set at the peptide and protein levels. The MaxQuant output file was used as a standard spectral library.

The raw data from DIA-MS were analyzed with the Spectronaut Pulsar 11.0 with default settings, which used the iRT peptides for retention time calibration [55]. Decoy generation was set to mutated, which is similar to scrambled but will only apply a random number of AA position swamps (min = 2, max = length/2). The false discovery rate (FDR) was estimated with the mProphet scoring algorithm [56] and set to 1% at the peptide and protein levels. Next, the R package MSstats [57] finished log2 transformation, normalization, and q-value (adjusted p-value) calculation. Differentially abundant proteins (DAPs) were also analyzed by MSstats, according to a fold change ≥ 2 and q-value < 0.05 (Student's t-test) as the screening criteria.

4.8. Bioinformatic Analysis

We annotated the function of proteins based on the following databases: Nr (NCBI nonredundant protein sequences), Swiss-Prot (A manually annotated and reviewed protein sequence database), KOG/COG (Clusters of Orthologous Groups of proteins), and KEGG. GO annotation was performed against the Nr database using the Blast2GO program [58]. The proteins were mapped to different biological pathways according to functional categories in the KEGG pathway database (http://www.genome.ad.jp/kegg/pathway.html). GO terms and KEGG pathways enrichment analysis were performed based on the presence of two proteins and the results of a hypergeometric test (p-value ≤ 0.05 indicated statistical significance). PCA was completed by psych package in R. Time series analysis was performed using the Mfuzz R package [59]. The heat maps of gene expression and protein abundance were drawn by MeV software (version 4.9.0). PPI network was constructed using the STRING online tool (https://string-db.org/) and visualized with the Cytoscape software (version 3.6.1) [60]. WGCNA was performed with an online analysis platform (http://www.ehbio.com/Cloud_Platform/front/#/). The core parameters for constructing WGCNA network were as follows: Power = 12, minimal module size = 25, network type = signed, merge cut height = 0.2, deep split = 2.

4.9. Quantitative Real-Time PCR (qRT-PCR) Analysis

The total RNA was extracted as described by Zou et al. [32]. Next, 1 μg RNA sample of each stage was used for reverse transcription with the RevertAid First Strand cDNA Synthesis Kit (Thermo Scientific, Waltham, MA, USA), according to the manufacturer's instructions. Then, qRT-PCR was performed with a StepOne Plus real-time PCR system (Applied Biosystems, Waltham, MA, USA) using a FastStart Universal SYBR Green Master Mix (Rox) (Roche, Germany). The reaction procedure was as follows: An initial denaturation at 95 °C for 10 min, followed by 40 cycles of 95 °C for 15 s, 60 °C for 1 min. The melting curve was determined for each primer pair to ensure specific amplification during the reaction process. Three replicates were made for each sample. The primers used in these experiments are shown in Table S11. *NnEF1a* gene (GenBank accession number AB491177.1) was used as the internal control, according to Zhu et al. [61], and the relative expression levels were normalized and calculated using the $2^{-\Delta\Delta Ct}$ method [62].

4.10. Western Blot Analysis

Equal protein samples were separated on SDS-PAGE and transferred to the nitrocellulose membrane. The membrane was blocked for 1 h at room temperature in TBST buffer (25 mM Tris, 137 mM NaCl, 2.7 mM KCl, 0.05% Tween, pH 7.4) containing 5% non-fat milk, and was then washed three times with TBST buffer. Primary antibodies were diluted in TBST buffer containing 3% non-fat milk (AOX at 1:1000; pUCP at 1:2000; CAT at 1:1000) and incubated with the membrane for 2 h at room temperature with agitation or overnight at 4 °C. After being rinsed three times with TBST buffer, the membrane was incubated with the secondary antibody (diluted at 1: 6000) for 1 h at room temperature, and then washed in TBST buffer three times. Finally, the prepared membranes were incubated with WesternBright™ ECL-HRP Substrate (Advansta, USA) for 1 to 3 min and put in a ChemiDoc™ Imaging System (Bio-Rad, Hercules, CA, USA) to visualize signal intensities. Western blot assays were repeated at least three times. Grayscale value analysis was performed using ImageJ software (National Institutes of Health, Bethesda, MD, USA).

The antibodies used for Western blot assays in the present study were as follows: (1) primary antibody: Anti-AOX1/2 (Agrisera, Vännäs, Sweden), anti-UCP (Agrisera, Vännäs, Sweden), anti-CAT (Agrisera, Vännäs, Sweden), and plant actin monoclonal antibody (ImmunoWay Biotechnology, Plano, TX, USA); (2) Secondary antibody: HRP-conjugated goat anti-rabbit/mouse IgG secondary antibodies (Proteintech Group Inc., Rosemont, IL, USA).

4.11. Statistical Analysis

Statistical significance of the data was determined using the Student's *t*-test to compare the two groups, and the one-way ANOVA test to compare multiple groups ($p < 0.05$).

5. Conclusions

In the present study, we reported the large-scale protein profiles related to floral thermogenesis through the next generation of label-free quantitative proteomics identification technology. A total of 6913 proteins were identified from receptacles at five different developmental stages, of which ten comparison groups with 3513 proteins were DAPs. Subsequently, we have identified 640 DAPs that were highly abundant during thermogenesis. The functional annotation of these proteins was mainly involved in pathways related to cellular respiration metabolism, such as the TCA cycle. Western blot analysis of AOX and pUCP further confirmed that pUCP was unlikely to participate in the floral thermogenesis of *N. nucifera*. Then, a protein module highly related to the thermogenic phenotype was identified by WGCNA analysis, which was mainly involved in metabolic processes, fatty acid degradation, and ubiquinone synthesis. In short, complex regulatory networks including TCA cycle metabolism, starch and sucrose metabolism, fatty acid degradation, and ubiquinone synthesis may participate in the regulation of floral thermogenesis in *N. nucifera*. The acquisition of the *N. nucifera* receptacle proteome will further enrich the research in the field of lotus proteomics and provide candidate proteins for future studies.

Supplementary Materials: The following are available online at https://www.mdpi.com/article/10.3390/ijms22158251-/s1, Figure S1: Thermogenesis pattern of *N. nucifera* receptacles at five developmental stages. The y-axis represents the difference between the receptacle temperature (Tr) and the ambient temperature (Ta). The error bars indicate the standard deviation of five replicates. Asterisks indicate a significant difference as determined by one-way ANOVA using stage 1 as a control (* $p < 0.05$; ** $p < 0.01$). Figure S2: The experimental strategy of DIA-based mass spectrometry identification technology in the present study. Figure S3: Heatmap of Pearson correlation between samples. Figure S4: Heatmap of the abundance of proteins in cluster 5 (A) and cluster 3 (B). Figure S5: Top five enriched GO terms of cluster 3 (A) and cluster 5 (B). The abscissa represents the number of proteins in each GO term, and the ordinate represents the GO terms of three categories. Figure S6: KOG annotation of the proteins in clusters 3 and 5. Figure S7: Temporal expression patterns of 'magenta', 'blue', 'black', and 'grey' modules. Figure S8: Omics correlation between transcriptome and proteome data. (A) DEGs/ DAPs that were highly abundant during thermogenesis. (B) DEGs/DAPs that were low-abundant during thermogenesis. Table S1: The detailed information of the peptides and proteins identified in the spectral library. Table S2: Details of the proteins identified and quantified by DIA-MS. Table S3: Annotation of all identified proteins. Table S4: The details of differentially abundant and non-differentially abundant proteins in each comparison group. Table S5: The protein members of six clusters classified by time series analysis. Table S6: The detailed functional annotation information of cluster 3 and cluster 5. Table S7: Ten modules identified by WGCNA analysis. Table S8: KEGG pathway enrichment analysis of the 'red' module. Table S9: Morphological characteristics of *N. nucifera* flowers in five sequential stages. Table S10: Measurement of receptacle temperature and ambient temperature. Table S11: The primers used in qRT-PCR assays.

Author Contributions: Conceptualization, Y.D., Y.S. and Y.Z.; validation, Y.S.; formal analysis, Y.S. and Z.L.; investigation, Y.S., Y.Z., H.C., Z.X., J.J., Y.W., Q.L.; resources, Y.D., J.P., and Y.Z.; writing—original draft preparation, Y.S. and Y.D.; writing—review and editing, Y.D., Y.S., J.J. and Q.Z.; visualization, Y.S. and Y.D.; supervision, Y.D.; project administration, Y.D.; funding acquisition, Y.D. All authors have read and agreed to the published version of the manuscript.

Funding: This research was funded by the National Natural Science Foundation of China (31271310).

Data Availability Statement: The mass spectrometry proteomics data have been deposited to the ProteomeXchange Consortium (http://proteomecentral.proteomexchange.org, 31 July 2021) via the iProX partner repository [63] with the dataset identifier PXD024269.

Acknowledgments: We thank BGI-Shenzhen Technology Co., Ltd. for the technical support in mass spectroscopy and bioinformatics analysis.

Conflicts of Interest: The authors declare no conflict of interest. The funders had no role in the design of the study; in the collection, analyses, or interpretation of data; in the writing of the manuscript, or in the decision to publish the results.

References

1. Chaffee, R.R.; Roberts, J.C. Temperature acclimation in birds and mammals. *Annu. Rev. Physiol.* **1971**, *33*, 155–202. [CrossRef]
2. Nagy, K.A.; Odell, D.K.; Seymour, R.S. Temperature regulation by the inflorescence of philodendron. *Science* **1972**, *178*, 1195–1197. [CrossRef] [PubMed]
3. Knutson, R.M. Heat production and temperature regulation in eastern skunk cabbage. *Science* **1974**, *186*, 746–747. [CrossRef]
4. Raskin, I.; Ehmann, A.; Melander, W.R.; Meeuse, B.J. Salicylic acid: A natural inducer of heat production in *Arum* lilies. *Science* **1987**, *237*, 1601–1602. [CrossRef]
5. Wagner, A.M.; Krab, K.; Wagner, M.J.; Moore, A.L. Regulation of thermogenesis in flowering Araceae: The role of the alternative oxidase. *Biochim. Biophys. Acta Bioenerg.* **2008**, *1777*, 993–1000. [CrossRef]

6. Seymour, R.S.; Schultze-Motel, P. Thermoregulating lotus flowers. *Nature* **1996**, *383*, 305. [CrossRef]
7. Dieringer, G.; Leticia Cabrera, R.; Mottaleb, M. Ecological relationship between floral thermogenesis and pollination in *Nelumbo lutea* (Nelumbonaceae). *Am. J. Bot.* **2014**, *101*, 357–364. [CrossRef]
8. Wang, R.; Liu, X.; Mou, S.; Xu, S.; Zhang, Z. Temperature regulation of floral buds and floral thermogenicity in *Magnolia denudata* (Magnoliaceae). *Trees* **2013**, *27*, 1755–1762. [CrossRef]
9. Ito-Inaba, Y.; Sato, M.; Sato, M.P.; Kurayama, Y.; Yamamoto, H.; Ohata, M.; Ogura, Y.; Hayashi, T.; Toyooka, K.; Inaba, T. Alternative oxidase capacity of mitochondria in microsporophylls may function in cycad thermogenesis. *Plant Physiol.* **2019**, *180*, 743–756. [CrossRef] [PubMed]
10. Li, J.K.; Huang, S.Q. Flower thermoregulation facilitates fertilization in Asian sacred lotus. *Ann. Bot.* **2009**, *103*, 1159–1163. [CrossRef] [PubMed]
11. Seymour, R.S.; Ito, Y.; Onda, Y.; Ito, K. Effects of floral thermogenesis on pollen function in Asian skunk cabbage *Symplocarpus renifolius*. *Biol. Lett.* **2009**, *5*, 568–570. [CrossRef]
12. Seymour, R.S.; Schultze-Motel, P. Heat-producing flowers. *Endeavour* **1997**, *21*, 125–129. [CrossRef]
13. Seymour, R.S.; White, C.R.; Gibernau, M. Heat reward for insect pollinators. *Nature* **2003**, *426*, 243–244. [CrossRef]
14. Jarmuszkiewicz, W.; Sluse-Goffart, C.M.; Vercesi, A.E.; Sluse, F.E. Alternative oxidase and uncoupling protein: Thermogenesis versus cell energy balance. *Biosci. Rep.* **2001**, *21*, 213–222. [CrossRef] [PubMed]
15. Watling, J.R.; Grant, N.M.; Miller, R.E.; Robinson, S.A. Mechanisms of thermoregulation in plants. *Plant Signal. Behav.* **2008**, *3*, 595–597. [CrossRef] [PubMed]
16. Moore, A.L.; Siedow, J.N. The regulation and nature of the cyanide-resistant alternative oxidase of plant mitochondria. *Biochim. Biophys. Acta Bioenerg.* **1991**, *1059*, 121–140. [CrossRef]
17. Vanlerberghe, G.C.; McLntosh, L. Signals regulating the expression of the nuclear gene encoding alternative oxidase of plant mitochondria. *Plant Physiol.* **1996**, *111*, 589–595. [CrossRef] [PubMed]
18. Rhoads, D.M.; Mcintosh, L. Salicylic acid regulation of respiration in higher plants: Alternative oxidase expression. *Plant Cell* **1992**, *4*, 1131–1139. [CrossRef] [PubMed]
19. Chivasa, S.; Berry, J.O.; Rees, T.A.; Carr, J.P. Changes in gene expression during development and thermogenesis in Arum. *Aust. J. Plant Physiol.* **1999**, *26*, 391–399. [CrossRef]
20. Grant, N.M.; Miller, R.E.; Watling, J.R.; Robinson, S.A. Synchronicity of thermogenic activity, alternative pathway respiratory flux, AOX protein content, and carbohydrates in receptacle tissues of sacred lotus during floral development. *J. Exp. Bot.* **2008**, *59*, 705–714. [CrossRef] [PubMed]
21. Ricquier, D.; Bouillaud, F. The uncoupling protein homologues: UCP1, UCP2, UCP3, StUCP and AtUCP. *Biochem. J.* **2000**, *345*, 161–179. [CrossRef]
22. Hourton-Cabassa, U.; Matos, A.R.; Zachowski, A.; Moreau, F. The plant uncoupling protein homologues: A new family of energy-dissipating proteins in plant mitochondria. *Plant Physiol. Biochem.* **2004**, *42*, 283–290. [CrossRef] [PubMed]
23. Vercesi, A.E.; Borecky, J.; Godoy Maia, I.D.; Arruda, P.; Cuccovia, I.M.; Chaimovich, H. Plant uncoupling mitochondrial proteins. *Annu. Rev. Plant Biol.* **2006**, *57*, 383–404. [CrossRef]
24. Ito-Inaba, Y.; Hida, Y.; Ichikawa, M.; Kato, Y.; Yamashita, T. Characterization of the plant uncoupling protein, SrUCPA, expressed in spadix mitochondria of the thermogenic skunk cabbage. *J. Exp. Bot.* **2008**, *59*, 995–1005. [CrossRef] [PubMed]
25. Seymour, R.S.; Schultze-Motel, P. Physiological temperature regulation by flowers of the sacred lotus. *Philos. Trans. R. Soc. Lond. B* **1998**, *353*, 935–943. [CrossRef]
26. Watling, J.R.; Robinson, S.A.; Seymour, R.S. Contribution of the alternative pathway to respiration during thermogenesis in flowers of the sacred lotus. *Plant Physiol.* **2006**, *140*, 1367–1373. [CrossRef]
27. Grant, N.; Onda, Y.; Kakizaki, Y.; Ito, K.; Watling, J.; Robinson, S. Two cys or not two cys? That is the question; alternative oxidase in the thermogenic plant sacred lotus. *Plant Physiol.* **2009**, *150*, 987–995. [CrossRef]
28. Raskin, I.; Turner, I.M.; Melander, W.R. Regulation of heat production in the inflorescences of an *Arum* lily by endogenous salicylic acid. *Proc. Natl. Acad. Sci. USA* **1989**, *86*, 2214–2218. [CrossRef] [PubMed]
29. Ito-Inaba, Y.; Hida, Y.; Inaba, T. What is critical for plant thermogenesis? Differences in mitochondrial activity and protein expression between thermogenic and non-thermogenic skunk cabbages. *Planta* **2009**, *231*, 121–130. [CrossRef] [PubMed]
30. Ito-Inaba, Y.; Hida, Y.; Matsumura, H.; Masuko, H.; Yazu, F.; Terauchi, R.; Watanabe, M.; Inaba, T. The gene expression landscape of thermogenic skunk cabbage suggests critical roles for mitochondrial and vacuolar metabolic pathways in the regulation of thermogenesis. *Plant Cell Environ.* **2012**, *35*, 554–566. [CrossRef]
31. Onda, Y.; Mochida, K.; Yoshida, T.; Sakurai, T.; Seymour, R.S.; Umekawa, Y.; Pirintsos, S.A.; Shinozaki, K.; Ito, K. Transcriptome analysis of thermogenic *Arum concinnatum* reveals the molecular components of floral scent production. *Sci. Rep.* **2015**, *5*, 8753. [CrossRef]
32. Zou, Y.; Chen, G.; Jin, J.; Wang, Y.; Xu, M.; Peng, J.; Ding, Y. Small RNA and transcriptome sequencing reveals miRNA regulation of floral thermogenesis in *Nelumbo nucifera*. *Int. J. Mol. Sci.* **2020**, *21*, 3324. [CrossRef]
33. Ludwig, C.; Gillet, L.; Rosenberger, G.; Amon, S.; Collins, B.C.; Aebersold, R. Data-independent acquisition-based SWATH-MS for quantitative proteomics: A tutorial. *Mol. Syst. Biol.* **2018**, *14*, e8126. [CrossRef]
34. Quan, J.; Kang, Y.; Li, L.; Zhao, G.; Sun, J.; Liu, Z. Proteome analysis of rainbow trout (*Oncorhynchus mykiss*) liver responses to chronic heat stress using DIA/SWATH. *J. Proteomics* **2021**, *233*, 104079. [CrossRef]

35. Meng, K.; Lu, S.; Yan, X.; Sun, Y.; Gao, J.; Wang, Y.; Yin, X.; Sun, Z.; He, Q.Y. Quantitative mitochondrial proteomics reveals ANXA7 as a crucial factor in mitophagy. *J. Proteome Res.* **2020**, *19*, 1275–1284. [CrossRef]
36. Miller, R.E.; Watling, J.R.; Robinson, S.A. Functional transition in the floral receptacle of the sacred lotus (*Nelumbo nucifera*): From thermogenesis to photosynthesis. *Funct. Plant Biol.* **2009**, *36*, 471–480. [CrossRef]
37. Zhang, B.; Horvath, S. A general framework for weighted gene co-expression network analysis. *Stat. Appl. Genet. Mol. Biol.* **2005**, *4*, 17. [CrossRef]
38. Langfelder, P.; Horvath, S. WGCNA: An R package for weighted correlation network analysis. *BMC Bioinform.* **2008**, *9*, 559. [CrossRef]
39. Vanlerberghe, G.C.; Day, D.A.; Wiskich, J.T.; Vanlerberghe, A.E.; McIntosh, L. Alternative oxidase activity in tobacco leaf mitochondria (dependence on tricarboxylic acid cycle-mediated redox regulation and pyruvate activation). *Plant Physiol.* **1995**, *109*, 353–361. [CrossRef] [PubMed]
40. Gray, G.R.; Maxwell, D.P.; Villarimo, A.R.; McIntosh, L. Mitochondria/nuclear signaling of alternative oxidase gene expression occurs through distinct pathways involving organic acids and reactive oxygen species. *Plant Cell Rep.* **2004**, *23*, 497–503. [CrossRef]
41. Finkemeier, I.; König, A.C.; Heard, W.; Nunes-Nesi, A.; Pham, P.A.; Leister, D.; Fernie, A.R.; Sweetlove, L.J. Transcriptomic analysis of the role of carboxylic acids in metabolite signaling in *Arabidopsis* leaves. *Plant Physiol.* **2013**, *162*, 239–253. [CrossRef]
42. Bhagavan, N.V.; Ha, C.-E. *Essentials of Medical Biochemistry*, 2nd ed.; Academic Press: San Diego, CA, USA, 2015; pp. 165–185.
43. Weitzman, P.D.; Danson, M.J. Citrate synthase. *Curr. Top. Cell. Regul.* **1976**, *10*, 161–204.
44. Akram, M. Citric acid cycle and role of its intermediates in metabolism. *Cell Biochem. Biophys.* **2014**, *68*, 475–478. [CrossRef]
45. Vanlerberghe, G.C.; McIntosh, L. Alternative oxidase: From gene to function. *Annu. Rev. Plant Physiol. Plant Mol. Biol.* **1997**, *48*, 703–734. [CrossRef]
46. Vanlerberghe, G.C.; Dahal, K.; Alber, N.A.; Chadee, A. Photosynthesis, respiration and growth: A carbon and energy balancing act for alternative oxidase. *Mitochondrion* **2020**, *52*, 197–211. [CrossRef]
47. Sluse, F.E.; Almeida, A.M.; Jarmuszkiewicz, W.; Vercesi, A.E. Free fatty acids regulate the uncoupling protein and alternative oxidase activities in plant mitochondria. *FEBS Lett.* **1998**, *433*, 237–240. [CrossRef]
48. Dry, I.B.; Moore, A.L.; Day, D.A.; Wiskich, J.T. Regulation of alternative pathway activity in plant mitochondria: Nonlinear relationship between electron flux and the redox poise of the quinone pool. *Arch. Biochem. Biophys.* **1989**, *273*, 148–157. [CrossRef]
49. Ribas-Carbo, M.; Wiskich, J.T.; Berry, J.A.; Siedow, J.N. Ubiquinone redox behavior in plant mitochondria during electron transport. *Arch. Biochem. Biophys.* **1995**, *317*, 156–160. [CrossRef]
50. Ng, S.; De Clercq, I.; Van Aken, O.; Law, S.R.; Ivanova, A.; Willems, P.; Giraud, E.; Van Breusegem, F.; Whelan, J. Anterograde and retrograde regulation of nuclear genes encoding mitochondrial proteins during growth, development, and stress. *Mol Plant* **2014**, *7*, 1075–1093. [CrossRef]
51. Bradford, M.M. A rapid and sensitive method for the quantitation of microgram quantities of protein utilizing the principle of protein-dye binding. *Anal. Biochem.* **1976**, *72*, 248–254. [CrossRef]
52. Cox, J.; Mann, M. MaxQuant enables high peptide identification rates, individualized p.p.b.-range mass accuracies and proteome-wide protein quantification. *Nat. Biotechnol.* **2008**, *26*, 1367–1372. [CrossRef]
53. Cox, J.; Neuhauser, N.; Michalski, A.; Scheltema, R.A.; Olsen, J.V.; Mann, M. Andromeda: A peptide search engine integrated into the MaxQuant environment. *J. Proteome Res.* **2011**, *10*, 1794–1805. [CrossRef]
54. Wang, K.; Deng, J.; Damaris, R.N.; Yang, M.; Xu, L.; Yang, P. LOTUS-DB: An integrative and interactive database for *Nelumbo nucifera* study. *Database* **2015**, *2015*, bav023. [CrossRef]
55. Bruderer, R.; Bernhardt, O.M.; Gandhi, T.; Miladinović, S.M.; Cheng, L.Y.; Messner, S.; Ehrenberger, T.; Zanotelli, V.; Butscheid, Y.; Escher, C.; et al. Extending the limits of quantitative proteome profiling with data-independent acquisition and application to acetaminophen-treated three-dimensional liver microtissues. *Mol. Cell. Proteomics* **2015**, *14*, 1400–1410. [CrossRef] [PubMed]
56. Reiter, L.; Rinner, O.; Picotti, P.; Hüttenhain, R.; Beck, M.; Brusniak, M.Y.; Hengartner, M.O.; Aebersold, R. mProphet: Automated data processing and statistical validation for large-scale SRM experiments. *Nat. Methods* **2011**, *8*, 430–435. [CrossRef]
57. Choi, M.; Chang, C.Y.; Clough, T.; Broudy, D.; Killeen, T.; MacLean, B.; Vitek, O. MSstats: An R package for statistical analysis of quantitative mass spectrometry-based proteomic experiments. *Bioinformatics* **2014**, *30*, 2524–2526. [CrossRef] [PubMed]
58. Conesa, A.; Götz, S.; García-Gómez, J.M.; Terol, J.; Talón, M.; Robles, M. Blast2GO: A universal tool for annotation, visualization and analysis in functional genomics research. *Bioinformatics* **2005**, *21*, 3674–3676. [CrossRef] [PubMed]
59. Kumar, L.; Futschik, M. Mfuzz: A software package for soft clustering of microarray data. *Bioinformation* **2007**, *2*, 5–7. [CrossRef]
60. Shannon, P.; Markiel, A.; Ozier, O.; Baliga, N.S.; Wang, J.T.; Ramage, D.; Amin, N.; Schwikowski, B.; Ideker, T. Cytoscape: A software environment for integrated models of biomolecular interaction networks. *Genome Res.* **2003**, *13*, 2498–2504. [CrossRef]
61. Zhu, Z.; Gui, S.; Jin, J.; Yi, R.; Wu, Z.; Qian, Q.; Ding, Y. The NnCenH3 protein and centromeric DNA sequence profiles of *Nelumbo nucifera* Gaertn. (sacred lotus) reveal the DNA structures and dynamics of centromeres in basal eudicots. *Plant J.* **2016**, *87*, 568–582. [CrossRef] [PubMed]
62. Livak, K.J.; Schmittgen, T.D. Analysis of relative gene expression data using real-time quantitative PCR and the 2 (-Delta Delta C(T)) Method. *Methods* **2001**, *25*, 402–408. [CrossRef] [PubMed]
63. Ma, J.; Chen, T.; Wu, S.; Yang, C.; Bai, M.; Shu, K.; Li, K.; Zhang, G.; Jin, Z.; He, F.; et al. iProX: An integrated proteome resource. *Nucleic. Acids. Res.* **2019**, *47*, D1211–D1217. [CrossRef] [PubMed]

International Journal of
Molecular Sciences

MDPI

Article

Proteomics Data Analysis for the Identification of Proteins and Derived Proteotypic Peptides of Potential Use as Putative Drought Tolerance Markers for *Quercus ilex*

Bonoso San-Eufrasio [1], Ezequiel Darío Bigatton [1,2], Victor M. Guerrero-Sánchez [1], Palak Chaturvedi [3], Jesús V. Jorrín-Novo [1], María-Dolores Rey [1] and María Ángeles Castillejo [1,*]

1 Agroforestry and Plant Biochemistry, Proteomics and Systems Biology, Department of Biochemistry and Molecular Biology, University of Cordoba, UCO-CeiA3, 14014 Cordoba, Spain; z82samab@uco.es (B.S.-E.); ezequielbigatton@gmail.com (E.D.B.); b12gusav@uco.es (V.M.G.-S.); bf1jonoj@uco.es (J.V.J.-N.); b52resam@uco.es (M.-D.R.)
2 Agricultural Microbiology, Faculty of Agricultural Science, National University of Cordoba, CONICET, 5001 Cordoba, Argentina
3 Molecular Systems Biology Lab (MOSYS), Department of Functional and Evolutionary Ecology, University of Vienna, Althanstrasse 14, A-1090 Vienna, Austria; palak.chaturvedi@univie.ac.at
* Correspondence: bb2casam@uco.es; Tel.: +34-957-218-609

Abstract: Drought is one of the main causes of mortality in holm oak (*Quercus ilex*) seedlings used in reforestation programs. Although this species shows high adaptability to the extreme climate conditions prevailing in Southern Spain, its intrinsic genetic variability may play a role in the differential response of some populations and individuals. The aim of this work was to identify proteins and derived proteotypic peptides potentially useful as putative markers for drought tolerance in holm oak by using a targeted post-acquisition proteomics approach. For this purpose, we used a set of proteins identified by shotgun (LC-MSMS) analysis in a drought experiment on *Q. ilex* seedlings from four different provenances (viz. the Andalusian provinces Granada, Huelva, Cadiz and Seville). A double strategy involving the quantification of proteins and target peptides by shotgun analysis and post-acquisition data analysis based on proteotypic peptides was used. To this end, an initial list of proteotypic peptides from proteins highly represented under drought conditions was compiled that was used in combination with the raw files from the shotgun experiment to quantify the relative abundance of the fragment's ion peaks with the software Skyline. The most abundant peptides under drought conditions in at least two populations were selected as putative markers of drought tolerance. A total of 30 proteins and 46 derived peptides belonging to the redox, stress-related, synthesis,-folding and degradation, and primary and secondary metabolism functional groups were thus identified. Two proteins (viz., subtilisin and chaperone GrpE protein) were found at increased levels in three populations, which make them especially interesting for validation drought tolerance markers in subsequent experiments.

Keywords: peptide markers; *Quercus ilex*; drought tolerance; targeted post-acquisition proteomics

Citation: San-Eufrasio, B.; Bigatton, E.D.; Guerrero-Sánchez, V.M.; Chaturvedi, P.; Jorrín-Novo, J.V.; Rey, M.-D.; Castillejo, M.Á. Proteomics Data Analysis for the Identification of Proteins and Derived Proteotypic Peptides of Potential Use as Putative Drought Tolerance Markers for *Quercus ilex*. *Int. J. Mol. Sci.* **2021**, *22*, 3191. https://doi.org/10.3390/ijms22063191

Academic Editor: Frank M. You

Received: 13 February 2021
Accepted: 18 March 2021
Published: 21 March 2021

1. Introduction

Holm oak (*Quercus ilex*) is the dominant tree species in natural forest ecosystems over the Western Mediterranean Basin, as well as in the agrosilvopastoral Spanish *"dehesa"*, which is environmentally, economically and socially important [1,2]. This species is highly adaptable to drought and to the high temperatures and irradiation typical of Southern Spain. However, the main cause of mortality in holm oak plantations is water deficiency, with drought stress acting as a major factor of decline [3,4]. This situation can be expected to worsen in a scenario of climate change where statistical models have predicted that 40% of the land areas with a high-density of *Q. ilex* will be unsuitable for its survival [5].

Studies on the genetic variability associated with both environmental factors and genotypes have shown high population variability and polymorphism in *Quercus* spp [6–9]. In addition, *Q. ilex* has been shown to exhibit high variability in traits associated with drought tolerance, both within and between populations [10]. Because this is a non-domesticated species with a very long-life cycle, it is not amenable to conventional plant breeding. Therefore, for management and conservation practices based on resilient, elite genotypes of holm oak trees to be effective, they must rely on a sound knowledge of their biology and molecular mechanisms of adaptation to adverse climatic conditions. The response of plants to stress-related situations may, in theory, be improved by characterizing their biodiversity and selecting elite genotypes based on specific molecular markers. This approach can be quite challenging with orphan species, such as holm oak, which has a still incompletely sequenced genome and largely unexplored molecular features [11–21]. Fortunately, omics approaches have enabled crucial advances in these directions. Thus, some multiomics studies have addressed *Q. ilex* [19]; also, a reference transcriptome for this species has been generated [16,17], and, more recently, the metabolome of the acorn was determined [22]. In any case, the greatest efforts have focused on the proteomics of *Q. ilex*. A number of proteomic studies have used 1D and 2D gel-based analysis to investigate drought tolerance in this species [11,12,23]. In addition, recent studies have addressed various aspects of its biology by using shotgun (LC–MS/MS) proteomic analysis [19,21,24]. In addition, species-specific improved databases, such as the recently compiled holm oak transcriptome database [16,17], and other sequenced *Quercus* species databases, such as those for *Q. robur* [25] and *Q. suber* [26], have substantially expanded available knowledge of holm oak biology.

Quantitative proteomics is providing increasingly powerful tools for identifying markers of complex traits. Thus, identifying target peptide signals against mass spectrometry libraries is an efficient method for protein identification and quantification [27]. Targeted proteomics, however, does not allow the identification of new proteins as it requires the prior measurement of the targeted proteins by discovery proteomics; rather, it is useful for detecting changes in the protein abundances from previously acquired information [28]. Therefore, this proteomics branch can be useful to characterize coordinated changes in protein abundance with a view to identifying or validating proteins as markers for specific traits. Recently, proteotypic peptides have proved useful for protein quantification [29].

In this work, we used a double strategy for proteins and peptides quantification by targeted post-acquisition data analysis against a species-specific *Q. ilex* database with a view to identifying proteotypic peptides of potential use as putative drought tolerance markers for holm oak. For this purpose, a set of raw data generated in a shotgun experiment performed after 17 and 24 days under drought conditions in *Q. ilex* seedlings from four different provinces in Andalusia was used. Based on previous studies of inter-population variability of this species [12,18,30,31], we selected two populations from the southeast (Cadiz, Granada) and two from the northwest (Huelva, Seville) of Andalusia, with the purpose of identifying changes in proteins and derived peptides persistent over time in response to drought in different populations. In this work, various proteins and peptides are proposed as putative markers of drought tolerance in holm oak that transcend not only the tolerant phenotype but also populations and are examined in biological terms.

2. Results

2.1. Qualitative and Quantitative Analysis of Drought Stress Responsive Proteins

Shotgun analysis allowed a total of 4470 proteins to be identified in the *Q. ilex* leaf proteome (Supplementary Table S1; data are available via ProteomeXchange with identifier PXD023782) of which 2920 fulfilled the following criterion for confident identification: XCorr ≥ 2 and at least two different peptides per protein. An overall 2261 proteins were deemed variable in accordance with the following confidence criteria: (a) consistent presence in all replicates, (b) statistical significance with false discovery rate (FDR) < 0.05, and (c) drought/control ratio ≥ 2 and ≤ 0.5. In this group, 1692 proteins exhibited qualitative changes in at least one population and sampling time and 569 quantitative changes, 1683 being more abundant in droughted seedlings. The initial dataset was screened to select those variable proteins most markedly represented under drought conditions at both sampling times in each population. A total of 380 proteins were thus screened, of which 48 were present in at least two populations and deemed markers candidates. A schematic view of the workflow, as well as the details of the experimental design, are shown in Figure 1 and Supplementary Figure S1.

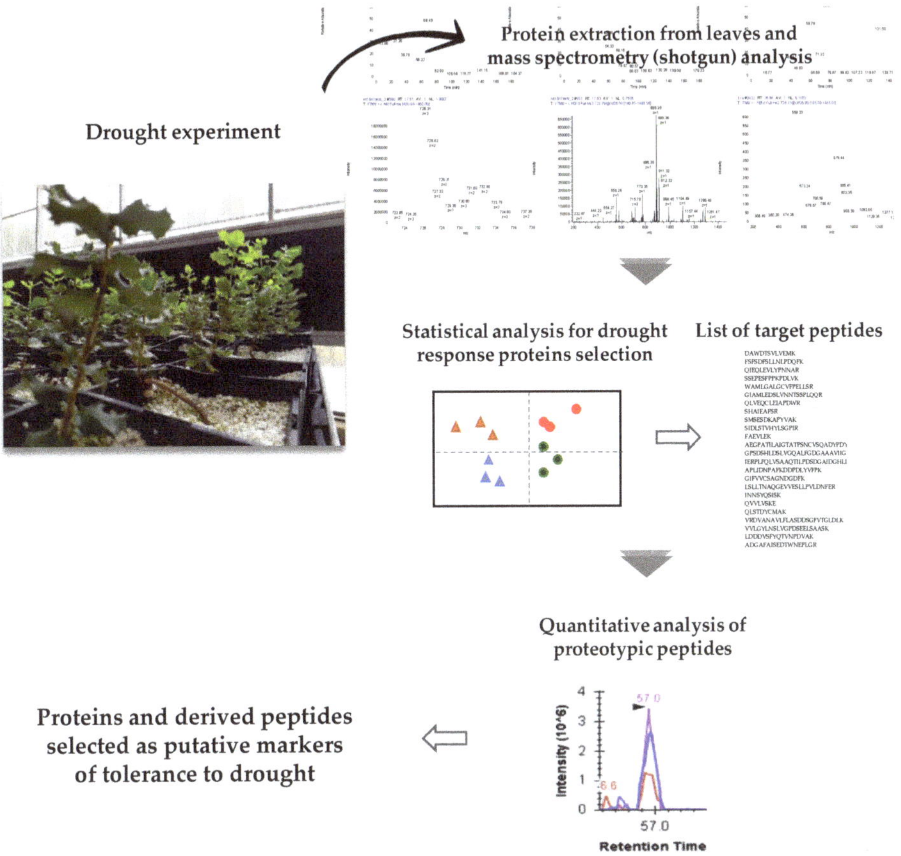

Figure 1. Schematic workflow for selection of putative markers of drought tolerance.

Multivariate analysis integrating the entire dataset is shown in Supplementary Figure S2A, where the first two components (25% of the total variability) separated populations. Hierarchical clustering was also performed and represented in a dendrogram (Supplementary Figure S2B), in which two main clusters were observed: one grouped Huelva and Granada populations, and the other grouped Seville and Cadiz, independently of the sampling time. Replicates of the same experimental condition (population, treatment and sampling time) were grouped, demonstrating reproducibility throughout the experiment. To check the effect of stress on the populations, the sampling times were analyzed separately by partial least-squares regression analysis (PLS-DA) (Figure 2). The first two components explained 20% (first sampling time) and 31% (second sampling time) of the total variability. Component 1 resolved Cadiz control plants, and component 2 resolved plants under drought for 17 days (first sampling) from control plants except in the Huelva population (Figure 2A). For the second sampling time (24 days), component 1 resolved the population pairs Huelva–Granada and Seville–Cadiz, and component 2 resolved treatments (Figure 2B). Based on this analysis, a clearer effect of drought treatment and populations can be appreciated at the second sampling time.

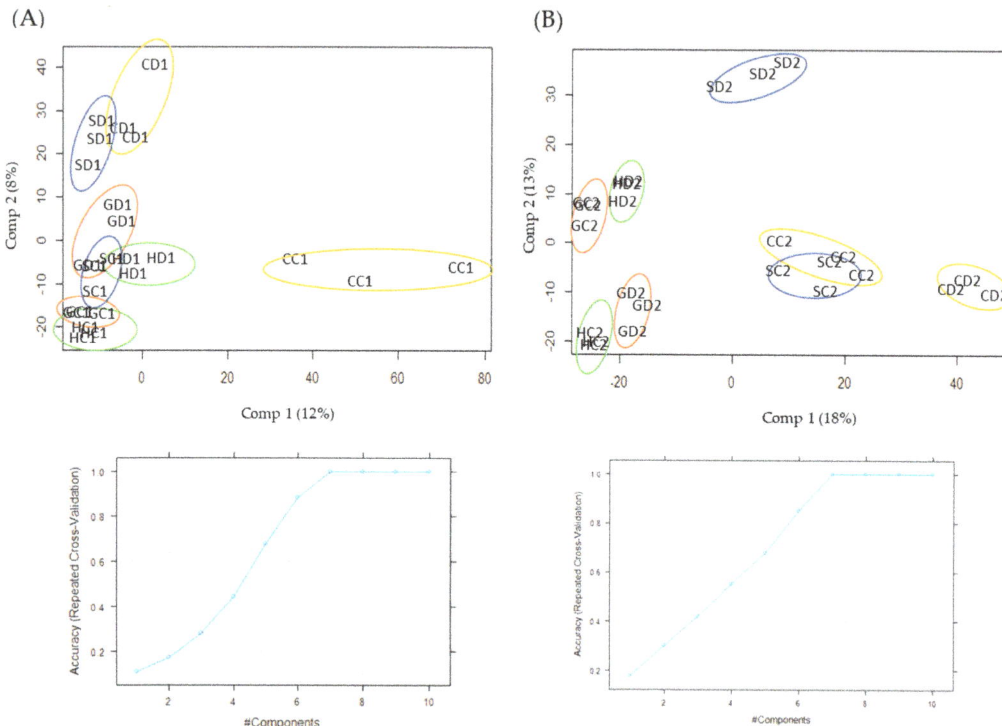

Figure 2. Partial least-squares discriminant analysis (PLS) of the entire dataset after 17 (**A**) and 24 days of drought (**B**) is shown in the upper part. The cumulative proportion of variance explained by components is shown below. C, Cadiz; G, Granada; H, Huelva; S, Seville. The letters following C, G, H, or S denote treatment (D, drought; C, control), the numbers before the underscore sampling time (1, 17 days; 2, 24 days) and that after it replicates (1, 2 or 3).

Figure 3A shows the 2261 variable proteins significantly (FDR < 0.05) up- or down-accumulated (twofold change) in the drought group, and Figure 3B a Venn diagram of the 380 variable proteins among them. The Granada and Huelva populations exhibited the greatest number of unique variable proteins (107 each), followed by Seville (71) and Cadiz (47). Huelva and Seville shared the largest number of variable proteins (16), followed by Granada–Huelva (14), Huelva–Cadiz (8), Seville–Cadiz (8), Granada–Seville (4) and Granada–Cadiz (2). Only two proteins changed significantly by the effect of drought in three populations; thus, GrpE protein (qilexprot_13677) changed in Huelva, Seville and Cadiz, and subtilisin-like protease (qilexprot_25223) in Granada, Seville and Cadiz.

The previous 380 variable proteins were characterized in functional terms by using Mercator [32] and GO enrichment (http://pantherdb.org/ accessed September 2020) for classification into 16 main groups (Figure 4A), namely: energy, carbohydrate, amino acid, lipid, hormone, coenzyme and secondary metabolism, other metabolic processes, cellular processes,-folding-sorting and degradation, synthesis (transcription/translation), structural, defense and response to stress, redox, signaling and transport. The best represented functional group was synthesis (62), followed by folding-sorting and degradation (42), defense and response to stress (37), carbohydrate metabolism (33) and redox (29), the previous five groups accounting for more than 50% of all identified proteins. Figure 4B compares the proteins whose abundance was altered by drought in each population as grouped by functional category.

(A)

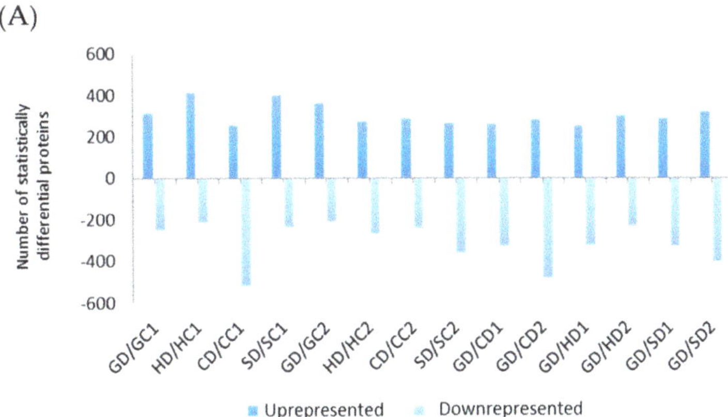

(B)

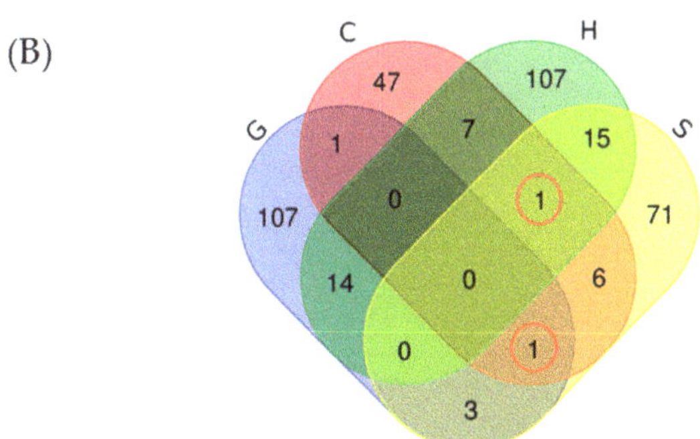

Figure 3. Variable proteins significantly (false discovery rate of 5%) uprepresented or downrepresented (twofold change) under drought conditions (**A**); C, Cadiz; G, Granada; H, Huelva; S, Seville. The letter following C, G, H or S denotes treatment (D, drought; C, control) and the number sampling time (1, 17 days; 2, 24 days). Venn diagram showing significantly up-represented proteins under drought conditions in each population (**B**).

2.2. *Targeted Data Analysis for Selection of Peptides as Putative Markers of Drought Tolerance*

A list of proteotypic peptides derived from the 48 selected proteins was compiled for subsequent targeted analysis. The list, which included 219 proteotypic peptides with a charge state of +2 or higher, was used for quantification with the software Skyline as described in Section 4.5. As confirmed by Supplementary Figure S3, using the above-described library allowed 159 peptides to be successfully integrated with a robust, reproducible, high-quality peak shape in all three replicates (Supplementary Table S2). A statistical analysis of variance (ANOVA) on normalized data revealed 71 peptides comprising 32 different proteins were significantly better represented in droughted seedlings. Those peptides and proteins better represented in at least two populations (46 peptides from 30 proteins) were selected as putative markers of drought tolerance (Table 1). Supplementary Figure S4 illustrates protein and peptide quantification graphically. The most representative protein functions in the marker panel (Table 1) were synthesis and mRNA processing with 9 proteins. Seven proteins belonged to the redox and response to functional stress groups; 3 to the folding, sorting and degradation group; and 2 to the transport group. Other metabolic functional groups, such as carbohydrate metabolism (2), secondary metabolism (1), photosynthesis (1) and other processes (5), were also represented. The Huelva population was that exhibiting the greatest number of proteins, most of which belonged to the synthesis and mRNA processing group (7) and the stress-related and secondary metabolism group (6). Other proteins associated with energy and metabolism (viz., photosynthesis, carbohydrate metabolism and other cellular processes) were also represented (5). The second population as regards protein changes was Seville, with proteins of the metabolism (6), synthesis (4),-folding and degradation (3), and stress-related (3) groups. The Granada population exhibited smaller numbers of changing proteins, and only in the synthesis (5) and metabolism (3) groups. Finally, the Cadiz population was that exhibiting the least changes and mainly in stress-related (3), metabolic processes (2) and folding and degradation proteins (2). A protein–protein interaction network among the 48 proteins previously selected was performed using the web-tool STRING10 (http://string-db.org accessed January 2021) (Figure 5). A strong connection between proteins of synthesis and those of folding and degradation was observed.

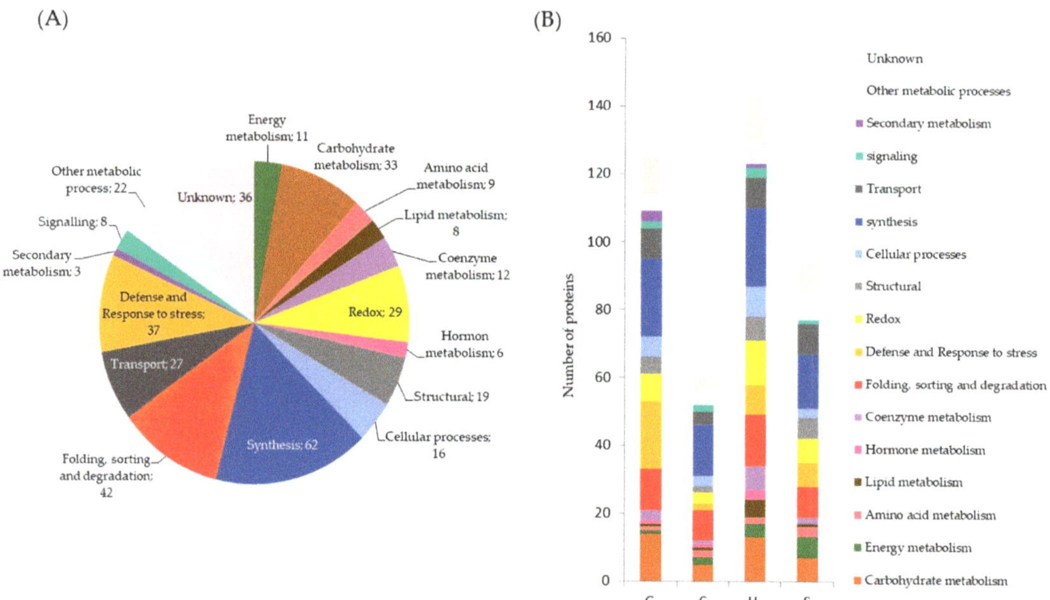

Figure 4. Functional categories of the 380 proteins. Total number of proteins significantly increasing in abundance after drought (**A**) and in each of the populations (**B**): C, Cadiz; G, Granada; H, Huelva; S, Seville.

Table 1. List of peptides and proteins selected as putative markers of tolerance to drought.

Protein ID	Peptide Sequences	Precursor *m/z*	Protein Description	Protein Function	Experimental Condition Showing Significant Change [a]			
qilexprot_45247	DAWDTSVLVEMK	697,333	Granule-bound starch synthase 1, chloroplastic/amyloplastic	Carbohydrate metabolism			S	G
	FSFSDFSLLNLPDQFK	952,971					S	G
	QIEQLEVLYPNNAR	843,940					S	G
qilexprot_71384	SSEPESFPPKPDLVK	828,924	Glycosyl hydrolase family protein with chitinase insertion domain	Carbohydrate metabolism		H	S	
qilexprot_18873	WAMLGALGCVFPELLSR	968,491	Chlorophyll a-b binding protein, chloroplastic	Photosynthesis		H	S	
qilexprot_32784	GIAMLEDSLVNNTSSPLQQR	1087,046	Mitochondrial fission 1 protein A	Cellular processes	C		S	
	QLVEQCLEIAPDWR	878,934			C		S	
qilexprot_49492	SHAIEAFSR	509,256	T-complex protein 1 subunit beta	Cellular processes			S	G
qilexprot_42362	SMSESDKAPYVAK	714,834	High mobility group B protein 4	Cellular processes		H	S	
qilexprot_14200	SIDLSTVHYLSGPIR	552,964	Formamidase	Other metabolic process	C	H		
qilexprot_39395	FAEVLEK	418,228	3-phosphoshikimate 1-carboxyvinyltransferase	Other metabolic process		H		G
qilexprot_29542	AEGPATILAIGTATPSNCVSQADYPDYYFR	1083,173	Chalcone synthase	Secondary metabolism		H		G
	GPSDSHLDSLVGQALFGDGAAAVIIGADPDTK	1031,844				H		G
	IERPLFQLVSAAQTILPDSDGAIDGHLR	1011,205				H		G
qilexprot_49771	APLIDNPAFKDDPDLYVFPK	1138,082	Calreticulin	Folding, sorting and degradation		H	S	
qilexprot_25223	GIFVVCSAGNDGDFK	793,366	Subtilisin-like protease	Folding, sorting and degradation	C		S	G
qilexprot_13677	LSLLTNAQGEVVESLLPVLDNFER	1328,700	GrpE protein	Folding, sorting and degradation	C	H	S	
	INNSYQSISK	577,300			C	H	S	
qilexprot_55000	QVVLVSKE	451,268	2-alkenal reductase (NADP(+)-dependent)	Redox	C	H		
qilexprot_1533	QLSTDYCMAK	616,764	Short-chain alcohol dehydrogenase A	Redox	C		S	
	VRDVANAVLFLASDDSGFVTGLDLK	874,459			C		S	

Table 1. *Cont.*

Protein ID	Peptide Sequences	Precursor m/z	Protein Description	Protein Function	Experimental Condition Showing Significant Change [a]			
qilexprot_55764	VVLGYLNSLVGPDSEELSAASK	1124,585	Protein disulfide-isomerase	Redox		H	S	
	LDDDVSFYQTVNPDVAK	963,456				H	S	
	ADGAFAISEDTWNEPLGR	974,952				H	S	
qilexprot_3345	EVTEEEYTK	564,255	Endoplasmin homolog	Response to stress		H	S	
	FYHSLAK	433,228				H	S	
	YLNFLMGLVDSDTLPLNVSR	1142,084				H	S	
	IAEEDPDEANDKDK	794,849				H	S	
qilexprot_72159	NKDEHETTTTTPGGNEGAVESK	801,364	Dehydrin	Response to stress		H		G
	EEEMASEFEK	614,752				H		G
qilexprot_48029	ELAENLFPEDDTVSQTPTLQSSENVLVR	1044,179	Senescence/dehydration-associated protein AT3g51250	Response to stress	C		S	
qilexprot_8871	KGCTPSQLALAWVHHQGK	673,013	Probable aldo-keto reductase 1	Response to stress		H	S	
qilexprot_19464	VNWAYASGQR	576,300	Oligouridylate-binding protein	mRNA processing	C			G
	SVVELTNGSSEDGK	711,300			C			G
qilexprot_70616	LITVTASENPDSR	701,900	BnaC03g49780D protein	mRNA processing		H		G
qilexprot_26698	DKPESDGADLANK	680,319	Zinc finger protein VAR3, chloroplastic	mRNA processing		H	S	
	SVASNAIEWTGNASGSSVPDK	524,745				H	S	
qilexprot_2527	IEDIDAYAPK	567,784	Myb domain containing transcription regulator	Synthesis	C		S	
qilexprot_7552	IVDVCEIGDSFIR	761,800	Proliferation-associated protein 2G4	Synthesis		H	S	
	ALQLVVSECKPK	686,383				H	S	
qilexprot_56656	ISFSGIDGKPEDVLNPK	605,650	Probable methionyl-tRNA synthetase	Synthesis		H		G
qilexprot_70881	VQDTYDTELAGK	670,319	Eukaryotic translation initiation factor 2c	Synthesis		H		G
qilexprot_71168	EDENRLDEVGYDDVGGVR	679,305	Transitional endoplasmic reticulum ATPase	Synthesis		H		G
qilexprot_68980	EFFGSENNSLVSAQVIFHENPR	840,737	RNA-binding family protein	Synthesis		H	S	
qilexprot_68567	SPPINEVVQSGVVPR	789,400	Importin subunit alpha	Transport		H	S	
qilexprot_4392	GLYENSGGGANVVNHGYTK	968,957	Aquaporin	Transport		H		G

[a] *Q. ilex* population in which significant changes occurred under drought stress: C, Cadiz; G, Granada; H, Huelva; S, Seville.

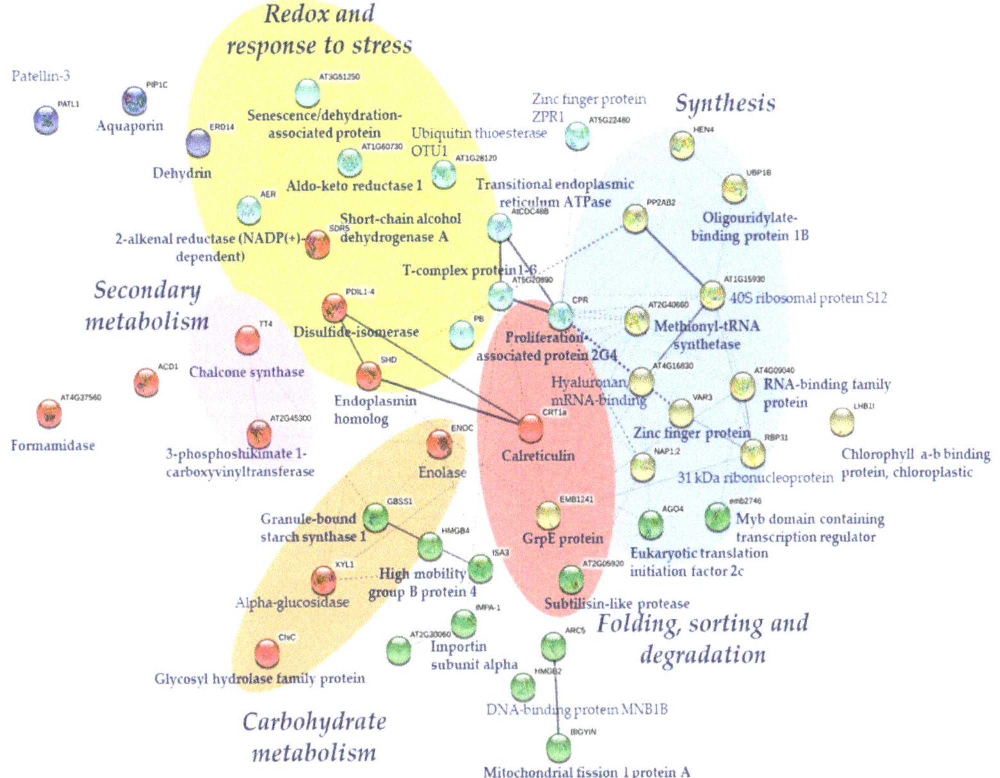

Figure 5. Analysis of the protein interaction network of the 48 proteins selected as putative markers of drought tolerance.

3. Discussion

In this work, we used a double proteomic strategy for protein and peptide quantification in order to identify putative protein markers associated with drought tolerance in *Quercus ilex*. For this purpose, a dataset obtained by shotgun proteomics analysis of four *Q. ilex* populations from different provinces of Andalusia, Southern Spain, under severe drought stress [31] was analyzed by using a double strategy, combining shotgun protein quantification of proteins and target peptides with post-acquisition analysis of data based on proteotypic peptides. The populations were selected based on previous studies of *Q. ilex* variability between eastern populations and western ones [30,31], which demonstrated a relationship between tolerance and provenance. However, intrapopulation variability was also observed as corroborated by the study of morpho-physiological and biochemical parameters [31]. For this reason, the aim of this work was the search for putative drought tolerance markers that transcend not only the tolerant phenotype but also populations.

The typically high complexity of proteomes makes protein identification by mass spectrometry irreproducible as a result of precursor ions being selected stochastically. In addition, forest plants exhibit enormous biological variability that results in even poorer reproducibility among samples. Targeted proteomics analyses can be used to identify, characterize and quantify small sets of proteins previously selected by mass spectrometry analysis [33]. The few targeted proteomic studies conducted so far to identify markers of important traits have focused on crops, such as potato [34], apple [35], grapevine [36] and tomato [37,38]. Some have examined gluten profile [39,40], but even fewer have dealt with forest species. Although selected reaction monitoring (SRM) and its variants are the current gold standard for quantitative estimation of proteins, the recently data-independent acquisition (DIA) method is increasingly being used for targeted proteomics. Unlike existing alternatives, DIA extracts specific information from previously acquired data [41]. This method has been used in combination with proteotypic peptides to identify peptide markers of resistance to *Peyronellaea pinodes* in pea [42]. The use of proteotypic peptides for protein quantification recently proved more accurate than methods based on algorithms (usually on the intensity of the strongest ion peaks) [29].

As shown here, the shotgun technique, in combination with proteotypic peptides extracted from previously acquired data, has great analytical potential. While not a targeted proteomic strategy proper, this approach allows one to select peptides and proteins closely associated with specific traits. This requires processing a dataset compiled from a properly designed and conducted experiment (viz., one using a large enough number of replicates or individuals). In this work, we compiled a list of peptides and proteins potentially useful as putative markers of drought tolerance in *Q. ilex* that are briefly discussed in biological terms below. The proteins in the marker panel were differentially represented among populations, with the greatest numbers of changes found in the Huelva population, followed by Seville, Granada and Cadiz. Many of the selected proteins are involved in synthesis or degradation processes or, to a lesser extent, in metabolic processes, such as carbohydrate metabolism, photosynthesis and other metabolic reactions. Stress response, redox and secondary metabolism proteins were also well represented.

Changes in synthesis and degradation of proteins under stress conditions, such as drought, can be interpreted as a mechanism of adaptation through the installation of the translational apparatus and protein synthesis by recycling available amino acids in plants through protein degradation. Thus, plants respond to drought by synthetizing protective proteins and repairing or degrading damaged proteins [43]. Considering the general changes observed in the proteome in response to drought, synthesis (ribosomal and transcription) was the most represented group of proteins showing qualitative and quantitative changes, followed by folding and degradation category. Many of the proteins selected as putative markers here are involved in synthesis processes; such is the case, for instance, with translation initiation factor, zinc finger protein VAR3, RNA-binding protein, and methionyl-tRNA synthetase. The marker panel also included folding and degradation proteins, such as the chaperones calreticulin and GrpE protein, and serine protease subtilisin. 2DE-MSMS proteomic analysis previously revealed a similar response involving some of the previous proteins in *Q. ilex* [12,23] and *Q. robur* [44] under drought and suggested active metabolic adjustment to stress.

By contrast, degradation of starch in response to stress has been often associated with improved tolerance and potentially limited photosynthesis [45,46]. Sugars resulting from starch degradation, and other derivative metabolites, help plants grow under stress and function as osmoprotectants and compatible solutes to mitigate the adverse effects of stress [47], as found in droughted *Q. robur* [44]. Although environmental factors are known to have strong effects on the starch synthesis, their regulatory mechanisms remain unclear [48]. Some studies reported increased starch accumulation under stress, mainly in response to high salinity or cold [49–51]. In this work, two proteins of carbohydrate metabolism of the marker panel (viz., granule-bound starch synthase 1, which is chloroplastic/amyloplastic, and the glycosyl hydrolase family protein with chitinase insertion domain) were found at increased levels in, mainly, the Seville population. However, several starch degradation proteins not selected as putative markers were significantly increased in response to drought in some of the experimental conditions studied, including phosphoglucan water dikinase, α-glucan phosphorylase and α-amylase. Although apparently contradictory, this response may be related to the presence of different types of starch, whether permanent or transitory and that of isoforms involved in their synthesis and mobilization [48,52,53]. On the other hand, the photosynthetic machinery was seemingly unaffected; in fact, only a few photosynthesis proteins exhibited any changes, and only one (chlorophyll a-b binding protein) was included in the marker panel. This result is consistent with those of physiological studies on *Q. ilex* populations under severe drought in Seville, Granada and Cadiz, which exhibited no significant changes in photosynthetic pigments [31].

The broadest group of proteins and derived peptides selected as putative markers consisted of redox (2-alkenal reductase NADP-dependent, short-chain alcohol dehydrogenase A, disulfide-isomerase) and stress response proteins (endoplasmin, dehydrin, senescence/dehydration-associated protein and aldo-keto reductase). Some were closely associated with drought in several studies on the genus *Quercus* [4,44] or with biotic stress caused by *Phytophthora cinnamomi* [54]. Furthermore, a representative number of redox proteins not included in the marker panel have been identified as being increased to a greater or lesser extent in some of the conditions studied in response to drought, including glutathione S-transferase, glutathione peroxidase, thioredoxin, peroxidase, superoxide dismutase, lipoxygenase, among others. Our marker panel also included two enzymes involved in the shikimate–phenolic biosynthetic pathways, namely: chalcone synthase (CHS) and 3-phosphoshikimate 1-carboxyvinyltransferase. One of the potential roles of phenolic compounds is to scavenge harmful reactive oxygen species [55]. Consistent with our results, *Q. ilex* [31,56] and other *Quercus* spp. [57,58] were previously found to exhibit increased total levels of phenolics.

Finally, transport proteins, such as the water channel protein aquaporins, have been associated with plant tolerance of biotic and abiotic stresses, to which they respond by regulating the movement of water and small molecules through plasma membranes and vacuoles [59]. Based on a proteomics strategy involving the identification of proteotypic peptides, some transport proteins have been proposed as markers of tolerance to drought [60] and resistance to *Ascochyta* blight [42] in pea. The proteins were assumed to induce signaling and transport processes as mechanisms to maintain homeostatic equilibrium and cope with stress. The proposed putative markers included the importin subunit alpha, aquaporin and mitochondrial fission 1 protein A, although other transport and signaling proteins, such as 14-3-3-like protein, lipocalin, outer envelope pore protein (OEP), voltage-dependent anion-selective channel (VDAC) and translocase of chloroplast 90 protein, were also more represented under drought in some of the experimental conditions in this study.

Despite it is outside the scope of this work, based on our results, a clear distinction in response to drought among the populations studied cannot be postulated, for, which a greater number of populations covering a wider area must be included. However, attending to the high number of the proteins that showed changes in the Huelva population also observed in Seville, we could speculate on a similar response pattern to drought in both, perhaps due to geographic proximity. The fact that the proteomic profile of Cadiz is different from the rest may also have a geographical explanation as has already been described by Fernandez i Marti et al. [18], suggesting that the Guadalquivir Valley has played an important role in determining population divergence. To the authors' knowledge, this is the first study aimed to identify proteins and derived peptides as putative markers of drought tolerance for a forest species, such as *Q. ilex*. Such markers may be useful with a view to selecting drought-tolerant genotypes or individuals.

4. Materials and Methods

4.1. Proteomic Dataset

The dataset used was compiled from the leaf proteome of *Q. ilex* seedlings in four different Andalusian populations, Southern Spain, namely: Granada (G), Huelva (H), Cadiz (C) and Seville (S). A detailed map of the Andalusian localizations from which samples were collected is shown in Supplementary Figure S5. All populations were under severe drought stress conditions, such as those imposed in summer under Mediterranean climate (Supplementary Table S3). Detailed information about specimen provenances (Supplementary Table S3), plant growth, stress conditions imposed, and physiological measurements can be found elsewhere [31]. Briefly, acorns were germinated and grown under greenhouse conditions in perlite, according to Simova-Stoilova et al. [23]. Severe drought was imposed by withholding water for 28 days at the 10-leaf stage under the following experimental conditions: 46/22 °C, 28 MJm^{-2}/day, 41% HR [31]. Two different sampling times based on measured leaf physiological parameters were used for proteomic analysis. Thus, leaves were collected after 17 and 24 days of drought, corresponding to an average drop in chlorophyll fluorescence of 20% and 40%, respectively, in droughted seedlings relative to well-watered seedlings in all populations [31] (Supplementary Figure S6). These sampling times were selected as representative of an early and later stage of the response to drought with photosynthetically active leaves. Supplementary Figure S7 shows visual damage symptoms observed in the seedlings 25 days after a drought. Healthy leaves from different seedlings under different conditions as regards population, treatment and sampling time were collected, frozen in liquid nitrogen and stored at −80 °C for subsequent proteomic analysis.

4.2. Protein Extraction and Mass Spectrometry Analysis

Five fully expanded (photosynthetically active) leaves per plant from each population (G, H, C, S), treatment (control well-watered, control and drought) and sampling time (17 and 24 days) were crushed with liquid nitrogen and used for protein extraction. Proteins from three independent biological replicates (200 mg of fresh tissue each) were extracted with TCA/acetone-phenol [61], solubilized in a solution containing 7 M urea, 2 M thiourea, 4% (*w/v*) CHAPS (3 [(3-cholamidopropyl) dimethylammonio]-1-propanesulfonate), 0.5% (*w/v*) Triton X-100 and 100 mM DTT, and quantified by the Bradford method [62] using bovine serum albumin (BSA) as standard.

Shotgun analysis was performed by using 90 µg of BSA protein equivalents per sample that were prefractionated in SDS–PAGE according to Valledor and Weckwerth [63]. The resulting unique bands were excised from the gels and digested with proteomics-grade trypsin (Promega) to a final concentration of 12.5 ng/µL according to Romero-Rodríguez et al. [21]. Digested peptides were desalted by passage through C18 cartridges from Scharlau (Barcelona, Spain), and eluted peptides were vacuum dried and dissolved in a mixture of 70:30 (*v/v*) acetonitrile (ACN)/water containing 0.1% trifluoroacetic acid. Mass spectrometry analyses were conducted at the Proteomics Facility for Research Support Central Service (SCAI) of the University of Cordoba (Spain), using a Dionex Ultimate 3000 nano UPLC instrument from Thermo Scientific (San Jose, CA, USA) coupled to a nanoelectrospray ionization source and a trihybrid analyzer Thermo Orbitrap Fusion mass spectrometer, also from Thermo Scientific, operating in the positive ion mode. The specific settings used in the LC–MS/MS analyses are described elsewhere [42].

4.3. Protein Identification and Quantification

The raw data obtained from the MS analysis were processed with the software Proteome Discoverer v. 2.1.0.81 from Thermo Scientific. MS2 spectra were searched with the SEQUEST engine against the FASTA database obtained from the *Q. ilex* transcriptome [16,17]. Precursor mass tolerance was set at 10 ppm and fragment ion mass tolerance fixed at 0.1 Da. Only those ions with a charge state of +2 or greater were used. In silico peptide lists were generated by theoretical tryptic digestion, allowing up to two missed cleavages, carbamidomethylation of cysteines as a fixed modification and oxidation of methionine as a variable modification. Peptides were classified into protein groups according to the law of parsimony and filtered to FDR = 5% and XCorr $\geq$ 2.

Proteins were quantified in relative terms from the peak areas for precursor ions (specifically, the average of the three strongest peptide ion signals) from the identified peptides were used [64]. Protein values were then normalized by using a method that accounts for variability between runs to normalize relative protein abundance between samples, using the sum of the peak area values for each sample. Only those values consistently present in the three biological replicates were considered for further statistical analysis. Proteins were categorized by function by using the protein FASTA sequences in the software MERCATOR (https://www.plabipd.de/portal/mercator4 accessed August 2020) [30], an online tool for batch classification of proteins or gene sequences into MapMan functional plant categories. In addition, nonannotated proteins were subjected to GO enrichment by using the Panther tool (http://pantherdb.org/ accessed September 2020).

The raw mass spectrometry data thus obtained were deposited on the ProteomeXchange Consortium (http://proteomecentral.proteomexchange.org on 25 January 2021) via the PRIDE partner repository [65] with the dataset identifier PXD023782.

4.4. Statistical Analysis of Data and Selection of Target Proteins

A partial least-squares-discriminant analysis (PLS-DA) of the entire dataset was performed to explain variance and correlation between the different variables, using the software RStudio v. 4.0.3 (Caret package v. 6.0-86 available at https://CRAN.R-project.org/package=caret) [66]. Target proteins were selected, and in silico data were analyzed by using the entire dataset provided by the shotgun analysis. All proteins selected met the following criteria: (a) the entire dataset was filtered by confidence parameters (score ≥ 2, at least 2 peptides per protein); and (b) they were the best represented in qualitative and/or quantitative terms ($p \leq 0.05$) under drought conditions at both sampling times in at least two populations. Figure 1 illustrates the experimental workflow.

4.5. Targeted Post-Acquisition Data Analysis for Selection of Putative Peptides Markers

A list of target peptides generated from the proteotypic peptides (specific peptide sequences from the selected proteins as far as existing annotation information allowed for) was compiled. Proteotypic peptides were searched by aligning the sequences on the entire *Q. ilex* database, using the bioinformatics tools Bourne-again shell (Bash) (available at http://ftp.gnu.org/gnu/bash/ accessed November 2020) [67] and the BLAST protein (BLASTp) tool for the open source software operating system Ubuntu (available at https://www.exoscale.com/syslog/blast/ accessed November 2020). Proteotypic peptides were quantified by integrating the areas of the chromatographic fragment ion peaks in Skyline software (available at https://skyline.ms). The parameters used for relative quantification of MS1 were 0.055 m/z mass tolerance for the instrument, 0.5 m/z for library peak integration and a resolution of 120,000 at m/z 200. The integration peak and retention time (RT) for each peptide were checked by hand in order to confirm the reproducibility of ions among samples. Peptide values were subsequently assessed for statistical significance by using the external tool MS Stats in Skyline [68]. Data were normalized by equalizing intensity medians and then subjected to log2-transformation, after which ANOVA analysis was used to select the best-represented proteins among those exhibiting significant changes ($p \leq 0.05$) under drought conditions. Supplementary Figure S4 provides a quantitative depiction of the proteins and precursor ions.

A protein interaction network of the selected proteins was generated by using the web tool STRING10 (http://string-db.org accessed January 2021). The protein homologs in *Arabidopsis* were analyzed by sequence BLASTing of the TAIR database (http://www.arabidopsis. org/Blast/index.jsp accessed January 2021), followed by application of STRING10 to develop a proteome-scale interaction network [69].

5. Conclusions

Methodologically, the proposed targeted strategy is aimed at identifying peptides associated with the response of *Q. ilex* to drought stress. As a supplement to shotgun analysis, using proteotypic peptides in addition to proteins allows putative markers enabling the identification of specific phenotypes, such as those best-resisting drought, to be selected. Our methodological workflow consisted of selecting those consistent and confident proteins whose proteotypic peptides were used for quantification. Of them, 46 peptides showed significant changes in response to drought stress in the same way that the protein they come from, which were proposed as putative markers.

Biologically, the results suggest that plants possess effective protective mechanisms for adaptation to drought through water loss prevention and protein protection and detoxification. However, small differences in response mechanisms may result in plant survival or adaptation to extreme conditions depending on the particular population or individual. As can be inferred from the composition of the marker panel, *Q. ilex* seedlings from four different populations responded differentially to drought, with the greatest number of changes being observed in the Huelva and Seville populations. Only two proteins (viz., the protease subtilisin and chaperone GrpE protein) were increased to a similar extent in three of the four populations. These proteins should be validated as biomarkers of drought tolerance in *Q. ilex* with further testing.

Our study constitutes a step forward in the molecular elucidation of this forest species. Advances in molecular techniques, including omics, in combination with physiological studies, can be expected to allow especially tolerant or resilient genotypes or individuals under stress conditions, such as those propitiated by a climate change scenario, to be selected in reforestation programs.

Supplementary Materials: The following are available online at https://www.mdpi.com/1422-0067/22/6/3191/s1, Figure S1: Schematic workflow for selection of putative tolerance drought markers in *Q. ilex* seedlings by targeted post-acquisition proteomic analysis. C, Cadiz; G, Granada; H, Huelva; S, Seville, Figure S2: multivariate statistical analysis (PCA) of the entire dataset (A). Dendrogram shows hierarchical clustering of experimental conditions (B). C, Cadiz; G, Granada; H, Huelva; S, Seville. The letters following C, G, H, or S denotes treatment (D, drought; C, control), the numbers sampling time (1, 17 days; 2, 24 days), Figure S3: Representative MS1 fragment ion, retention times (RT) and normalized peak area of a target precursor ion given by the software Skyline, Figure S4: Quantitative protein (left) and abundance of precursor ions (right), Figure S5: Localization of all Andalusian Q. ilex provenances used in this study. Andalusia is delimited by the blue outline, Figure S6: Measurements of quantum yield of photosystem II (Fv/Fm) in dark-adapted leaves of Q. ilex of Andalusian provenances (Cadiz, C; Granada, G; Huelva, H; Seville, S) under drought conditions throughout the experiment. Values are means ± standard errors (SE). Arrows indicate the sampling times when the mean of leaf fluorescence had decreased by 20% after 17 days and 40% after 24 days relative to control seedlings, Figure S7: Visual damage symptoms in four Andalusian Q. ilex provenances (C, Cadiz; H, Huelva; S, Seville; G, Granada; the letters following C, H, S or G denotes treatment: D, drought; C, control) 25 days after drought treatment. Control seedlings did not show damage symptoms throughout the experiment. In contrast, some seedlings did show clear damage symptoms related to drought. Only asymptomatic seedlings were used in this study. Table S1: Raw data, normalized data and significant changes (FDR < 0.05) as calculated from the shotgun proteomics experiment, Table S2: List of proteotypic peptides derived from the 48 selected proteins, Table S3: Locations and environmental conditions of the four Q. ilex Andalusian populations.

Author Contributions: Conceptualization, J.V.J.-N. and M.Á.C.; methodology, M.Á.C., B.S.-E. and V.M.G.-S.; software, M.Á.C., B.S.-E. and V.M.G.-S.; formal analysis, M.Á.C., B.S.-E., P.C. and V.M.G.-S.; investigation, M.Á.C. and E.D.B.; writing—original draft preparation, M.Á.C., writing—review and editing, M.Á.C., M.-D.R. and J.V.J.-N., supervision, M.Á.C. and J.V.J.-N.; funding acquisition, J.V.J.-N. and M.Á.C. All authors have read and agreed to the published version of the manuscript.

Funding: This research was funded by the Spanish BIO2015-64737-R and PID2019-10908RB-100 projects.

Institutional Review Board Statement: Not applicable.

Informed Consent Statement: Not applicable.

Data Availability Statement: The raw mass spectrometry data were deposited to the ProteomeXchange Consortium (http://proteomecentral.proteomexchange.org on 25 January 2021) via the PRIDE partner repository with the dataset identifier PXD023782.

Acknowledgments: M.Á.C. and M.-D.R. are grateful for awards of a Ramón y Cajal (RYC-2017-23706) and a Juan de la Cierva-Incorporación (IJC2018-035272-I) contract, respectively, by the Spanish Ministry of Science, Innovation and Universities. E.D.B. is grateful for the fellowship of the Argentinian National Research Council (CONICET).

Conflicts of Interest: The authors declare no conflict of interest.

References

1. Olea, L.; San Miguel-Ayanz, A. The Spanish dehesa. A traditional Mediterranean silvopastoral system linking production and nature conservation. In Proceedings of the 21st General Meeting of the European Grassland Federation, Badajoz, Spain, 3–6 April 2006; pp. 1–15.
2. Abril, N.; Gion, J.M.; Kerner, R.; Muller-Starck, G.; Cerrillo, R.M.; Plomion, C.; Renaut, J.; Valledor, L.; Jorrin-Novo, J.V. Proteomics research on forest trees, the most recalcitrant and orphan plant species. *Phytochemistry* **2011**, *72*, 1219–1242. [CrossRef]
3. Crescente, M.F.; Gratani, L.; Larcher, W. Shoot growth efficiency and production of *Quercus ilex* L. in different climates. *Flora Morphol. Distrib. Funct. Ecol. Plants* **2002**, *197*, 2–9. [CrossRef]
4. Echevarria-Zomeno, S.; Ariza, D.; Jorge, I.; Lenz, C.; Del Campo, A.; Jorrin, J.V.; Navarro, R.M. Changes in the protein profile of *Quercus ilex* leaves in response to drought stress and recovery. *J. Plant. Physiol.* **2009**, *166*, 233–245. [CrossRef] [PubMed]
5. Keenan, T.; Maria Serra, J.; Lloret, F.; Ninyerola, M.; Sabate, S. Predicting the future of forests in the Mediterranean under climate change, with niche- and process-based models: CO2 matters! *Glob. Chang. Biol.* **2011**, *17*, 565–579. [CrossRef]
6. Castro-Diez, P.; Villar-Salvador, P.; Pérez-Rontomé, C.; Maestro-Martínez, M.; Montserrat-Martí, G. Leaf morphology and leaf chemical composition in three *Quercus* (Fagaceae) species along a rainfall gradient in NE Spain. *Trees* **1997**, *11*, 127–134. [CrossRef]
7. Lumaret, R.; López de Heredia, U.; Soto, A. Origin and genetic variability. In *Cork Oak Woodlands on the Edge: Ecology, Adaptive Management and Restoration*; Island Press: Washington, DC, USA, 2009; pp. 25–32.
8. Valero-Galván, J.; Navarro Cerrillo, R.M.; Romero-Rodríguez, M.C.; Ariza-Mateos, D.; Jorrín-Novo, J.V. Estudio de la respuesta al estrés hídrico en dos poblaciones de encina (*Quercus ilex* subsp. *ballota* (Desf.) Samp.) mediante una aproximación de proteómica comparativa basada en electroforesis bidimensional. In *Proteómica*; Jorrín, J.V., Vázquez, J., Eds.; SEPROT; Argos Press: Córdoba, Spain, 2010; Volume 5, pp. 156–157. Available online: http://www.seprot.es/wp-content/uploads/2016/05/Proteomica_Vol5.pdf (accessed on 1 March 2021).
9. Valero Galvan, J.; Valledor, L.; Navarro Cerrillo, R.M.; Gil Pelegrin, E.; Jorrin-Novo, J.V. Studies of variability in Holm oak (*Quercus ilex* subsp. *ballota* [Desf.] Samp.) through acorn protein profile analysis. *J. Proteom.* **2011**, *74*, 1244–1255. [CrossRef] [PubMed]
10. Ramirez-Valiente, J.A.; Lorenzo, Z.; Soto, A.; Valladares, F.; Gil, L.; Aranda, I. Elucidating the role of genetic drift and natural selection in cork oak differentiation regarding drought tolerance. *Mol. Ecol.* **2009**, *18*, 3803–3815. [CrossRef] [PubMed]
11. Jorge, I.; Navarro, R.M.; Lenz, C.; Ariza, D.; Jorrin, J. Variation in the holm oak leaf proteome at different plant developmental stages, between provenances and in response to drought stress. *Proteomics* **2006**, *6* (Suppl. S1), S207–S214. [CrossRef]
12. Valero-Galvan, J.; Gonzalez-Fernandez, R.; Navarro-Cerrillo, R.M.; Gil-Pelegrin, E.; Jorrin-Novo, J.V. Physiological and proteomic analyses of drought stress response in Holm oak provenances. *J. Proteome Res.* **2013**, *12*, 5110–5123. [CrossRef]
13. Jorrín-Novo, J.; Navarro-Cerrillo, R.M. Variabilidad y respuesta a distintos estreses en poblaciones de encina (*Quercus ilex* L.) en Andalucía mediante una aproximación proteómica. *Ecosistemas* **2014**, *23*, 99–107. [CrossRef]
14. Rico, L.; Ogaya, R.; Terradas, J.; Penuelas, J. Community structures of N2 -fixing bacteria associated with the phyllosphere of a Holm oak forest and their response to drought. *Plant. Biol.* **2014**, *16*, 586–593. [CrossRef] [PubMed]
15. Rivas-Ubach, A.; Barbeta, A.; Sardans, J.; Guenther, A.; Ogaya, R.; Oravec, M.; Urban, O.; Peñuelas, J. Topsoil depth substantially influences the responses to drought of the foliar metabolomes of Mediterranean forests. *Perspect Plant. Ecol. Evol. Syst.* **2016**, *21*, 41–54. [CrossRef]
16. Guerrero-Sanchez, V.M.; Maldonado-Alconada, A.M.; Amil-Ruiz, F.; Jorrin-Novo, J.V. Holm Oak (*Quercus ilex*) Transcriptome. De novo Sequencing and Assembly Analysis. *Front. Mol. Biosci.* **2017**, *4*, 70. [CrossRef] [PubMed]
17. Guerrero-Sanchez, V.M.; Maldonado-Alconada, A.M.; Amil-Ruiz, F.; Verardi, A.; Jorrin-Novo, J.V.; Rey, M.D. Ion Torrent and Illumina, two complementary RNA-seq platforms for constructing the holm oak (*Quercus ilex*) transcriptome. *PLoS ONE* **2019**, *14*, e0210356. [CrossRef] [PubMed]
18. Fernández i Marti, A.; Romero-Rodríguez, C.; Navarro-Cerrillo, R.; Abril, N.; Jorrín-Novo, J.; Dodd, R. Population Genetic Diversity of *Quercus ilex* subsp. *ballota* (Desf.) Samp. Reveals Divergence in Recent and Evolutionary Migration Rates in the Spanish Dehesas. *Forests* **2018**, *9*, 337. [CrossRef]
19. Lopez-Hidalgo, C.; Guerrero-Sanchez, V.M.; Gomez-Galvez, I.; Sanchez-Lucas, R.; Castillejo-Sanchez, M.A.; Maldonado-Alconada, A.M.; Valledor, L.; Jorrin-Novo, J.V. A Multi-Omics Analysis Pipeline for the Metabolic Pathway Reconstruction in the Orphan Species *Quercus ilex*. *Front. Plant. Sci.* **2018**, *9*, 935. [CrossRef]
20. Natali, L.; Vangelisti, A.; Guidi, L.; Remorini, D.; Cotrozzi, L.; Lorenzini, G.; Nali, C.; Pellegrini, E.; Trivellini, A.; Vernieri, P.; et al. How *Quercus ilex* L. saplings face combined salt and ozone stress: A transcriptome analysis. *BMC Genom.* **2018**, *19*, 872. [CrossRef]
21. Romero-Rodriguez, M.C.; Jorrin-Novo, J.V.; Castillejo, M.A. Toward characterizing germination and early growth in the non-orthodox forest tree species *Quercus ilex* through complementary gel and gel-free proteomic analysis of embryo and seedlings. *J. Proteom.* **2019**, *197*, 60–70. [CrossRef]
22. Lopez-Hidalgo, C.; Trigueros, M.; Menendez, M.; Jorrin-Novo, J.V. Phytochemical composition and variability in *Quercus ilex* acorn morphotypes as determined by NIRS and MS-based approaches. *Food Chem.* **2021**, *338*, 127803. [CrossRef]

23. Simova-Stoilova, L.P.; Romero-Rodriguez, M.C.; Sanchez-Lucas, R.; Navarro-Cerrillo, R.M.; Medina-Aunon, J.A.; Jorrin-Novo, J.V. 2-DE proteomics analysis of drought treated seedlings of *Quercus ilex* supports a root active strategy for metabolic adaptation in response to water shortage. *Front. Plant. Sci.* **2015**, *6*, 627. [CrossRef]
24. Gomez-Galvez, I.; Sanchez-Lucas, R.; San-Eufrasio, B.; de Francisco, L.E.R.; Maldonado-Alconada, A.M.; Fuentes-Almagro, C.; Castillejo, M.A. Optimizing Shotgun Proteomics Analysis for a Confident Protein Identification and Quantitation in Orphan Plant Species: The Case of Holm Oak (*Quercus ilex*). *Methods Mol. Biol.* **2020**, *2139*, 157–168. [CrossRef]
25. Plomion, C.; Aury, J.M.; Amselem, J.; Alaeitabar, T.; Barbe, V.; Belser, C.; Berges, H.; Bodenes, C.; Boudet, N.; Boury, C.; et al. Decoding the oak genome: Public release of sequence data, assembly, annotation and publication strategies. *Mol. Ecol. Resour.* **2016**, *16*, 254–265. [CrossRef] [PubMed]
26. Ramos, A.M.; Usie, A.; Barbosa, P.; Barros, P.M.; Capote, T.; Chaves, I.; Simoes, F.; Abreu, I.; Carrasquinho, I.; Faro, C.; et al. The draft genome sequence of cork oak. *Sci. Data* **2018**, *5*, 180069. [CrossRef]
27. Gillet, L.C.; Navarro, P.; Tate, S.; Rost, H.; Selevsek, N.; Reiter, L.; Bonner, R.; Aebersold, R. Targeted data extraction of the MS/MS spectra generated by data-independent acquisition: A new concept for consistent and accurate proteome analysis. *Mol. Cell Proteom.* **2012**, *11*. [CrossRef] [PubMed]
28. Domon, B.; Aebersold, R. Options and considerations when selecting a quantitative proteomics strategy. *Nat. Biotechnol.* **2010**, *28*, 710–721. [CrossRef]
29. Escandon, M.; Jorrin-Novo, J.V.; Castillejo, M.A. Application and optimization of label-free shotgun approaches in the study of *Quercus ilex*. *J. Proteom.* **2021**, *233*, 104082. [CrossRef]
30. Navarro-Cerrillo, R.M.; Ruiz Gomez, F.J.; Cabrera-Puerto, R.J.; Sánchez-Cuesta, R.; Palacios Rodriguez, G.; Quero Pérez, J.L. Growth and physiological sapling responses of eleven *Quercus ilex* ecotypes under identical environmental conditions. *Ecol. Manag.* **2018**, *415–416*, 58–69. [CrossRef]
31. San-Eufrasio, B.; Sánchez-Lucas, R.; López-Hidalgo, C.; Guerrero-Sánchez, V.M.; Castillejo, M.Á.; Maldonado-Alconada, A.M.; Jorrín-Novo, J.V.; Rey, M.-D. Responses and Differences in Tolerance to Water Shortage under Climatic Dryness Conditions in Seedlings from *Quercus* spp. and Andalusian *Q. ilex* Populations. *Forests* **2020**, *11*, 707. [CrossRef]
32. Lohse, M.; Nagel, A.; Herter, T.; May, P.; Schroda, M.; Zrenner, R.; Tohge, T.; Fernie, A.R.; Stitt, M.; Usadel, B. Mercator: A fast and simple web server for genome scale functional annotation of plant sequence data. *Plant. Cell Environ.* **2014**, *37*, 1250–1258. [CrossRef]
33. Rodiger, A.; Baginsky, S. Tailored Use of Targeted Proteomics in Plant-Specific Applications. *Front. Plant. Sci.* **2018**, *9*, 1204. [CrossRef]
34. Chawade, A.; Alexandersson, E.; Bengtsson, T.; Andreasson, E.; Levander, F. Targeted Proteomics Approach for Precision Plant Breeding. *J. Proteome Res.* **2016**, *15*, 638–646. [CrossRef]
35. Buts, K.; Michielssens, S.; Hertog, M.L.; Hayakawa, E.; Cordewener, J.; America, A.H.; Nicolai, B.M.; Carpentier, S.C. Improving the identification rate of data independent label-free quantitative proteomics experiments on non-model crops: A case study on apple fruit. *J. Proteom.* **2014**, *105*, 31–45. [CrossRef] [PubMed]
36. Riebel, M.; Fronk, P.; Distler, U.; Tenzer, S.; Decker, H. Proteomic profiling of German Dornfelder grape berries using data-independent acquisition. *Plant. Physiol. Biochem.* **2017**, *118*, 64–70. [CrossRef]
37. Martin, L.B.; Sherwood, R.W.; Nicklay, J.J.; Yang, Y.; Muratore-Schroeder, T.L.; Anderson, E.T.; Thannhauser, T.W.; Rose, J.K.; Zhang, S. Application of wide selected-ion monitoring data-independent acquisition to identify tomato fruit proteins regulated by the CUTIN DEFICIENT2 transcription factor. *Proteomics* **2016**, *16*, 2081–2094. [CrossRef] [PubMed]
38. Mata, C.I.; Fabre, B.; Parsons, H.T.; Hertog, M.; Van Raemdonck, G.; Baggerman, G.; Van de Poel, B.; Lilley, K.S.; Nicolai, B.M. Ethylene Receptors, CTRs and EIN2 Target Protein Identification and Quantification Through Parallel Reaction Monitoring During Tomato Fruit Ripening. *Front. Plant. Sci.* **2018**, *9*, 1626. [CrossRef]
39. Bose, U.; Byrne, K.; Howitt, C.A.; Colgrave, M.L. Targeted proteomics to monitor the extraction efficiency and levels of barley alpha-amylase trypsin inhibitors that are implicated in non-coeliac gluten sensitivity. *J. Chromatogr. A* **2019**, *1600*, 55–64. [CrossRef] [PubMed]
40. Bromilow, S.N.; Gethings, L.A.; Langridge, J.I.; Shewry, P.R.; Buckley, M.; Bromley, M.J.; Mills, E.N. Comprehensive Proteomic Profiling of Wheat Gluten Using a Combination of Data-Independent and Data-Dependent Acquisition. *Front. Plant. Sci.* **2016**, *7*, 2020. [CrossRef] [PubMed]
41. Meyer, J.G.; Schilling, B. Clinical applications of quantitative proteomics using targeted and untargeted data-independent acquisition techniques. *Expert. Rev. Proteom.* **2017**, *14*, 419–429. [CrossRef] [PubMed]
42. Castillejo, M.A.; Fondevilla-Aparicio, S.; Fuentes-Almagro, C.; Rubiales, D. Quantitative Analysis of Target Peptides Related to Resistance Against Ascochyta Blight (*Peyronellaea pinodes*) in Pea. *J. Proteome Res.* **2020**, *19*, 1000–1012. [CrossRef] [PubMed]
43. Vaseva, I.; Sabotič, J.; Šuštar-Vozlič, J.; Meglič, V.; Kidrič, M.; Demirevska, K.; Simova-Stoilova, L. The response of plants to drought stress: The role of dehydrins, chaperones, proteases and protease inhibitors in maintaining cellular protein function. In *Droughts: New Research*; Neves, D.F., Sanz, J.D., Eds.; Nova Science Publishers, Inc.: New York, NY, USA, 2012; Volume 1, pp. 1–45.
44. Sergeant, K.; Spiess, N.; Renaut, J.; Wilhelm, E.; Hausman, J.F. One dry summer: A leaf proteome study on the response of oak to drought exposure. *J. Proteom.* **2011**, *74*, 1385–1395. [CrossRef]
45. González-Cruz, J.; Pastenes, C. Water-stress-induced thermotolerance of photosynthesis in bean (*Phaseolus vulgaris* L.) plants: The possible involvement of lipid composition and xanthophyll cycle pigments. *Environ. Exp. Bot.* **2012**, *77*, 127–140. [CrossRef]

46. Cuellar-Ortiz, S.M.; De La Paz Arrieta-Montiel, M.; Acosta-Gallegos, J.; Covarrubias, A.A. Relationship between carbohydrate partitioning and drought resistance in common bean. *Plant. Cell Environ.* **2008**, *31*, 1399–1409. [CrossRef] [PubMed]
47. Krasensky, J.; Jonak, C. Drought, salt, and temperature stress-induced metabolic rearrangements and regulatory networks. *J. Exp. Bot.* **2012**, *63*, 1593–1608. [CrossRef] [PubMed]
48. Thalmann, M.; Santelia, D. Starch as a determinant of plant fitness under abiotic stress. *New Phytol.* **2017**, *214*, 943–951. [CrossRef]
49. Kaplan, F.; Guy, C.L. RNA interference of Arabidopsis beta-amylase8 prevents maltose accumulation upon cold shock and increases sensitivity of PSII photochemical efficiency to freezing stress. *Plant. J.* **2005**, *44*, 730–743. [CrossRef]
50. Yin, Y.G.; Kobayashi, Y.; Sanuki, A.; Kondo, S.; Fukuda, N.; Ezura, H.; Sugaya, S.; Matsukura, C. Salinity induces carbohydrate accumulation and sugar-regulated starch biosynthetic genes in tomato (*Solanum lycopersicum* L. cv. 'Micro-Tom') fruits in an ABA- and osmotic stress-independent manner. *J. Exp. Bot.* **2010**, *61*, 563–574. [CrossRef] [PubMed]
51. Skirycz, A.; De Bodt, S.; Obata, T.; De Clercq, I.; Claeys, H.; De Rycke, R.; Andriankaja, M.; Van Aken, O.; Van Breusegem, F.; Fernie, A.R.; et al. Developmental stage specificity and the role of mitochondrial metabolism in the response of Arabidopsis leaves to prolonged mild osmotic stress. *Plant. Physiol.* **2010**, *152*, 226–244. [CrossRef]
52. Wang, S.J.; Liu, L.F.; Chen, C.K.; Chen, L.W. Regulations of granule-bound starch synthase I gene expression in rice leaves by temperature and drought stress. *Biol. Plant.* **2006**, *50*, 537–541. [CrossRef]
53. Prathap, V.; Tyagi, A. Correlation between expression and activity of ADP glucose pyrophosphorylase and starch synthase and their role in starch accumulation during grain filling under drought stress in rice. *Plant. Physiol. Biochem.* **2020**, *157*, 239–243. [CrossRef]
54. Sghaier-Hammami, B.; Valero-Galvan, J.; Romero-Rodriguez, M.C.; Navarro-Cerrillo, R.M.; Abdelly, C.; Jorrin-Novo, J. Physiological and proteomics analyses of Holm oak (*Quercus ilex* subsp. *ballota* [Desf.] Samp.) responses to *Phytophthora cinnamomi*. *Plant. Physiol. Biochem.* **2013**, *71*, 191–202. [CrossRef]
55. Sharma, A.; Shahzad, B.; Rehman, A.; Bhardwaj, R.; Landi, M.; Zheng, B. Response of Phenylpropanoid Pathway and the Role of Polyphenols in Plants under Abiotic Stress. *Molecules* **2019**, *24*, 2452. [CrossRef]
56. Nogués, I.; Llusià, J.; Ogaya, R.; Munné-Bosch, S.; Sardans, J.; Peñuelas, J.; Loreto, F. Physiological and antioxidant responses of *Quercus ilex* to drought in two different seasons. *Plant. Biosyst. Int. J. Deal. All Asp. Plant Biol.* **2013**, *148*, 268–278. [CrossRef]
57. Jafarnia, S.; Akbarinia, M.; Hosseinpour, B.; Modarres Sanavi, S.A.M.; Salami, S.A. Effect of drought stress on some growth, morphological, physiological, and biochemical parameters of two different populations of *Quercus brantii*. *Iforest Biogeosciences For.* **2018**, *11*, 212–220. [CrossRef]
58. Ghanbary, E.; Tabari Kouchaksaraei, M.; Zarafshar, M.; Bader, K.M.; Mirabolfathy, M.; Ziaei, M. Differential physiological and biochemical responses of *Quercus infectoria* and *Q. libani* to drought and charcoal disease. *Physiol. Plant* **2020**, *168*, 876–892. [CrossRef] [PubMed]
59. Li, J.; Ban, L.; Wen, H.; Wang, Z.; Dzyubenko, N.; Chapurin, V.; Gao, H.; Wang, X. An aquaporin protein is associated with drought stress tolerance. *Biochem. Biophys. Res. Commun.* **2015**, *459*, 208–213. [CrossRef] [PubMed]
60. Castillejo, M.A.; Iglesias-Garcia, R.; Wienkoop, S.; Rubiales, D. Label-free quantitative proteomic analysis of tolerance to drought in *Pisum sativum*. *Proteomics* **2016**, *16*, 2776–2787. [CrossRef] [PubMed]
61. Wang, W.; Vignani, R.; Scali, M.; Cresti, M. A universal and rapid protocol for protein extraction from recalcitrant plant tissues for proteomic analysis. *Electrophoresis* **2006**, *27*, 2782–2786. [CrossRef] [PubMed]
62. Bradford, M.M. A rapid and sensitive method for the quantitation of microgram quantities of proteins utilizing the principle of protein-dye binding. *Anal. Biochem.* **1976**, *72*, 248–254. [CrossRef]
63. Valledor, L.; Weckwerth, W. An Improved Detergent-Compatible Gel-Fractionation LC-LTQ-Orbitrap-MS Workflow for Plant and Microbial Proteomics. In *Plant Proteomics. Methods in Molecular Biology (Methods and Protocols)*; Jorrin-Novo, J.K.S., Weckwerth, W., Wienkoop, S., Eds.; Humana Press: Totowa, NJ, USA, 2014; Volume 1072. [CrossRef]
64. Al Shweiki, M.R.; Monchgesang, S.; Majovsky, P.; Thieme, D.; Trutschel, D.; Hoehenwarter, W. Assessment of Label-Free Quantification in Discovery Proteomics and Impact of Technological Factors and Natural Variability of Protein Abundance. *J. Proteome Res.* **2017**, *16*, 1410–1424. [CrossRef]
65. Perez-Riverol, Y.; Csordas, A.; Bai, J.; Bernal-Llinares, M.; Hewapathirana, S.; Kundu, D.J.; Inuganti, A.; Griss, J.; Mayer, G.; Eisenacher, M.; et al. The PRIDE database and related tools and resources in 2019: Improving support for quantification data. *Nucleic Acids Res.* **2019**, *47*, D442–D450. [CrossRef] [PubMed]
66. Kuhn, M. Caret: Classification and Regression Training. R package version 6.0-86. 2020. Available online: https://CRAN.R-project.org/package=caret (accessed on 1 March 2021).
67. Altschu, S.F.; Gish, W.; Miller, W.; Myers, E.W.; Lipman, D.J. Basic Local Alignment Search Tool. *J. Mol. Biol.* **1990**, *215*, 403–410. [CrossRef]
68. Chang, C.-Y.; Picotti, P.; Hüttenhain, R.; Heinzelmann-Schwarz, V.; Jovanovic, M.; Aebersold, R.; Vitek, O. Protein Significance Analysis in Selected Reaction Monitoring (SRM) Measurements. *Mol. Cell Proteom.* **2012**, *11*. [CrossRef] [PubMed]
69. Suo, J.; Zhao, Q.; Zhang, Z.; Chen, S.; Cao, J.; Liu, G.; Wei, X.; Wang, T.; Yang, C.; Dai, S. Cytological and Proteomic Analyses of *Osmunda cinnamomea* Germinating Spores Reveal Characteristics of Fern Spore Germination and Rhizoid Tip Growth. *Mol. Cell Proteom.* **2015**, *14*, 2510–2534. [CrossRef] [PubMed]

International Journal of
Molecular Sciences

MDPI

Article

Proteomic and Biochemical Analyses of the Mechanism of Tolerance in Mutant Soybean Responding to Flooding Stress

Setsuko Komatsu [1,*], Hisateru Yamaguchi [2], Keisuke Hitachi [3], Kunihiro Tsuchida [3], Yuhi Kono [4] and Minoru Nishimura [5]

1 Faculty of Environment and Information Sciences, Fukui University of Technology, Fukui 910-8505, Japan
2 Department of Medical Technology, Yokkaichi Nursing and Medical Care University, Yokkaichi 512-8045, Japan; h-yamaguchi@y-nm.ac.jp
3 Institute for Comprehensive Medical Science, Fujita Health University, Toyoake 470-1192, Japan; hkeisuke@fujita-hu.ac.jp (K.H.); tsuchida@fujita-hu.ac.jp (K.T.)
4 Central Region Agricultural Research Center, National Agriculture and Food Research Organization, Joetsu 943-0193, Japan; k41523@affrc.go.jp
5 Graduate School of Science and Technology, Niigata University, Niigata 950-2181, Japan; nisimura@agr.niigata-u.ac.jp
* Correspondence: skomatsu@fukui-ut.ac.jp; Tel.: +81-276-29-2466

Abstract: To investigate the mechanism of flooding tolerance of soybean, flooding-tolerant mutants derived from gamma-ray irradiated soybean were crossed with parent cultivar Enrei for removal of other factors besides the genes related to flooding tolerance in primary generated mutant soybean. Although the growth of the wild type was significantly suppressed by flooding compared with the non-flooding condition, that of the mutant lines was better than that of the wild type even if it was treated with flooding. A two-day-old mutant line was subjected to flooding for 2 days and proteins were analyzed using a gel-free/label-free proteomic technique. Oppositely changed proteins in abundance between the wild type and mutant line under flooding stress were associated in endoplasmic reticulum according to gene-ontology categorization. Immunoblot analysis confirmed that calnexin accumulation increased in both the wild type and mutant line; however, calreticulin accumulated in only the mutant line under flooding stress. Furthermore, although glycoproteins in the wild type decreased by flooding compared with the non-flooding condition, those in the mutant line increased even if it was under flooding stress. Alcohol dehydrogenase accumulated in the wild type and mutant line; however, this enzyme activity significantly increased and mildly increased in the wild type and mutant line, respectively, under flooding stress compared with the non-flooding condition. Cell death increased and decreased in the wild type and mutant line, respectively, by flooding stress. These results suggest that the regulation of cell death through the fermentation system and glycoprotein folding might be an important factor for the acquisition of flooding tolerance in mutant soybean.

Keywords: proteomics; mutant soybean; flooding; glycoprotein folding; fermentation; cell death

Citation: Komatsu, S.; Yamaguchi, H.; Hitachi, K.; Tsuchida, K.; Kono, Y.; Nishimura, M. Proteomic and Biochemical Analyses of the Mechanism of Tolerance in Mutant Soybean Responding to Flooding Stress. *Int. J. Mol. Sci.* **2021**, *22*, 9046. https://doi.org/10.3390/ijms22169046

Academic Editor: Jiří Fajkus

Received: 9 July 2021
Accepted: 19 August 2021
Published: 22 August 2021

1. Introduction

Along with climate changes, the increased tendency of flooding has triggered severe crop reductions in yield and quality around the world [1]. Flooding is classified into two forms, which are waterlogging and submergence, depending on water depth [2]. Waterlogging is the condition that water exists on the soil surface and only plant roots are surrounded by water, while submergence is the state that the whole plant partially or completely immerses in water [3]. Flooding is a major threat causing substantial yield decline of crops and is expected to be even more serious in many parts of the world due to climatic anomalies in the future. In Japan, more than 80% of soybeans are cultivated in fields converted from paddy fields. During the rainy season, from June to the middle of July, soybean seedlings are damaged due to submergence in fields with poor drainage. Understanding the mechanisms of plants coping with unanticipated flooding is crucial for developing new flooding-tolerant crop varieties [4]. These previous findings indicated that the development of flooding-tolerant crop is an important task for improvement of crop yield.

In rice, which is one of the few flooding-tolerant crops, seedlings cannot grow under extended periods of complete submergence [5]. One of the two adaptive strategies for flooding tolerance is the *SNORKEL 1/2* dependent escape strategy [6]. It promotes the internode elongation through the stimulation of gibberellin biosynthesis in deep-water rice under flooding stress, and thereby enabling rice grows upward to the water surface for air exchange [6]. Another adaptive strategy is the *SUBMERGENCE1* dependent quiescence strategy [7]. In a few rice-tolerant varieties against submergence, the elongation of stem and leaf is inhibited by suppressing the increase in ethylene concentration, which decreases the sensitivity of plants to gibberellin [7]. Although the mechanisms of flooding tolerance in rice are well reported, those of soybean are still unclear, because the flooding-tolerant materials such as mutant lines or cultivars are not enough.

Although soybean is one of the major agricultural crops, it is particularly sensitive to flooding stress [8]. Plant growth and grain yield of soybean are markedly reduced in flooded soil [8]. When soybean was exposed to flooding at the vegetative growth stage or reproductive growth stage, grain yield and quality were reduced [9]. In addition, secondary aerenchyma was formed and worked as an oxygen pathway under flooded conditions [10]. Furthermore, flooding stress impaired plant growth by inhibiting root elongation and reducing hypocotyl pigmentation [11]. Under flooding, soybean seedlings showed differential regulation of proteins involved in signal transduction, hormonal signaling, transcriptional control, glucose degradation/sucrose accumulation, alcohol fermentation, gamma-aminobutyric acid shunt, suppression of reactive-oxygen species scavenging, mitochondrial impairment, ubiquitin/proteasome-mediated proteolysis, and cell-wall loosening [12–15]. Although flooding-response mechanisms in soybean were reported, characterization of the mechanism of flooding tolerance is needed regarding to agricultural usage.

To characterize the mechanism of flooding tolerance in soybean, flooding-tolerant lines were generated by physical mutagenesis of gamma-ray irradiation [16]. For preparation of the flooding-tolerant mutant [16], flooding-tolerant tests were repeated five times using gamma-ray irradiated soybeans, whose root growth (M6 stage) was not suppressed even if it was under flooding stress. Using the primary generated flooding-tolerant mutant, gel-based proteomic analysis was performed, indicating that activation of the fermentation system in the early stages of flooding is an important factor for the acquisition of flooding tolerance in soybean [16]. Using this mutant soybean and abscisic-acid treated soybean, which show flooding tolerance, proteomics [17], transcriptomics [18], and metabolomics [19] were performed at the initial stage of flooding stress. In this study, to remove other factors besides the genes related to flooding tolerance in primary generated mutant soybean, it was crossed with parent cultivar Enrei and flooding-tolerant tests were repeated two times. Using the progeny of this cross, morphological analysis was performed under flooding stress. Based on the morphological results, gel-free/label-free proteomic analysis was carried out to explore the mechanism of tolerance for the positive effects on growth of mutant soybean treated with flooding. Proteomic results were subsequently confirmed by immunoblot, enzyme activity, and physiological analyses.

2. Results

2.1. Morphological Analysis of Mutant Soybean under Flooding Stress

To remove other factors besides the genes related to flooding tolerance in primary generated mutant soybean [16], they were crossed with parent cultivar Enrei. Mutant lines with flooding tolerance selected from progeny were used in this study. As mutant lines, 1386-6 (G2), 1386-9 (G3), 1387-12 (G4) lines were used for morphological analysis.

To investigate the morphology of mutant soybean under flooding stress, morphological changes of soybean were analyzed (Figure 1). To induce flooding stress, water was added to immerse 2-day-old soybeans, and samples were collected at 6 days after 5-days flooding (Figure 1). The length/weight of stem/root of the mutant lines did not change compared with the wild type without flooding stress (Figure 2). The length/weight of stem/root of the wild type soybean was significantly suppressed by flooding; however, those of the 1386-6 (G2) and 1386-9 (G3) lines were better than those of the wild type even if they were treated with flooding. By flooding, length and weight of hypocotyl did not change between wild type and mutants; however, those of epicotyls increased in the mutant compared with wild type (Figure 2). Based on morphological results, 1386-6 (G2) line of mutant was used for proteomic analysis.

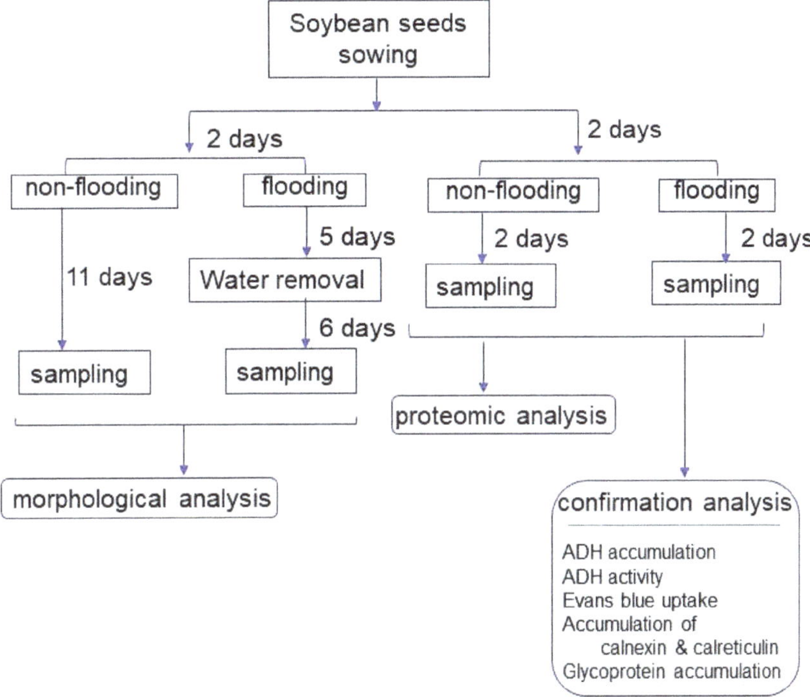

Figure 1. Experimental design for investigation of the mechanism of flooding tolerance in mutant soybean. As mutant lines, 1386-6 (G2), 1386-9 (G3), and 1387-12 (G4) lines were used. Soybean seedlings were analyzed with morphological and proteomics. Proteomic results were subsequently confirmed by immunoblot, enzyme activity, and physiological analyses. All experiments were performed with 3 independent biological replicates.

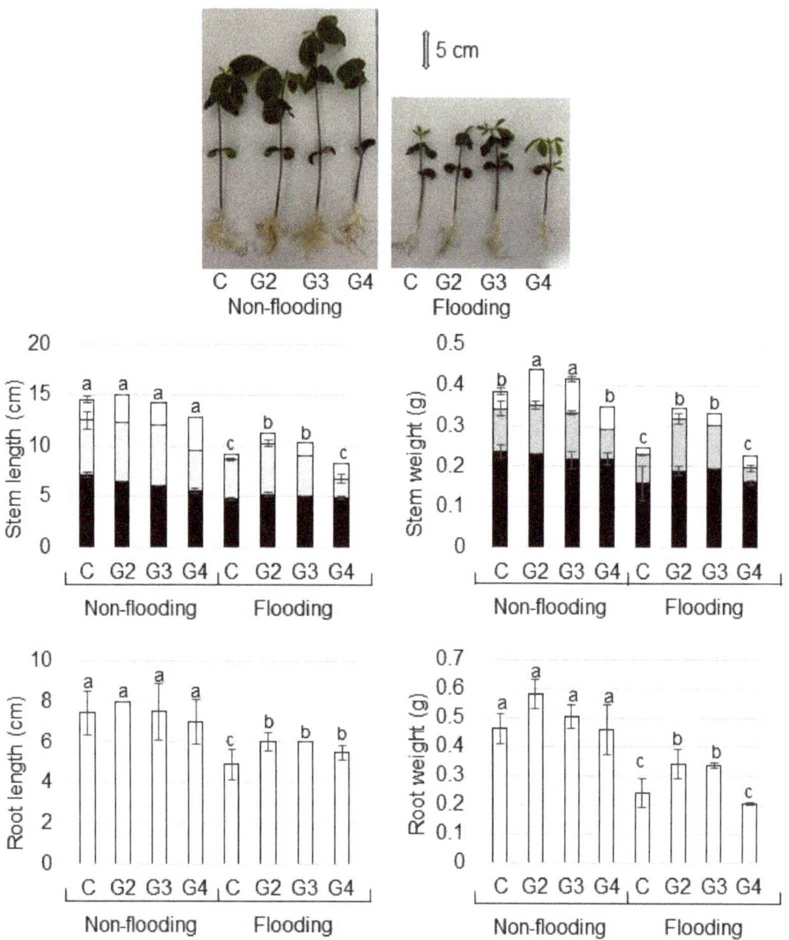

Figure 2. Morphological effect of flooding stress in mutant soybean. Wild type (C) and mutant lines, which are 1386-6 (G2), 1386-9 (G3), 1387-12 (G4), were used. Two-day-old soybeans were flooded for 5 days and water was removed. After 6 days of water removal, root and hypocotyl were collected. For non-flooded group, samples were collected at 13 days after sowing. Photograph shows 13-day-old soybeans. Hypocotyl (black column)/epicotyl (gray column)/2nd internode (white column) length, main root length, hypocotyl/epicotyl/2nd internode weight, and total root weight were measured as morphological parameters. In the graph, hypocotyl/epicotyl/2nd internode was summarized as stem. Data are shown as means ± SD from 3 independent biological replicates. The different letters indicate significant changes according to one-way ANOVA followed by Tukey's multiple comparisons ($p < 0.05$).

2.2. Identification and Functional Investigation of Proteins in Mutant Soybean under Flooding Stress

In order to explore the cellular mechanism on growth of mutant soybean, a gel-free/label-free proteomic technique was used. Proteins were extracted from the root including hypocotyl after 2-days flooding of 2-day-old soybeans (Figure 1). Totally, 7889 proteins were detected by MS analysis. Four kinds of group, which were wild type/mutant line under flooding/non-flooding, were prepared. The criteria for significantly changed proteins were 2 or more than 2 matched peptides with a p-value less than 0.05 (Tables S1–S4).

Relative abundance of 6967 proteins from wild type under flooding was compared with that from non-flooding. Among the differentially changed 986 proteins, 514 proteins increased and 472 proteins decreased in wild type under flooding stress compared with under non-flooding (Table S1). Furthermore, relative abundance of 6967 proteins from mutant under flooding was compared with that from non-flooding. Among the differentially changed 833 proteins, 350 proteins increased and 483 proteins decreased in the mutant line under flooding stress compared with non-flooding (Table S2). The proteomic data of all samples from different groups were compared by PCA, which indicated the different accumulation patterns of proteins from different treatment (Figure S1). Flooding stress largely affected soybean proteins in both wild type and mutant line (Figure 3).

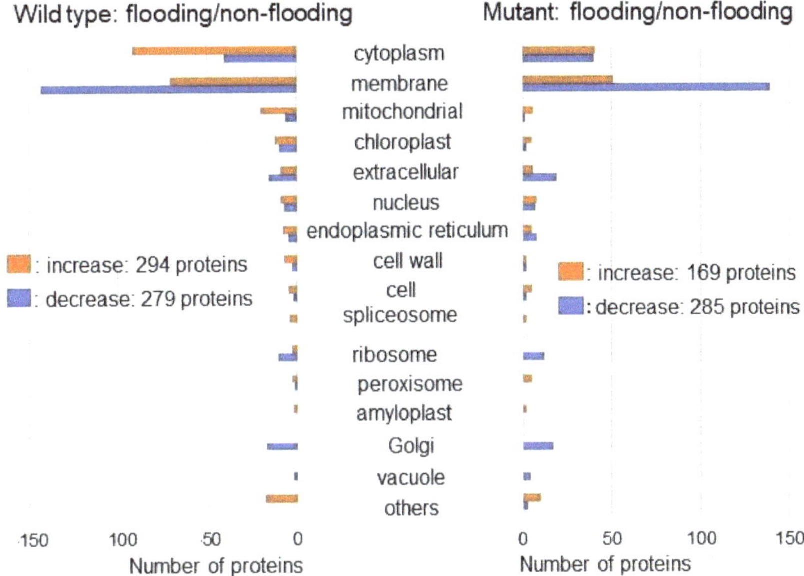

Figure 3. Gene-ontology categories of proteins with differential abundance in the mutant line and wild type treated with flooding. Two-day-old soybeans of the mutant line and wild type were exposed without (non-flooded) or with (flooded) flooding stress. After proteomic analysis, significantly changed proteins ($p < 0.05$) in the mutant line and wild type were compared between flooding and non-flooding conditions. Functional categories of changed proteins were determined using gene-ontology analysis (Tables S1 and S2). Red and blue columns show increased and decreased proteins, respectively.

Furthermore, the abundance of proteins differentially changed with fold change $\geq$1.5 and $\leq$1/3 in wild type or mutant line under flooding stress compared with non-flooding were categorized using cellular component of gene-ontology analysis (Figure 3). Among the differentially changed 573 proteins, 294 proteins increased and 279 proteins decreased in wild type under flooding stress compared with non-flooding (Figure 3 left). Among the differentially changed 454 proteins, 169 proteins increased and 285 proteins decreased in the mutant line under flooding stress compared with non-flooding (Figure 3 right). Differentially increased proteins were mainly located in cytoplasm and decreased proteins were in membrane. Endoplasmic reticulum related proteins were oppositely changed between the wild type and mutant line under flooding condition (Figure 3). Based on proteomic results, proteins in oppositely changed subcellular category, which is endoplasmic reticulum, were confirmed by biochemical and physiological analyses.

2.3. Immunoblot Analysis of Proteins Involved in Endoplasmic Reticulum in Mutant Soybean under Flooding Stress

To better uncover the change of proteins from different treatments, immunoblot analysis of proteins in endoplasmic reticulum was performed (Figure 4). Proteins were extracted from root including hypocotyl of soybeans, which were wild type and mutant line treated with or without flooding. Coomassie brilliant blue staining pattern was used as loading control (Figure S2). To investigate the change of proteins in endoplasmic reticulum, the accumulation of calnexin and calreticulin was analyzed (Figure S3). Immunoblot analysis confirmed that calnexin accumulation increased in both wild type and mutant line; however, calreticulin accumulated in only mutant line under flooding stress (Figure 4A,B). Furthermore, Concanavalin A antibody was used for lectin blot (Figure S4); and the accumulation of glycoproteins decreased in wild type by flooding and that in the mutant line under flooding stress even increased in the level of the non-flooding (Figure 4C,D). In N-linked glycoproteins, exostosin domain-containing protein (I1JXR7) was significantly decreased in wild type by flooding stress (Table S1); however, it was increased in the mutant line even if it was under flooding stress (Table S2). These results indicated that folding of glycoproteins was improved in the mutant line, even if it was flooding condition.

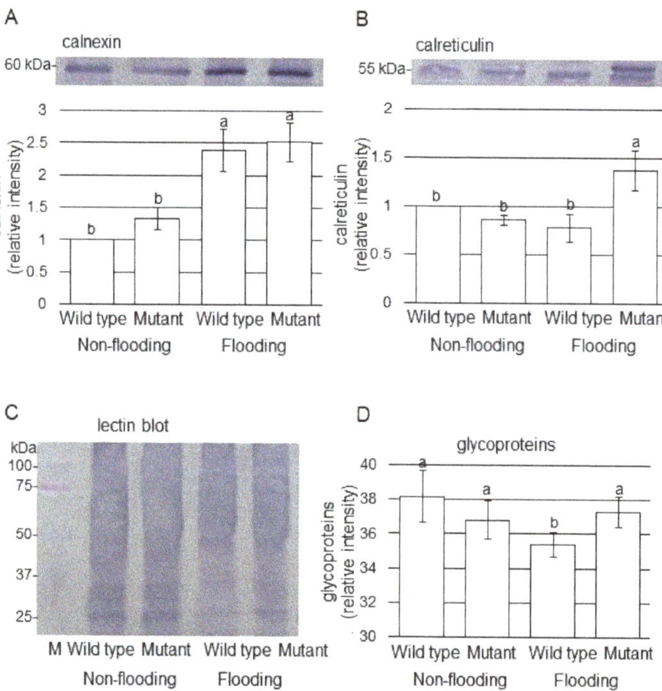

Figure 4. Immunoblot analysis of proteins involved in endoplasmic reticulum. Proteins were extracted from root including hypocotyl, separated on SDS-polyacrylamide gel by electrophoresis, and transferred onto membrane. The membrane was cross-reacted with anti-calnexin (**A**) and calreticulin (**B**) antibodies. In the case of glycoproteins, lectin-blot (**C**,**D**) was performed. Coomassie brilliant blue staining pattern was used as loading control (Figure S2). The integrated densities of bands were calculated using ImageJ software. Data are shown as the means ± SD from 3 independent biological replicates (Figures S3 and S4). The different letters indicate significant changes according to one-way ANOVA followed by Tukey's multiple comparisons ($p < 0.05$).

2.4. Alcohol Dehydrogenase (ADH) Accumulation and Activity in Mutant Soybean under Flooding Stress

To better uncover the change of fermentation in different treatments, change of ADH was analyzed. Proteins were extracted from root including hypocotyl of soybeans, which were wild type and mutant line treated with or without flooding. Coomassie brilliant blue staining pattern was used as loading control (Figure S2), and the accumulation of ADH was analyzed with immunoblot (Figure S5). ADH accumulated in wild type and mutant line by flooding stress (Figure 5A). Furthermore, in order to understand the molecular mechanism of the mutant line under flooding stress, ADH activity was analyzed. Enzyme activity of ADH significantly increased and mildly increased in wild type and mutant line, respectively, under flooding stress (Figure 5B). These results indicated that the fermentation was caused in flooded seedlings and it was recovered in the mutant line, even if it was flooding condition.

A

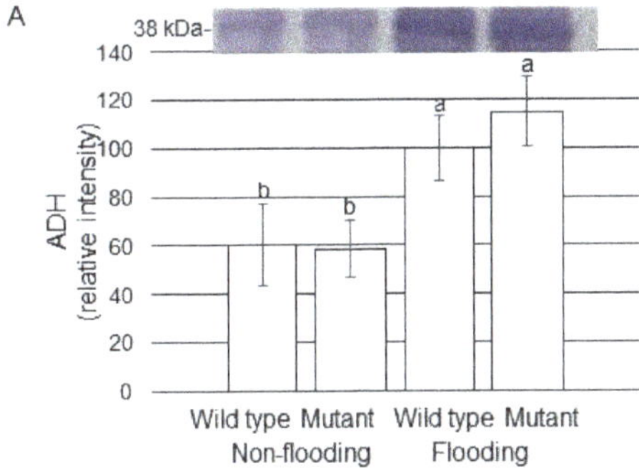

B

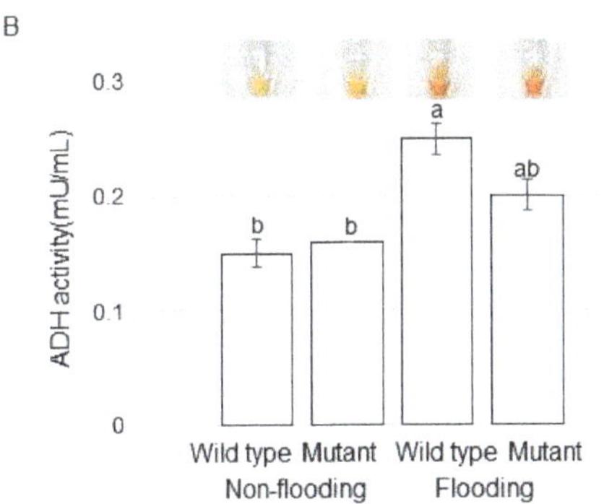

Figure 5. Immunoblot analysis and activity assay of proteins involved in fermentation. (**A**) Proteins were extracted from root including hypocotyl, separated on SDS-polyacrylamide gel by electrophoresis, and transferred onto membrane. The membrane was cross-reacted with anti-ADH antibody. Coomassie brilliant blue staining pattern was used as loading control (Figure S2). The integrated densities of bands were calculated using ImageJ software with 3 independent biological replicates (Figure S5). (**B**) Proteins were extracted from root including hypocotyl and ADH activity assay was performed. Data are shown as means ± SD from 3 independent biological replicates. The different letters indicate significant changes according to one-way ANOVA followed by Tukey's multiple comparisons ($p < 0.05$).

2.5. Evans-Blue Staining with Mutant Soybean under Flooding Stress

To better understand suppression of root growth, the cell death of root-tip was evaluated by Evans-blue staining (Figure 6). Seedlings after 2 days of flooding treatment were stained with Evans blue, and the Evans blue was extracted into the solvent from the stained root tips and measured. Evans-blue uptake in root tip increased in wild type by flooding stress and decreased in the mutant line (Figure 6). These results indicated that the suppression of root growth by flooding was caused by cell death of root tip induced by flooding stress in wild type and it has not happened in the mutant line.

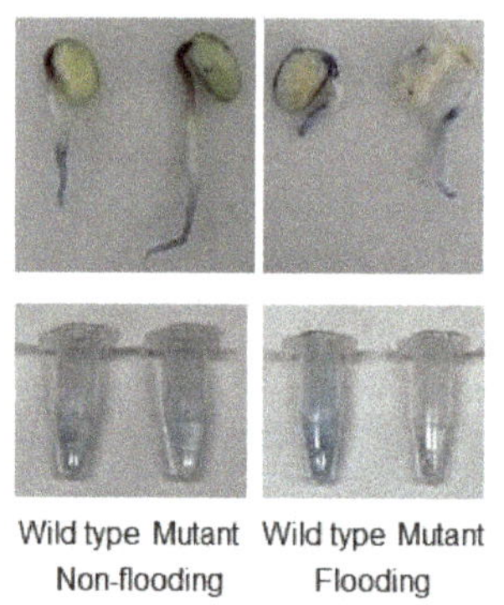

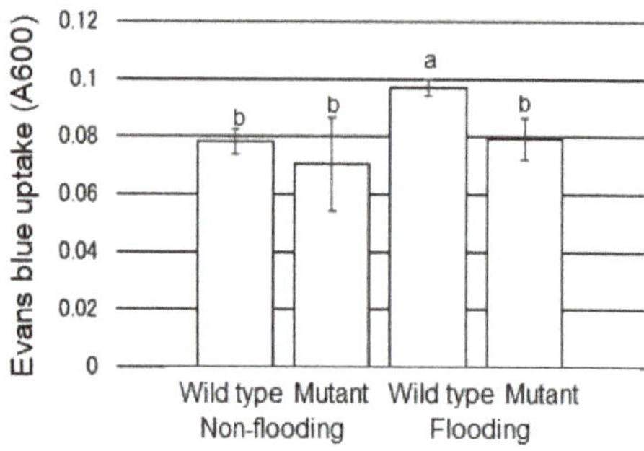

Figure 6. Evaluation of cell death in soybean with Evans-blue staining. After flooding, soybean seedlings were stained with Evans-blue dye. The Evans-blue dye was extracted from the root tip and absorbance was measured at 600 nm. Data are shown as means ± SD from 3 independent biological replicates. The different letters indicate significant changes according to one-way ANOVA followed by Tukey's multiple comparisons ($p < 0.05$).

3. Discussion

3.1. Mutation in Flooding-Tolerant Soybean Has Positive Effects on the Growth of Plant under Flooding Stress

To investigate the difference of mutant soybean before and after being crossed with parent cultivar Enrei, morphological changes of the mutant lines were analyzed (Figure 2) and compared with previous results [16]. Before being crossed with cultivar Enrei, these mutant soybean lines exhibited restricted growth when exposed to flooding conditions for 7 days, whereas no wild-type plants survived this condition [16]. When the water was removed, however, the mutant lines started to grow again [16]. In this study, although the growth of the wild type was significantly suppressed by flooding compared with non-flooding condition, that of mutant lines was better than that of the wild type even if it was treated with flooding (Figure 2). Taken together, this study in conjunction with the previous report [16] indicates that mutant lines keep the morphological improvement after being crossed with parent cultivar Enrei.

Bottlenecks in genetic diversity challenge germplasm screening for flood-tolerant cultivars of soybean, while proteomics and metabolomics assist in the elucidation of flooding tolerance in soybean. Nanjo et al. [20] reported that RNA binding/processing-related proteins and flooding stress indicator proteins correlated with flood-tolerant index. In addition, Li et al. [21] reported that secondary metabolites such as isoflavonoid improved flooding tolerance cultivar. However, these factors did not exist in flooding-tolerant mutant in this study. In the case of rice, there are two types showing tolerance against flooding stress, which are an escape strategy [6,22] and a quiescence strategy [7,23]. Because *SNORKEL 1/2* and *SUBMERGENCE 1* genes in rice do not exist in soybean, the mechanism of flooding tolerance might be completely different in soybean.

3.2. Regulation of Alcohol-Formation Pathway Is Related to the Mechanism of Flooding Tolerance

In the flooding environment, the energy to maintain plant vitality mainly relies on the ethanol metabolic pathway in glycolysis to degrade glucose and glycogen accompanied by ATP generation. ADH and pyruvate decarboxylase are key to the establishment of the fermentative metabolism in plants during oxygen shortage [24,25]. The flooding tolerance of plants was proportional to the change in ADH activity in response to flooding. Due to the oxygen-depleted environment arising from flooding, a crisis in ATP availability occurred; however, to manage the energy crisis, the balance of glycolysis aided in soybean survival from submergence [17,21] and enhanced fermentation was necessary for the acquisition of flooding tolerance [16]. These findings indicate that activated fermentation and glycolysis confer soybean-flooding tolerance to ensure survival.

The respective overexpression of *ThADH1* and *ThADH4* in *Populus* [26] and/or *GmAdh2* in soybean [27] improved the growth of transgenic soybeans under flooding stress. *ADH* genes were regarded as important candidates for genetic manipulation to achieve flooding tolerance in plants, through improving the adaptability to hypoxia [28]. In this study, 3 proteins having ADH activities were identified (Tables S1 and S2), which were "uncharacterized protein" by search with UniProtKB Glycine max (SwissProt TreEMBL). ADH accumulation increased in both wild type and mutant line under flooding stress; however, ADH activity significantly increased and mildly increased in wild type and mutant line, respectively, under flooding stress (Figure 5). These results suggest that mutant line can survive from flooding through the mild activation of the alcohol-fermentation pathway.

3.3. Glycoprotein Folding Is Related to the Mechanism of Flooding Tolerance

Based on proteomic results, oppositely changed proteins determined in the endoplasmic reticulum (Figure 3) were confirmed by immunoblot analysis (Figure 4). Endoplasmic reticulum mediates protein folding and assembly through a well-coordinated system of chaperones such as calnexin, protein disulfide isomerase, and heat shock proteins [29]. Among these proteins, calnexin is involved in protein folding and quality control [30]. Calreticulin, which is a major calcium binding chaperone in the endoplasmic reticulum, is a key component of the calnexin/calreticulin cycle [31]. Furthermore, two calreticulins existed in rice [32] and phosphorylated calreticulin showed the molecular weight of 56 kDa [28]. The calnexin/calreticulin cycle is responsible for the folding of newly synthesized proteins, especially glycoproteins, for quality control and stability in the endoplasmic reticulum [31]. In this study, heat shock 70 kDa proteins as well as calreticulin and calnexin were identified by proteomic analysis (Tables S1 and S2). The present results with previous reports indicate that the activation by the phosphorylation of both calreticulin and calnexin as well as heat shock protein 70 is needed for glycoprotein folding. This study suggests that glycoproteins might not be folded under flooding stress, because only calnexin increased in calnexin/calreticulin cycle.

In soybean, calnexin and calreticulin decreased, which led to the reduced accumulation of glycoproteins and disruption of endoplasmic reticulum homeostasis under flooding and drought stresses [33]. The accumulation of calnexin/calreticulin and glycoproteins significantly increased in soybean under flooding with silver nano particles with other chemicals, which improved soybean growth [34]. In this study, lectin legB domain-containing protein and glycosyltransferase were also found by proteomic analysis (Tables S1 and S2). Furthermore, in N-linked glycoproteins, exostosin domain-containing protein significantly decreased in wild type by flooding stress; however, it was recovered in the mutant line even if it was under flooding stress (Tables S1–3). Exostosin domain-containing protein is in Golgi apparatus and related to secondary cell-wall biogenesis. The root and hypocotyl of soybean caused the suppression of lignification through the decrease in glycoproteins by downregulation of reactive oxygen species and jasmonate biosynthesis under flooding stress [35]. Because the abundances of calnexin, calreticulin, and glycoproteins increased in mutant soybean under flooding (Figure 4), the folding of glycoproteins might be improved in the mutant line with the increase in both calnexin and calreticulin, even if it was flooding condition. These results with previous reports suggest that the folding of glycoproteins is essential for recovery from damage caused by flooding stress.

3.4. Cell Death in Soybean Root Is Related to Flooding Tolerance

At the initial stage of flooding, proteomic studies using flood-tolerant materials of mutant and ABA-treated soybeans showed that protein synthesis and RNA regulation-related proteins triggered soybean tolerance [17]. Together with RNA regulation and protein metabolism, hormone response contributed to initial flooding tolerance through the inhibition of cytochrome P450 77A1 [18]. Integrated omic data derived from proteomics and metabolomics indicated that fructose was a critical metabolite to assist in soybean flooding tolerance at initial stages via regulation of hexokinase and phosphofructokinase [19]. Taken together, proteomics, in combination with transcriptomics and metabolomics, improves the capability to uncover prospective flood tolerance responses in soybean as for RNA modification, protein synthesis, and sugar catabolism at the initial stage of flooding. However, 4-day-old mutant soybean with 2-days flooding stress [36], did not show same mechanisms compared with flooding stress at the initial stage.

The loss of root tips by flooded seedlings was caused by flooding-induced root tip cell death and it was recovered in the mutant line (Figure 6). Within cells, protein synthesis and degradation are well balanced, because a small decrease in synthesis or increase in degradation results in cell death [37]. The ubiquitin proteasome system responds to the stress conditions by regulating the degradation of misfolded proteins to restrict their accumulation [38]. Autophagic programmed cell death of the meristematic cells has been implicated in root-tip death of several species, including pea and maize exposed to severe stress conditions [39]. The death program was triggered by an imbalance between folded and unfolded proteins in the endoplasmic reticulum, which represents a common cellular stress [40]. In this study, glycoproteins did not fold under flooding stress; however, the folding of glycoproteins was improved in the mutant line with the increase in both calnexin and calreticulin (Figure 6). These results with previous reports suggest that increased accumulation of misfolded related proteins causes cell death, which might reduce the growth of soybean under flooding stress. On the other hand, mutation in flooding-tolerant soybean might improve the quality of proteins; as a result, cell death is suppressed.

4. Materials and Methods

4.1. Plant Material and Experimental Design

Experimental design for investigation of the mechanism of flooding tolerance in mutant soybean was summarized in Figure 1. Flooding-tolerant founder mutant soybeans [16] were crossed with wild type soybean (*Glycine max* L. cultivar Enrei). Mutant lines with flooding-tolerant selected from progeny were used. As mutant lines, 1386-6 (G2), 1386-9 (G3), 1387-12 (G4) lines were used in this study.

Seeds were sterilized with 2% sodium hypochlorite solution, rinsed twice in water, and sown in 400 mL of silica sand in a seedling case. Soybeans were grown at 25 °C and 60% humidity under white fluorescent light (160 µmol m^{-2} s^{-1}, 16 h light period/day). To induce flooding stress, water was added above the soil surface to immerse 2-day-old soybeans. For morphological analysis, root and hypocotyl were collected from at 6 days after 5-days flooding. For proteomic analysis and other confirmation experiments, roots including hypocotyl were collected from 2-days flooded soybeans.

Three independent experiments were performed as biological replications for all experiments, meaning that the seeds were sown on different days. A total of 14 seeds were sown evenly in each seedling case. The statistical significance of multiple groups was evaluated by one-way ANOVA test. SPSS 20.0 (IBM, Chicago, IL, USA) statistical software was used for the evaluation of the results. A p-value of less than 0.05 was considered as statistically significant.

4.2. Protein Extraction, Protein Enrichment, Reduction, Alkylation, and Digestion

Protein extraction was performed with methods described in the previous study [41]. The method of Bradford [42] was used to determine the protein concentration with bovine serum albumin used as the standard. Extracted proteins (100 µg) were adjusted to a final volume of 100 µL. Protein enrichment, reduction, alkylation, and digestion were performed with methods described in the previous study [41].

4.3. Protein Identification Using Nano-Liquid Chromatography Mass Spectrometry

The peptides were loaded onto the LC system (EASY-nLC 1000; Thermo Fisher Scientific, San Jose, CA, USA) equipped with a trap column (Acclaim PepMap 100 C18 LC column, 3 µm, 75 µm ID × 20 mm; Thermo Fisher Scientific). The liquid chromatography (LC) conditions as well as the mass spectrometry (MS) acquisition conditions were described in the previous study [43].

4.4. Mass Spectrometry Data Analysis

The MS/MS searches were carried out using MASCOT (version 2.6.1, Matrix Science, London, UK) and SEQUEST HT search algorithms against the UniprotKB *Glycine max* (SwissProt TreEMBL, TaxID = 3847, version 2020-09-18) using Proteome Discoverer (PD) 2.2 (version 2.2.0.388; Thermo Fisher Scientific). The workflow for both algorithms included spectrum files RC, spectrum selector, MASCOT, SEQUEST HT search nodes, percolator, ptmRS, and minor feature detector nodes. Oxidation of methionine was set as a variable modification and carbamidomethylation of cysteine was set as a fixed modification. Mass tolerances in MS and MS/MS were set at 10 ppm and 0.6 Da, respectively. Trypsin was specified as protease and a maximum of 2 missed cleavage was allowed. Target-decoy database searches used for calculation of false discovery rate (FDR) and for peptide identification FDR was set at 1%.

4.5. Differential Analysis of Proteins Using Mass Spectrometry Data

Label-free quantification was also performed with PD 2.2 using precursor ions quantifiler nodes. For differential analysis of the relative abundance of peptides and proteins between samples, the free software PERSEUS (version 1.6.14.0, Max Planck Institute of biochemistry, Martinsried, Germany) [44] was used. The condition of analysis is described in the previous study [43]. Principal component analysis (PCA) was performed with PD 2.2.

4.6. Bioinformatic Analyses of Protein Functional Categorization

The sequences of the differentially accumulated proteins were subjected to a BLAST query against the gene-ontology database (http://www.geneontology.org/ (accessed on 19 September 2019)).

4.7. Immunoblot Analysis

SDS-sample buffer consisting of 60 mM Tris-HCl (pH 6.8), 2% SDS, 10% glycerol, and 5% dithiothreitol was added to protein extracts. As protein marker, Precision Plus Protein Standard (Bio-Rad, Hercules, CA, USA) was used. Quantified proteins (10 µg) were separated by electrophoresis on a 10% SDS-polyacrylamide gel and transferred onto a polyvinylidene difluoride membrane using a semidry transfer blotter (Nippon Eido, Tokyo, Japan). The blotted membrane was blocked for 5 min in Bullet Blocking One reagent (Nacalai Tesque, Kyoto, Japan). After blocking, the membrane was cross-reacted with a 1: 1000 dilution of the primary antibodies for 1 h at room temperature. As primary antibodies, anti- ADH [45], calnexin [46], and calreticulin [47] antibodies were used. Anti-rabbit IgG conjugated with horseradish peroxidase (Bio-Rad) was used as the secondary antibody. For lectin blot, peroxidase-Concanavalin A antibody (Seikagaku, Tokyo, Japan) was used. After 1 h incubation, signals were detected using TMB Membrane Peroxidase Substrate Kit (Seracare, Milford, MA, USA) following protocol from the manufacturer. Coomassie brilliant blue staining was used as loading control. Raw data were saved TIFF format (16-bit gray scale image) and the integrated densities of bands were calculated using Image J software (version 1.8; National Institutes of Health, Bethesda, MD, USA).

4.8. ADH Activity Assay

ADH activity was analyzed using ADH Activity Colorimetric Assay Kit (BioVision, Milpitas, CA, USA). A portion (50 mg) of samples was homogenized in 200 μL of ice-cold assay buffer and centrifuged at 13,000× g for 10 min at 4 °C to remove insoluble material. Extracts (50 μL) were added in 100 μL of reaction mixture containing 82 μL of ADH assay buffer, 8 μL of developer, and 10 μL of substrate. Mixture was incubated for 2 and 10 min at 37 °C and the absorbance of mixture was measured at 450 nm.

4.9. Evaluation of Cell Death Using Evans-Blue Dye

Cell death was evaluated by Evans-blue staining as described by Jacyn-Baker and Mock [48].

5. Conclusions

To remove other factors besides the genes related to flooding tolerance in previous isolated from gamma-ray irradiated soybean [16], mutants were crossed with the parent cultivar Enrei. To investigate the mechanism of flooding tolerance of soybean, flooding-tolerant mutants generated in this study were analyzed using a gel-free/label-free proteomic technique. Furthermore, proteomic results were confirmed with biochemical techniques. The main findings are as follows: (i) the growth of the mutant line was better than that of the wild type even if it was treated with flooding; (ii) oppositely changed proteins between the wild type and mutant lines under flooding stress were associated in endoplasmic reticulum; (iii) calnexin accumulation increased in the wild type and mutant line but calreticulin accumulated in only the mutant line under flooding stress; (iv) glycoproteins in the mutant line increased compared with those of the wild type even if it was under flooding stress; (v) ADH accumulated in the wild type and mutant line but enzyme activity significantly increased and mildly increased in the wild type and mutant line, respectively, under flooding stress; (vi) cell death increased and decreased in the wild type and mutant line, respectively, by flooding stress. These results suggest that regulation of cell death through the fermentation system and glycoprotein folding under flooding might be an important factor for the acquisition of flooding tolerance in soybean.

Supplementary Materials: The following are available online at https://www.mdpi.com/article/10.3390/ijms22169046/s1.

Author Contributions: Y.K. and M.N. cultivated mutant soybeans; S.K. performed morphological measurement and protein preparation; H.Y., K.H. and K.T. performed MS analysis; S.K. performed the immunoblot analysis and other confirmation analyses; S.K. analyzed the data and wrote the paper. All authors have read and agreed to the published version of the manuscript.

Funding: This work was supported by the program of "Breeding of soybean varieties with new trait in Niigata prefecture", Niigata, Japan for S.K.

Institutional Review Board Statement: Not applicable.

Informed Consent Statement: Not applicable.

Data Availability Statement: For MS data, RAW data, peak lists and result files have been deposited in the ProteomeXchange Consortium [49] via the jPOST [50] partner repository under data-set identifiers PXD024711.

Acknowledgments: Authors are grateful to Shunya Higashiyama and Atsuya Tamura of Fukui University of Technology for helping molecular biology research.

Conflicts of Interest: The authors declare no conflict of interest.

References

1. Tanoue, M.; Hirabayashi, Y.; Ikeuchi, H. Global-scale river flood vulnerability in the last 50 years. *Sci. Rep.* **2016**, *6*, 36021. [CrossRef]
2. Fukao, T.; Barrera-Figueroa, B.E.; Juntawong, P.; Pena-Castro, J.M. Submergence and waterlogging stress in plants: A review highlighting research opportunities and understudied aspects. *Front. Plant Sci.* **2019**, *10*, 340. [CrossRef]
3. Setter, T.L.; Waters, I. Review of prospects for germplasm improvement for waterlogging tolerance in wheat, barley and oats. *Plant Soil* **2013**, *253*, 1–34. [CrossRef]
4. Jia, W.; Ma, M.; Chen, J.; Wu, S. Plant morphological, physiological and anatomical adaption to flooding stress and the underlying molecular mechanisms. *Int. J. Mol. Sci.* **2021**, *22*, 1088. [CrossRef]
5. Ma, M.; Cen, W.; Li, R.; Wang, S.; Luo, J. The molecular regulatory pathways and metabolic adaptation in the seed germination and early seedling growth of rice in response to low O_2 stress. *Plants* **2020**, *9*, 1363. [CrossRef]
6. Hattori, Y.; Nagai, K.; Furukawa, S.; Song, X.J.; Kawano, R.; Sakakibara, H.; Wu, J.; Matsumoto, T.; Yoshimura, A.; Kitano, H.; et al. The ethylene response factors *SNORKEL1* and *SNORKEL2* allow rice to adapt to deep water. *Nature* **2009**, *460*, 1026–1030. [CrossRef] [PubMed]
7. Fukao, T.; Xu, K.; Ronald, P.C.; Bailey-Serres, J. A variable cluster of ethylene response factor-like genes regulates metabolic and developmental acclimation responses to submergence in rice. *Plant Cell* **2006**, *18*, 2021–2034. [CrossRef]
8. Githiri, S.M.; Watanabe, S.; Harada, K.; Takahashi, R. QTL analysis of flooding tolerance in soybean at an early vegetative growth stage. *Plant Breed.* **2006**, *125*, 613–618. [CrossRef]
9. Oosterhuis, D.M.; Scott, H.D.; Hampton, R.E.; Wullschleter, S.D. Physiological response of two soybean (*Glycine max* L. Merr) cultivars to short-term flooding. *Environ. Exp. Bot.* **1990**, *30*, 85–92. [CrossRef]
10. Shimamura, S.; Mochizuki, T.; Nada, Y.; Fukuyama, M. Formation and function of secondary aerenchyma in hypocotyl, roots and nodules of soybean (*Glycine max*) under flooded conditions. *Plant Soil* **2003**, *251*, 351–359. [CrossRef]
11. Komatsu, S.; Hiraga, S.; Yanagawa, Y. Proteomics techniques for the development of flood tolerant crops. *J. Proteome Res.* **2012**, *11*, 68–78. [CrossRef]
12. Komatsu, S.; Shirasaka, N.; Sakata, K. 'Omics' techniques for identifying flooding-response mechanisms in soybean. *J. Proteom.* **2013**, *93*, 169–178. [CrossRef]
13. Komatsu, S.; Sakata, K.; Nanjo, Y. 'Omics' techniques and their use to identify how soybean responds to flooding. *J. Anal. Sci. Technol.* **2015**, *6*, 9. [CrossRef]
14. Wang, X.; Komatsu, S. Proteomic approaches to uncover the flooding and drought stress response mechanisms in soybean. *J. Proteom.* **2018**, *172*, 201–215. [CrossRef] [PubMed]
15. Wang, X.; Komatsu, S. Review: Proteomic techniques for the development of flood-tolerant soybean. *Int. J. Mol. Sci.* **2020**, *21*, 7497. [CrossRef]
16. Komatsu, S.; Nanjo, Y.; Nishimura, M. Proteomic analysis of the flooding tolerance mechanism in mutant soybean. *J. Proteom.* **2013**, *79*, 231–250. [CrossRef]
17. Yin, X.; Nishimura, M.; Hajika, M.; Komatsu, S. Quantitative proteomics reveals the flooding-tolerance mechanism in mutant and abscisic acid-treated soybean. *J. Proteome Res.* **2016**, *15*, 2008–2025. [CrossRef]
18. Yin, X.; Hiraga, S.; Hajika, M.; Nishimura, M.; Komatsu, S. Transcriptomic analysis reveals the flooding progeny in flooding tolerant line and abscisic acid treated soybean. *Plant Mol. Biol.* **2017**, *93*, 479–496. [CrossRef]
19. Wang, X.; Zhu, W.; Hashiguchi, A.; Nishimura, M.; Tian, J.; Komatsu, S. Metabolic profiles of flooding-tolerant mechanism in early-stage soybean responding to initial stress. *Plant Mol. Biol.* **2017**, *94*, 669–685. [CrossRef]
20. Nanjo, Y.; Jang, H.Y.; Kim, H.S.; Hiraga, S.; Woo, S.H.; Komatsu, S. Analyses of flooding tolerance of soybean varieties at emergence and varietal di_erences in their proteomes. *Phytochemistry* **2014**, *106*, 25–36. [CrossRef]
21. Lin, Y.; Li, W.; Zhang, Y.; Xia, C.; Liu, Y.; Wang, C.; Xu, R.; Zhang, L. Identification of Genes/Proteins Related to Submergence Tolerance by Transcriptome and Proteome Analyses in Soybean. *Sci. Rep.* **2019**, *9*, 14688. [CrossRef]
22. Nagai, K.; Hattori, Y.; Ashikari, M. Stunt or elongate? Two opposite strategies by which rice adapts to floods. *J. Plant Res.* **2010**, *123*, 303–309. [CrossRef]
23. Xu, K.; Xu, X.; Fukao, T.; Canlas, P.; Maghirang-Rodriguez, R.; Heuer, S.; Ismail, A.M.; Bailey-Serres, J.; Ronald, P.C.; Mackill, D.J. Sub1A is an ethylene-response-factor-like gene that confers submergence tolerance to rice. *Nature* **2006**, *442*, 705–708. [CrossRef]
24. Schwartz, D. Genetic control of alcohol dehydrogenase—A competition model for regulation of gene action. *Genetics* **1971**, *67*, 411–425. [CrossRef]

25. Ventura, I.; Brunello, L.; Iacopino, S.; Valeri, M.C.; Novi, G.; Dornbusch, T.; Perata, P.; Loreti, E. Arabidopsis phenotyping reveals the importance of alcohol dehydrogenase and pyruvate decarboxylase for aerobic plant growth. *Sci. Rep.* **2020**, *10*, 16669. [CrossRef]
26. Xuan, L.; Hua, J.; Zhang, F.; Wang, Z.; Pei, X.; Yang, Y.; Yin, Y.; Creech, D.L. Identification and functional analysis of ThADH1 and ThADH4 genes involved in tolerance to waterlogging stress in taxodium hybrid 'Zhongshanshan 406'. *Genes* **2021**, *12*, 225. [CrossRef]
27. Tougou, M.; Hashiguchi, A.; Yukawa, K.; Nanjo, Y.; Hiraga, S.; Nakamura, T.; Nishizawa, K.; Komatsu, S. Responses to flooding stress in soybean seedling with the alcohol dehydrogenase transgene. *Plant Biotech.* **2012**, *29*, 301–306. [CrossRef]
28. Takahashi, H.; Greenway, H.; Matsumura, H.; Tsutsumi, N.; Nakazono, M. Rice alcohol dehydrogenase 1 promotes survival and has a major impact on carbohydrate metabolism in the embryo and endosperm when seeds are germinated in partially oxygenated water. *Ann. Bot.* **2014**, *113*, 851–859. [CrossRef]
29. Healy, S.J.; Verfaillie, T.; Jäger, R.; Agostinis, P.; Samali, A. Biology of the endoplasmic reticulum. In *Endoplasmic Reticulum Stress in Health and Disease*; Agostinis, P., Samali, A., Eds.; Springer: Dordrecht, The Netherlands, 2012; pp. 3–22.
30. Bergeron, J.J.; Brenner, M.B.; Thomas, D.Y.; Williams, D.B. Calnexin: A membrane-bound chaperone of the endoplasmic reticulum. *Trends Biochem. Sci.* **1994**, *19*, 124–128. [CrossRef]
31. Michalak, M.; Robert Parker, J.M.; Opas, M. Ca2+ signaling and calcium binding chaperones of the endoplasmic reticulum. *Cell Calcium* **2002**, *32*, 269–278. [CrossRef]
32. Tanaka, N.; Mitsui, S.; Nobori, H.; Yanagi, K.; Komatsu, S. Expression and Function of Proteins during Development of the Basal Region in Rice Seedlings. *Mol. Cell. Proteom.* **2005**, *4*, 796–808. [CrossRef]
33. Wang, X.; Komatsu, S. Gel-free/label-free proteomic analysis of endoplasmic reticulum proteins in soybean root tips under flooding and drought stresses. *J. Proteome Res.* **2016**, *15*, 2211–2227. [CrossRef]
34. Hashimoto, T.; Mustafa, G.; Nishiuchi, T.; Komatsu, S. Comparative analysis of the effect of inorganic and organic chemicals with silver nanoparticles on soybean under flooding stress. *Int. J. Mol. Sci.* **2020**, *21*, 1300. [CrossRef]
35. Komatsu, S.; Kobayashi, Y.; Nishizawa, K.; Nanjo, Y.; Furukawa, K. Comparative proteomics analysis of differentially expressed proteins in soybean cell wall during flooding stress. *Amino Acids* **2010**, *39*, 1435–1449. [CrossRef]
36. Wang, X.; Sakata, K.; Komatsu, S. An integrated approach of proteomics and computational genetic modification effectiveness analysis to uncover the mechanisms of flood tolerance in soybeans. *Int. J. Mol Sci.* **2018**, *19*, 1301. [CrossRef]
37. Mitch, W.E.; Goldberg, A.L. Mechanisms of muscle wasting. The role of the ubiquitin-proteasome pathway. *N. Engl. J. Med.* **1996**, *335*, 1897–1905. [CrossRef]
38. Dielen, A.S.; Badaoui, S.; Candresse, T.; German-Retana, S. The ubiquitin/26S proteasome system in plant-pathogen interactions: A never-ending hide-and-seek game. *Mol. Plant Pathol.* **2010**, *11*, 293–308. [CrossRef]
39. Subbaiah, C.C.; Sachs, M.M. Molecular and cellular adaptations of maize to flooding stress. *Ann. Bot.* **2003**, *91*, 119–127. [CrossRef]
40. Williams, B.; Verchot, J.; Dickman, M.B. When supply does not meet demand-ER stress and plant programmed cell death. *Front. Plant Sci.* **2014**, *5*, 211. [CrossRef]
41. Komatsu, S.; Han, C.; Nanjo, Y.; Altaf-Un-Nahar, M.; Wang, K.; He, D.; Yang, P. Label-free quantitative proteomic analysis of abscisic acid effect in early-stage soybean under flooding. *J. Proteome Res.* **2013**, *12*, 4769–4784. [CrossRef]
42. Bradford, M.M. A rapid and sensitive method for the quantitation of microgram quantities of protein utilizing the principle of protein-dye binding. *Anal. Biochem.* **1976**, *72*, 248–254. [CrossRef]
43. Li, X.; Rehman, S.U.; Yamaguchi, H.; Hitachi, K.; Tsuchida, K.; Yamaguchi, T.; Sunohara, Y.; Matsumoto, H.; Komatsu, S. Proteomic analysis of the effect of plant-derived smoke on soybean during recovery from flooding stress. *J. Proteom.* **2018**, *181*, 238–248. [CrossRef] [PubMed]
44. Tyanova, S.; Temu, T.; Sinitcyn, P.; Carlson, A.; Hein, M.Y.; Geiger, T.; Mann, M.; Cox, J. The Perseus computational platform for comprehensive analysis of (prote) omics data. *Nat. Methods* **2016**, *13*, 731–740. [CrossRef]
45. Komatsu, S.; Deschamps, T.; Hiraga, S.; Kato, M.; Chiba, M.; Hashiguchi, A.; Tougou, M.; Shimamura, S.; Yasue, H. Characterization of a novel flooding stress-responsive alcohol dehydrogenase expressed in soybean roots. *Plant Mol. Biol.* **2011**, *77*, 309–322. [CrossRef]
46. Nouri, M.Z.; Hiraga, S.; Yanagawa, Y.; Sunohara, Y.; Matsumoto, H.; Komatsu, S. Characterization of calnexin in soybean roots and hypocotyls under osmotic stress. *Phytochemistry* **2012**, *74*, 20–29. [CrossRef] [PubMed]
47. Komatsu, S.; Masuda, T.; Abe, K. Phosphorylation of a protein (pp56) is related to the regeneration of rice cultured suspension cells. *Plant Cell Physiol.* **1996**, *37*, 748–753. [CrossRef] [PubMed]
48. Jacyn Baker, C.; Mock, N.M. An improved method for monitoring cell death in cell suspension and leaf disc assays using Evans blue. *Plant Cell Tissue Organ Cult.* **1994**, *39*, 7–12. [CrossRef]
49. Vizcaíno, J.A.; Côté, R.G.; Csordas, A.; Dianes, J.A.; Fabregat, A.; Foster, J.M.; Griss, J.; Alpi, E.; Birim, M.; Contell, J.; et al. The PRoteomics IDEntifications (PRIDE) database and associated tools: Status in 2013. *Nucleic Acids Res.* **2013**, *41*, D1063–D1069. [CrossRef]
50. Okuda, S.; Watanabe, Y.; Moriya, Y.; Kawano, S.; Yamamoto, T.; Matsumoto, M.; Takami, T.; Kobayashi, D.; Araki, N.; Yoshizawa, A.C.; et al. jPOSTrepo: An international standard data repository for proteomes. *Nucleic Acids Res.* **2017**, *45*, D1107–D1111. [CrossRef]

International Journal of
Molecular Sciences

MDPI

Article

Endoplasmic Reticulum Subproteome Analysis Reveals Underlying Defense Mechanisms of Wheat Seedling Leaves under Salt Stress

Junwei Zhang †, Dongmiao Liu †, Dong Zhu, Nannan Liu and Yueming Yan *

College of Life Science, Capital Normal University, Beijing 100048, China; zhangjunwei_22@163.com (J.Z.); liudongmiao1230@163.com (D.L.); ZD19920804@163.com (D.Z.); liunnCNU@163.com (N.L.)
* Correspondence: yanym@cnu.edu.cn; Tel./Fax: +86-10-68902777
† These authors contributed equally to this work.

Abstract: Salt stress is the second most important abiotic stress factor in the world, which seriously affects crop growth, development and grain production. In this study, we performed the first integrated physiological and endoplasmic reticulum (ER) proteome analysis of wheat seedling leaves under salt stress using a label-free-based quantitative proteomic approach. Salt stress caused significant decrease in seedling height, root length, relative water content and chlorophyll content of wheat seedling leaves, indicating that wheat seedling growth was significantly inhibited under salt stress. The ER proteome analysis identified 233 ER-localized differentially accumulated proteins (DAPs) in response to salt stress, including 202 upregulated and 31 downregulated proteins. The upregulated proteins were mainly involved in the oxidation-reduction process, transmembrane transport, the carboxylic acid metabolic process, stress response, the arbohydrate metabolic process and proteolysis, while the downregulated proteins mainly participated in the metabolic process, biological regulation and the cellular process. In particular, salt stress induced significant upregulation of protein disulfide isomerase-like proteins and heat shock proteins and significant downregulation of ribosomal protein abundance. Further transcript expression analysis revealed that half of the detected DAP genes showed a consistent pattern with their protein levels under salt stress. A putative metabolic pathway of ER subproteome of wheat seedling leaves in response to salt stress was proposed, which reveals the potential roles of wheat ER proteome in salt stress response and defense.

Keywords: wheat; salt stress; physiological characteristics; ER proteome; label-free quantitation

Citation: Zhang, J.; Liu, D.; Zhu, D.; Liu, N.; Yan, Y. Endoplasmic Reticulum Subproteome Analysis Reveals Underlying Defense Mechanisms of Wheat Seedling Leaves under Salt Stress. *Int. J. Mol. Sci.* **2021**, *22*, 4840. https://doi.org/10.3390/ijms22094840

Academic Editor: Sixue Chen

Received: 1 March 2021
Accepted: 30 April 2021
Published: 3 May 2021

1. Introduction

As one of the three major crops in the world, wheat (*Triticum aestivum* L.) is a staple source of food and protein in the human diet. However, global food security is facing great challenges, including arable land reduction, extreme climate change, and environmental pollution. Plants are often exposed to a variety of abiotic stress environments, among which salt stress significantly impacts crop growth and yield formation. High salt levels affect an estimated 20% of arable land, globally. [1]. The expansion of saline soil damages crop production at the cost of US $27.3 billion annually seriously threatens the world's human populations [2,3].

The most obvious effect of salt injury on plants includes the decreased growth rate of new leaves that is proportional to the osmotic pressure of roots. The high salt content in leaves is a direct cause of plant death [4]. The primary impacts of elevated ionic strength are osmotic stress and ion toxicity. Osmotic stress first causes plant hormone abscisic acid to accumulate, leading to a series of adaptive responses [5]. Excessive sodium ion leads to ion toxicity and triggers a series of reactions that further activate the expression of *ENA1* and other target genes. *ENA1* encodes Na^+-ATPase formation, urging the cell to pump excess sodium ions out of the cell, and finally restoring the ion homeostasis inside and outside the cell. Plants use calcium-dependent protein kinase pathways, also known as salt hypersensitivity pathways or salt overly sensitive (SOS) pathways for salt stress signal transduction [5]. In Arabidopsis, research has focused on the interaction and ion homeostasis mechanism of the three main proteins of the SOS pathways SOS1, SOS2 and SOS3 [6].

As an important organelle unique to eukaryotes, endoplasmic reticulum (ER) can form into tubules, vesicles and tubular interconnected reticular membrane systems in the cells. There are two kinds of ER present that are distinct in structural features: rough ER and smooth ER. The rough ER is involved in protein synthesis, processing and transport, while smooth ER participates in synthesizing lipids, phospholipids and steroids [7]. When subjected to endogenous and exogenous stresses, the accumulation of the misfolded proteins activates an ER stress response in which cells produce a series of responses to achieve a new ER homeostasis. One can detect the misfolded proteins using an ER quality control (ERQC) system and degrade them through an ER-associated degradation (ERAD) system. Plant ER stress response is an important mechanism for abiotic stress defense, which enhances stress resistance in plants and alleviates damages that various stresses cause [8]. The ER plays the role of protein factory and calcium bank, and it is an environment for protein quality monitoring mediated by various signal pathways and protein folding auxiliaries. Unfolded or misfolded proteins accumulate under adverse environmental conditions and leads to ER stress and cell death [9].

Generally, proteomics is a powerful approach performed at the large scale level while currently enabling also high-throughput procedures in an automatic and therefore more precise and safe manner [10,11]. Among different goals, proteomics also enables one to provide insight concerning plant responses and defense mechanisms under various biotic and abiotic stresses [12]. In recent years there has been great progress in plant proteomics, and subcellular proteomics research has become an important field to decipher plant cell responses during development and upon exposure to environmental stresses [13]. To date as of 2021, researchers have performed subcellular proteomic analyses in various plant species, such as soybean cell wall proteome, in response to flooding stress [14]; chloroplast proteome has been involved in photosynthesis and abiotic stress in rice and mangrove [15,16]; mitochondrial proteome in Arabidopsis tissues [17]; and finally, soybean plasma membrane proteome in response to osmotic stress [18]. In terms of plant ER subproteome studies, there has been relatively little work. The authors of [19] identified the flooding response proteins rich in ER by using non-gel and 1D gel-based proteomic techniques, finding that flood stress mainly affected the protein synthesis and glycosylation function of ER in soybean root tips. Through a gel-free/label-free proteomic approach, [20] found that the increase of cellular solute calcium levels induced by drought and flood stress might disturb the ER environment and affect protein folding in soybean root tips. In wheat, the subcellular proteomic studies are mainly performed in chloroplast [21–23], mitochondria [24,25] and plasma membrane [26]. This is because these subcellular structures are relatively easy to isolate intact and remove contamination of proteins from other organelles. However, other subcellular structures that are diffuse and amorphic such as ER and Golgi apparatus are very difficult to isolate and characterize. Concurrent to the development of subcellular proteomics, powerful methods have been established to effectively isolate the ER and characterize its proteome. These include various hydrodynamic techniques such as differential centrifugation and density gradient centrifugation. In more recent studies, purity has been assessed via electron microscopy and/or Western blotting for known ER-located proteins using calreticulin, BiP and calnexin as common ER markers. Considering that wheat has a huge genome and proteome, isolating and purifying wheat ER components face a significant challenge. Therefore, the ER proteome characteristics in wheat and its roles in abiotic stress response remain unclear.

Along with the development of subcellular proteomics, some powerful methods have been established to effectively isolate ER and characterize ER proteome such as ultracentrifugation combined with differential centrifugation and density gradient centrifugation [27]. In the present work, we performed the first ER proteomic analysis of wheat seedling leaves under salt stress using a label-free quantitative proteome approach. We aimed to reveal the ER proteome profiling and the potential roles of the key ER proteins involved in salt stress response. Our results provided new insights into the underlying mechanisms of plant subcellular organelles in response to abiotic stress.

2. Results

2.1. Phenotype and Physiological Changes of Wheat Seedlings under Salt Stress

The morphological and physiological characteristics of wheat seedlings showed obvious changes under salt stress (Figure 1). Compared to the control group, the seedlings experienced leaf curl, lodging, wilting and yellowing with the increase of salt stress time, but particularly after 96 h of salt treatment (Figure 1A). Seedling height and root length significantly decreased (Figure 1B–C), indicating that salt stress severely inhibited plant growth.

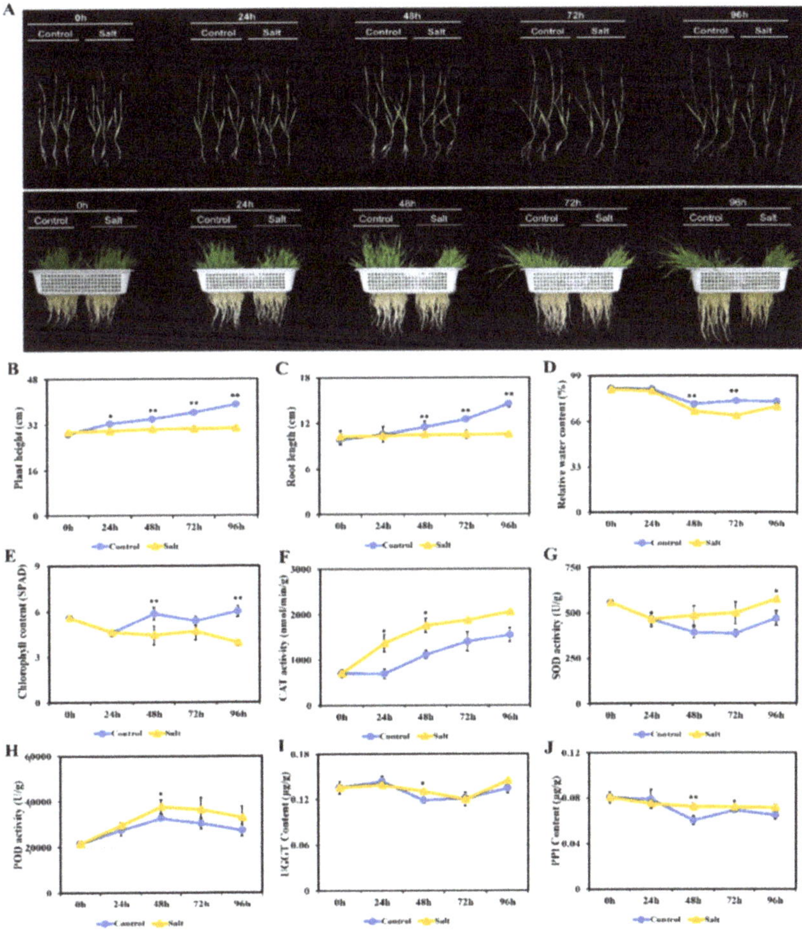

Figure 1. Morphological and physiological changes of wheat cultivar Zhongmai 175 seedling leaves under normal and salt stress conditions. (**A**) Morphology of individual seedling and bunch seedling. (**B**) Plant height. (**C**) Root length. (**D**) Relative water content. (**E**) Chlorophyll content. (**F**) CAT activity. (**G**) SOD activity. (**H**) POD activity. (**I**) UGGT content. (**J**) PPI content. Statistically significant differences were calculated based on an independent Student's t-test: * $p < 0.05$; ** $p < 0.01$.

Salt stress also significantly altered the physiological characteristics of seedling leaves. Compared with the control group, both relative water content and chlorophyll content gradually decreased under salt stress, though there was more significant reduction at 48–96 h (Figure 1D,E). On the contrary, the activities of catalase (CAT), superoxide dismutase (SOD) and peroxidase (POD), which are involved in plant reactive oxygen species (ROS) scavenging, significantly increased while spending time under salt stress (Figure 1F–H). Plants usually produce significant amounts of ROS in the wake of salt stress, and the increased activities of these enzymes enhance plant stress tolerance through a ROS scavenging mechanism. Meanwhile, UDP-glycosyltransferase and peptidyl prolyl cis-trans isomerase activities are closely related to ER stress response; this also increased after salt treatment, particularly at 48 h (Figure 1I,J). The process may promote plant salt tolerance through accelerating the misfolded proteins' degradation.

2.2. Quality Assessment of the Isolated ER Proteome

In this study, researchers extracted the components of intima media except plasma membrane and Golgi apparatus from lower layers through two phase separation systems (aqueous polymer phase), exhibited in Figure 2A. ER was further differentiated from other endometrial systems through a series of sucrose density gradients in 0.6–2 M. The sucrose density gradient between 0.6 M and 1.3 M was endoplasmic reticulum (Figure 2B). To test the purity of ER proteins, we used five polyclonal antibodies against different organelle-specific proteins using Western blotting, including plasma membrane specific protein antibody H^+-ATPase, ER specific protein antibody BiP, mitochondrial specific protein antibody AOX1/2, chloroplast specific protein antibody PsbA and nuclear specific protein antibody H3. As shown in Figure 2C, one extremely strong ER-specific protein BiP band was observed when compared against leaf ER proteome, indicating that the ER protein was effectively enriched. The remaining four organelle-specific protein antibodies had no obvious signals (Figure 2C). These results demonstrated that the ER extracted proteins from wheat leaves show high purity without clear protein contaminations from other organelles.

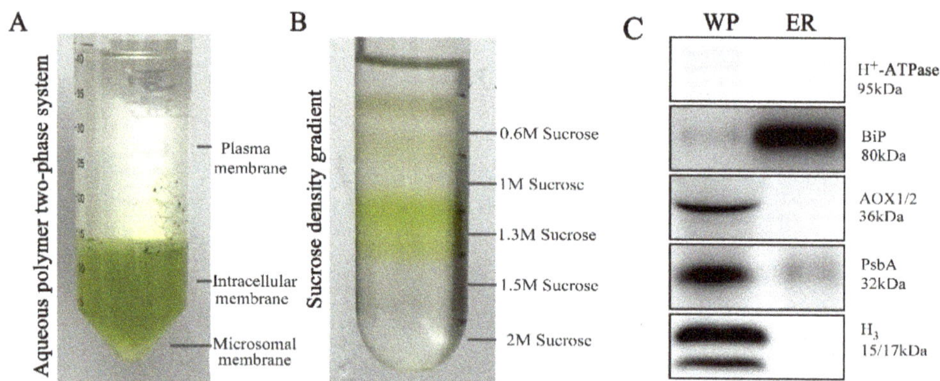

Figure 2. Purity assessment and immunoblotting analysis of the extracted ER proteins. (**A**) Discontinuous Percoll gradient. (**B**) Morphological observation of isolated chloroplasts. (**C**) Immunoblotting analysis using antibodies against plant cell compartment markers, including H+-ATPase (Plasma membrane), BiP (Endoplasmic reticulum), AOX1/2 (Mitochondrion), PsbA (Chloroplast) and anti-H3 (Nuclear). WP: whole protein of wheat leaves.

2.3. Salt Stress Responsive DAPs in ER Subproteome and Subcellular Localization

Label-free quantitative ER proteome analysis of wheat seedling leaves under normal conditions and salt stress identified 34,732 peptides corresponding to 6663 unique proteins with a high confidence (Table S1 and S2). Among them, salt stress changed the expression of 968 proteins by more than double compared with the control. Of these proteins, 516 were upregulated and 452 were downregulated (Table S3). In particular, 107 ribosomal proteins were identified, including 1 upregulated protein and 106 downregulated proteins. The subcellular localization prediction of the other 861 DAPs showed that there were 233 proteins (27.06%) in ER while the remaining 633 proteins were mainly located in the chloroplast, cytoplasm and nucleus (Table S3, Figure 3A).

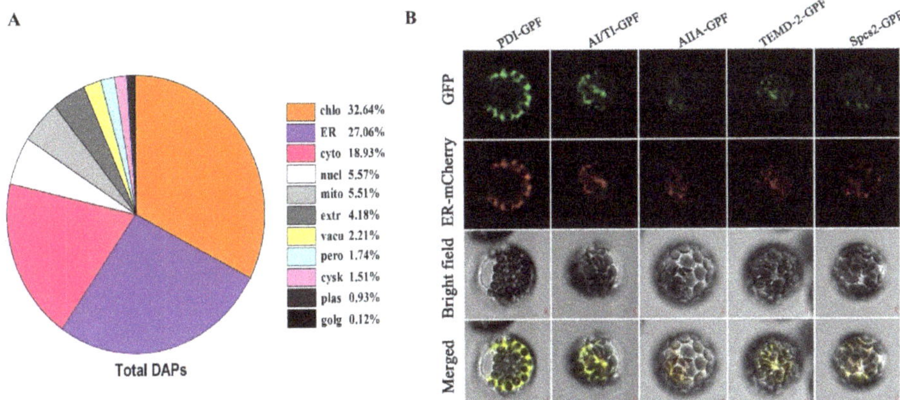

Figure 3. Subcellular localization of the proteins identified in wheat ER proteome under salt stresses. (**A**) Subcellular localization prediction of total proteins identified in the wheat endoplasmic reticulum. Chlo, chloroplast; ER, endoplasmic reticulum; Cyto, cytoplasm; Nucl, nucleus; Mito, mitochondria; Pero, peroxisome; Plas, plasma membrane; Extr, extracellular; Cysk, cytoskeleton; Vacu, vacuole. (**B**) Subcellular localization assay of five reprehensive proteins by wheat protoplast transformation. GFP, green fluorescent protein; ER-cherry; bright field, the field of bright light; merged, merged GFP fluorescence; ER-cherry, and the field of bright light; 16,318, empty vector; PDI: protein disulfide-isomerase; AI/TI: alpha-amylase/trypsin inhibitor; AIIA: probable ureidoglycolate hydrolase; TMED2:GOLD domain-containing protein; Spcs2: probable signal peptidase complex subunit 2; bar = 5 μm.

To verify the prediction results of subcellular localization, we selected five representative ER-located DAPs using online prediction to further perform a subcellular localization assay, including protein disulfide isomerase (PDI), α-amylase/trypsin inhibitor (AI/TI), urea glycolate hydrolase (AIIA), GOLD domain-containing protein (TMED2) and signal peptidase complex subunit 2 (Spcs2). We cloned five DAP genes and transformed them into Arabidopsis protoplasts for transient expression, then observed them using confocal microscopes. The ER suborganelle is a relatively large membrane system that connects with the plasma membrane and nucleus and evenly distributes in the cytoplasm. Considering that Arabidopsis protoplast cells are generally small and endoplasmic reticulum subcellular organelles are a large membrane system, the ER proteins may be interfered as connected with other organelle components. We therefore selected ER-specific calreticulin to mark and co-locate the target proteins in Arabidopsis protoplasts, which allowed us to then eliminate the interference of other components. The ER mark protein calreticulin gene carried an RFP fluorescent tag while we cloned the target genes with GFP fluorescent tags that used different and specific primers (Table S4). The results showed that five DAPs were all co-located with ER mark protein calreticulin, verifying their location in ER (Figure 3B). These experimental results are consistent with the website-based prediction results (Figure 3A).

2.4. Function Classification of the Salt Stress Responsive DAPs from ER

Among 515 upregulated and 346 downregulated DAPs induced by salt stress, 202 (39.22%) and 31 (8.96%) were respectively localized in the ER (Figure 4A,B). The function classification of 233 ER DAPs responsive to salt stress was performed using the blast2go software functional annotation. The upregulated ER proteins were mainly involved in organic substance transport, cellular response to stimulus, the phosphate-containing compound metabolic process, stress response, transmembrane transport, the carbohydrate metabolic carboxylic acid metabolic process and the oxidation reduction process (Table S5, Figure 4C). The downregulated ER proteins mainly participated in the metabolic process, biological regulation and the cellular process (Table S5, Figure 4D). Most upregulated ER proteins under salt stress were distributed in ER membrane, as well as ER and an integral component of membrane, which generally had transmembrane transporter activity, protein binding activity, transferase activity and oxidoreductase activity. This indicates that many proteins synthesized in ER play important roles in salt stress response. The downregulated ER proteins primarily exhibit catalytic activity and binding ability in the cellular anatomical entity.

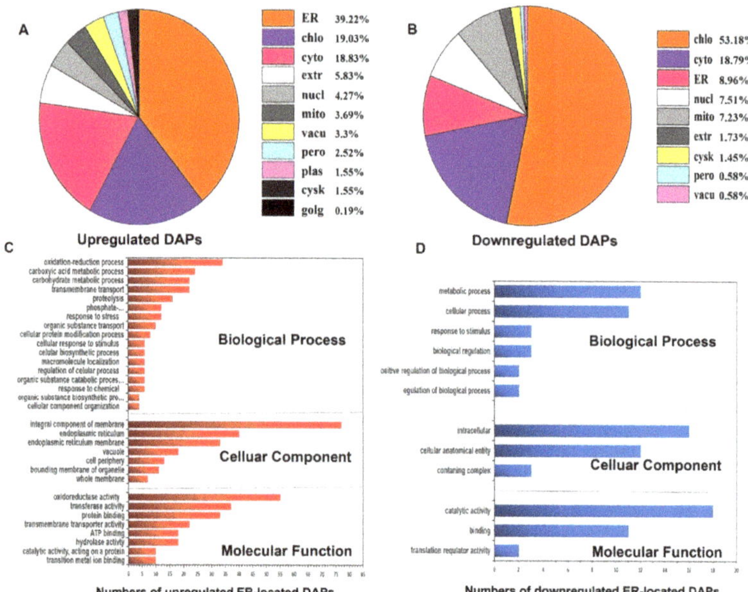

Figure 4. Subcellular localization and functional classification of the identified differentially accumulated proteins (DAPs) in wheat ER under salt stress. (**A**) Subcellular localization of 515 upregulated DAPs identified from wheat ER under salt stress. Chlo, chloroplast; cyto, cytoplasm; mito, mitochondrial; nucl, nucleus; vac, vacuole. (**B**) Subcellular localization of 346 downregulated DAPs identified in wheat ER under salt stress. Chlo, chloroplast; cyto, cytoplasm; mito, mitochondrial; nucl, nucleus; per, peroxisome. (**C**) Functional classification of the ER-localized upregulated DAPs under salt stress. (**D**) Functional classification of the ER-localized downregulated DAPs under salt stress.

2.5. Transcription Expression Analysis of the ER-DAP Genes under Salt Stress

We selected eight DAP genes closely related to ER stress with significant upregulation or downregulation under 0–96 h of salt stress to further detect their dynamic transcriptional expression characteristics under different salt stress treatments. These protein genes included: protein disulfide isomerase (PDI), cytochrome P450 (Cyto P450), calcium-dependent protein kinase (CDPK), calcitonin homologue (CNX), dolichyl-diphosphooligosaccharide-protein glycosyltransferase subunit 1 (RPN1), calcium transport ATP enzyme (CTA), signal recognition particle receptor subunit beta (SRPRB) and 50 s ribosomal L6 protein (RPL6). Table S6 lists the specific primers. As Figure 5 presents, *PDI* and *Cyto P450* genes were significantly upregulated after 48 h of salt stress, while *CTA*, *CNX*, *RPN1*, *CDPK* and *RPL6* genes were significantly downregulated after 24 h and 48 h of salt stress. *SRPRB* showed significant upregulation at 24 h after salt stress, but gradually downregulated after 48 h. Compared to the protein expression patterns, five DAP genes (*PDI, Cyto P450, CTA, RPL6* and *SRPRB*) showed high consistency or similar patterns. The remaining three DAP genes (*CNX, RPN1* and *CDPK*) showed poor consistency between transcriptional and translational expression, perhaps because of time-space span between transcription and translation as well as various posttranscriptional and posttranslational modifications [28,29].

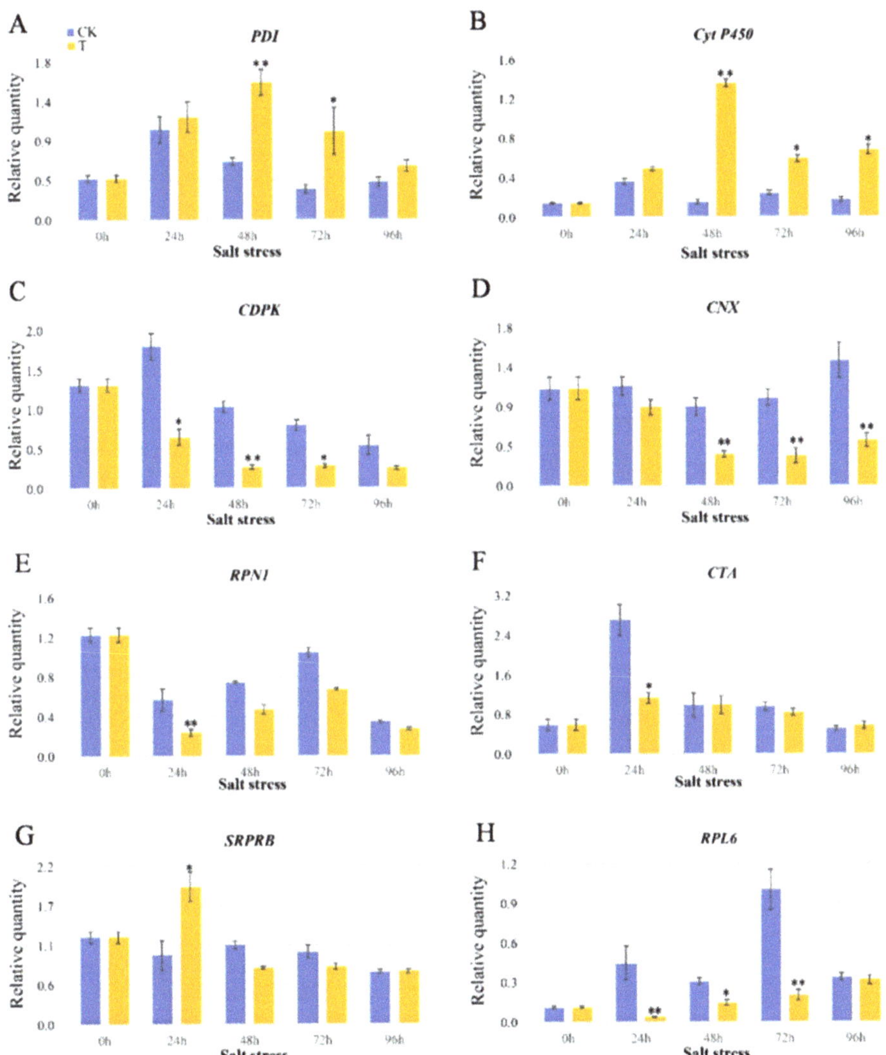

Figure 5. Transcript expression patterns of eight DAP genes in wheat ER under salt stress via RT-qPCR. (**A**) Protein disulfide-isomerase; (**B**) cytochrome P450; (**C**) calcium-dependent protein kinase; (**D**) calnexin homolog; (**E**) dolichyl-diphosphooligosaccharide-protein glycosyltransferase subunit 1 (RPN1); (**F**) calcium-transporting ATPase; (**G**) signal recognition particle receptor subunit beta (SRPRB); (**H**) 50S ribosomal protein L6. Statistically significant differences were calculated based on an independent Student's *t*-test: * $p < 0.05$; ** $p < 0.01$.

3. Discussion

Salt stress leads to plants producing and accumulating a large number of ROS, causing oxidative damage to cells [30] and inhibiting plant growth (Figure 1A–D). Meanwhile, a large amount of ROS can also act as a signal of stress response [31,32], inducing the activity increase of plant ROS scavenging-related enzymes such as SOD, CAT and POD (Figure 1F–H). SOD can convert superoxide radicals into oxygen and hydrogen peroxide, and CAT can catalyze the conversion of hydrogen peroxide into water and oxygen. POD can catalyze the oxidation of substrate with hydrogen peroxide as the electron acceptor. These are the key antioxidant enzymes in plant ROS scavenging systems [33].

ROS accumulation can cause continuous oxidization and reduction of disulfide bonds [34], which accounts for most of the nicotinamide adenine dinucleotide phosphate (NADPH) in the cell and results in misfolded proteins accumulating. Meanwhile, if unfolded or misfolded proteins accumulate in the ER lumen, the cells face ER stress that induces the unfolded protein response (UPR) to relieve ER stress [35]. The UPR pathway reestablishes ER homeostasis and protein synthesis, including initiating expression of chaperones and foldases for promoting protein folding, attenuating translation and removing unfolded proteins through proteasome degradation [36]. When these misfolded proteins are improperly treated, plant growth inhibition and cell death occur [37]. The current study found that the upregulated ER-DAPs involved in UPR mainly participated in the redox process, cellular stimulus response, phosphate-containing compound metabolic process, stress response and transmembrane transport (Figure 4C). The upregulation of these proteins alleviated the stress pressure of ER under salt stress conditions.

ER can anchor the heat shock 40 protein (ERdj3) to form a complex with the molecular chaperone proteins, binding directly to the hydrophobic region of the newborn protein, preventing protein aggregation, and assisting protein folding [38]. We identified two upregulated molecular chaperone proteins (R9W6A6 and A0A3B6QFL1) under salt stress (Table S3). As a kind of subcellular organelle, ER not only participates in maturing and folding protein, but also stores intracellular calcium ions that regulate calcium dependent protease activity in the ER lumen. When calcium ions unbalanced, it affects the folding ability and activity of these proteins, also causing stress response. After calcium ions complete messengers in the cytoplasm, they are repumped into ER storage by calcium-transporting ATPase on the ER. This not only ensures the difference of calcium concentration, but also maintains molecular chaperon proteins' activity. In a previous report [20], we found that two calcium-transporting ATPases (A0A3B6HQ49 and A0A3B5ZQ67) were significantly upregulated while under salt stress (Table S3, Figure 5). The ER lumen has unique oxidizing potential that supports disulfide bond formation during protein folding, as well as high protein concentration to form a gel-like protein matrix of chaperones and folding enzymes [26]. As a versatile protein folding factory, ER contains a specialized set of folding enzymes including PDI family (A0A3B6IQT3, W5BSJ0 and D8L9B3), PPI (A0A3B6PP75 and A0A3B6TIS2). We detected many such proteins as abundant proteins in the current study (Table S3, Figure 6).

Another UPR mechanism is mainly involved in attenuating translation as a response to salt stress. Ribosomal proteins play an integral role in generating rRNA structure and forming protein synthesizing machinery. They are also crucial in the growth and development of all organisms [39]. In the current study, we found that the expression of ribosomal related proteins was significantly downregulated such as RPL (A0A1D6BC85), A0A1D5U807, A0A3B5YZT7, 60sRPL (A0A0C4BIR6), A0A3B6C6A1, and A0A3B6RED0 (Table S3). This corresponded to their transcription level (Figure 5). It is possible that this could reduce ER protein concentration under salt stress. Consistent with the previous report on soybeans by [18], protein synthesis was impaired under salt conditions due to the decreased abundance of ribosomal proteins. The slowdown of the protein translation process could help alleviate ER load under salt stress conditions.

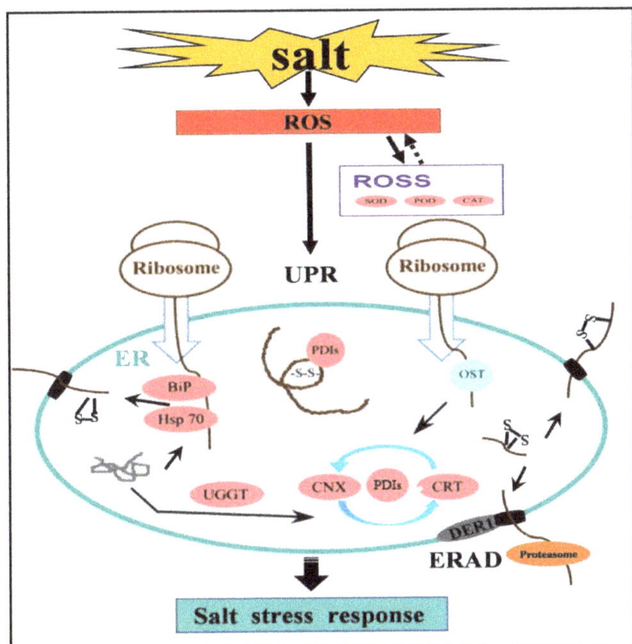

Figure 6. A putative synergistic responsive network of wheat ER proteome to salt stress. The red circle represents upregulated proteins. *BiP*, binding protein; *CAT*, catalase; *CNX*, calnexin; *CRT*, calreticulin; *ERAD*, endoplasmic reticulum associated degradation; *DER1*, degradation in the endoplasmic reticulum protein 1; *Hsp70*, 70 kilodalton heat shock proteins; *ROS*, reactive oxygen species; *ROSS*, reactive oxygen species scavenging; *OST*, oligosaccharide transferase; *PDIs*, protein disulfide-isomerase; *POD*, peroxidase; *SOD*, superoxide dismutase; *UGGT*, UDP-glycosyl transferases; *UPR*, unfolded protein reaction.

The ERAD mechanism is important as a means to remove unfolded proteins in the UPR, but mainly in the CNX/CRT folding cycle. Its basic process primarily includes four steps: target protein recognition, protein ubiquitination, reverse transcriptase transport and proteasome degradation [40]. The intermediate products of protein folding and the final misfolded proteins present a few structural similarities. The hydrophobic region embedded in the protein is exposed to the outside, easily leading to protein aggregation and misfolding. Molecular chaperones such as the heat shock protein 70 family can promote the folding of polypeptide chains by binding to hydrophobic regions. When the target protein interacts with Hsp70, the E3 ubiquitin ligase Hrd1 complex in the ERAD pathway can recognize it [41]. The misfolded proteins were ubiquitinated by E3 ubiquitin ligase on the cytoplasmic sol surface of the ER and then entered 26S proteasome to be degraded. Newborn peptides with glycosylation sites glycosylate when oligosaccharide transferase (OST) enters them. This process transfers pre-combined lipids (polyterpenol)-linked oligosaccharides to glycoprotein-dependent asparagine residues [42]. Monoglycosylated oligosaccharides are recognized by lectin molecular chaperone calcitonin, which is a kind of membrane-anchored protein, but they are also recognized by lumen protein calcium reticulin. These two proteins form a folding cage, thereby folding the target protein. PDI family not only catalyzes the formation of disulfide bonds, but also participates in the CNX/CRT cycle [43]. Moreover, UDP-glucosyltransferase (UGGT) and re-glycosylated detect misfolded proteins to reenter the calcitonin/calreticulin-mediated folding cycle and complete correct folding. UGGT can identify hydrophobic residues clusters exposed on the surface of spherical conformation isomers in the structural domain of unfolded proteins, acting as decision maker in ERQC [43–45]. We found that plants significantly upregulated UGGT in response to salt stress (Figure 2I). Several protein family members were also significantly upregulated under salt stress in the ERAD mechanism, including Hsp70 family (A0A3B6GVY6 and A0A3B5XXV7) and calcium (A0A3B6NLQ9 and A0A3B6NMA6) (Table S3). This indicated that the rapid accumulation of misfolded ER proteins could activate the degradation reaction mechanism of ER misfolded proteins under salt stress.

According to our results and previous reports, we proposed a putative metabolic network of wheat ER proteome responsive to salt stress (Figure 6). When plants were subjected to salt stress, ROS was rapidly accumulated. With continuous salt stress, excessive ROS activated the related ROS scavenging (ROSS) to maintain the balance of redox in cells. Meanwhile, ROS promoted the continuous redox of protein disulfide isomerase and accelerated the folding and removal of unfolded and misfolded proteins in the ER. When salt stress continued, excessive accumulation of misfolded proteins in the ER lumen induced significant upregulation of chaperone proteins as a means of reducing the ER load. The upregulation of calcium-transporting ATPasesbalanced the homeostasis of calcium concentration difference, ensured the activity and folding of calcium-dependent proteins and molecular chaperones and reduced the misfolded proteins. The excessive accumulation of misfolded proteins caused ER stress response, resulting in ER-ERAD activation. The unfolded and misfolded proteins accumulated in the ER cavity continue to fold through the cycle of calcitonin and calcium reticulum. Misfolded proteins correctly entered the ER autodegradation system. On the other hand, ER proteins were no longer transported into the ER lumen to relieve ER lumen pressure, and ribosomal proteins greatly reduced, thereby inhibiting the cell process.

4. Materials and methods

4.1. Wheat Materials and Salt Stress Treatments

Elite Chinese winter wheat cultivar Zhongmai 175 (*Triticum aestivum* L.) was used in this study. The mature seeds with similar size were surfacely sterilizedwith 70% (v/v) alcohol and 15% (v/v) sodium hypochlorite and rinsed four times with sterile distilled water. Then, the sterilized seeds were transferred onto wet filter paper and germinated at room temperature. After 48 h, uniformly germinated seeds were picked out and further grown in half strength Hoagland culture solution. The salt stress treatments were applied to wheat seedlings at the three-leaf stage by adding 200 mM NaCl to culture solution. The control group was grown in normal culture solution. Both treatment and control groups included three biological replicates (each with 300 seedlings), and the samples of whole seedlings at 0, 24, 48, 72 and 96 h in salt treatment and control were collected and stored in −80 °C prior to analysis.

4.2. Seedling Morphology Observation and Physiological Parameter Measurement

The measurement of plant height and root length were performed on 30 seedlings from the control and treatment group after 0, 24, 48, 72 and 96 h. The relative water content (RWC) and chlorophyll content of wheat seedling leaves were measured based on the method of Lv et al. [46]. The enzyme activities were demonstrated, including catalase (CAT), superoxide dismutase (SOD) and peroxidase (POD) related to reactive oxygen scavenging pathway, using a kit purchased from Suzhou Keming Biotechnology Co., LTD. The activates of UDP-glycosyl transferases (UGGT) and peptidyl-prolyl cis-trans isomerase (PPI) related to ER stress response were demonstrated using an ELISA kit (Jianglai Biological Co., LTD, Shanghai,China) based on the kit instructions. All measurements included three biological replicates to minimize experimental error. Statistical significances of the differences between the control and treatments were determined using a Student's t-test using SPSS 17.0 software (SPSS Institute Ltd., Armonk, NY, USA).

4.3. Enrichment of ER Components

In the current study, isolating and purifying ER from wheat seedling leaves was based on the method of Wang et al. [47] with some modifications. Wheat seedling leaves (25 g, about 500 seedlings) from each replicate were collected from each replicate after 48 h of treatment. Fresh leaves were cut into 1 cm pieces and ground using a tissue homogenizer (Ultra Turrax-T18, IKA, Staufen, Germany) with isolation buffer I containing PBS (pH 7.8), 250 mM sucrose, 50 mM HEPES, 5 mM EDTA-2Na, 0.2% caseinhydrolysate, 10% PEG, 0.6% PVPP, 0.2% BSA, 10% glycerin, 5 mM vitamin C tablets, 5 mM DTT, 15 mM CsCl, 0.2 mm PMSF and 1× Protease Inhibitor Cocktail (1 tablet/10 mL; Roche, Basel, Switzerland). The homogenate was filtered through three layers of Miracloth (Darmstadt, Germany) with two repeats. Then, the filtrate was centrifuged at 200× *g* for 10 min at 4 °C, after which the supernatant was collected and centrifuged at 3000× *g* for 10 min at 4 °C three times. Further, a high-speed centrifugation at 10,000× *g* and 4 °C for 20 min was carried out on the supernatant with two repeats. Then, ultra-high-speed centrifugation with 100,000× *g* at 4 °C for 1 h was conducted immediately, and the precipitate was resuscitated with 6 mL resuspension buffer (PBS [pH 7.8], 250 mM SUR, 50 mM HEPES and 2 mM DTT). Subsequently, the resuspended precipitate was loaded onto discontinuous sucrose gradients (8 mL 21.5% sucrose and 5 mL 37% sucrose) and centrifuged at 65,000× *g* for 30 min at 4 °C in a swing-out rotor. The inner membrane system components containing plasma membrane were collected from 21.5% sucrose gradient. A two-phase separation system (water polymer phase: 40 mL PBS buffer pH 7.8, 2.52 g dextran T-500, 2.52 g polyethylene glycol 4000, 250 mM sucrose, 0.8 M sodium chloride, 0.004 g DTT) was used to separate the inner membrane system and plasma membrane, and the inner membrane system was collected from the lower layer. The ER components were separated from the inner membrane system by a series of discontinuous sucrose gradients (3 mL 2 M sucrose, 5 mL 1.5 M sucrose, 4 mL 1.3 M sucrose, 4 mL 1 M sucrose, 8 mL 0.6 M sucrose), and smooth and rough ER were respectively collected from the interface between 0.6 and 1.3 M sucrose gradients and the precipitation. The mixture of smooth and rough ER were diluted ten times with resuspension buffer and centrifuged at 100,000× *g* for 1 h at 4 °C. The precipitation was immediately used for protein extraction.

4.4. Protein Extraction

The extraction of ER proteins was performed based on the previous study [48] with some modifications. Briefly, the ER components were suspended in 7 mL of extraction buffer (0.9 m sucrose, 2% TritonX-100 (*v*/*v*), 0.1 M Tris-HCl, pH 7.5, 50 mm EDTA-2Na, 2% SDS, 1% PVPP, 20 mm DTT, 1 mm PMSF, 1× Protease Inhibitor Cocktail) and ultrasound runs for 2 s and stops for 9 s, which is a cycle. Ultrasound lasts for 10 min with 500 W and the temperature was set as not exceeding 25 °C. An equal volume of Tris-balanced phenol (pH 7.5) was added to the mixture and further completely mixed by grounding in a mortar for 10 min. All the homogenate was transferred to a clean centrifuge tube, centrifuged at 14,000× *g* for 30 min at 4 °C, and the supernatant phenol phase was put into a new tube. Proteins were precipitated by adding four volumes of 100 mM ammonium acetate and remained at −20 °C overnight. On the next day, protein pellets were collected via centrifugations at 14,000× *g* for 30 min and rinsed once with pre-cooling methyl alcohol containing 20 mM β-mercaptoethanol, two times with pre-cooling acetone containing 20 mM β-mercaptoethanol and finally freeze-dried in a vacuum for subsequent experiments.

4.5. Trypsin Digestion and HPLC Fractionation

Sample was sonicated three times on ice using a high intensity ultrasonic processor (Scientz, Ningbo, China) in 500 µL lysis buffer (8 M urea, 1× Protease Inhibitor Cocktail). The remaining debris was removed via centrifugation at 12,000× *g* at 4 °C for 10 min. Finally, the supernatant was collected and the protein concentration was determined with BCA kit [49]. For digestion, the protein solution was reduced with 5 mM DTT for 30 min at 56 °C and alkylated with 11 mM iodoacetamide for 15 min at room temperature in darkness. The protein sample was then diluted by adding 100 mM NH_4HCO_3 to urea with a concentration of less than 2 M. Finally, 4 µg trypsin (Promega, Madison, WI, USA) was added to a final enzyme/protein ratio of 1:50 (*w*/*w*) at 37 °C overnight according to a previous study [50]. The tryptic peptides were fractionated into fractions via high pH reverse-phase HPLC using Agilent 300 Extend C18 column (5 µm particles, 4.6 mm ID, 250 mm length). The peptides were gradient eluted with 8-32% acetonitrile (Na_2CO_3-$NaHCO_3$ buffer, pH 9) and collected for later analysis.

4.6. NanoUPLC and Mass Spectrometric Analysis

The peptides were dissolved in mobile phase A and separated using EASY-nLC 1000 (Thermo/Finnigan, San jose, CA, USA). Mobile phases A and B contained 0.1 % formic acid in water and 0.2% formic acid in 90% acetonitrile, respectively. The gradient was set to 6–23% from 0 to 40 min, 23–35% from 40 to 54 min, 35–80% from 54 to 57 min and 80% from 57 to 60 min, with flow rate setting at 0.40 µL min^{-1}. Meanwhile, the LC system was equipped withorbitrap Q Exactive mass spectrometry (Thermo/Finnigan). The m/z scan range was 350 to 1800 for a full scan, and intact peptides were detected in the Orbitrap at a resolution of 70,000. Automatic gain control (AGC) was set at 5×10^4 Peptides were selected for MS/MS using a NCE setting of 28 and the fragments were detected in the Orbitrap at a resolution of 17,500. The electrospray voltage applied was 2.0 kV.

4.7. Sequence Database Search and Data Analysis

Maxquant (v1.5.2.8 Max Planck Institute for Biochemistry, Martinsried, Germany) was used to complete the retrieval of secondary mass spectrometry data from UniProt *Triticum aestivum* containing 116,790 sequences. The method of trypsin digestion was set to the maximum number of modified peptides to 5 and the minimum length to 7 amino acid residues, the parameter of omission site to 2, and the mass tolerance for precursor ions was set as 20 ppm in the first search and 5 ppm in the main search, and the mass tolerance for fragment ions was set as 0.02 Da. The fixed modification was set to cysteine alkylation, the oxidation of methionine and the acetylation of protein N-terminal were set to variable modification, and the false positive rates of protein identification was set to 1% [51,52]. In this study, the quantitative values of each sample in three replicates were obtained via LFQ intensity. The ratio of the mean LFQ intensity between the two samples represents the protein fold change. To calculate the significant p value of differential expression between two samples, LFQ intensity was taken as log2 transform. Then, when the protein is quantified at least twice in the two compared samples, the p value is calculated using the two-tailed t test. When the p value < 0.05 and protein ratio > 2, it was regarded as up-regulation. When the p value < 0.05 and protein ratio < 0.5, it was regarded as down-regulation. (Student's t-test, $p < 0.05$).

4.8. Western Blotting

For immunoblotting analysis, approximately 10 µg of total proteins and chloroplast proteins were prepared and separated using 12% sodium dodecyl sulfate-polyacrylamide gel electrophoresis (SDS-PAGE). The separated proteins on the gel were transferred to polyvinylidene fluoride (PVDF) membrane via semi-dry transfer imprinting and incubated in blocking buffer containing 20 mM of trihydrochloric acid (pH 7.5), 500 mM of sodium chloride and 5% skim milk, and then further incubated with 1 RV 5000 diluted polyclonal antibody (Swedish Agrisera, Vännäs, Sweden) at room temperature for 1 h. Anti-rabbit or mouse antibodies (Bio-Rad, Hercules, CA, USA) conjugated with horseradish peroxidase or anti-mouse antibodies (Bio-Rad) conjugated with horseradish peroxidase were used as secondary antibodies. After incubating with the second antibody for 1 h, the signals were detected using an ECL plus Western blotting kit (General Electric Healthcare, Piscataway, NJ, USA) according to the manufacturer's instruction.

4.9. Subcellular Localization

The subcellular localization predication of the identified proteins were performed according to the combination of the predicated results from WoLF PSORT (https://wolfpsort.hgc.jp/, accessed on 28 June 2020), TargetP-2.0 (http://www.cbs.dtu.dk/services/TargetP/, accessed on 28 June 2020), Plant-mPLoc (http://www.csbio.sjtu.edu.cn/bioinf/plant-multi/, accessed on 29 June 2020), CELLO (http://cello. life.nctu.edu.tw/, accessed on 29 June 2020) and UniProtKB (https://www.uniprot.org/help/uniprotkb/, accessed on 29 June 2020). Then, further subcellular localization assay by transforming Arabidopsis protoplasts was performed to verify the predicated results according to [53]. The amplified target fragment was reconstructed onto pSAT1-GFP-N (Pe3449) and PSAT1-RFP-N (Pe3449) vectors. Psat1-gfp-n (Pe3449) and PSAT1-YFP-N (Pe3449) carried a green fluorescent protein (GFP) gene and a red fluorescent protein (RFP) gene, respectively.

4.10. Total mRNA Extraction and Real-Time Quantitative Polymerase Chain Reaction (RT-qPCR)

RT-qPCR was used to detect the dynamic transcript levels of the key DAPs genes in response to salt stress. Total RNA was isolated from seedling leaves of control and salt treatment groups using TRIZOL Reagent (Invitrogen, Carlsbad, CA, USA). Genomic DNA was removed and then the reverse transcription reactions were performed using a PrimeScript® RT Reagent Kit with gDNA Eraser (TaKaRa, Shiga, Japan) according to [54]. Gene-specific primers for selected genes were designed using online Primer3Plus (www.bioinformatics.nl/cgi-bin/primer3plus/primer3plus.cgi). Ubiquitin was used as the reference for normalization. RT-qPCR was conducted using a CFX96 Real-Time PCRD detection system (Bio-Rad), and all data were analyzed with CFX Manager Software (Bio-Rad). Three independent replications were conducted for each sample.

5. Conclusions

Salt stress significantly inhibited the growth of wheat seedlings and resulted in phenotypic, physiological and biochemical changes, including the decrease of plant height, root length, relative water content and chlorophyll content. Label-free quantitative proteomic analysis identified 234 ER-localized DAPs in response to salt stress, including 203 upregulated and 31 downregulated proteins. The upregulated proteins were mainly involved in protein folding and quality control, ER stress response, unfolded protein response and ER-related degradation, while the downregulated proteins mainly participated in basic plant metabolic processes such as protein synthesis and translation. Through two main pathways of UPR and ERAD regulating ER stress in plants, the synergistic response of these ER proteins could play important roles in plant salt stress defense.

Supplementary Materials: The following are available online at https://www.mdpi.com/article/10.3390/ijms22094840/s1, Table S1: The list of peptides identified in wheat endoplasmic reticulum under salt stresses, Table S2: The list of total proteins identified in wheat endoplasmic reticulum under salt stresses, Table S3: The list of differentially accumulated proteins (DAPs) identified from extracted endoplasmic reticulum of wheat seedling leaves under salt stress, Table S4: List of specific primers for gene cloning and vector construction, Table S5: GO annotation of ER-localized differentially accumulated proteins (DAPs) identified in wheat endoplasmic reticulum under salt stresses, Table S6: List of specific primers for RT-qPCR analysis.

Author Contributions: J.Z. and D.L. performed all of the experiments, data analysis and wrote the paper. D.Z. and N.L. performed experimental treatments and protein extraction. Y.Y. designed and supervised experiments. All authors have read and agreed to the published version of the manuscript.

Funding: This research was financially supported by grants from the National Natural Science Foundation of China (31971931, 31771773).

Acknowledgments: All the mass spectrometry data were deposited in the Proteome X change Consortium (http://proteomecentral.proteomexchange.org, accessed on 3 September 2020) via the PRIDE partner repository. The dataset identifier number is PXD019453.

Conflicts of Interest: The authors have no conflicts of interest relevant to this article.

Abbreviation

AGPase	ADP glucose pyrophosphorylase
AI/TI	Alpha-amylase/trypsin inhibitor
AIIA	Probable ureidoglycolate hydrolase
AOX1/2	Plant alternative oxidase 1 and 2
BiP	Binding protein
CAT	Catalase
CNX	Calnexin
CDPK	Calcium-dependent protein kinase
CRT	Calreticulin
CTAB	Hexadecyl trimethyl ammonium bromide
DAPs	Differentially accumulated proteins
DPA	Day post anthesis
ER	Endoplasmic reticulum
ERAD	Endoplasmic reticulum associated degradation
ERQC	Endoplasmic reticulum quality control
GO	Gene ontology
H3	Histones 3
HSP	Heat shock protein
NADPH	Nicotinamide adenine dinucleotide phosphate
OST	Oligosaccharide transferase
PDI	Protein disulfide isomerase
PEG	Polyethyleneglycol
PMSF	Phenylmethylsulfonyl fluoride
POD	Peroxidase
PPI	Peptidyl-prolyl cis-trans isomerase
RT-qPCR	Real-time quantitative polymerase chain reaction
RFP	Red fluorescent protein
ROS	Reactive oxygen species
ROSS	Reactive oxygen species scavenging
RP	Ribosomal proteins
RP-HPLC	Reversed-phase high performance liquid chromatogram
RT-PCR	Reverse transcription polymerase chain reaction
RWC	Relative water content
SDS	Sodium dodecyl sulfate
SDS-PAGE	SDS-polyacrylamide gel electrophoresis
SOD	Superoxide dismutase
SOS	Salt overly sensitive
Spcs2	Probable signal peptidase complex subunit 2
UGGT	UDP-glycosyl transferase
UPR	Unfolded protein reaction

References

1. Shrivastava, P.; Kumar, R. Soil salinity: A serious environmental issue and plant growth promoting bacteria as one of the tools for its alleviation. *Saudi. J. Biol. Sci.* **2015**, *22*, 123–131. [CrossRef]
2. Jesus, J.M.; Danko, A.S.; Fiúza, A.; Borges, M.T. Phytoremediation of salt-affected soils: A review of processes, applicability, and the impact of climate change. *Environ. Sci. Pollut. Res.* **2015**, *22*, 6511–6525. [CrossRef]
3. Qadir, M.; Quillérou, E.; Nangia, V.; Murtaza, G.; Singh, M.; Thomas, R.J.; Noble, A.D. Economics of salt-induced land degradation and restoration. *Nat. Resour. Forum* **2014**, *38*, 282–295. [CrossRef]
4. Munns, R.; Tester, M. Mechanisms of salinity tolerance. *Annu. Rev. Plant. Biol.* **2008**, *59*, 651–681. [CrossRef]
5. Zhu, J.K. Salt and drought stress signal transduction in plants. *Annu Rev. Plant Biol.* **2002**, *53*, 247–273. [CrossRef] [PubMed]
6. Zhu, J.K. Abiotic stress signaling and responses in plants. *Cell* **2016**, *167*, 313–324. [CrossRef] [PubMed]
7. Healy, S.J.; Verfaillie, T.; Jager, R. A biology of the endoplasmic reticulum. In *Endoplasmic Reticulum Stress in Health and Disease*; Agostinis, P., Samali, A., Eds.; Springer: Dordrecht, The Netherlands, 2012; pp. 3–22.
8. Howell, S.H. Endoplasmic reticulum stress responses in plants. *Annu. Rev. Plant. Biol.* **2013**, *64*, 477–499. [CrossRef]

9. Wang, X.; Komatsu, S. Plant subcellular proteomics: Application for exploring optimal cell function in soybean. *J. Proteom.* **2016**, *143*, 45–56. [CrossRef] [PubMed]
10. Lv, D.W.; Zhu, G.R.; Zhu, D.; Bian, Y.W.; Liang, X.N.; Cheng, Z.W.; Yan, Y.M. Proteomic and phosphoproteomic analysis reveals the response and defense mechanism in leaves of diploid wheat *T. monococcum* under salt stress and recovery. *J. Proteom.* **2016**, *128*, 388–402. [CrossRef] [PubMed]
11. Ma, A.; Plu, B.; Ht, C.; Js, A. Recent advances in robotic protein sample preparation for clinical analysis and other biomedical applications. *Clin. Chim. Acta* **2020**, *507*, 104–116.
12. Hasan, M. High-throughput proteomics and metabolomic studies guide re-engineering of metabolic pathways in eukaryotic microalgae: A review. *Bioresour. Technol.* **2020**, *321*, 124495.
13. Wang, W.Q.; Liu, S.J.; Song, S.Q. Proteomics of seed development, desiccation tolerance, germination and vigor. *Plant. Physiol. Biochem.* **2015**, *86*, 1–15. [CrossRef] [PubMed]
14. Komatsu, S.; Kobayashi, Y.; Nishizawa, K.; Nanjo, Y.; Furukawa, K. Comparative proteomics analysis of differentially expressed proteins in soybean cell wall during flooding stress. *Amino Acids* **2010**, *39*, 1435–1449. [CrossRef] [PubMed]
15. Gayen, D.; Barua, P.; Varshney, S.; Sengupta, S.; Chakraborty, S. Dehydration-responsive alterations in the chloroplast proteome and cell metabolomic profile of rice reveals key stress adaptation responses. *Environ. Exp. Bot.* **2019**, *160*, 16–24. [CrossRef]
16. Wang, L.; Liang, W.; Xing, J.; Tan, F.; Chen, Y.; Huang, L.; Cheng, C.L.; Chen, W. Dynamics of chloroplast proteome in salt-stressed mangrove *Kandelia candel* (L.) Druce. *J. Proteome. Res.* **2013**, *12*, 5124–5136. [CrossRef] [PubMed]
17. Lee, C.P.; Eubel, H.; O'Toole, N.; Millar, A.H. Combining proteomics of root and shoot mitochondria and transcript analysis to define constitutive and variable components in plant mitochondria. *Phytochemistry* **2011**, *72*, 1092–1108. [CrossRef]
18. Nouri, M.Z.; Komatsu, S. Comparative analysis of soybean plasma membrane proteins under osmotic stress using gel-based and LC MS/MS-based proteomics approaches. *Proteomics* **2010**, *10*, 1930–1945. [CrossRef]
19. Komatsu, S.; Kuji, R.; Nanjo, Y.; Hiraga, S.; Furukawa, K. Comprehensive analysis of endoplasmic reticulum-enriched fraction in root tips of soybean under flooding stess using proteomics techniques. *J. Proteom.* **2012**, *77*, 531–560. [CrossRef]
20. Wang, X.; Komatsu, S. Gel-Free/Label-Free Proteomic Analysis of Endoplasmic Reticulum Proteins in Soybean Root Tips under Flooding and Drought Stresses. *J. Proteom. Res.* **2016**, *15*, 2211–2227. [CrossRef]
21. He, Z.H.; Li, H.W.; Shen, Y.; Li, Z.S.; Mi, H. Comparative analysis of the chloroplast proteomes of a wheat (*Triticum aestivum* L.) single seed descent line and its parents. *J. Plant. Physiol.* **2013**, *170*, 1139–1147. [CrossRef]
22. Luo, Y.; Liu, H.Y.; Fan, Y.Z.; Wang, W.; Zhao, Y.Y. Comparative chloroplast proteome analysis of exogenously supplied trehalose to wheat seedlings under heat stress. *Photosynthetica* **2018**, *56*, 1123–1133. [CrossRef]
23. Zhu, D.; Luo, F.; Zou, R.; Liu, J.X.; Yan, Y.M. Integrated physiological and chloroplast proteome analysis of wheat seedling leaves under salt and osmotic stresses. *J. Proteom.* **2021**, *234*, 104097. [CrossRef] [PubMed]
24. Chen, R.H.; Liu, W.; Zhang, G.S.; Ye, J.X. Mitochondrial proteomic analysis of cytoplasmic male sterility line and its maintainer in wheat (*Triticum aestivum* L.). *Agric. Sci. China* **2010**, *9*, 771–782. [CrossRef]
25. Liu, X.; Gan, F.; Wang, Z.; Gao, Y.; Peng, Y.; Dang, C.; Yang, Z. Characterization of mitochondrial proteomic changes in resistant wheat near-isogenic line after inoculation with powdery mildew. *J. Phytopathol.* **2013**, *161*, 215–223. [CrossRef]
26. Shen, Y.; Du, J.; Yue, L.; Zhan, X. Proteomic analysis of plasma membrane proteins in wheat roots exposed to phenanthrene. *Environ. Sci. Pollut. Res.* **2016**, *23*, 10863–10871. [CrossRef]
27. Chen, X.; Karnovsky, A.; Sans, M.D.; Andrews, P.C.; Williams, J.A. Molecular characterization of the endoplasmic reticulum: Insights from proteomic studies. *Proteomics* **2010**, *10*, 4040–4052. [CrossRef] [PubMed]
28. Guo, G.F.; Lv, D.W.; Yan, X.; Subburaj, S.; Ge, P.; Li, X.H.; Yan, Y.M. Proteome characterization of developing grains in bread wheat cultivars (*Triticum aestivum* L.). *BMC Plant. Biol.* **2012**, *12*, 147. [CrossRef]
29. Zhu, D.; Zhu, G.R.; Zhang, Z.; Wang, Z.; Yan, Y.M. Effects of independent and combined water-deficit and high-nitrogen treatments on flag leaf proteomes during wheat grain development. *Int. J. Mol. Sci.* **2020**, *21*, 2098. [CrossRef]
30. Kim, C.; Meskauskiene, R.; Apel, K.; Laloi, C. No single way to understand singlet oxygen signalling in plants. *EMBO Rep.* **2008**, *9*, 435–439. [CrossRef]
31. Ashraf, M. Biotechnological approach of improving plant salt tolerance using antioxidants as markers. *Biotechnol. Adv.* **2009**, *27*, 84–93. [CrossRef]
32. Huang, C.; He, W.; Guo, J.; Chang, X.; Su, P.; Zhang, L. Increased sensitivity to salt stress in an ascorbate-deficient Arabidopsis mutant. *J. Exp. Bot.* **2005**, *56*, 3041–3049. [CrossRef]
33. Yadav, N.M.; Bagdi, D.L.; Kakralya, B.L. Effect of salt stress on physiological, biochemical, growth and yield variables of wheat (*Triticum aestivum* L.). *Agric. Sci. Dig.* **2011**, *31*, 247–253.
34. Tu, B.P.; Weissman, J.S. Oxidative protein folding in eukaryotes: Mechanisms and consequences. *J. Cell. Biol.* **2004**, *164*, 341–346. [CrossRef]
35. Walter, P.; Ron, D. The unfolded protein response: From stress pathway to homeostatic regulation. *Science* **2011**, *334*, 1081–1086. [CrossRef] [PubMed]
36. Liu, J.X.; Howell, S.H. Endoplasmic Reticulum Protein Quality Control and Its Relationship to Environmental Stress Responses in Plants. *Plant. Cell* **2010**, *22*, 2930–2942. [CrossRef] [PubMed]
37. Liu, L.; Cui, F.; Li, Q.; Yin, B.; Zhang, H.; Lin, B.; Xie, Q. The endoplasmic reticulum-associated degradation is necessary for plant salt tolerance. *Cell Res.* **2011**, *21*, 957–969. [CrossRef]

38. Jin, Y.; Zhuang, M.; Hendershot, L.M. ERdj3, a luminal ER DnaJ homologue, binds directly to unfolded proteins in the mammalian ER: Identification of critical residues. *Biochemistry* **2009**, *48*, 41–49. [CrossRef] [PubMed]
39. Ishii, K.; Washio, T.; Uechi, T.; Yoshihama, M.; Kenmochi, N.; Tomita, M. Characteristics and clustering of human ribosomal protein genes. *BMC Genom.* **2006**, *7*, 37. [CrossRef]
40. Çakır Aydemir, B.; Yüksel Özmen, C.; Kibar, U.; Mutaf, F.; Buyuk, P.B.; Bakır, M.; Ali, E. Salt stress induces endoplasmic reticulum stress-responsive genes in a grapevine rootstock. *PLoS ONE* **2020**, *15*, e0236424. [CrossRef]
41. Izawa, T.; Nagai, H.; Endo, T.; Nishikawa, S.-i. Yos9p and Hrd1p mediate ER retention of misfolded proteins for ER-associated degradation. *Mol. Biol. Cell* **2012**, *23*, 1283–1293. [CrossRef]
42. Pattison, R.J.; Amtmann, A. N-glycan production in the endoplasmic reticulum of plants. *Trends Plant. Sci.* **2009**, *14*, 92–99. [CrossRef] [PubMed]
43. Caramelo, J.J.; Castro, O.A.; Alonso, L.G.; Prat-Gay, G.; Parodi, A.J. UDP-Glc:glycoprotein glucosyltransferase recognizes structured and solvent accessible hydrophobic patches in molten globule-like folding intermediates. *Proc. Natl. Acad. Sci. USA* **2003**, *100*, 86–91. [CrossRef]
44. Sousa, M.; Parodi, A.J. The molecular basis for the recognition of misfoldhed glycoproteins by the UDP-Glc: Glycoprotein glucosyltransferase. *EMBO J.* **1995**, *14*, 4196–4203. [CrossRef] [PubMed]
45. Taylor, S.C.; Thibault, P.; Tessier, D.C.; Bergeron, J.J.; Thomas, D.Y. Glycopeptide specificity of the secretory protein folding sensor UDP-glucose glycoprotein: Glucosyltransferase. *EMBO Rep.* **2003**, *4*, 405–411. [CrossRef]
46. Lv, D.W.; Subburaj, S.; Cao, M.; Yan, X.; Li, X.H.; Appels, R.; Sun, D.; Ma, W.J.; Yan, Y.M. Proteome and phosphor proteome characterization reveals new response and defense mechanisms of *Brachypodium distachyon* leaves under salt stress. *Mol. Cell. Proteom.* **2014**, *13*, 632–652. [CrossRef] [PubMed]
47. Wang, G.; Wang, F.; Wang, G.; Wang, F.; Zhang, X.; Zhong, M.; Song, R. Opaque1 encodes a myosin XI motor protein that is required for endoplasmic reticulum motility and protein body formation in maize endosperm. *Plant. Cell* **2012**, *24*, 3447–3462. [CrossRef]
48. Wang, X.C.; Li, X.F.; Deng, X.; Han, H.P.; Shi, W.L.; Li, Y.X. A protein extraction method compatible with proteomic analysis for the euhalophyte salicornia europea. *Electrophoresis* **2007**, *28*, 3976–3987. [CrossRef]
49. Walker, J.M. The bicinchoninic acid (BCA) assay for protein quantitation. *Methods Mol. Biol.* **1994**, *32*, 5–8. [PubMed]
50. Shen, Z.; Li, P.; Ni, R.J.; Ritchel, M.; Yang, C.P.; Liu, G.F.; Ma, W.; Liu, G.J.; Ma, L.; Li, S.J.; et al. Label-free quantitative proteomics analysis of etiolated maize seedling leaves during greening. *Mol. Cell Proteom.* **2009**, *8*, 2443–2460. [CrossRef]
51. Elias, J.E.; Gygi, S.P. Target-decoy search strategy for increased confidence in large-scale protein identifications by mass spectrometry. *Nat. Methods.* **2007**, *4*, 207–214. [CrossRef] [PubMed]
52. Chick, J.M.; Kolippakkam, D.; Nusinow, D.P.; Zhai, B.; Rad, R.; Huttlin, E.L. A mass-tolerant database search identifies a large proportion of unassigned spectra in shotgun proteomics as modified peptides. *Nat. Biotechnol.* **2015**, *33*, 743–749. [CrossRef] [PubMed]
53. Zhen, S.M.; Deng, X.; Li, M.F.; Zhu, D.; Yan, Y.M. 2D-DIGE comparative proteomic analysis of developing wheat grains under high-nitrogen fertilization revealed key differentially accumulated proteins that promote storage protein and starch biosyntheses. *Anal. Bioanal. Chem.* **2018**, *410*, 6219–6235. [CrossRef] [PubMed]
54. Yu, Y.L.; Zhu, D.; Ma, C.Y.; Cao, H.; Wang, Y.P.; Xu, Y.; Yan, Y.M. Transcriptome analysis reveals key differentially expressed genes involved in wheat grain development. *Crop. J.* **2016**, *4*, 92–106. [CrossRef]

International Journal of
Molecular Sciences

MDPI

Article

Proteomics of Homeobox7 Enhanced Salt Tolerance in *Mesembryanthemum crystallinum*

Xuemei Zhang [1,2], Bowen Tan [2], Dan Zhu [2,3], Daniel Dufresne [4], Tingbo Jiang [1,*] and Sixue Chen [2,5,6,*]

1 State Key Laboratory of Tree Genetics and Breeding, Northeast Forestry University, Harbin 150040, China; zhangxuemei199111@gmail.com
2 Department of Biology, Genetics Institute, University of Florida, Gainesville, FL 32610, USA; tanbowen@ufl.edu (B.T.); zhudan2014dora@163.com (D.Z.)
3 College of Life Sciences, Qingdao Agricultural University, Qingdao 266109, China
4 Department of Chemistry, Florida Atlantic University, Boca Raton, FL 33431, USA; dufresne71@yahoo.com
5 Plant Molecular and Cellular Biology Program, University of Florida, Gainesville, FL 32610, USA
6 Proteomics and Mass Spectrometry, Interdisciplinary Center for Biotechnology Research, University of Florida, Gainesville, FL 32610, USA
* Correspondence: tbjiang@yahoo.com (T.J.); schen@ufl.edu (S.C.)

Abstract: *Mesembryanthemum crystallinum* (common ice plant) is a halophyte species that has adapted to extreme conditions. In this study, we cloned a *McHB7* transcription factor gene from the ice plant. The expression of *McHB7* was significantly induced by 500 mM NaCl and it reached the peak under salt treatment for 7 days. The *McHB7* protein was targeted to the nucleus. *McHB7*-overexpressing in ice plant leaves through *Agrobacterium*-mediated transformation led to 25 times more *McHB7* transcripts than the non-transformed wild type (WT). After 500 mM NaCl treatment for 7 days, the activities of superoxide dismutase (SOD) and peroxidase (POD) and water content of the transgenic plants were higher than the WT, while malondialdehyde (MDA) was decreased in the transgenic plants. A total of 1082 and 1072 proteins were profiled by proteomics under control and salt treatment, respectively, with 22 and 11 proteins uniquely identified under control and salt stress, respectively. Among the 11 proteins, 7 were increased and 4 were decreased after salt treatment. Most of the proteins whose expression increased in the *McHB7* overexpression (OE) ice plants under high salinity were involved in transport regulation, catalytic activities, biosynthesis of secondary metabolites, and response to stimulus. The results demonstrate that the *McHB7* transcription factor plays a positive role in improving plant salt tolerance.

Keywords: *Mesembryanthemum crystallinum*; *McHB7*; transcription factor; salt stress; proteomics

Citation: Zhang, X.; Tan, B.; Zhu, D.; Dufresne, D.; Jiang, T.; Chen, S. Proteomics of Homeobox7 Enhanced Salt Tolerance in *Mesembryanthemum crystallinum*. *Int. J. Mol. Sci.* **2021**, *22*, 6390. https://doi.org/10.3390/ijms22126390

Academic Editor: Karl-Josef Dietz

Received: 14 May 2021
Accepted: 13 June 2021
Published: 15 June 2021

Publisher's Note: MDPI stays neutral with regard to jurisdictional claims in published maps and institutional affiliations.

1. Introduction

Mesembryanthemum crystallinum (common ice plant) is a well-known halophyte plant native to Africa, southern Europe, and widely naturalized elsewhere [1]. It has been used as a health-promoting vegetable against diabetes [2]. Also, ice plant has been increasingly used as a model for studying abiotic stress responses [1] because it can shift from C3 photosynthesis to crassulacean acid metabolism (CAM) [3]. Halophytes, such as ice plants, tolerate high salinity by accumulating osmolytes in specific locations and balancing ion homeostasis, thereby maintaining osmolarity [4]. In ice plants, the special epidermal bladder cells (EBCs) on the leaf surface help to sequester salt [5]. The mesophyll cells and diurnal regulation of stomatal guard cells also improve water use efficiency of the ice plants [6].

Recent studies have revealed changes in ice plants at transcriptomic [7,8], intracellular ion [9], metabolite [10], and protein levels [11] in response to salt stress, with many candidate genes, proteins and metabolites of relevance having been identified. For example, *McSnRK1* encodes a SNF1-related protein kinase, which functions as transcriptional modulator to control metabolic adaption, regulate Na^+ flux and maintain Na/K homeostasis under salt stress [12]. Transgenic tobacco co-expressing an ice plant inositol methyl transferase gene, *McIMT1*, and a wild halophytic rice gene, *PcINO1*, accumulated high levels of inositol and methylated inositol. These metabolites may help to improve plant tolerance to high degrees of salt stress [13]. A sodium transporter, *McHKT2*, was reported to participate in plasma membrane Na^+ transport and induce significant salt tolerance in transgenic *Arabidopsis thaliana* (*A. thaliana*) seedlings [14]. In addition, Kong et al. found that the transition from C3 to CAM photosynthesis takes place between day 5 and day 7 after treating ice plants with 500 mM NaCl. A total of 495 transcripts were significantly changed in guard cells based on RNA-Seq data analysis [15]. Among them, 18 transcription factors (TFs) were identified, including NAC (comp21521_c0_seq1 and Mcr002150.001), GRAS (Contig16446), WRKY (Contig20720), and homeobox family TFs (Contig9771, *McHB7*).

Homeobox genes have been found in all eukaryotic organisms [16]. They have a highly conserved homeodomain (HD), which contains a specific 60 amino acids (aa) DNA-binding motif. The HD forms a helix-turn-helix structure that regulates the expression of target genes [17,18]. The HD is always followed by a leucine zipper motif (35–42 aa), termed as homeobox associated leucine zipper (HALZ), which functions as a dimerization motif associated with the HD [19]. Homeobox genes are master regulators of plant development [20], meristem regulation [20], hormone mediation [21], and response to various environmental stimuli—including high salinity, drought, and extreme temperature [22]. For example, STIMPY/AtWOX9 was found to be important to the growth of vegetative shoot apical meristems and the maintenance of cell division in the shoot and root apex [23]. Overexpressing a poplar WUSCHEL-related homeobox gene, *PagWOX11/12a*, increased plant root biomass and enhanced drought tolerance [24,25]. *WOX6/HOS9* is important for low temperature response in Arabidopsis by mediating the expression of genes independent of element-binding factor pathway [26]. In upland cotton, *GhWOX10_Dt*, *GhWOX13b_At/Dt*, and *GhWOX13a_At/Dt* were significantly induced by salt stress, and they may play vital roles in improving salt stress response of plants [27].

In Kong et al. [15], it was reported that *McHB7* in ice plant was induced by 500 mM NaCl treatment for 7 days during the transition from C3 to CAM. Here we report additional results on *McHB7* functions. Although it is common to analyze gene functions by genetic transformation, transformation of ice plants to study the functions of salinity-related genes is rare. The objective of this study is to analyze the ice plant TF *McHB7* and its function in salt stress tolerance. We cloned the *McHB7* from ice plant leaves and examined the molecular changes in transiently transformed plants compared to WT under control and high salinity conditions.

2. Results

2.1. Bioinformatic Analysis of McHB7

As shown in Supplementary Figure S1, the cloned *McHB7* gene is 804 bp and it encodes a TF of 267 amino acids. According to NCBI blast, 10 homologous proteins from other species shared high homology, i.e., XP_021748183.1 (64.1%, *Chenopodium quinoa*), XP_021767274.1 (64.0%, *Chenopodium quinoa*), XP_021846961.1 (61.1%, *Spinacia oleracea*), XP_010692437.1 (61.6%, *Beta vulgaris subsp. vulgaris*), KNA21535.1 (61.1%, *Spinacia oleracea*), KMT20244.1 (61.7%, *Beta vulgaris subsp. vulgaris*), ACZ05048.1 (54.8%, *Phytolaccaacinosa*), PSS26255.1 (48.4%, *Actinidia chinensis var. chinensis*), QEQ92614.1 (48.9%, *Populusussuriensis*), and XP_002320889.1 (48.9%, *Populustrichocarpa*). Phylogenetic analysis (Figure 1A) and sequence alignment (Figure 1B) showed that these proteins contain two highly conserved domains: homeobox and homeobox associated HALZ. The basic region homeobox at the N terminal (Figure 1C) consists of three α-helices that form a helix-turn-helix DNA-binding motif, followed by a HALZ that is markedly different from those homeobox TFs in animal systems [19].

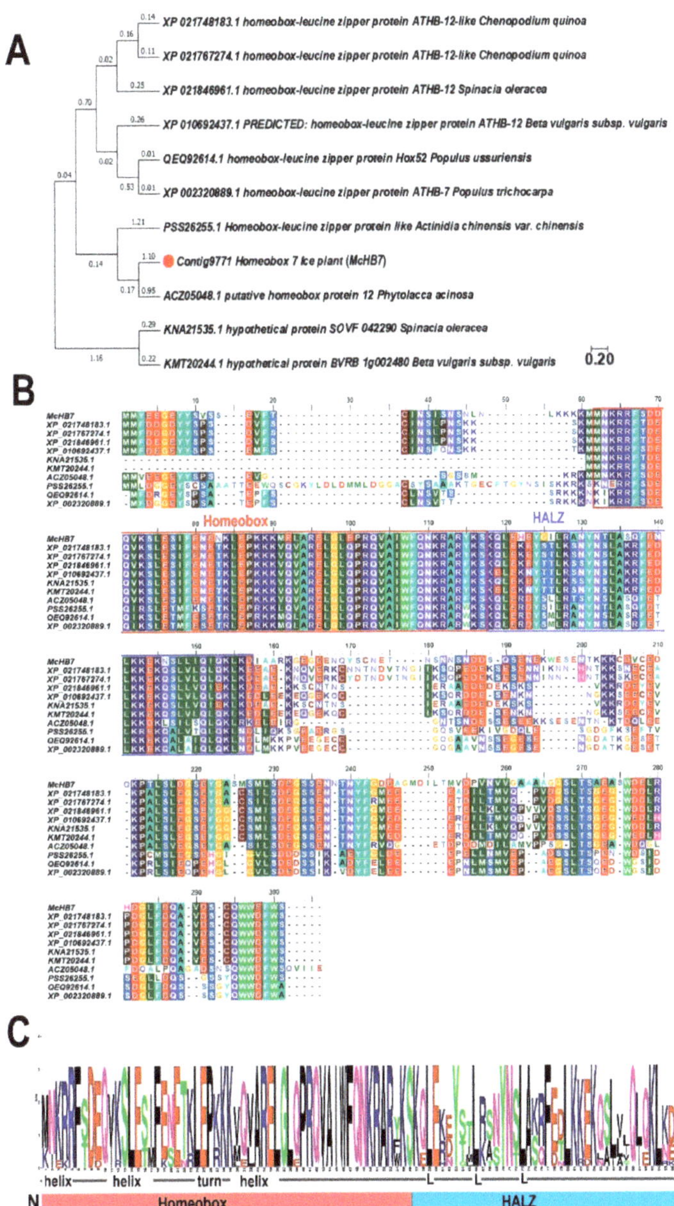

Figure 1. Bioinformatic analysis of *McHB7* and homologous TF proteins. (**A**) Phylogenetic tree of XP_021748183.1, XP_021767274.1, XP_021846961.1, XP_010692437.1, KNA21535.1, KMT20244.1, ACZ05048.1, PSS26255.1, QEQ92614.1, and XP_002320889.1. The red dot indicates the *McHB7* TF. (**B**) Amino acid sequence alignment of XP_021748183.1, XP_021767274.1, XP_021846961.1, XP_010692437.1, KNA21535.1, KMT20244.1, ACZ05048.1, PSS26255.1, QEQ92614.1, and XP_002320889.1. (**C**) Two domains of homeobox TFs: conserved homeobox domain (HD) and variable HALZ domain.

2.2. *McHB7 TF Protein Localized to the Nucleus*

CELLO2GO prediction showed the localization of *McHB7* protein to the nucleus (Supplementary Figure S2). To validate this predicted result, *A. tumefaciens* GV3101 containing an *McHB7-GFP* construct (Figure 2A) and the positive control pCAMBIA1302-*GFP* were transiently expressed in tobacco leaves. As shown in Figure 2B, *GFP* signal of the positive control was observed in the cytosol and nucleus of whole epidermal cell, while the *GFP* signal of *McHB7-GFP* only existed in the nucleus. This result clearly showed the nuclear localization of the *McHB7* TF.

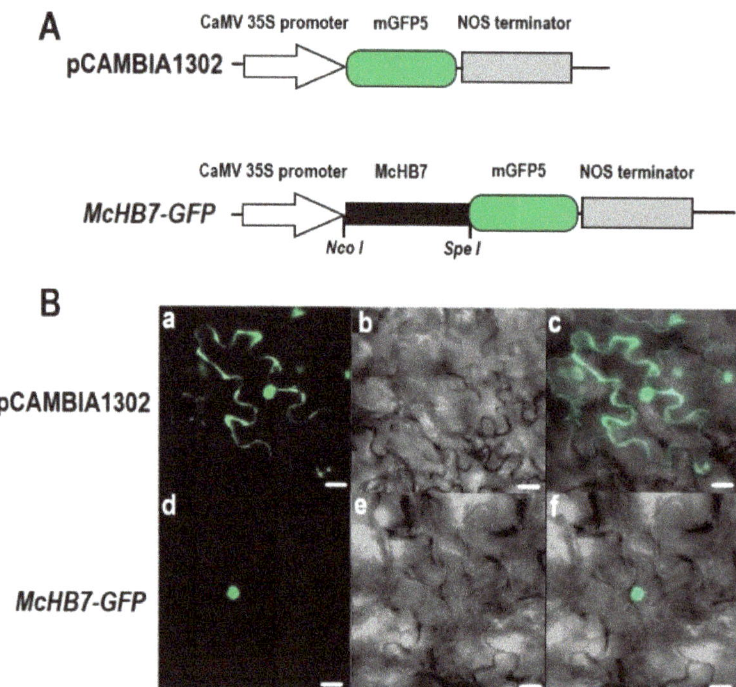

Figure 2. Subcellular localization of *McHB7* protein. (**A**) Schematic maps of the positive control pCAMBIA1302-*GFP* and the *McHB7-GFP* construct in the pCAMBIA1302vector. (**B**) The *McHB7-GFP* fusion construct and the positive control pCAMBIA1302were introduced into tobacco leaves by infiltration. *GFP* fluorescence was observed by confocal laser scanning microscopy. (**a**,**d**) fluorescence images observed in a dark field (green); (**b**,**e**) light images observed in bright field; (**c**,**f**) merged images of dark field and bright field. Scale bar = 30 μm.

2.3. *Relative Expression of McHB7 in Ice Plant Leaves after Salt Stress Treatment*

To investigate the expression pattern of *McHB7* under high salinity, we treated the four-week-old ice plant seedlings with 500 mM NaCl and collected leaf samples each day for 14 days. When ice plants were salt-stressed for seven or 14 days, the growth was severely inhibited and the leaves were smaller when compared to the control seedlings (Figure 3A), and fresh weight of salt-treated leaves for seven days and 14 days was 49.9% and 24.9% that of control leaves, respectively (Supplementary Figure S3). According to the RT-qPCR results in Figure 3B, the relative expression of *McHB7* in ice plant leaves was significantly induced by salt stress, reached the peak at day 7, up to almost 12 times higher than untreated leaves.

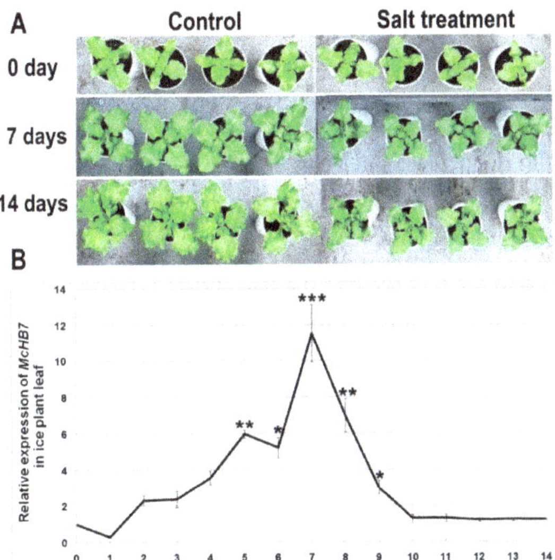

Figure 3. Ice plant growth phenotype and *McHB7* expression after high salinity treatment. (**A**) Phenotype of ice plant seedlings under 500 mM NaCl treatment for 0, 7, and 14 days. (**B**) Relative expression of *McHB7* after 500 mM NaCl treatment (compared with the control condition). Student's *t*-test: *t*: * $p < 0.05$; ** $p < 0.01$, *** $p < 0.001$, error bars indicate mean ± standard deviation (SD) ($n = 4$).

2.4. Overexpressing McHB7 in Transgenic Ice Plants

To verify the OE of *McHB7* in the ice plant leaves, we performed RT-qPCR with *McHB7*-overexpressing (OE) plants and WT leaves. As shown in Figure 4A, 7 days after infiltration, the relative expression of *McHB7* in OE ice plant was significantly higher (~25 times) than WT. According to the Western blot analysis, the Flag signal only existed in the *McHB7*-infiltrated leaves (Figure 4B). These data clearly showed successful OE of *McHB7* at both the transcriptional and translational levels. However, there was no significant difference in SOD and POD activity or water content between OE plants and WT. MDA content of OE plants was a little lower than WT, but not statistically significant (Supplementary Figure S4).

2.5. Changes in Biochemical Parameters in Ice Plants after Salt Stress Treatment

One week after infiltration, the OE ice plant seedlings and WT were treated with 500 mM NaCl solution for another week (14 days after infiltration). As shown in Figure 5A, there were no significantly differences between the *McHB7*-overexpressingplants and WT under control conditions. After high salinity treatment for one week, the leaves of OE plants grew better than WT, with a fresh weight 1.2 times higher (Supplementary Figure S5). To further test whether the transgene is still functional, we carried out the Western blot with transgenic leaves under control and salt-treated conditions. The results showed that the Flag signal still existed in OE leaves 14 days after infiltration (Figure 5B). The control and OE leaves were collected for physiological analyses and histochemical staining. As described in Figure 5C–F, SOD activity, POD activity, and water content in OE were 1.2 ± 0.2, 1.5 ± 0.8, and 1.1 ± 0.5 times higher than in WT, respectively. Nitrotetrazoliumblue chloride (NBT) and3, 3′-diaminobenzidine (DAB) were used to analyze the levels of superoxide anions (O_2^-) and hydrogen peroxide (H_2O_2), respectively [28]. Evans blue was used to check cell death due to plasma membrane damage [29]. The staining intensities of OE and WT leaves were similar under control conditions (Figure 5G–I). After 500 mM NaCl treatment for seven days, the staining of WT was darker than the OE plants, and the relative intensity of NBT and DAB in WT was significantly higher than that of OE (Supplementary Figure S6).

The result indicates overexpressing *McHB7* led to lower accumulation of reactive oxygen species (ROS) and stronger capability for ROS removal.

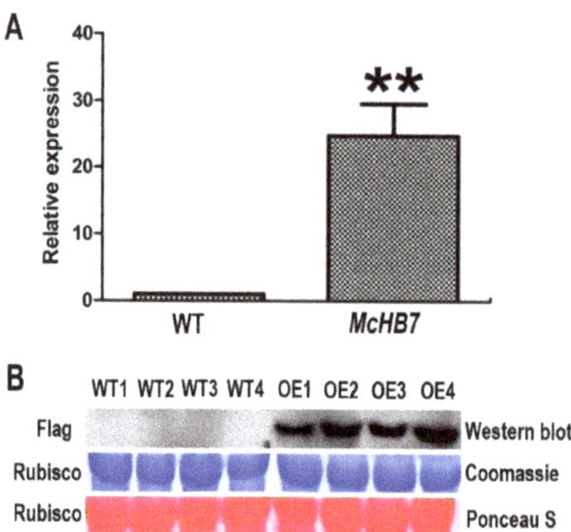

Figure 4. Overexpression of *McHB7* in ice plants at day 7 after transformation. (**A**) The relative expression of *McHB7* in the overexpression (OE) plants compared to WT; Student's *t*-test, ** $p < 0.01$. Error bars indicate mean ±SD ($n = 4$). (**B**) Images of Western blot showing Flag signal of *McHB7* protein level in the OE plants. The protein loading was assessed by Coomassie staining and Ponceau S staining of Rubisco, showing equal loading. WT1–WT4 are four replicates of WT, OE1–OE4 are four replicates of *McHB7*-over-expressing seedlings.

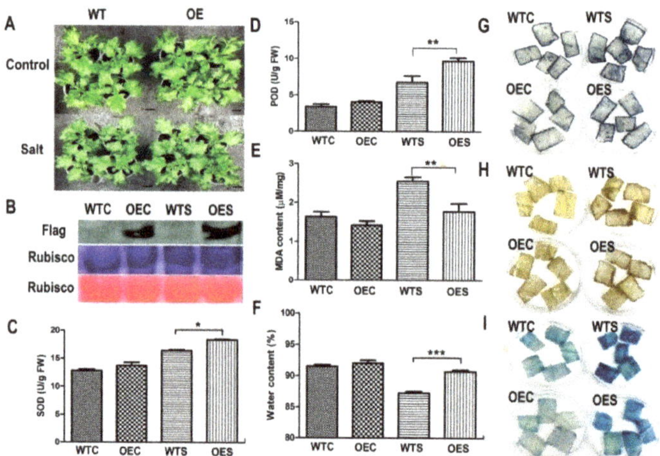

Figure 5. *McHB7*-overexpressing (OE) plants under control and salt stress conditions. (**A**) Phenotype of OE plants and WT under control and salt stress conditions, bar = 2 cm. (**B**) Validation of OE plants after salt stress treatment. The OE plants were at day 14 after transformation; Student's *t*-test, * $p< 0.05$, ** $p < 0.01$, *** $p < 0.001$. Error bars indicate mean ± SD ($n = 4$). (**C**) SOD activity of OE plants and WT under control and salt stress conditions. (**D**) POD activity. (**E**) MDA content. (**F**) Water content. (**G**) NBT staining. (**H**) DAB staining. (**I**) Evans blue staining. WTC, WT under control conditions; OEC, transgenic plants under control conditions; WTS, WT after salt stress treatment; OES, transgenic plants after salt stress treatment.

2.6. Proteomic Changes Attributed to McHB7 Overexpression and Salinity Treatment

Total proteins of OE and WT plant leaves were isolated and label-free quantitative proteomics was carried out using LC-MS/MS. Under control and salt stress conditions, 1082 and 1072 proteins were identified, respectively (Figure 6A,B). GO functional classifications including biological process, cellular component and molecular function were generated by Proteome Discoverer. Of the biological processes, most proteins were involved in metabolic process, transport, stimulus response or regulation (Figure 6C). Most of them were categorized to the cytoplasm, membrane, ribosome, mitochondrion, and nucleus (Figure 6D). Many proteins played a role in catalytic activity, nucleotide, and metal ion binding. Some of them were involved in protein and RNA binding (Figure 6E). Among these identified proteins, 34 proteins had phosphorylation modifications under control and salt stress conditions (Supplementary Table S2). Most of them were chlorophyll a-b binding proteins, which belong to the light-harvesting complex and could be photo-regulated through reversible phosphorylation. These were reported to mediate the distribution of excitation energy in the photosystems I and II [30].

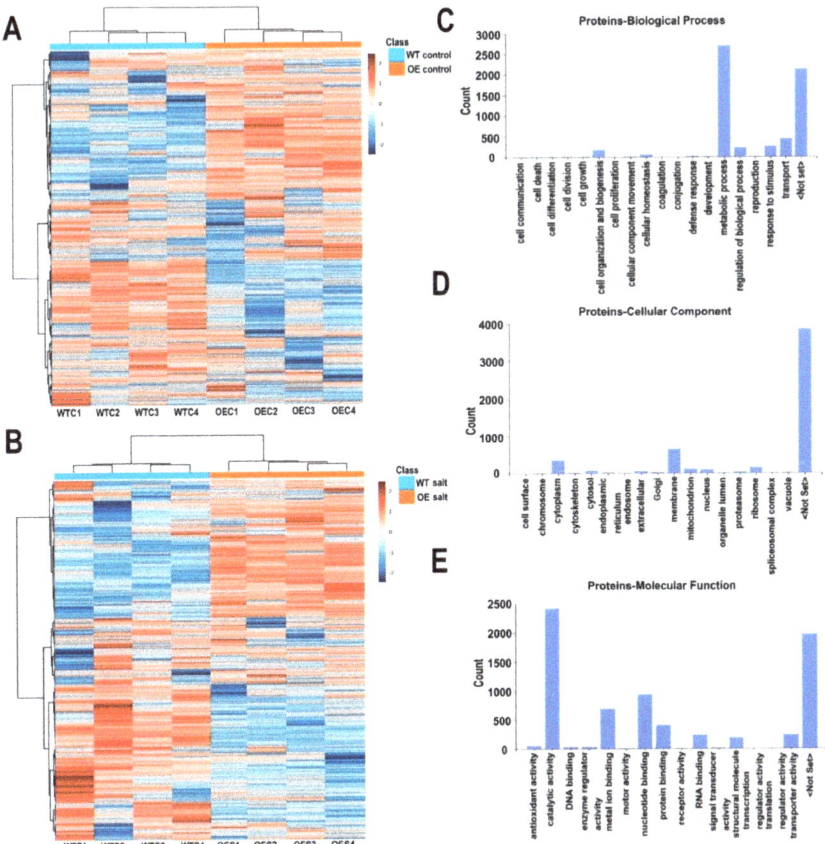

Figure 6. Heatmap of identified proteins under control and high salinity conditions and GO functional categorization of the proteins. (**A**) Identified proteins under control conditions. WTC1–WTC4, four biological replicates of WT under control conditions; OEC1–OEC4, four replicates of OE plants under control conditions. (**B**) Identified proteins under salt stress conditions. WTS1–WTS4, four biological replicates of WT after salt stress treatment; OES1–OES4, four biological replicates of transgenic plants after salt stress treatment. (**C**) Biological process of identified proteins. (**D**) Cellular component of identified proteins. (**E**) Molecular function of identified proteins.

2.7. Differentially Expressed Proteins under Control and Salt Stress Treatment

To address the molecular mechanisms underlying *McHB7* function, we conducted proteomics on WT and OE plants under control and salt stress conditions. Principal component analysis (PCA) is an unsupervised multivariate statistics-based detection method to reveal the differences and relationships between samples [31]. In this study, four replicates of each sample were grouped together, but different samples were classified into distinct clusters (Figure 7A). The results showed that samples from OE and WT leaves under control and salt treatment, respectively, occupied relatively independent spaces in the distribution map, especially OES was separated and far away from WTS.

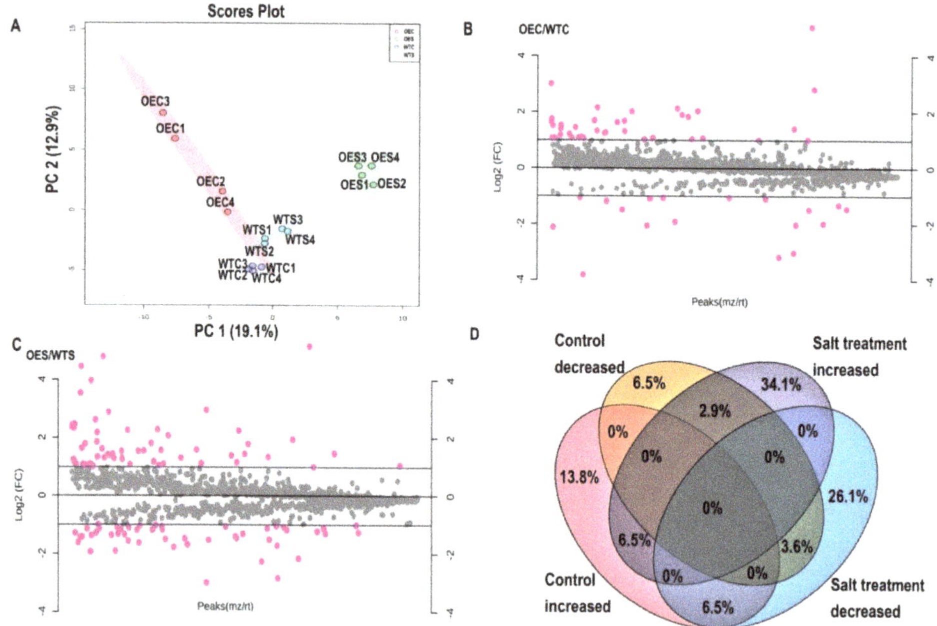

Figure 7. Differentially expressed proteins (DEPs) under control and salt stress treatment. WTC1-WTC4, four biological replicates of WT under control conditions; OEC1-OEC4, four replicates of OE plants under control conditions; WTS1-WTS4, four biological replicates of WT after salt stress treatment; OES1-OES4, four biological replicates of transgenic plants after salt stress treatment. (**A**) PCA result of all the control and salt stress samples of WT and *McHB7* OE. (**B**) DEPs under control conditions. (**C**) DEPs under salt stress treatment. (**D**) Venn diagram of DEPs under control and salt stress treatment.

Based on the identified proteins from OE and WT leaves, differentially expressed proteins (DEPs) were analyzed. There were 55 proteins—37 increased and 18 decreased under control conditions (Figure 7B; Supplementary Figure S7A). Among these, seven proteins were experienced fold change (OEC/WTC) greater than four magnitudes. Notably, K9NCW5, a ribulose bisphosphate carboxylase/oxygenase (Rubisco) homologue was about 33 times higher in OE. Other increased proteins like A0A022RVT2, A0A1S3 × 8Z6, B9GEL5, A0A200R0A6, A0A072UU34, and M1C498 were involved in nucleotide binding that may interact with the nuclear localized *McHB7*. After high salinity treatment, there were 110 proteins including 60 increased and 50 decreased (Figure 7C; Supplementary Figure S7B). Among these increased proteins, six including D7KSJ9, A0A1J3DHL5, F4Y5A9, A0A1D6LMX1, A9TQU1, and A0A1Q3C6B9 were responsive to stimulus and were all involved in metabolic process. Besides, some proteins like A2TJU5 were proteins sensitive to high salinity and played a role in transport activities. W8E1S1 and A0A0F7GYT3 are photosystem II CP43 reaction center protein and photosystem II reaction center PsbP family protein, respectively. Their induction by salt may help maintain photosynthesis activity under the stress condition. A0A022RMX1, A0A1J3DHL5 and A9PAY7 were involved in regulation of biological activities, A0A022RMX1 and A0A1J3DHL5 took a part in the cell organization and biogenesis, A9PAY7 worked in cellular homeostasis. A0A0D2Q010 and M0U9P1 were involved in biosynthesis of secondary metabolites, A0A0D2Q010 played a role in citrate acid cycle, and M0U9P1was involved in glycine, serine, and threonine metabolism. A Venn Diagram shows that nine proteins were increased in *McHB7*-overexpressing leaves and after high salinity treatment (Figure 7D). Of these, the relative expression of K9NCW5, A0A178UQY1, A0A2P5EU89, A0A022RVT2, A0A078G853, and D7KSJ9 was 33.09, 8.04, 3.24, 2.89, 2.31, and 2.06 times higher in OE vs. WT, respectively, under control conditions. However, they were elevated to 35.65, 21.84, 6.1298, 3.62, 3.46, and 11.62 times, respectively, under high salinity. Interestingly, A0A075M528, A0A022RT96, A0A2J6MJ80, and A0A2R4KXR9 were decreased in transgenic plant leaves while being significantly increased after the salt stress treatment.

Also, we found that among the 1082 and 1072 proteins profiled under control and salt stress, 22 (2%) proteins were uniquely found under control conditions, while 11 (1%) were uniquely found after salt stress treatment (Supplementary Figure S8). Among these 11 proteins, G3XDJ4, A0A1Z1CK67, and C8CS23 were adenosine triphosphate (ATP) synthase subunits that are important for the formation of energy storage molecule ATP [32]. Besides, 7 of the 11 proteins including A0A2P5PX19, I1IJ13, A0A1Z1CK67, E9LWD6, I1INX4, and A0A1S3BNT8 were increased after salt stress treatment.

3. Discussion

In this study, we cloned a salt-stress responsive transcription factor *McHB7* from an ice plant with a highly conserved HD and a variable HALZ domain. Using a popular *Agrobacterium*-mediated transformation method [33,34], we were able to obtain transient transgenic ice plants and confirmed OE of the transgene *McHB7* by both qPCR and Western blot after infiltration for 7 and 14 days (Figures 4 and 5). Western blot analysis is important to show increase of *McHB7* without affecting the total protein level [35]. This work was necessary to demonstrate transformation efficiency and genetic manipulation of ice plant. The ability to test gene functions through reverse genetics in calcitrant native system is a breakthrough. Many studies have used heterologous systems—e.g., *A. thaliana* and *Nicotiana benthamiana* (*N. benthamiana*)—for testing gene functions [36–38]. While heterologous systems are useful, in vivo research is always more applicable. Interestingly, the transgene expression in the ice plant appears to be long lasting, in contrast to the *N. benthamiana* system that often lasts a short time [39]. Here, OE of the *McHB7* gene in ice plant leaves prolonged for two weeks after infiltration (Figure 5).

Salinity is a major abiotic stress factor that can induce water deficit [40], leading to an accumulation of ROS (e.g., O_2^-, H_2O_2, and hydroxyl radical (OH^-)) that damage cell membranes and essential molecules like DNA, proteins, and lipids [41,42]. With the OE ice plants, we examined the different physiological parameters including SOD, POD, MDA, and water content in the WT and OE plants under salt stress treatment. SOD is known to function as a primary O_2^- scavenger [43]. POD is an oxidoreductase that detoxifies H_2O_2 [44]. MDA is often used as an indicator of oxidative membrane lipid damage [45]. Water content is used to reflect general stress tolerance and water use efficiency [46]. The results clearly showed that OE of *McHB7* effectively increased the activities of SOD and POD, which reduce ROS levels and damage under salt stress. *McHB7* OE improved plant development and growth under salinity stress by maintaining ion homeostasis and water balance.

Proteomics analysis suggested that many proteins were increased or decreased in the *McHB7* overexpressing plants compared to WT. For example, K9NCW5 (Rubisco), a key enzyme involved in the first step of carbon fixation in photosynthesis [47]. It was reported that the small subunit of Rubisco was coded by nuclear genes [48,49]. Here we showed that *McHB7* protein was localized within the nucleus and that OE of *McHB7* in ice plant increased the abundance of Rubisco, indicating that *McHB7* is likely a positive regulator of carbon fixation. The abundance of Rubisco in OE plants was significantly elevated under both control and stress conditions, which showed *McHB7* may function in the process of photosynthesis by mediating the expression of the proteins. More experiments are needed to uncover the relationship between *McHB7* and Rubisco. D7KSJ9 is known to play an important role in response to stimulus and regulation [50]. Its protein level is two times higher in OE plants than in the WT before salt stress. After salt stress, its protein level increased about 12 times in the OE plants compared to WT. A0A0D2Q010 is an ATP-citrate synthase beta chain protein and known to participate in the citrate cycle [51]. It was also increased in OE plants after salt treatment. These results suggest that *McHB7* is involved in stress response through regulating stress and energy-related proteins as a master switch.

Since the ice plant genome has not been published, we used the corresponding Arabidopsis homolog to determine potential HB7 binding sites in the 5-kb upstream promoter of *AtHB7* (AT2G46680.1) by searching PlantCARE (http://bioinformatics.psb.ugent.be/webtools/plantcare/html/) (accessed on 1 May 2021). Several *cis*-acting elements related to light-regulation like the LAMP-element, ACE, GT1-motif, 3-AF1 binding site, Sp1, and chs-CMA1a were found (Supplementary Table S3), which could be bound by numerous proteins [52]. Among these elements GT1-motif was reported in *Rubisco*, which was regulated at the transcriptional level through light-responsive pathway [53]. In addition, numerous stress-responsive elements like an oxidative stress-responsive element *as*-1 were found in the promoter region of *AtHB7*. The binding activity of *as*-1 could be stimulated by oxidative stress [54]. However, salt stress was considered to induce osmotic stress and enhance accumulation of ROS in plant cells [55]. Surprisingly, several proteins that were related to oxidative phosphorylation like AOA078JRB4, B9HTA1, and I1J2V1 were identified in OE plants and the abundances of these proteins were increased compared to WT. This indicates *McHB7* might function as a stress response TF induced by high salinity to enhance the stress tolerance of plants through regulation of oxidative-related proteins. MBS, LTR, and TC-rich repeats were also reported to respond to various abiotic stresses [56]. Some hormone-responsive *cis*-elements including ABRE, GARE-motif, TGACG-motif, P-box, TATC-box, and TCA-element were identified which were known to be responsive to abscisic acid, gibberellin, methyl jasmonate, or salicylic acid [56]. The results suggest that *McHB7* is a signal for stresses dependent on the hormonal signaling pathway. Interestingly, according to our proteomics, we found the protein A0AOJ8BHN7 was identified under control and salt stress conditions and elevated in the OE plants. It is also involved in plant hormone signal transduction based on GO analysis.

Protein functions may be informed by their interacting networks. The homologous gene of *McHB7* in Arabidopsis is *AtHB7*. They share 41.3% identity at the amino acid level (Supplementary Figure S9A) with high similarity in the HD and HAZL domains. According to String network analysis, four proteins—*At*HB5, RD26, ABI2, and ABF3—were found to be tightly associated with *At*HB7 at high confidence (confidence > 0.7) (Supplementary Figure S9B). Among them, *At*HB5 acts as a positive regulator of ABA-responsiveness and mediates the inhibitory effect of ABA on growth [57]. RD26 encodes a NAC transcription factor and participates in ABA-mediated dehydration response [58]. ABI2 encodes a phosphatase 2C protein and regulates numerous ABA responses [59]. ABF3 encodes an ABA-responsive element-binding protein and can respond to stress and abscisic acid [60]. *At*HB7 is also transcriptionally regulated in an ABA-dependent pathway [61]. It may interact with these ABA-related proteins and mediate the response to adverse environmental effects. In addition, it was reported that ABA-induced *AtHB7* can promote leaf development, chlorophyll synthesis, and reduce stomatal conductance in mature plants [62]. Also, *AtHB12* functions as a positive transcriptional regulator of PP2C genes and plays a vital role in response to water deficit [63]. The proteomics data and String network analysis provides a functional context of *McHB7*, which can be explored further in the future studies.

4. Materials and Methods

4.1. Plant Materials and Salt Stress Treatment

Mesembryanthemum crystallinum seeds were sowed into the moist soil in a growth chamber at 12-h (26 °C) light/12-h dark (18 °C) cycle. Seven days later, the seedlings with four leaves were transplanted to the 946 mL foam cups and watered with 50 mL 0.5× Hoagland's solution each day. One-month-old ice plants were used for transformation and salt treatment. For high salinity stress, the plants were irrigated with 50 mL 500 mM NaCl in 0.5× Hoagland's solution every day. The control plants were irrigated with 50 mL 0.5× Hoagland's solution.

4.2. Cloning and Sequence Analysis of McHB7 Gene

A total of 100 mg leaf material from an individual ice plant was collected and ground in liquid nitrogen into a fine powder. Total RNA was extracted according to the instructions of a RNeasy® Plant Mini Kit (QIAGEN, Germantown, MD, USA). The RNA was then reversely transcribed into cDNA following a protocol of ProtoScript® II First Strand cDNA Synthesis Kit (New England BioLabs, Ipswich, MA, USA). Based on the sequence information obtained from RNA-Seq [15], a pair of specific primers *cMcHB7* (Supplementary Table S1) were designed for PCR amplification of the *McHB7* open-reading-frame (ORF). With the NCBI database, the amino acid of *McHB7* was used to blast for homologous proteins. A phylogenetic tree was constructed with MEGA 7 and multi-sequence alignment was analyzed by BioEdit (Version 7.2). Pfam database (http://pfam.xfam.org, accessed on 8 May 2021) was applied for conserved motifs prediction, and online software WebLogo (https://weblogo.berkeley.edu/logo.cgi, accessed on 8 May 2021) was used for motif visualization.

4.3. Subcellular Localization of McHB7 Protein

The subcellular localization of *McHB7* protein was predicted by online software CELLO2GO (http://cello.life.nctu.edu.tw/cello2go, accessed on 8 May 2021). Briefly, the amino acid sequence of *McHB7* was pasted to the software in FASTA format, and blast-searched in Eukaryote (E-value 0.001). To confirm the predicted result, the ORF without stop codon of *McHB7* was PCR amplified using a pair of primers *gMcHB7* (Supplementary Table S1), which contains *NcoI* and *Spe I* restriction sites, respectively. Then the PCR fragment was ligated to a plant expression vector pCAMBIA1302 with a green fluorescent protein mGFP5. The recombinant plasmid called *McHB7-GFP* was transformed into *Agrobacterium tumefaciens* GV3101. *McHB7-GFP* and the positive control pCAMBIA1302 agrobacteria were cultured in a LB broth liquid medium containing 50 mg/mL kanamycin and 25 mg/mL rifampicin to an OD_{600} of 0.8–1.0. The agrobacteria were collected by centrifugation and resuspended in a MES solution containing 100 mM MES (pH 5.8), 100 mM $MgCl_2$, and 100 µM acetosyringone (As) and used for transformation. *N. benthamiana* seedlings were grown in the growth chamber conditions, and two-month-old tobacco leaves were used for infiltration with blunt-end micro-syringe [33]. Each leaf was injected with 500 µL *Agrobacterium*, and four replicate experiments were done. Two days later, the GFP signal was observed using a confocal laser scanning microscope (Zeiss, Jena, Germany).

4.4. McHB7 Gene Expression Analysis

To investigate the relative expression of *McHB7* in the leaves of ice plant under high salinity conditions, one-month-old ice plant seedlings were irrigated with 50 mL 500 mM NaCl solution each day for 14 days. The second pair of leaves were harvested and midveins discarded for RNA extraction. Each sample has four biological replicates and three technical replicates. Plasma membrane intrinsic protein 1; 2 (*McPIP1; 2*) was used as the internal reference. The 2× SYBR Green qPCR Master Mix kit (Bimake, Houston, TX, USA) was used for RT-qPCR using a CFX96 171 Touch™ Real-Time PCR Detection System (Bio-Rad, Hercules, CA, USA). Real-time primers are listed in Supplemental Table S1. The relative expression in different samples was calculated using the $2^{-\Delta\Delta Ct}$ method [64]. Based on Student's *t*-test, bars in the figures correspond to standard deviation. A star indicates *p*-value < 0.05, two stars indicate *p*-value < 0.01, and three stars indicate *p*-value < 0.001.

4.5. *McHB7 Overexpression Vector Contruction and Plant Transient Transformation*

Based on the *McHB7* sequence, a pair of primers *pMcHB7* contains *Bam HI* and *Xba I*, respectively, was designed (Supplementary Table S1). The reverse primer contained the sequence of 3× Flag tag. The ORF without stop codon was ligated to an *Agrobacterium* binary vector pCAMBIA1300, which confers kanamycin (Kan) resistance. The recombinant plasmid was transformed into *Agrobacterium tumefaciens* GV3101 for plant transformation and selected on LB medium with 50 mg/mL Kan. A modified *Agrobacterium*-mediated transformation method was used to transform the leaves of four-week-old ice plants [34]. GV3101 with the recombinant plasmid *pMcHB7* was grown in 25 mL LB liquid medium containing 50 mg/mL Kan and 25 mg/mL rifampicin overnight at 28 °C. The cells were collected by centrifugation at 10,000 rpm for 10 min at room temperature, and then resuspended into 154 mM NaCl solution with 100 µM As. The OD_{600} was adjusted to 0.8–1.0 for infection. The second pair of expanded leaves was injected with 500 µL of the agrobacterial cultures using a needle-less syringe. The control wild type (WT) plants were injected with the same solution without agrobacteria. Seven days later, the injected leaves were harvested for further analyses. Each sample had four replicates.

4.6. *Protein Extraction and Validation of Transgenic Ice Plant*

A phenol method was used for protein extraction [6]. Infiltrated ice plant leaves (1 g fresh weight) were ground into powder by motor and pestle. A total of 3 mL extraction buffer (100 mM Tris-HCl, pH 8.8, 10 mM EDTA, 0.9 M Sucrose, 20 mM 2-Mercaptoethanol, 1× protease inhibitor cocktail, and 1 mM PMSF) and 3 mL Tris-saturated phenol were added into the protein powder, and then transferred to a centrifugation tube, and vortexed at 4 °C for 1 h. The samples were centrifuged at 4 °C, 15,000× *g* for 15 min. The upper layer was transferred to a new tube carefully, then five times volume of 0.1 M ammonium acetate in 100% methanol was added for protein precipitation. The samples were placed in −20 °C overnight. After centrifugation at 4 °C, 15,000× *g* for 15 min, the supernatant was removed. The precipitate was washed, resuspended in 80% acetone, and transferred to a new 2 mL centrifuge tube, placed in −20 °C for 20 min. After centrifugation at 4 °C, 13,000 rpm for 15 min, and washing again with 100% acetone, the protein pellet was dissolved in 150 µL dissolution buffer (6 M Urea, 1 mM EDTA, 1% SDS, and 50 mM Tris-HCl, pH 8.5). Protein quantification was determined by using a Bradford assay (Thermo Scientific, San Jose, CA, USA) following manufacturer's instructions. Equal amounts of proteins were loaded on a 10% SDS-PAGE gel. Four biological replicates were conducted for each sample.

For Western blot analysis [65,66], proteins in the SDS-PAGE gel were transferred onto a nitrocellulose immobilization membrane (PerkinElmer Life Sciences, Boston, MA, USA) at constant amperage (0.01 A overnight at 4 °C). Protein loading was assessed by Ponceau S staining. The membrane was washed several times by TBST buffer (50 mM Tris base, 150 mM NaCl, 0.1% Tween 20, pH7.4) and blocked by blocking buffer (5% non-fat milk in TBST buffer) for 1 h. Then, the membrane was incubated with a primary anti-Flag monoclonal antibody (Sigma-Aldrich, Saint Louis, MO, USA) that was diluted into 1 mL TBST buffer (at 1:5000 dilution) and 30 mg no-fat milk for 2 h. After washing the membrane five times by TBST, 5 min. each time, the membrane was incubated by a secondary anti-IgG antibody that was diluted into 1 mL TBST (at 1:10,000) for 1 h. The membrane was washed five times in TBST, then incubated with Western blot signal enhancer reagent (Thermo Scientific™ Pierce™ Western Blot Signal Enhancer, Walham, MA, USA) for 5 min and detected by chemiluminescence using an Amersham Imager 600 (GE Healthcare, Marlborough, MA 01752, USA).

4.7. Biochemical Analysis of Antioxidative System Components and ROS

Physiological parameters including superoxide SOD, POD, MDA, and water content of transgenic ice plant seedlings and WT control were measured according to previously published methods [67]. Histochemical staining is an important technique for visualizing biological structures and detection of ROS, including O_2^-, H_2O_2, hydroxyl radical (OH^-), singlet oxygen, and lipid hydroperoxides [68]. In this study, DAB, NBT, and Evans blue were carried out according to published methods [69,70]. Four biological replicates were conducted for each treatment and a Student's *t*-test was used for statistical analysis.

4.8. Liquid Chromatography Mass Spectrometry (LC-MS/MS) and Data Analysis

The protein samples from four biological replicates of WT control and four biological replicates of transgenic plants were digested with trypsin as previously described [6]. Liquid chromatography tandem mass spectrometry (LC−MS/MS) was carried out on an Easy-nLC 1200 system (Thermo Fisher Scientific Inc., Germering, DE, USA) coupled with a Q-Exactive HF Orbitrap mass spectrometer (Thermo FisherScientific Inc., San Jose, CA, USA). The peptides were separated by an Acclaim PepMap100 C18 column (250 mm × 75 μm; 2 μm-C18) (Thermo Fisher Scientific Inc., San Jose, CA, USA) with 3 h gradient at the flow rate of 0.35 μL/min. The LC gradient was: 0–5 min, 2% B; 140 min, 35% B; 160 min, 100% B; 165 min, 100% B; 170 min, 2% B; 180 min, 2% B, run stop. The MS was operated between MS scan and MS/MS scan automatically with a cycle time of 3 s. Eluted peptides were detected in the Orbitrap MS at a resolution of 120 K with a scan range of 350–1800 m/z, and the most abundant ions bearing 2–7 charges were selected for MS/MS analysis. Automatic gain control (AGC) for the full MS scan was set as 200,000 with maximum injection time (MIT) as 50 ms, and AGC Target of 10,000 and MIT of 35 ms were set for the MS/MS scan. The MS/MS scan used quadrupole isolation mode, high energy collision-induced dissociation (HCD) activation energy, and 35% collision energy, and Orbitrap detection. A dynamic exclusion time of 30 s was applied to prevent repeated sequencing of the most abundant peptides.

Proteome Discoverer™ 2.4 (Thermo Fisher Scientific, Bremen, Germany) was used for protein identification. The SEQUEST algorithm in the Proteome Discoverer was used to process raw data files. Spectra were searched using the TAIR10 protein database with the following parameters: 10 ppm mass tolerance for MS1 and 0.02 Retention Time tolerance as mass tolerance for MS2, two maximum missed tryptic cleavage sites, a fixed modification of carbamidomethylation (+57.021) on cysteine residues, dynamic modifications of (oxidation of methionine (+15.996) and phosphorylation (+79.966) on tyrosine, serine, and threonine. Search results were filtered at 1% false discovery rate (FDR) and peptide confidence level was set for at least two unique peptides per protein for protein identification. Relative protein abundance in the samples was measured using label-free quantification in the Proteome Discoverer 2.4. Proteins identified and quantified in all 4 out of 4 biological samples were used, and no imputation was performed. Peptides in the control and transgenic samples were quantified as area under the chromatogram peak. The data were normalized by medium and generalized logarithm transformation (Log2). The heatmap was generated using online software MetaboAnalyst 5.0 (https://www.metaboanalyst.ca, accessed on 15 April 2021). Functional categorization was built by Proteome Discoverer software and venn diagram was made by Venny 2.1.0 (https://bioinfogp.cnb.csic.es/tools/venny, accessed on 1 May 2021). Protein network analysis was conducted using String (https://string-db.org/cgi/network, accessed on 1 May 2021).

5. Conclusions

In this study, we cloned *McHB7* TF gene from ice plant leaves, and successfully obtained transient *McHB7* OE plants. Under salt stress, ROS and cell death tend to increase in WT. Overexpressing *McHB7* helped to counteract these changes and maintain ROS homeostasis. The results of biochemical and proteomic analyses showed *McHB7* TF exerts a positive effect on the response to high salinity in ice plant. This work provides a quick and effective method for testing gene functions in the native plant species. Identification of genes directly regulated by this salinity responsive TF, as well as its molecular complexes, are interesting future research directions. The findings about *McHB7* may inform molecular breeding and biotechnological efforts towards enhancing crop resilience and productivity.

Supplementary Materials: The following are available online at https://www.mdpi.com/article/10.3390/ijms22126390/s1. Figure S1: Cloning *McHB7* from ice plant leaves, Figure S2: Subcellular localization prediction with CELLO2GO, Figure S3: Leaf fresh weight measurement of WT and *McHB7* OE plants, Figure S4: Physiological parameters in of WT and *McHB7* OE plants, Figure S5: Leaf fresh weight measurement of WT and *McHB7* OE plants, Figure S6: Relative quantification of levels of H2O2 and O_2^- in ice plant leaves based on NBT and DAB staining, Figure S7: Significantly increased proteins in transgenic ice plant leaves under control and salt stress treatment, Figure S8: Identified proteins under control and salt stress conditions, Table S1: List of primers sequences for cloning and quantitative real-time PCR, Table S2: List of phosphorylated proteins, Table S3: Cis-acting elements in the upstream promoter of AtHB7.

Author Contributions: S.C. and T.J. designed research. X.Z. conducted experiments and wrote the manuscript draft. B.T. and D.Z. performed proteomic data analysis. D.D. acquired LC-MS/MS data and analyzed the data. S.C. and T.J. finalized the manuscript for submission. All authors have read and agreed to the published version of the manuscript.

Funding: This work was supported by the University of Florida and China Scholarship Council (201906600017).

Institutional Review Board Statement: Not applicable.

Informed Consent Statement: Not applicable.

Data Availability Statement: The proteomics data have been deposited to the ProteomeXchange Consortium via the PRIDE partner repository with the data set identifier PXD025961 (userID: reviewer_pxd025961@ebi.ac.uk; password: M1rQA2wo).

Acknowledgments: The authors thank Aneirin Lott from Plant Molecular and Cellular Biology Program at University of Florida for critical reading and editing of the manuscript.

Conflicts of Interest: The authors have no conflicts of interest to declare.

References

1. Bohnert, H.J.; Cushman, J.C. The ice plant cometh: Lessons in abiotic stress tolerance. *J. Plant Growth Regul.* **2000**, *19*, 334–346. [CrossRef]
2. Kim, Y.J.; Kim, H.M.; Kim, H.M.; Lee, H.R.; Jeong, B.R.; Lee, H.J.; Kim, H.J.; Hwang, S.J. Growth and phytochemicals of ice plant (*Mesembryanthemum crystallinum* L.) as affected by various combined ratios of red and blue LEDs in a closed-type plant production system. *J. Appl. Res. Med. Aromat. Plants* **2021**, *20*, 100267. [CrossRef]
3. Kim, Y.J.; Kim, H.M.; Kim, H.M.; Jeong, B.R.; Lee, H.J.; Kim, H.J.; Hwang, S.J. Ice plant growth and phytochemical concentrations are affected by light quality and intensity of monochromatic light-emitting diodes. *Hortic. Environ. Biotechnol.* **2018**, *59*, 529–536. [CrossRef]
4. Agarie, S.; Shimoda, T.; Shimizu, Y.; Baumann, K.; Sunagawa, H.; Kondo, A.; Ueno, O.; Nakahara, T.; Nose, A.; Cushman, J.C. Salt tolerance, salt accumulation, and ionic homeostasis in an epidermal bladder-cell-less mutant of the common ice plant *Mesembryanthemum crystallinum*. *J. Exp. Bot.* **2007**, *58*, 1957–1967. [CrossRef] [PubMed]
5. Winter, K.; Holtum, J.A. The effects of salinity, crassulacean acid metabolism and plant age on the carbon isotope composition of *Mesembryanthemum crystallinum* L., a halophytic C_3-CAM species. *Planta* **2005**, *222*, 201–209. [CrossRef] [PubMed]
6. Guan, Q.; Kong, W.; Zhu, D.; Zhu, W.; Dufresne, C.; Tian, J.; Chen, S. Comparative proteomics of *Mesembryanthemum crystallinum* guard cells and mesophyll cells in transition from C_3 to CAM. *J. Proteom.* **2021**, *231*, 104019. [CrossRef]
7. Tsukagoshi, H.; Suzuki, T.; Nishikawa, K.; Agarie, S.; Ishiguro, S.; Higashiyama, T. RNA-seq analysis of the response of the halophyte, *Mesembryanthemum crystallinum* (ice plant) to high salinity. *PLoS ONE* **2015**, *10*, e0118339. [CrossRef]
8. Oh, D.H.; Barkla, B.J.; Vera-Estrella, R.; Pantoja, O.; Lee, S.Y.; Bohnert, H.J.; Dassanayake, M. Cell type-specific responses to salinity–the epidermal bladder cell transcriptome of *Mesembryanthemum crystallinum*. *New Phytol.* **2015**, *207*, 627–644. [CrossRef]
9. Barkla, B.J.; Vera-Estrella, R.; Raymond, C. Single-cell-type quantitative proteomic and ionomic analysis of epidermal bladder cells from the halophyte model plant *Mesembryanthemum crystallinum* to identify salt-responsive proteins. *BMC Plant. Biol.* **2016**, *16*, 110. [CrossRef]

10. Barkla, B.J.; Vera-Estrella, R. Single-cell comparative metabolomics of epidermal bladder cells from the halophyte *Mesembryanthemum crystallinum*. *Front. Plant. Sci.* **2015**, *6*, 435. [CrossRef]
11. Cosentino, C.; di Silvestre, D.; Fischer-Schliebs, E.; Homann, U.; de Palma, A.; Comunian, C.; Mauri, P.L.; Thiel, G. Proteomic analysis of *Mesembryanthemum crystallinum* leaf microsomal fractions finds an imbalance in V-ATPase stoichiometry during the salt-induced transition from C_3 to CAM. *Biochem. J.* **2013**, *450*, 407–415. [CrossRef] [PubMed]
12. Hwang, H.H.; Wang, C.H.; Huang, H.W.; Chiang, C.P.; Chi, S.F.; Huang, F.C.; Yen, H.E. Functional analysis of McSnRK1 (SNF1-related protein kinase 1) in regulating Na/K homeostasis in transgenic cultured cells and roots of halophyte *Mesembryanthemum crystallinum*. *Plant Cell Rep.* **2019**, *38*, 915–926. [CrossRef]
13. Patra, B.; Ray, S.; Richter, A.; Majumder, A.L. Enhanced salt tolerance of transgenic tobacco plants by co-expression of *PcINO1* and *McIMT1* is accompanied by increased level of myo-inositol and methylated inositol. *Protoplasma* **2010**, *245*, 143–152. [CrossRef] [PubMed]
14. Nishijima, T.; Furuhashi, M.; Sakaoka, S.; Morikami, A.; Tsukagoshi, H. Ectopic expression of *Mesembryanthemum crystallinum* sodium transporter *McHKT2* provides salt stress tolerance in *Arabidopsis thaliana*. *Biosci. Biotechnol. Biochem.* **2017**, *81*, 2139–2144. [CrossRef] [PubMed]
15. Kong, W.; Yoo, M.J.; Zhu, D.; Noble, J.D.; Kelley, T.M.; Kirst, M.; Assmann, S.M.; Chen, S. Molecular changes in *Mesembryanthemum crystallinum* guard cells underlying the C_3 to CAM transition. *Plant. Mol. Biol.* **2020**, *103*, 653–667. [CrossRef]
16. Mukherjee, K.; Brocchieri, L.; Bürglin, T.R. A comprehensive classification and evolutionary analysis of plant homeobox genes. *Mol. Biol. Evol.* **2009**, *26*, 2775–2794. [CrossRef]
17. Ariel, F.D.; Manavella, P.A.; Dezar, C.A.; Chan, R.L. The true story of the HD-Zip family. *Trends Plant. Sci.* **2007**, *12*, 419–426. [CrossRef]
18. Li, Y.; Zhu, Y.; Yao, J.; Zhang, S.; Wang, L.; Guo, C.; Nocker, S.; Wang, X. Genome-wide identification and expression analyses of the homeobox transcription factor family during ovule development in seedless and seeded grapes. *Sci. Rep.* **2017**, *7*, 1–16. [CrossRef]
19. Jain, M.; Tyagi, A.K.; Khurana, J.P. Genome-wide identification, classification, evolutionary expansion and expression analyses of homeobox genes in rice. *FEBS J.* **2008**, *275*, 2845–2861. [CrossRef]
20. Mittempergher, J.B.; Morelli, G.; Ruberti, I. A dynamic balance between gene activation and repression. *Genes Dev.* **2005**, *19*, 2811–2815.
21. Nakamura, M.; Katsumata, H.; Abe, M.; Yabe, N.; Komeda, Y.; Yamamoto, K.T.; Takahashi, T. Characterization of the class IV homeodomain-leucine zipper gene family in Arabidopsis. *Plant Physiol.* **2006**, *141*, 1363–1375. [CrossRef] [PubMed]
22. Shen, W.; Li, H.; Teng, R.; Wang, Y.; Wang, W.; Zhuang, J. Genomic and transcriptomic analyses of HD-Zip family transcription factors and their responses to abiotic stress in tea plant (*Camellia sinensis*). *Genomics* **2019**, *111*, 1142–1151. [CrossRef] [PubMed]
23. Wu, X.; Dabi, T.; Weigel, D. Requirement of homeobox gene *STIMPY/WOX9* for Arabidopsis meristem growth and maintenance. *Curr. Biol.* **2005**, *15*, 436–440. [CrossRef] [PubMed]
24. Wang, L.; Li, Z.; Wen, S.; Wang, J.; Zhao, S.; Lu, M. WUSCHEL-related homeobox gene *PagWOX11/12a* responds to drought stress by enhancing root elongation and biomass growth in poplar. *J. Exp. Bot.* **2020**, *71*, 1503–1513. [CrossRef]
25. Liu, R.; Wang, R.; Lu, M.; Wang, L. WUSCHEL-related homeobox gene *PagWOX11/12a* is involved in drought tolerance through modulating reactive oxygen species scavenging in poplar. *Plant Signal. Behav.* **2020**, *16*, 1866312. [CrossRef]
26. Zhu, J.; Shi, H.; Lee, B.; Damsz, B.; Cheng, S.; Stirm, V.; Zhu, J.K.; Hasegawa, P.M.; Bressan, R.A. An Arabidopsis homeodomain transcription factor gene, HOS9, mediates cold tolerance through a CBF-independent pathway. *Proc. Natl. Acad. Sci. USA* **2004**, *101*, 9873–9878. [CrossRef] [PubMed]
27. Yang, Z.; Gong, Q.; Qin, W.; Yang, Z.; Cheng, Y.; Lu, L.; Ge, X.; Zhang, C.; Wu, Z.; Li, F. Genome-wide analysis of WOX genes in upland cotton and their expression pattern under different stresses. *BMC Plant Biol.* **2017**, *17*, 1–17. [CrossRef]
28. Wang, C.F.; Huang, L.L.; Buchenauer, H.; Han, Q.M.; Zhang, H.C.; Kang, Z.S. Histochemical studies on the accumulation of reactive oxygen species (O_2^- and H_2O_2) in the incompatible and compatible interaction of wheat-*Puccinia striiformis* f. sp. tritici. *Physiol. Mol. Plant Pathol.* **2007**, *71*, 230–239. [CrossRef]
29. Cohen, J.A.; Edwards, T.N.; Liu, A.W.; Hirai, T.; Jones, M.R.; Wu, J.; Li, Y.; Zhang, S.; Ho, J.; Davis, B.M.; et al. Cutaneous TRPV1$^+$ neurons trigger protective innate type 17 anticipatory immunity. *Cell* **2019**, *178*, 919–932.e14. [CrossRef]
30. Yang, D.H.; Paulsen, H.; Andersson, B. The N-terminal domain of the light-harvesting chlorophyll a/b-binding protein complex (LHCII) is essential for its acclimative proteolysis. *FEBS Lett.* **2000**, *466*, 385–388. [CrossRef]
31. Yang, Y.; Wang, B.; Fu, Y.; Shi, Y.; Chen, F.; Guan, H.; Liu, L.; Zhang, C.; Zhu, P.; Liu, Y.; et al. HS-GC-IMS with PCA to analyze volatile flavor compounds across different production stages of fermented soybean whey tofu. *Food Chem.* **2021**, *346*, 128880. [CrossRef] [PubMed]
32. Duan, L.; He, Q.; Wang, K.; Yan, X.; Cui, Y.; Möhwald, H.; Li, J. Adenosine triphosphate biosynthesis catalyzed by FoF1 ATP synthase assembled in polymer microcapsules. *Angew. Chem.* **2007**, *119*, 7126–7130. [CrossRef]
33. Wydro, M.; Kozubek, E.; Lehmann, P. Optimization of transient Agrobacterium-mediated gene expression system in leaves of *Nicotiana benthamiana*. *Acta Biochim. Pol.* **2006**, *53*, 289–298. [CrossRef] [PubMed]
34. Hwang, H.H.; Wang, C.H.; Chen, H.H.; Ho, J.F.; Chi, S.F.; Huang, F.C.; Yen, H.E. Effective Agrobacterium-mediated transformation protocols for callus and roots of halophyte ice plant (*Mesembryanthemum crystallinum*). *Bot. Stud.* **2019**, *60*, 1–15. [CrossRef]
35. Mahmood, T.; Yang, P.C. Western blot: Technique, theory, and trouble shooting. *N. Am. J. Med. Sci.* **2012**, *4*, 429.

36. Zhang, T.; Chhajed, S.; Schneider, J.D.; Feng, G.; Silveira, J.A.; Song, W.; Chen, S. Proteomic characterization of MPK4 signaling network and putative substrates. *Plant Mol. Biol.* **2019**, *101*, 325–339. [CrossRef]
37. Zhang, T.; Schneider, J.D.; Lin, C.; Koh, J.; Chen, S. Proteomics data of SNF1-related protein kinase 2.4 interacting proteins revealed by immunoprecipitation-mass spectrometry. *Data Brief* **2020**, *32*, 106326. [CrossRef]
38. Fang, P.; Yan, H.; Chen, F.; Peng, Y. Overexpression of maize *ZmHDZIV14* increases abscisic acid sensitivity and mediates drought and salt stress in Arabidopsis and tobacco. *Plant Mol. Biol. Rep.* **2020**, *39*, 275–287. [CrossRef]
39. Zhang, T.; Lin, C.; Schneider, J.D.; Geng, S.; Ma, T.; Lawrence, S.R.; Dufresne, C.P.; Harmon, A.C.; Chen, S. MPK4 phosphorylation dynamics and interacting proteins in plant immunity. *J. Proteome Res.* **2019**, *18*, 826–840. [CrossRef]
40. Sairam, R.K.; Srivastava, G.C. Changes in antioxidant activity in sub-cellular fractions of tolerant and susceptible wheat genotypes in response to long term salt stress. *Plant Sci.* **2002**, *162*, 897–904. [CrossRef]
41. Hernandez, J.A.; Corpas, F.J.; Gomez, M.; del Rio, L.A.; Sevilla, F. Salt-induced oxidative stress mediated by activated oxygen species in pea leaf mitochondria. *Physiol. Plant.* **1993**, *89*, 103–110. [CrossRef]
42. Fadzilla, N.M.; Finch, R.P.; Burdon, R.H. Salinity, oxidative stress and antioxidant responses in shoot cultures of rice. *J. Exp. Bot.* **1997**, *48*, 325–331. [CrossRef]
43. Esfandiari, E.; Shekari, F.; Shekari, F.; Esfandiari, M. The effect of salt stress on antioxidant enzymes' activity and lipid peroxide ation on the wheat seedling. *Not. Bot. Horti Agrobot. Cluj-Napoca* **2007**, *35*, 48.
44. Baque, M.A.; Hahn, E.J.; Paek, K. Induction mechanism of adventitious root from leaf explants of *Morindacitrifolia* as affected by auxin and light quality. *In Vitro Cell. Dev. Biol. Plant* **2010**, *46*, 71–80. [CrossRef]
45. Kong, W.; Liu, F.; Zhang, C.; Zhang, J.; Feng, H. Non-destructive determination of Malondialdehyde (MDA) distribution in oilseed rape leaves by laboratory scale NIR hyperspectral imaging. *Sci. Rep.* **2016**, *6*, 1–8. [CrossRef]
46. González, L.; González-Vilar, M. Determination of relative water content. In *Handbook of Plant Ecophysiology Techniques*; Springer: Dordrecht, The Netherlands, 2001; pp. 207–212.
47. Darabi, M.; Seddigh, S. Computational study of biochemical properties of Ribulose-1,5-Bisphosphate Carboxylase/Oxygenase (RuBisCO) enzyme in C_3 plants. *J. Plant Biol.* **2017**, *60*, 35–47. [CrossRef]
48. Spreitzer, R.J. *Role of the Rubisco Small Subunit*; University of Nebraska: Lincoln, NE, USA, 2016.
49. Andersson, I.; Backlund, A. Structure and function of Rubisco. *Plant. Physiol. Biochem.* **2008**, *46*, 275–291. [CrossRef] [PubMed]
50. Kaur, A.; Pati, P.K.; Pati, A.M.; Nagpal, A.K. Physico-chemical characterization and topological analysis of pathogenesis-related proteins from *Arabidopsis thaliana* and *Oryza sativa* using in-silico approaches. *PLoS ONE* **2020**, *15*, e0239836. [CrossRef]
51. Verschueren, K.H.G.; Blanchet, C.; Felix, J.; Vos, D.D.; Bloch, Y.; Beeumen, J.V.; Suergun, D.; Gutsche, I.; Savvides, S.N.; Verstraete, K. Structure of ATP citrate lyase and the origin of citrate synthase in the Krebs cycle. *Nature* **2019**, *568*, 571–575. [CrossRef] [PubMed]
52. Terzaghi, W.B.; Cashmore, A.R. Light-regulated transcription. *Annu. Rev. Plant. Biol.* **1995**, *46*, 445–474. [CrossRef]
53. Lam, E.; Chua, N.H. GT-1 binding site confers light responsive expression in transgenic tobacco. *Science* **1990**, *248*, 471–474. [CrossRef]
54. Pang, C.H.; Wang, B.S. Oxidative stress and salt tolerance in plants. In *Progress in Botany*; Springer: Berlin/Heidelberg, Germany, 2008; pp. 231–245.
55. Garretón, V.; Carpinelli, J.; Jordana, X.; Holuigue, L. The as-1 promoter element is an oxidative stress-responsive element and salicylic acid activates it via oxidative species. *Plant Physiol.* **2002**, *130*, 1516–1526. [CrossRef] [PubMed]
56. Tyagi, S.; Sembi, J.K.; Upadhyay, S.K. Gene architecture and expression analyses provide insights into the role of glutathione peroxidases (GPXs) in bread wheat (*Triticum aestivum* L.). *J. Plant Physiol.* **2018**, *223*, 19–31. [CrossRef] [PubMed]
57. Johannesson, H.; Wang, Y.; Hanson, J.; Engström, P. The *Arabidopsis thaliana* homeobox gene ATHB5 is a potential regulator of abscisic acid responsiveness in developing seedlings. *Plant Mol. Biol.* **2003**, *51*, 719–729. [CrossRef] [PubMed]
58. Fujita, M.; Fujita, Y.; Maruyama, K.; Seki, M.; Hiratsu, K.; Ohme-Takagi, M.; Tran, L.P. Yamaguchi-Shinozaki, K.; Shinozaki, K. A dehydration-induced NAC protein, RD26, is involved in a novel ABA-dependent stress-signaling pathway. *Plant J.* **2004**, *39*, 863–876. [CrossRef]
59. Leung, J.; Merlot, S.; Giraudat, J. The Arabidopsis ABSCISIC ACID-INSENSITIVE2 (*ABI2*) and *ABI1* genes encode homologous protein phosphatases 2C involved in abscisic acid signal transduction. *Plant Cell* **1997**, *9*, 759–771.
60. Yoshida, T.; Fujita, Y.; Sayama, H.; Kidokoro, S.; Maruyama, K.; Mizoi, J.; Shinozaki, K.; Yamaguchi-Shinozaki, K. *AREB1*, *AREB2*, and *ABF3* are master transcription factors that cooperatively regulate ABRE-dependent ABA signaling involved in drought stress tolerance and require ABA for full activation. *Plant J.* **2010**, *61*, 672–685. [CrossRef]
61. Söderman, E.; Mattsson, J.; Engström, P. The Arabidopsis homeobox gene *ATHB-7* is induced by water deficit and by abscisic acid. *Plant J.* **1996**, *10*, 375–381. [CrossRef]
62. Ré, D.A.; Capella, M.; Bonaventure, G.; Chan, R.L. Arabidopsis *AtHB7* and *AtHB12* evolved divergently to fine tune processes associated with growth and responses to water stress. *BMC Plant Biol.* **2014**, *14*, 1–14. [CrossRef]
63. Valdés, A.E.; Övernäs, E.; Johansson, H.; Rada-Iglesias, A.; Engström, P. The homeodomain-leucine zipper (HD-Zip) class I transcription factors *ATHB7* and *ATHB12* modulate abscisic acid signalling by regulating protein phosphatase 2C and abscisic acid receptor gene activities. *Plant Mol. Biol.* **2012**, *80*, 405–418. [CrossRef]
64. Willems, E.; Leyns, L.; Vandesompele, J. Standardization of real-time PCR gene expression data from independent biological replicates. *Anal. Biochem.* **2008**, *379*, 127–129. [CrossRef] [PubMed]

65. Ausubel, F.M.; Brent, R.; Kingston, R.E.; Moore, D.D.; Smith, J.A.; Seidman, J.G.; Smith, J.A.; Struhl, K. Preparation and Analysis of DNA: Preparation of Genomic DNA from Bacteria. In *Current Protocols in Molecular Biology*; John Wiley Sons Inc.: Hoboken, NJ, USA, 2003.
66. Gallo-Oller, G.; Ordonez, R.; Dotor, J. A new background subtraction method for Western blot densitometry band quantification through image analysis software. *J. Immunol. Methods* **2018**, *457*, 1–5. [CrossRef] [PubMed]
67. Yao, W.; Wang, S.; Zhou, B.; Jiang, T. Transgenic poplar overexpressing the endogenous transcription factor *ERF76* gene improves salinity tolerance. *Tree Physiol.* **2016**, *36*, 896–908. [CrossRef]
68. Jambunathan, N. Determination and detection of reactive oxygen species (ROS), lipid peroxidation, and electrolyte leakage in plants. In *Plant Stress Tolerance*; Humana Press: Totowa, NJ, USA, 2010; pp. 291–297.
69. Huang, X.S.; Luo, T.; Fu, X.Z.; Fan, Q.J.; Liu, J.H. Cloning and molecular characterization of a mitogen-activated protein kinase gene from *Poncirus trifoliata* whose ectopic expression confers dehydration/drought tolerance in transgenic tobacco. *J. Exp. Bot.* **2011**, *62*, 5191–5206. [CrossRef]
70. Cheng, Z.; Zhang, X.; Zhao, K.; Zhou, B.; Jiang, T. Ectopic expression of a poplar gene *NAC13* confers enhanced tolerance to salinity stress in transgenic *Nicotiana tabacum*. *J. Plant Res.* **2020**, *133*, 727–737. [CrossRef] [PubMed]

International Journal of
Molecular Sciences

MDPI

Article

Quantitative Proteomics Reveals the Dynamic Regulation of the Tomato Proteome in Response to *Phytophthora infestans*

Kai-Ting Fan [1,†], Yang Hsu [1,†], Ching-Fang Yeh [1], Chi-Hsin Chang [1,2,3], Wei-Hung Chang [1] and Yet-Ran Chen [1,2,3,*]

1. Agricultural Biotechnology Research Center, Academia Sinica, Taipei 11529, Taiwan; kaitingfan@sinica.edu.tw (K.-T.F.); sheuyoung@gmail.com (Y.H.); cfyeh@gate.sinica.edu.tw (C.-F.Y.); as0190450@gate.sinica.edu.tw (C.-H.C.); whchang@gate.sinica.edu.tw (W.-H.C.)
2. Molecular and Biological Agricultural Sciences Program, Taiwan International Graduate Program, Academia Sinica, Taipei 11529, Taiwan
3. Graduate Institute of Biotechnology, National Chung Hsing University, Taichung 402, Taiwan

* Correspondence: yetran@gate.sinica.edu.tw; Tel.: +886-02-2787-2050

† Contributed equally to this work.

Abstract: Late blight (LB) disease is a major threat to potato and tomato production. It is caused by the hemibiotrophic pathogen, *Phytophthora infestans. P. infestans* can destroy all of the major organs in plants of susceptible crops and result in a total loss of productivity. At the early pathogenesis stage, this hemibiotrophic oomycete pathogen causes an asymptomatic biotrophic infection in hosts, which then progresses to a necrotrophic phase at the later infection stage. In this study, to examine how the tomato proteome is regulated by *P. infestans* at different stages of pathogenesis, a data-independent acquisition (DIA) proteomics approach was used to trace the dynamics of the protein regulation. A comprehensive picture of the regulation of tomato proteins functioning in the immunity, signaling, defense, and metabolism pathways at different stages of *P. infestans* infection is revealed. Among the regulated proteins, several involved in mediating plant defense responses were found to be differentially regulated at the transcriptional or translational levels across different pathogenesis phases. This study increases understanding of the pathogenesis of *P. infestans* in tomato and also identifies key transcriptional and translational events possibly targeted by the pathogen during different phases of its life cycle, thus providing novel insights for developing a new strategy towards better control of LB disease in tomato.

Keywords: plant pathogenesis responses; quantitative proteomics; data-independent acquisition; *Phytophthora infestans*

Citation: Fan, K.-T.; Hsu, Y.; Yeh, C.-F.; Chang, C.-H.; Chang, W.-H.; Chen, Y.-R. Quantitative Proteomics Reveals the Dynamic Regulation of the Tomato Proteome in Response to *Phytophthora infestans*. *Int. J. Mol. Sci.* **2021**, *22*, 4174. https://doi.org/10.3390/ijms22084174

Academic Editor: Setsuko Komatsu

Received: 7 March 2021
Accepted: 14 April 2021
Published: 17 April 2021

Publisher's Note: MDPI stays neutral with regard to jurisdictional claims in published maps and institutional affiliations.

1. Introduction

Late blight (LB), caused by the oomycete *Phytophthora infestans*, is one of the most notorious plant diseases to afflict solanaceous plants [1]. LB is difficult to control and causes severe losses of up to 100% in the production of potato and tomato crops [2]. LB was the major factor contributing to severe crop loss in Ireland in early 1840, causing the Great Famine and resulting in the population of the area falling by 20–25%. Until now, this disease remains one of the biggest threats to tomato and potato production. In the United States, LB has caused up to 7% yield losses in tomato crops and approximately 3.5% yield losses in potato crops over the last two decades [2]. The worldwide economic losses due to LB and the cost of controlling this disease are estimated to exceed $5 billion in tomato crops [3] and about $7 billion in potato crops [2] annually.

P. infestans destroys all the leaves, roots, tubers, and fruit of susceptible plants [4]. After infection of susceptible hosts by this pathogen, local lesions pervade the whole plant within a few days, eventually causing the death of the plant. The sporangia produced on an infected plant organ can be effectively dispersed by wind or by splashes from raindrops [5]. *P. infestans* exhibits a two-phase life cycle; it is initially hemibiotrophic, showing an asymptomatic biotrophic phase of infection. This is followed by a necrotrophic phase that is characterized by the degradation of the host tissue [6]. During the biotrophic stage, *P. infestans* forms appressoria, primary and secondary hyphae and a specialized structure called haustoria within living plant cells and obtains nutrients without killing the plant [7]. In this stage of pathogenesis, *P. infestans* suppresses the plant immune system and the programmed cell death (PCD) responses that limit biotrophic infection. At the necrotrophic stage, the pathogen hyphae ramify throughout the plant and toxins from the pathogen are injected into the host cells causing necrosis of the infected tissue [8].

To gain a comprehensive insight into the dynamic plant molecular mechanisms that respond to the hemibiotrophic infection of *P. infestans*, several time-course quantitative transcriptomics studies have been performed on tomato or potato infected by this pathogen [9–11]. In potato, massive transcriptional reprogramming was observed at the end of the early biotrophic infection stage but the pathogenesis-related (PR) and hypersensitive response (HR) genes were mainly upregulated at the later necrotrophic stage [9]. Using microarray analysis, genes encoding transcription factors and components of signaling pathways were shown to be upregulated by *P. infestans* in a resistant tomato cultivar, but suppressed in a susceptible one [10]. Due to the accuracy of the microarray technique, five genes considered to be involved in plant resistance in the microarray-based transcriptomics study were validated by real-time quantitative reverse transcription polymerase chain reaction (qRT-PCR) analysis: endochitinase 3 precursor, glyceraldehyde 3-phosphate dehydrogenase B, probable GST, carbonic anhydrase, and basic glucanase. More recently, a more accurate transcriptomics analysis using the next-generation sequencing technique was performed to observe the transcriptomic change in tomato from the well-defined biotrophic to the necrotrophic pathogenesis stages of *P. infestans* infection [11].

Furthermore, proteomics was used to investigate the pathogenic regulation dynamics of the plant proteome by this pathogen. The interaction of potatoes and *P. infestans* was examined by quantifying the changes in the potato proteome after infection, both in a temporal manner and between the resistant and susceptible cultivars [12–14]. These studies provided further direct evidence about the production of key proteins that can directly interact with this pathogen and proceed with the metabolic reaction for the defense metabolites or energy homeostasis. However, to date, there has been no comprehensive study reporting the dynamics of the tomato proteome in response to *P. infestans*. In a previous proteomic study that used two-dimensional electrophoresis (2-DE) and matrix-assisted laser desorption/ionization time-of-flight/time-of-flight (MALDI-TOF/TOF) for protein identification, only 56 proteins were identified in the proteome of resistant and susceptible tomato cultivars in response to *P. infestans* infection [15].

In this study, a data-independent acquisition (DIA) proteomics approach was applied to reveal the dynamics of the proteome change in tomato from the early (biotrophic phase) to the late stage (necrotrophic phase) of *P. infestans* infection. By establishing a DIA spectra library from two-dimensional-liquid chromatography with tandem mass spectrometry (LC-MS/MS) analysis operated in the data-dependent acquisition (DDA) mode, more proteins were identified compared to the use of a spectral library established by single-dimensional LC-MS/MS. This approach enabled the profiling of the tomato proteome in a high-throughput manner, and can potentially be used to facilitate the analysis of the dynamics of different pathogenic responses in plants.

2. Results and Discussion

2.1. *P. infestans* Pathogenesis Assessment in Tomato Leaves

A time-series phenotypic assessment of the leaf from a susceptible tomato (*Solanum lycopersicum*) cultivar (CL5915) challenged with *P. infestans* was subjected to microscopic and macroscopic examination (Figure 1A,B). Leaflets detached from 35-d-old tomato plants were inoculated with droplets of *P. infestans* sporangia suspension and collected at 6, 16, 24, 48, 96, and 120-h post-inoculation (hpi). Three biological replicates were used at each time point in the phenotypic assessment. Because there was no obvious disease phenotype at the early time points, the pathogen development on leaflets at 6, 16, and 24 hpi were stained with trypan blue. After the staining, the growth phenotype of the pathogen was recorded by picking thirty *P. infestans* spores randomly. Microscopy showed that 20% (6/30) sporangia started to germinate and 80% (24/30) sporangia remained at the cyst stage at 6 hpi (Figure 1A). At 16 hpi, about 90% of the resting cysts of *P. infestans* germinated and developed geminating tubes, and only ~3% (1/30) of the sporangia remained as cysts. The appressorium structures were first observed at 16 hpi and accounted for only ~7% (2/30) of the total cysts. At 24 hpi, appressoria appeared in ~73% (22/30) of the observed cysts, and primary hyphae could be observed extending through the host tissue. No disease symptoms on the leaf were observed macroscopically at 24 hpi and only little inconspicuous black spots were discovered at 48 hpi (Figure 1B). Water-soaked lesions were not obvious until 96 hpi and the average area of the lesions observed at this time point was 0.71 ± 0.12 cm^2. At 120 hpi, the lesion area had covered over 80% of the leaf surfaces and aerial mycelium were visually observable surrounding the center of the droplets, indicating a well-established necrotrophic phase.

To better assess the pathogenesis phase at the molecular level, the expressions of *P. infestans* genes that mark different pathogenesis stages were analyzed by reverse transcription-polymerase chain reaction (RT-PCR) in the leaflets at 24, 48, 96, and 120 hpi (Figure 1C and Figure S1). Two *P. infestans* RXLR effector genes *SNE1* and *IpiO* (also named *Avr-blb1*) were used as marker genes for infection at the biotrophic stage. *SNE1* has been reported to be specifically expressed at the transcriptional level during the biotrophic stage in infected tomato [16], and *IpiO* has been detected in the early stage of infection, but not in sporangia or old mycelia [17]. Another gene *PiNPP1.1* which encodes a cell death-inducing protein and thus marks the necrotrophic stage of infection [18] was used in this study. The expression of *SEN1* and *IpiO* was detected from the early stage of infection (24 hpi) and continually accumulated during the biotrophic phase until 96 hpi. The expression of both biotrophic markers decreased at 120 hpi, suggesting that the pathogen transited from biotrophy to necrotrophy at 96 hpi. In contrast, *PiNPP1.1* expression was detected at low abundance until 96 hpi and was mainly expressed at the later stages of the interaction. The expression of *P. infestans* actin was used to evaluate pathogen growth. Expression of this gene was continuously increased from 24 hpi to 120 hpi. Taken together, *P. infestans* exhibited a biotrophic invasion in tomato cv. CL5915 at 24 hpi by initiating the penetration. The time points 48 and 96 hpi represented the transition phase (the change from biotrophic to necrotrophic growth), and the necrotrophic phase, respectively.

2.2. Proteomics Analysis and Quantification Using the DIA Approach

To study the plant pathogenesis responses at the protein level across the early biotrophic to transition and later necrotrophic phases, tomato leaflets at 24, 48, and 96 hpi with *P. infestans* were subjected to proteomic analysis using data-independent acquisition (DIA). A DIA spectral library containing 11,563 proteins and 65,763 peptide transition groups from tomato and *P. infestans* was constructed by the analysis of the pooled tryptic peptide samples with or without fractionation by high-pH reversed-phase liquid chromatography (RPLC) using LC-MS/MS operated in data-dependent acquisition (DDA) mode. A total of 7324 proteins, 6631 tomato proteins, 678 *P. infestans* proteins (with at least 1 unique peptide), and 15 protein groups shared by tomato and *P. infestans* were identified in the DIA proteomics analysis of the time-series experiments. About 67% of tomato proteins were identified in samples collected at all time points (Supplemental Tables S1 and S2) and ~4–11% of the proteins were exclusively identified at a single time point (Figure 2A). To identify the tomato proteins regulated by this pathogen, the abundance of each protein in the plants inoculated with the pathogen was compared to the abundance in the mock-treated plants at the same time point. An average fold change of protein abundance in the pathogen and mock-treated plants from three biological replicates of ≥ 1.5 or ≤ 0.67 with significance ($p < 0.05$) was considered as up- or downregulated by *P. infestans*, respectively (Figure 2B). In all, 3165, 3345 and 2973 tomato proteins were quantified at 24, 48, and 96 hpi, respectively. In the quantitative analysis, 8, 21, and 270 tomato proteins were determined to be upregulated at 24, 48 and 96 hpi, respectively. In addition, 4, 13, and 110 tomato proteins were found to be downregulated at 24, 48 and 96 hpi, respectively. No protein was found to have differential up- or downregulation at all the time points and only ~4% of upregulated and ~2% of downregulated proteins were identified at both 48 and 96 hpi. The data, therefore, showed that the tomato proteome was barely changed during the biotrophic stage of infection at 24 hpi; in contrast, only certain proteins were quantified as regulated at the transition stage at 48 hpi; and more than 10% of the quantified proteins were regulated at the late necrotrophic stage at 96 hpi.

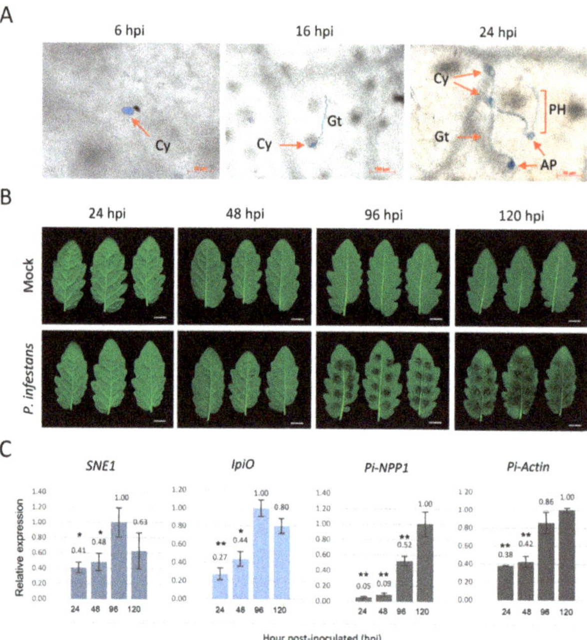

Figure 1. Inoculation of detached tomato leaflets by *P. infestans*. Detached leaflets of five-week-old tomato plants were inoculated with *P. infestans* by placing a 20 μL drop of inoculum at eight spots on the abaxial surface of the leaflets. (**A**) Microscopic assessment of *P. infestans* developing on tomato leaflets at different time points post-inoculation after staining with trypan blue. At 6-h post-inoculation (hpi), the pathogens remained as cysts and started to germinate at 16 hpi. At 24 hpi, the primary hyphae (PH) were developed and the appressorium (Ap) structure was generated, indicating that the pathogen was ready to invade the plant cells. (**B**) Example leaflets showing *P. infestans* lesion development at 24, 48, 96 and 120 hpi. The white bar represents 1 cm. (**C**) Bar graphs showing the relative band intensity of reverse transcription-polymerase chain reaction (RT-PCR) products of *P. infestans* marker genes, *SNE1*, *IpiO*, and *PiNPP1.1* is shown in Figure S1. The RNA samples were extracted from tomato leaflets at 24, 48, 96, and 120 hpi. The relative transcript abundance is shown as a proportion of the most intense band in each gel image. The expression level of *P. infestans* actin was used to evaluate the biomass of the pathogen in tomato leaves. The expression levels of biotrophic stage markers (*SNE1* and *IpiO*) were reduced after 96 hpi, and the necrotrophic stage marker (*PiNPP1.1*) was continuously increased from 24 hpi to 120 hpi. The error bars are standard deviations and the graph represents the combined data from three biological replicates (n = 3). Differentially regulated genes with a p-value of less than 0.05 or a p-value of less than 0.01 are marked with single or double asterisks, respectively. Cy, cyst; Gt, germ tube; AP, appressorium; PH, primary hyphae.

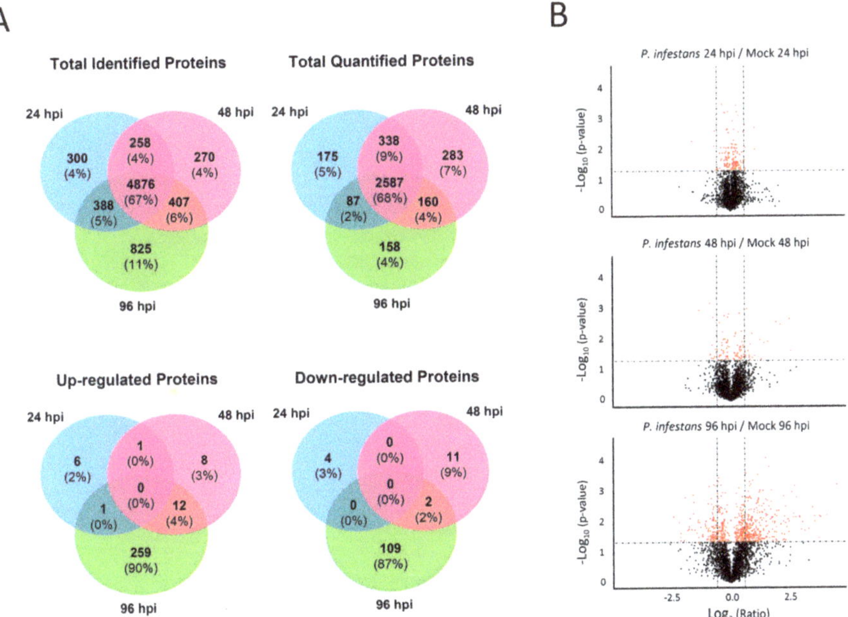

Figure 2. Proteome changes in tomato leaf between 24, 48 and 96 h post-inoculation with *P. infestans* compared to post-mock-inoculation at 24, 48, and 96 h, respectively. (**A**) Venn diagrams showing the unique and shared tomato or *P. infestans* proteins identified, quantified in three biological replicates, or showing a significant difference in abundance with a fold change greater than 1.5 or less than 0.67 in quantity ($p < 0.05$) due to inoculation. (**B**) Volcano plots showing the protein abundance ratio of *P. infestans*-inoculated over the mock group at 24, 48 and 96 hpi. Following liquid chromatography-mass spectrometry (LC-MS) analysis and data-independent acquisition (DIA) quantification, *t*-test-based significance values ($\log_{10}$ (*p*-value)) were plotted versus $\log_2$ (protein quantity ratio for all proteins between infected and mock). Differentially regulated proteins with $p < 0.05$ are plotted in red. A level of protein abundance change of 1.5 or 0.67-fold is marked with a dashed line.

2.3. Functional Classification of the Proteins Regulated by P. infestans

To understand the implications of the differentially regulated proteins in this study, the quantified proteins were subjected to functional categorization analysis using Gene Ontology (GO). In this analysis, we used The Arabidopsis Information Resource (TAIR) database combined with Protein Basic Local Alignment Search Tool (BLASTP) to identify the protein functions that showed a significant change in the *P. infestans*-treated and mock-treated samples. The upregulated and downregulated proteins were classified into 15 and 10 enriched categories, respectively (Figure 3). At 24 hpi, the enriched functional categories for the upregulated proteins were "single-organism metabolic process" and "single-organism cellular process"; the downregulated proteins were "response to other organism", "response to external stimulus", "response to biotic stimulus", and "response to abiotic stimulus". At 48 hpi, the enriched functional categories for the upregulated proteins were "developmental process involved in reproduction", "response to other organism", "response to external stimulus" and "response to biotic stimulus" but there were no enriched categories for the downregulated proteins. At 96 hpi, the enriched functional categories for the upregulated proteins were "catabolic process", "immune response", "interspecies interaction between organisms", "macromolecule localization", "multi-multicellular organism process", "protein folding", "response to abiotic stimulus", "response to biotic stimulus", "response to chemical", "response to external stimulus", "response to other organism", "response to stress", "single-organism cellular process" and "single-organism metabolic process", and the enriched functional categories for the downregulated proteins were "biosynthetic process", "cellular component biogenesis", "cellular metabolic process", "methylation", "nitrogen compound metabolic process" and "single-organism metabolic process".

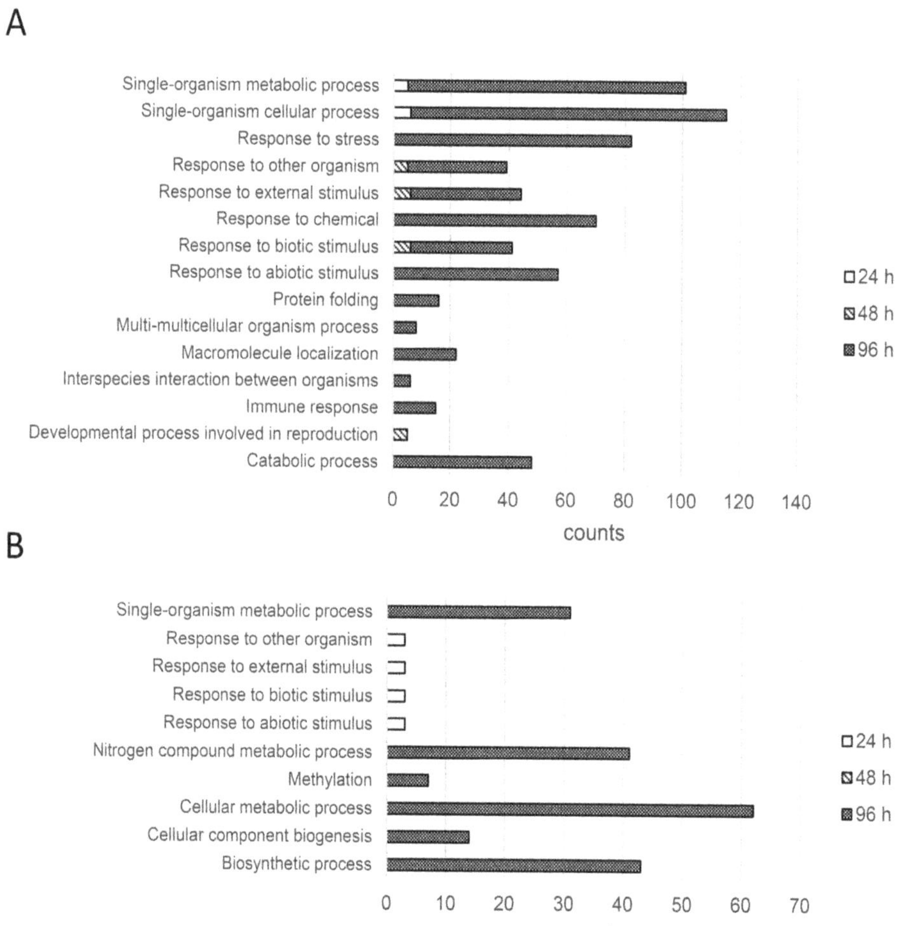

Figure 3. Functional analysis of biological processes for tomato proteins (**A**) upregulated or (**B**) downregulated at 24, 48, or 96 h post-inoculation with *P. infestans*. Tomato proteins up- or downregulated by *P. infestans* were matched to the Arabidopsis homolog proteins by Protein Basic Local Alignment Search Tool (BLASTP) against The Arabidopsis Information Resource (TAIR) database. The counts of these Arabidopsis homolog proteins were compared at different time points in the enriched function group with Gene Ontology (GO) categorization using the Database for Annotation, Visualization and Integrated Discovery (DAVID) v.6.8 ($p < 0.05$).

The upregulated proteins that are pathogen resistance-associated fall into enriched function categories like "immune response", "interspecies interaction between organisms", "response to biotic and abiotic stimulus", "response to external stimulus", "response to other organism", and "response to stress", suggesting that these functions in the plant were positively regulated during the tomato-*P. infestans* interaction. Conversely, proteins related to the metabolic pathways or cellular biosynthesis processes were mostly downregulated, suggesting that metabolic flux may be altered by the attack of this pathogen.

2.4. Changes Associated with Direct Defense

Tomato proteins involved in different pathogen-prohibiting functions were found to be regulated by *P. infestans* (Table 1 and Table S1). Three tomato beta-glucosidases (Solyc01g008620, Solyc10g079860, and a protein group of Solyc01g059965 and Solyc01g060020) for hydrolyzing the major constitutes (1→3)-β-D-glucans, (1→6)-β-D-glucans and cellulose of the oomycete cell wall [19] were observed to be regulated by *P. infestans*. Among these beta-glucosidases, the expression of one of the protein groups (Solyc01g059965 and Solyc01g060020) that encodes a putative beta-1,3-endoglucanase sequence was downregulated with a fold change of 0.62 at 24 hpi, but upregulated with a fold change of 3.67 at 96 hpi. Two other quantified beta-glucosidases Solyc01g008620 and Solyc10g079860 did not have a significant change in abundance at 24 hpi, but were increased by 5.91 and 16.62-fold, respectively, at 96 hpi. Four proteins with chitinase activity (Solyc10g055800, Solyc04g072000, Solyc10g055810, Solyc02g082920) for hydrolyzing the chitin of the fungal cell wall were all found to be upregulated only at 96 hpi with a fold change from 3.52 to 7.90. PR-5 protein (Solyc08g080670), which can bind to the mannose phosphate groups of the fungal cell walls and exhibit a broad spectrum in fungal resistance [20], was downregulated with a fold change of 0.32 at 24 hpi but was not regulated at later time points. The PR-5 family proteins (also called thaumatin-like proteins) which include the closely related proteins permatin, osmotin and zeamatin, may cause fungal hyphae leak and rupture likely via disrupting the fungal membrane permeabilization [21] and overexpressing PR-5 causes enhanced resistance against necrotrophic fungus in multiple plant species [22]. In addition, several proteins that may have roles in defense but for which direct studies to examine their biological functions are lacking were also found to be regulated. One PR-10 family protein named norcoclaurine synthase (NCS; Solyc07g005370) was identified to have a fold change of 0.59 at 96 hpi. This protein belongs to the Bet v 1 family which have RNase activity [23,24] and cysteine protease inhibition activities [25], and participate in anti-bacterial, anti-fungal, anti-viral, and anti-nematode activity. However, three other PR-10 family proteins with accession numbers Solyc09g090990, Solyc09g091000, and Solyc09g090980 showed a 6.26, 10.98, and 22.30-fold increase at 96 hpi, respectively. Two of them Solyc09g090990 and Solyc09g090980 were also, respectively, 5.45 and 2.31-fold upregulated at 48 hpi. PR-4 (Solyc01g09724) and PR-4b (Solyc01g097280) proteins which are bifunctional enzymes with both RNase and DNase activity and involved in regulating HR [26] were also shown to be increased at 96 hpi by 2.14 to 3.0-fold.

Of the defense proteins identified to be regulated by *P. infestans*, the beta-glucosidases and PR-5 protein involved in the disruption and blocking of the pathogen cell wall were repressed at the early stage of the pathogenesis. This repression may facilitate the early development of the pathogen, and establish successful penetration into plant cells at the biotrophic phase. In the *P. infestans*-resistant potato cultivar, endo-1,3-beta-glucosidase was upregulated as early as 6 hpi but not in the susceptible potato cultivar [13]. Furthermore, the previously claimed resistant potato cultivar became susceptible to another *P. infestans* isolate that caused severe lesion areas after inoculation had upregulated endo-1,3-beta-glucosidase at a later time point [14]. Therefore, the early induction of these hydrolytic enzymes in the resistance cultivar could be the key to plant resistance to this pathogen. In our study, tomato beta-glucosidases were not regulated by *P. infestans* at the early time point, which may be related to the plant susceptibility for this pathogen. However, at the later time point after *P. infestans*-inoculation, tomato proteins related to the disruption of the glucans were highly expressed. This may reflect enhanced plant resistance by the degradation of the hyphal cell wall of the oomycete, which not only renders it susceptible to cell lysis but also releases beta-1,3-glucans which serve as elicitors [27] to initiate a wide range of localized and systemic defense responses. Moreover, the proteins involved in regulating HR were induced at 96 hpi; this response may contribute to the triggering of rapid cell death in the local region surrounding the infection area to prevent the spread of the pathogen.

2.5. Changes Associated with Immune Regulation

Several tomato proteins involved in regulating immune responses were identified to have an abundance change in response to the pathogenesis of *P. infestans* (Table 2 and Table S1). Three PR-1 family proteins (Solyc00g174340, Solyc09g007010, and Solyc01g106620) were upregulated at 96 hpi with fold-changes from 4.91 to 7.47. The protein encoded by the gene Solyc00g174340, named pathogenesis-related protein 1b (PR-1b), is the precursor of the peptide hormone, cysteine-rich secretory protein, antigen 5 and PR-1 (CAP)-derived peptide 1 (CAPE1). PR-1b was upregulated at 48 and 96 hpi with a fold change of 1.51 and 7.47, respectively. CAPE1 derived from PR-1b has been demonstrated to activate tomato antiherbivore and antipathogen responses via jasmonic acid (JA) and salicylic acid (SA)-regulated responses [28]. The production of CAPE1 has been suggested to balance excessive JA production and facilitate the biosynthesis of SA in plants [29]. On the other hand, the sterol-binding activity of tobacco PR-1b contributes to the depletion of the lipids required for oomycete *Phytophthora brassicae* to inhibit the growth of this pathogen [30]. Several proteins involved in the effector-triggered immunity (ETI)-triggered programmed cell death (PCD) were all upregulated at 96 hpi, including phytophthora-inhibited protease 1 (PIP1; Solyc02g077040) with a 3.0-fold change, hypersensitive-induced response protein (HIR; Solyc06g071050) with 1.61-fold change and response to desiccation 21a (RD21a; Solyc04g078540) with 1.66-fold change. PIP1 is a tomato apoplastic cysteine protease that triggers cell death and fungal resistance [31]. The enzyme activity of PIP1 has been shown to be inhibited by an effector EPIC2B produced from *P. infestans* [32]. The function of HIR protein is associated with HR and cell death, positively contributing to the inhibition of hemibiotrophs [33,34]. HIR proteins have been shown to help the recognition of the *Pseudomonas syringae* effector AvrRpt2 by interacting with the R protein *Pseudomonas syringae* 2 (RPS2) via increasing the local concentration of RPS2 at plasma membrane microdomains [34]. RD21a contributes to the resistance to necrotrophic fungi in Arabidopsis and functions as a PCD-promoting protease which is released from the ER body or vacuoles to the cytoplasm [35]. This protein has also been shown to control stomata closure and its degradation by ubiquitin E3 ligase SINAT4 was recently found to be triggered by the bacterial type III effector AvrRxo1 [36].

Among all the quantified proteins, carbonic anhydrase (CA; Solyc02g067750) which is essential for various biological processes including stomatal aperture, respiration, pH regulation, and CO_2 homeostasis in plants [37] was upregulated with the greatest fold change at 96 hpi (~320-fold change). This protein is involved in attenuating flg22-triggered immunity [38], and functions in the *Pto:avrPto* mediated-HR in plant-*P. synrigae* interaction [39]. The gene expression of *CA* was more suppressed in the compatible interaction than the incompatible interaction between *P. infestans* and potato (*Solanum tuberosum*) [40]. Silencing this plastic *CA* gene results in higher susceptibility to *P. infestans* in tobacco (*Nicotiana benthamiana*) than the wild type [40].

Being a more general immune regulator, enhanced disease susceptibility 1 (EDS1; Solyc06g071280) was 1.91-fold upregulated at 96 hpi. EDS1 functions as a SA regulator and key ETI mediator for several toll-interleukin-1 receptor (TIR) NB-LRR resistance proteins against virulent pathogens in different plant species [41–45]. This protein enhances the SA-dependent defense response against the oomycete *Phytophthora parasitica* [44,46]. It has been proposed that EDS1 and SA signaling pathways are operated in parallel in defense responses. The EDS1-phytoalexin deficient4 (PAD4) signaling is distinct from the known SA-compensatory route involving MAPK signaling and can regulate both SA-dependent and SA-independent gene expression sectors [47]. In contrast to proteins initiating the ETI-triggered PCD, Kunitz-type protease inhibitor (KTI; Solyc03g098730) which antagonizes the pathogen-associated PCD [48] was 1.91-fold upregulated at 96 hpi of *P. infestans*. The expression of *KTI* is induced by SA and some PCD-eliciting toxins from necrotrophic fungi [48] and knocking out the *KTI* gene causes more H_2O_2 accumulation in plants [49].

Table 1. Proteins involved in defense mechanisms with a significant change in abundance at 24, 48 and 96 h post-inoculation with *P. infestans*.

			24 h			48 h			96 h		
Gene Accession	Protein Description	Arabidopsis Homolog	# Rep [a]	log_2 Ratio [b]	p [c]	# Rep [a]	log_2 Ratio [b]	p [c]	# rep [a]	log_2 Ratio [b]	p [c]
Solyc10g055800	chitinase	AT3G12500	3	−0.121	0.804	3	0.454	0.347	3	1.816	0.023
Solyc02g082920	class II chitinase	AT3G12500	3	0.140	0.326	3	0.556	0.009	3	2.982	0.019
Solyc01g008620	glucan endo-1,3-beta-glucosidase	AT4G16260	3	0.072	0.133	3	−0.022	0.933	3	2.563	0.001
Solyc10g079860	LEQB L.esculentum TomQ'b beta (1,3) glucanase	AT3G57270	3	0.089	0.496	3	0.612	0.344	3	4.055	0.007
Solyc09g090990	major allergen Pru ar.1 (PR-10 family)	AT5G45860	3	0.755	0.162	3	2.446	0.002	3	4.479	0.001
Solyc07g005370	norcoclaurine synthase (NCS; PR-10/Bet v 1 family)	AT2G26040	3	0.135	0.642	3	0.375	0.153	3	−0.756	0.033
Solyc08g080670	pathogenesis-related 5-like protein (PR-5)	AT4G11650	3	−1.633	0.005	3	−0.298	0.623	3	0.394	0.436
Solyc01g097240	pathogenesis-related protein 4 (PR-4)	AT3G04720	3	0.199	0.027	3	0.522	0.058	3	3.745	0.007
Solyc09g090980	pathogenesis-related protein STH-2-like (PR-10 family)	AT1G24020	3	0.185	0.009	3	1.206	0.010	3	3.456	0.004

[a] Number of biological replicates in which the protein was quantifiable. [b] The average log_2 ratio of protein quantity representing (inoculated/mock) from three biological replicates. [c] The *p*-value was calculated by Student's *t*-test if the protein could be quantified from three biological replicates. Color codes: red, significant quantity change with greater than 0.58 of log_2 ratio ($p < 0.05$); blue, significant quantity change less than −0.58 of log_2 ratio; grey, no significant change between the inoculated and mock group ($p \geq 0.05$) or not quantified in all three replicates (NQ); white, the quantity change is between 0.58 and −0.58 of log_2 ratio ($p < 0.05$).

Table 2. Proteins involved in immune regulation with a significant change in abundance at 24, 48 and 96 h post-inoculation with *P. infestans*.

			24 h			48 h			96 h		
Gene Accession	Protein Description	Arabidopsis Homolog	# Rep [a]	log_2 Ratio [b]	p [c]	# Rep [a]	log_2 Ratio [b]	p [c]	# Rep [a]	log_2 Ratio [b]	p [c]
Solyc02g077880	auxin-repressed protein (ARP)	AT2G33830	3	−0.404	0.202	3	−1.360	0.072	3	−2.441	0.007
Solyc03g033790	P-loop containing nucleoside triphosphate hydrolases superfamily protein	AT3G50930	1	0.083	NQ	2	0.304	NQ	3	1.340	0.021
Solyc06g068840	calcium-dependent phospholipid-binding copine family protein	AT5G61900	2	0.262	NQ	3	0.534	0.181	3	1.015	0.030
Solyc02g067750	carbonic anhydrase (CA)	AT3G01500	3	1.884	0.719	3	−1.904	0.374	3	8.330	0.012
Solyc04g078540	cysteine proteinase RD21a	AT1G47128	3	0.331	0.102	3	−0.029	0.888	3	0.716	0.036
Solyc06g071280	enhanced disease susceptibility 1 (EDS1)	AT3G48090	3	−0.015	0.100	3	0.092	0.594	3	0.930	0.037
Solyc06g071050	hypersensitive-induced response protein (HIR)	AT5G62740	2	−0.334	NQ	3	−0.430	0.525	3	0.686	0.016
Solyc03g098730	Kunitz-type protease inhibitor (KTI)	AT1G17860	3	−0.607	0.220	3	0.076	0.744	3	2.740	0.008
Solyc07g043320	myosin heavy chain, embryonic smooth protein	AT2G32240	3	0.129	0.276	3	0.578	0.188	3	1.050	0.030
Solyc00g174340	pathogenesis-related protein 1b (PR-1b)	AT4G33720	3	0.058	0.484	3	0.591	0.025	3	2.902	0.008
Solyc09g007010	pathogenesis-related protein 1 (PR-1)	AT4G33720	3	0.097	0.354	3	0.411	0.302	3	2.296	0.043
Solyc02g077040	phytophthora-inhibited protease 1 (PIP1)	AT3G49340	3	0.183	0.052	3	0.291	0.316	3	1.585	0.024

[a] Number of biological replicates in which the protein was quantifiable. [b] The average log_2 ratio of protein quantity representing (inoculated/mock) from three biological replicates. [c] The *p*-value was calculated by Student's *t*-test if the protein could be quantified from the three biological replicates. Color codes: red, significant quantity change with greater than 0.58 of log_2 ratio ($p < 0.05$); blue, significant quantity change less than −0.58 of log_2 ratio; grey, no significant change between the inoculated and mock group ($p \geq 0.05$) or not quantified in all three replicates (NQ); white, the quantity change is between 0.58 and −0.58 of log_2 ratio ($p < 0.05$).

In summary, *P. infestans* regulates the PR-1 protein family for JA and SA immune signaling and defense responses at 48 and 96 hpi. After 96 h of pathogen inoculation, CA protein, which functions to suppress the PTI responses and the activation HR development, was highly induced. In the meantime, the induction of EDS1 for more general deference responses may further counteract the immune suppression of pathogen effectors. At 96 hpi, we also observed upregulated proteins that are involved in the activation of PCD by ETI. In addition, the KTI protein, which may be involved in the suppression of the H_2O_2 accumulation, was also induced at a late time point, suggesting a fine-tuning mechanism in this pathogenesis stage to control the H_2O_2 mediated HR and PCD responses.

2.6. Changes Associated with the Regulation of Phytohormones

Several proteins involved in the biosynthetic or signaling pathways of different phytohormones, including SA, ethylene (ET), and abscisic acid (ABA), were identified in our study (Table 3). Three lipoxygenases involved in JA biosynthesis were regulated in various ways at different post-inoculation time points. Tomato lipoxygenase D (LoxD; Solyc03g122340) was downregulated with a fold change of 0.65 at 24 hpi but not significantly regulated at either 48 or 96 hpi. LoxD was one of the first proteins to be identified from the JA biosynthetic pathway, and mediates defense against fungal and bacterial pathogens [50,51]. In contrast, the other two lipoxygenase proteins (Solyc01g099160 and Solyc08g029000) were upregulated with a 3.7 to 4.19-fold increase at 48 hpi or a 13.97-fold increase at 96 hpi, respectively. In addition to the proteins involved in JA biosynthesis, topless 3 (TPL3; Solyc01g100050), a transcriptional corepressor of the JA responsive genes [52], was upregulated by 1.76-fold at 96 hpi. It has been speculated that TPL protein is involved in multiple pathways and interacts with numerous transcriptional repressors [53]. Notably, the repression of JA responses by TPL3 is dependent on its interaction with another repressor, Novel Interactor of JAZ (NINJA). TPL3 and NINJA are recruited together by the jasmonate ZIM domain (JAZ) protein to inhibit the expression of the early JA-responsive genes [52]. The transcriptional regulation by TPL and family members has been shown to be a crucial part of the immune signaling for defense against both biotrophic oomycetes and necrotrophic fungus [54]. 1-aminocyclopropane-1-carboxylate (ACC) oxidase (ACO1; Solyc07g049530), functioning as the rate-limiting enzyme of the ET biosynthesis, was upregulated by *P. infestans* at 48 and 96 hpi with 1.84 and 4.88-fold increases, respectively. CobN/magnesium chelatase (CHLH/ABAR; Solyc04g015750) was upregulated in abundance by 1.55-fold at 24 hpi, but downregulated with a fold change of 0.42 at 96 hpi. CHLH/ABAR is involved in the plastid-to-nucleus retrograde signaling for chlorophyll development, and stomatal response to ABA, interacting with a group of WRKY transcription factors such as WRKY40 thus attenuating their negative regulation on ABA-responsive genes [55–57].

Taken together, repression of the biosynthetic pathway of JA and enhancement of the ABA signaling pathway were observed at 24 hpi with *P. infestans*. The repression of JA production may facilitate the activation of SA-related responses to defeat the biotrophic infection of *P. infestans*. This JA repression may facilitate *P. infestans* transition from biotroph to necrotroph. The enhanced ABA signaling may benefit *P. infestans* infection at the early time point as ABA induces disease susceptibility to various biotrophic or necrotrophic pathogens in a wide range of plant species [58]. *P. infestans* also suppressed the JA responsive genes at the later time point of 96 hpi, while plants may increase the level of JA and ET from 48 hpi to 96 hpi as well as downregulate ABA signaling at 96 hpi. Since the activation of JA and ET production is involved in the enhanced plant resistance against necrotrophic pathogens [6]; these results suggest that plants try to trigger the pathogen resistance from the biotrophic to necrotrophic pathogenesis stage via adjusting the signaling and/or biosynthesis of JA, ET, and ABA.

2.7. Changes Associated with Reactive Oxygen Species and Oxidation–Reduction Reactions

In this study, we found that 25 proteins involved with reactive oxygen species (ROS) or redox function were differentially regulated in response to *P. infestans* pathogenesis (Table 4 and Table S1). The NADPH/respiratory burst oxidase protein D (RbohD) homolog, whitefly-induced p91-phox (GP91^{phox}, Solyc03g117980) protein was upregulated at 48 and 96 hpi by 1.70 and 5.08-fold, respectively. The plant NADPH oxidase RbohD is a primary player responsible for the ROS burst after the pathogen-associated molecular pattern (PAMP) perception and thus regulates the HR in and around the infection site [59]. Twelve enzymes working as ROS scavengers were up- or downregulated in this study, including 9 proteins with peroxidase (POX) enzyme activities (Solyc08g069040, Solyc04g073990, Solyc01g006300, Solyc10g076240, Solyc04g071900, Solyc02g079500, Solyc02g092580, Solyc04g071890, and Solyc03g006700). Among them, only one POX (Solyc08g069040) had downregulation at 48 hpi with a fold change of 0.56 while the rest were upregulated at 96 hpi with a fold change from 1.61 to 13.28. The other ROS scavenger enzymes regulated at 96 hpi were catalase (CAT; Solyc04g082460) with a fold-change of 0.56, superoxide dismutase 1 (CSD1; Solyc01g067740) with a 2.16-fold increase, and glutathione peroxidase-like encoding 1 (GPX-1; Solyc08g080940) with a 1.94-fold increase. In addition, another protein in the antioxidant redoxin family, thioredoxin (TRX; Solyc04g081970), was downregulated with a fold change of 0.47 at 96 hpi. Thioredoxin reductase (TRXR; Solyc02g082250) for catalyzing the reduction of TRX was upregulated by 1.52-fold. Another protein predicted to be involved in TRX regulation, CBS domain-containing protein-like (CDCP-like; Solyc01g107860) was 2.10-fold upregulated at 96 hpi. Seven proteins participating in the regulation of the homeostasis of the antioxidant glutathione (GSH) showed different patterns of regulation across the pathogen growth stages. Among them, four glutathione S-transferases (GST) or GST-like proteins (Solyc06g009020, Solyc08g080900, Solyc10g084400, and Solyc09g011590) for conjugation of the reduced form of GST to the xenobiotic substrate were increased at 96 hpi by 1.58 to 7.42-fold while two GSTs (Solyc06g083770 and Solyc02g081430) were downregulated with a fold change of 0.41 and 0.58, respectively. On the other hand, gamma-glutamylcyclotransferase (GGCG; Solyc11g012910) which is responsible for the degradation of GSH in the cytosol [60] was downregulated with a fold change of 0.68 and 0.33 at 24 hpi and 96 hpi, respectively.

In summary, of all the proteins involved in the ROS/redox homeostasis regulation, the majority of the increase in ROS burst likely occurs from 48 to 96 hpi. *P. infestans*-inoculated tomato plants also showed upregulated antioxidants and ROS scavenger enzymes at 96 hpi. The enhancement of GST proteins may be correlated with the detoxification of the effectors produced by the pathogen and the suppression of the GSH catabolic proteins may help increase the antioxidant and GST capabilities in the cytosol. During the necrotrophic pathogenesis stage, the specifically downregulated CAT, POX, or TRX could be host targets manipulated by the pathogen to favor its own growth.

2.8. Differentially Regulated Proteins Involved in Carbohydrate and Energy Metabolism

Proteins that are involved in carbohydrate and energy metabolism were significantly upregulated at 96 hpi (Table 5 and Table S1). Proteins participating in the glycolysis pathways, including ATP-dependent 6-phosphofructokinase (PFK; Solyc07g045160), glyceraldehyde-3-phosphate dehydrogenase (GAPDH; Solyc10g005510), and pyruvate dehydrogenase (PDH; Solyc03g097680) were all significantly upregulated at 96 hpi by 2.07 to 2.38 fold. In contrast, enolase (ENO1; Solyc03g114500) was increased earlier, at 24 hpi, with a fold change of 1.52. This enzyme catalyzes the generation of the immediate precursor of pyruvate in the glycolytic pathway and a branch point to the shikimic acid pathway. Three proteins that are involved in the pentose phosphate pathway (PPP) for producing NAPDH, transaldolase (TAL; Solyc11g033288), ribose 5-phosphate isomerase A (Rpi; Solyc05g008370), and 6-phosphogluconate dehydrogenase (6PGD; Solyc04g005160) were upregulated at 96 hpi by 1.59 to 3.70-fold. Proteins participating in fatty acid/lipid metabolism, acyl-CoA dehydrogenase (ACAD; Solyc10g076600) and enoyl-CoA hydratases (ECH; Solyc01g059830 and Solyc12g094450), were upregulated at 96 hpi with fold changes between 1.68 and 4.24. Seven proteins that function in the TCA cycle, pyruvate dehydrogenase complex (PDC; Solyc11g007720), malic enzyme (ME; Solyc08g066360), citrate synthase (CSY; Solyc01g073740), aconitases (ACO; Solyc07g052350 and Solyc12g005860), isocitrate dehydrogenase (IDH; Solyc01g005560), oxoglutarate dehydrogenase complexes (ODC; Solyc07g064800 and Solyc12g005080), succinyl-CoA ligases (SCoAL; Solyc06g083790 and Solyc01g007910) and fumarase (FUM; Solyc09g075450), were all upregulated at 96 hpi with a fold change ranging from 1.52 to 2.00. In contrast, all the proteins involved in carbon fixation and photosynthesis progress (Table 6 and Table S1), such as chlorophyll a-b binding protein (LHCB; Solyc06g063370), cytochrome b6-f complex iron-sulfur subunit (Solyc12g005630), photosystem I (PSI) P700 chlorophyll a apoprotein (Solyc06g009940), PSI reaction center subunit III (Solyc02g069450), PSI reaction center subunit N (Solyc08g013670), PSII 22 kDa protein (Solyc06g060340) and sedoheptulose-1,7-bisphosphatase (SBPase; Solyc05g052600), were downregulated at 96 hpi with fold changes between 0.54 and 0.66.

The upregulation of PFK, GAPDH, ENO1, and PDH for glycolysis together with ACAD and ECH in the fatty acid and lipid metabolisms may enhance the synthesis of acetyl-CoA for feeding the TCA cycle. TAL, Rpi, and 6PGD that function in the PPP were enhanced, and thus were likely to produce more NADPH and ribose 5-phosphate (R5P) for the alternative progress of glycolysis. The proteins involved in the TCA cycle to generate more NADH and $FADH_2$ were also significantly upregulated. The increase of NAPDH by these pathways may be related to the enhancement of the oxidative phosphorylation pathway in order to generate more energy. These results suggest that plants may attempt to generate more ATP by the glycolysis/TCA/PPP and oxidative pathways during the pathogenesis process. The carbohydrate metabolites produced from the glycolysis and fatty acid biosynthetic pathways may also be used as the signals for the activation of defense responses [61]. In contrast, proteins participating in photosynthesis and energy storage were downregulated in this process, suggesting a switch in the primary metabolisms wherein carbon is obtained from the sink tissue, rather than the source.

2.9. Changes Associated with Secondary Metabolites

Three of the proteins involved in the secondary metabolite biosynthesis pathway were upregulated at 96 hpi (Table 7 and Table S1). Two proteins involved in the phenylpropanoid biosynthetic pathway, cinnamate 4-hydroxylase (C4H; Solyc06g150137) and 4-coumarate:CoA ligase (4CL; Solyc03g117870) were increased by 2.62- and 4.90-fold at 96 hpi, respectively. In addition, 5-enolpyruvylshikimate-3-phosphate synthase (ESPS; Solyc01g091190), which is involved in the synthesis of phenylalanine starting the phenylpropanoid pathway, was increased by 2.24-fold at 96 hpi.

We further observed that *P. infestans* induced the proteins responsible for the synthesis of the upstream and intermediate metabolites of the phenylpropanoid pathway during the necrotrophic pathogenesis stage. These upregulated proteins may enhance the level of phenylpropanoids in plants to fight against pathogens [62]. Furthermore, lignin, suberin, or flavonoids which are derived from this pathway also contribute to the basal immunity response of the plant [63].

2.10. Novel P. infestans-Regulated Tomato Reponses Revealed by Time-Lapse Proteomics Studies

To date, only a limited number of studies have focused on the tomato proteome in response to *P. infestans*. A previous proteomics study discovered a total of 56 tomato proteins regulated by *P. infestans* infection, of which 39 and 17 were found in the resistant and susceptible tomato genotype, respectively [15]. In the resistant tomato genotype, six proteins associated with immune and defense mechanisms were observed with enhanced abundance after pathogen inoculation, which is comprised of three peroxiredoxins, one PR protein, carboxypeptidase, and predicted R protein. Two proteinase inhibitors and a glucosidase were observed to have suppressed abundance in the resistant tomato genotype. In the susceptible genotype, among the stress and defense-related proteins, one enzyme participating in the urea cycle and metabolism of amino groups was suppressed whereas one ROS scavenger enzyme and one glucosidase were both enhanced at the protein level after infection. Compared to the previous proteomics study, we identified 25-fold more tomato proteins regulated by this pathogen; thus most of the *P. infestans*-regulated mechanisms were first identified and dynamically quantified at the protein level. To the best of our knowledge, this study provides the most comprehensive proteome information available to date, revealing the dynamics of tomato proteome regulation at the defined key stages of *P. infestans* infection. With regard to the dynamics of tomato defense responses, we discovered that PR-5 protein was downregulated at the biotrophic stage, and multiple chitinases were induced to conduct the defense activity by degrading the pathogen cell wall in the necrotrophic stage. In addition, tomato proteins for direct antipathogen activity, biosynthesis and signaling of immune regulatory hormones, ETI responses, and biosynthesis of the secondary metabolites were also exclusively determined to be regulated by *P. infestans* infection in this study.

Although the response to *P. infestans* has been previously analyzed in tomato in well-defined stages of pathogenesis using a RNA-Seq-based transcriptomics approach [11], this study lacks a mock-treated sample as a control at each time point; this may lead to identifying non-pathogenetic responses. In addition, the regulation at the protein level also requires investigation to further clarify the roles of these gene candidates. With the mock-treated sample as the control at each time point, many of the *P. infestans*-responsive genes identified in the previous RNA-Seq analysis were statistically proved to be regulated at the protein level in this study. However, it is well-known that gene expression is not always proportional to protein expression, and accordingly we discovered there were some proteins differentially regulated by *P. infestans* infection that were reported differently in the previous transcriptomic study. Among them, nine proteins identified in this study were either not identified (C4H), not significantly expressed (TPL3), or showed opposite regulation patterns at the transcript level (PR-5, RD21a, EDS1, HIR, PIP1, CHLH/ABAR, and ENO1). These proteins have molecular functions in direct antifungal activity (PR-5), cysteine protease (RD21a and PIP1), R protein helper inducing HR (HIR), SA/ETI-mediator required for basal and NB-LRR protein-induced resistance (EDS1), JA/ET signaling (TPL3), ABA signaling (CHLH/ABAR), glycolysis (ENO1), and the phenylpropanoid biosynthetic pathway (C4H). For these selected proteins, their gene expressions were analyzed by qRT-PCR in the same biological samples used in this proteomics study (Figure 4).

Among those nine proteins, three proteins, RD21a, HIR, and C4H showed positively correlated regulations between the transcript and protein levels. Four proteins, PR-5, EDS1, PIP1, and ENO1 showed similar trends of direction in gene and protein regulation. In the case of PIP1, differential expression at the protein level was observed at a later time point (96 hpi) than the transcript level (24 and 48 hpi). There could be more variation in the expression of these four genes; thus the changes of transcript levels during the pathogen infection did not reject the null hypothesis in the significant test. The higher variation in transcript regulation than protein regulation might also explain the non-correlation of results in our proteome analysis and the previous RNA-Seq analysis. As the previous transcriptomic analysis only used one biological replicate, the actual changes in the transcript levels after pathogen infection may not be accurately presented. The non-correlation of the results of the transcriptome and proteome analysis could be due to the biosynthesis activity and stability of RNA and protein under differential regulation [64]. We found that two proteins TPL3 and CHLH/ABAR showed opposite directions of transcriptional and translational/post-translational regulation after pathogen infection. This opposite regulation trend could result from multiple factors beyond the transcript concentration, which influence the establishment of the protein expression level. These factors could include the modulation of translation rate by mechanisms like regulator elements/miRNA and constraints of machinery/resources for protein biosynthesis, the modulation of protein half-life by the ubiquitin-proteasome or autophagy pathway, the temporal delay in protein synthesis, and the protein export resulting in spatial changes of proteins [65].

Table 3. Proteins involved in hormone biosynthesis and signaling with significant changes in abundance at 24, 48 and 96 h post-inoculation with *P. infestans*.

			24 h			48 h			96 h		
Gene Accession	Protein Description	Arabidopsis Homolog	# Rep [a]	log_2 Ratio [b]	p [c]	# Rep [a]	log_2 Ratio [b]	p [c]	# Rep [a]	log_2 Ratio [b]	p [c]
Solyc07g049530	1-aminocyclopropane-1-carboxylate oxidase 1 (ACO1)	AT1G05010	3	0.109	0.521	3	0.877	0.007	3	2.288	0.012
Solyc04g015750	cobN/magnesium chelatase (CHLH/ABAR)	AT5G13630	3	0.629	0.042	3	0.155	0.625	3	−1.239	0.031
Solyc01g099160	lipoxygenase (Lox)	AT1G55020	0	ND	ND	3	2.067	0.003	3	3.063	0.055
Solyc08g029000	lipoxygenase (Lox)	AT1G55020	3	−0.291	0.527	3	1.889	0.003	3	3.804	0.008
Solyc03g122340	lipoxygenase D (LoxD)	AT1G17420	3	−0.611	0.046	3	−0.453	0.394	1	−1.177	NQ
Solyc12g062290	protease do-like 9 (DEGP9)	AT5G40200	3	0.133	0.268	3	−0.588	0.011	2	0.000	NQ
Solyc01g100050	topless 3 (TPL3)	AT5G27030	3	0.205	0.067	3	0.307	0.188	3	0.817	0.004

[a] Number of biological replicates in which the protein was quantifiable. [b] The average log_2 ratio of protein quantity representing (inoculated/mock) from three biological replicates. [c] The p-value was calculated by Student's t-test if the protein could be quantified from three biological replicates. Color codes: red, significant quantity change with greater than 0.58 of log_2 ratio ($p < 0.05$); blue, significant quantity change less than −0.58 of log_2 ratio; grey, no significant change between the inoculated and mock group ($p \geq 0.05$) or not quantified in all three replicates (NQ); white, the quantity change is between 0.58 and −0.58 of log_2 ratio ($p < 0.05$).

Table 4. Proteins involved in ROS homeostasis and redox regulation with a significant change in abundance at 24, 48 and 96 h post-inoculation with *P. infestans*.

			24 h			48 h			96 h		
Gene Accession	Protein Description	Arabidopsis Homolog	# Rep [a]	log_2 Ratio [b]	p [c]	# Rep [a]	log_2 Ratio [b]	p [c]	# Rep [a]	log_2 Ratio [b]	p [c]
Solyc04g073990	annexin p34 (ANXP34)	AT1G35720	3	−0.122	0.603	3	0.107	0.883	3	0.690	0.002
Solyc04g082460	catalase (CAT)	AT4G35090	3	0.223	0.122	3	−0.182	0.724	3	−0.845	0.046
Solyc01g107860	CBS domain-containing protein-like (CDCP-like)	AT4G36910	3	−0.026	0.345	3	−0.046	0.777	3	1.068	0.010
Solyc11g012910	gamma-glutamylcyclotransferase (GGCG)	AT1G44790	3	−0.564	0.046	3	0.369	0.418	3	−1.588	0.003
Solyc08g080940	glutathione peroxidase -like encoding 1 (GPX-1)	AT4G11600	3	−0.037	0.498	3	0.241	0.319	3	0.957	0.019
Solyc06g083770	glutathione S-transferase (GST)	AT5G44000	3	−0.008	0.967	3	−0.260	0.366	3	−1.300	0.016
Solyc10g084400	glutathione S-transferase (GST)	AT5G02790	3	−0.099	0.596	3	0.221	0.100	3	1.572	0.000
Solyc09g011590	glutathione S-transferase-like protein (GST-like)	AT3G09270	3	−1.121	0.316	3	0.987	0.272	3	2.891	0.007
Solyc02g081430	microsomal Glutathione s-Transferase (MGST)	AT1G65820	3	0.082	0.263	3	0.201	0.644	3	−0.790	0.006
Solyc01g096430	NADPH:quinone oxidoreductase-like (NQR-like)	AT3G27890	2	−0.040	NQ	3	0.140	0.814	3	1.794	0.034
Solyc03g111720	peptide methionine sulfoxide reductase (MSR)	AT5G61640	3	−0.440	0.079	3	0.135	0.396	3	0.829	0.010
Solyc10g076240	peroxidase (POX)	AT5G05340	0	NQ	NQ	0	NQ	NQ	3	3.731	0.037
Solyc08g069040	peroxidase (POX)	AT4G37530	2	−0.165	NQ	3	−0.831	0.041	2	−0.428	NQ
Solyc01g067740	superoxide dismutase 1 (CSD1)	AT1G08830	3	0.320	0.060	3	0.074	0.820	3	1.113	0.038
Solyc04g081970	thioredoxin (TRX)	AT1G76080	3	0.021	0.805	3	0.000	1.000	3	−1.087	0.011
Solyc02g082250	thioredoxin reductase (TRXR)	AT2G17420	3	0.015	0.883	3	0.217	0.262	3	0.607	0.013
Solyc03g117980	whitefly-induced gp91-phox (GP91phox)	AT5G47910	3	0.039	0.777	3	0.765	0.038	3	2.345	0.016

[a] Number of biological replicates in which the protein was quantifiable. [b] The average log_2 ratio of protein quantity representing (inoculated/mock) from three biological replicates. [c] The p-value was calculated by Student's t-test if the protein could be quantified from three biological replicates. Color codes: red, significant quantity change with greater than 0.58 of log_2 ratio ($p < 0.05$); blue, significant quantity change less than −0.58 of log_2 ratio; grey, no significant change between the inoculated and mock group ($p \geq 0.05$) or not quantified in all three replicates (NQ); white, the quantity change is between 0.58 and −0.58 of log_2 ratio ($p < 0.05$).

Table 5. Proteins involved in energy metabolisms with a significant change in abundance at 24, 48 and 96 h post-inoculation with *P. infestans*.

Gene Accession	Protein Description	Arabidopsis Homolog	24 h			48 h			96 h		
			# Rep [a]	$\log_2$ Ratio [b]	p [c]	# Rep [a]	$\log_2$ Ratio [b]	p [c]	# Rep [a]	$\log_2$ Ratio [b]	p [c]
Metabolism-Primary-Energy Metabolisms											
Solyc11g007720	pyruvate dehydrogenase complex (PDC)	AT3G52200	3	−0.121	0.358	3	0.082	0.842	3	0.645	0.017
Solyc01g073740	citrate synthase (CSY)	AT2G44350	3	−0.112	0.410	3	−0.198	0.283	3	0.639	0.017
Solyc07g052350	aconitate hydratase (ACO)	AT2G05710	3	0.019	0.872	3	0.110	0.544	3	0.708	0.047
Solyc12g005860	aconitate hydratase (ACO)	AT2G05710	3	−0.084	0.054	3	−0.100	0.605	3	0.968	0.023
Solyc01g005560	isocitrate dehydrogenase (IDH)	AT1G65930	3	−0.125	0.650	3	0.311	0.369	3	0.753	0.049
Solyc07g064800	oxoglutarate dehydrogenase complex (ODC)	AT4G26910	3	0.049	0.497	3	0.150	0.700	3	0.607	0.046
Solyc12g005080	oxoglutarate dehydrogenase complex (ODC)	AT5G55070	3	−0.073	0.661	3	0.107	0.043	3	1.003	0.023
Solyc06g083790	succinyl-CoA ligase (SCoAL)	AT2G20420	3	0.058	0.107	3	0.039	0.888	3	0.602	0.016
Solyc01g007910	succinyl-CoA ligase (SCoAL)	AT5G23250	3	−0.131	0.132	3	0.295	0.529	3	0.630	0.035
Solyc09g075450	fumarase (FUM)	AT2G47510	3	0.088	0.362	3	0.091	0.691	3	0.953	0.004
Solyc08g066360	malic enzyme (ME)	AT1G79750	3	−0.007	0.978	3	−0.016	0.957	3	0.903	0.045
Metabolism-Primary-Glycolysis											
Solyc07g045160	ATP-dependent 6-phosphofructokinase (PFK)	AT4G26270	3	0.217	0.106	3	0.107	0.497	3	1.052	0.037
Solyc10g005510	glyceraldehyde-3-phosphate dehydrogenase (GAPDH)	AT1G16300	3	0.097	0.413	3	0.220	0.713	3	1.096	0.045
Solyc03g114500	enolase (ENO1)	AT1G74030	3	0.599	0.049	3	−1.053	0.430	2	1.073	NQ
Solyc03g097680	pyruvate dehydrogenase 1 (PDH)	AT5G50850	3	−0.048	0.857	3	−0.270	0.724	3	1.251	0.038
Metabolism-Primary-Carbohydrate Metabolisms-PPP											
Solyc04g005160	6-phosphogluconate dehydrogenase (6PGD)	AT3G02360	3	−0.046	0.033	3	0.524	0.031	3	1.886	0.016
Solyc05g008370	ribose 5-phosphate isomerase A (Rpi)	AT1G71100	3	0.068	0.366	3	0.211	0.715	3	0.790	0.036
Solyc11g033288	transaldolase (TAL)	AT5G13420	3	0.029	0.751	3	−0.206	0.729	3	0.667	0.032
Metabolism-Primary-Fatty Acid/Lipids											
Solyc10g076600	acyl-CoA oxidase/dehydrogenase (ACAD)	AT3G51840	3	−0.057	0.872	3	0.050	0.928	3	2.085	0.014
Solyc01g059830	enoyl-CoA hydratase (ECH)	AT4G16210	3	0.469	0.386	2	−0.156	NQ	3	0.750	0.041
Solyc12g094450	enoyl-CoA hydratase (ECH)	AT1G76150	2	−0.278	NQ	3	0.202	0.216	3	1.663	0.002

[a] Number of biological replicates in which the protein was quantifiable. [b] The average $\log_2$ ratio of protein quantity representing (inoculated/mock) from three biological replicates. [c] The p-value was calculated by Student's t-test if the protein could be quantified from three biological replicates. Color codes: red, significant quantity change with greater than 0.58 of $\log_2$ ratio ($p < 0.05$); blue, significant quantity change less than −0.58 of $\log_2$ ratio; grey, no significant change between the inoculated and mock group ($p \geq 0.05$) or not quantified in all three replicates (NQ); white, the quantity change is between 0.58 and −0.58 of $\log_2$ ratio ($p < 0.05$).

Table 6. Proteins involved in carbon fixation with a significant change in abundance at 24, 48 and 96 h post-inoculation with *P. infestans*.

			24 h			48 h			96 h		
Gene Accession	Protein Description	Arabidopsis Homolog	# Rep [a]	log_2 Ratio [b]	p [c]	# Rep [a]	log_2 Ratio [b]	p [c]	# Rep [a]	log_2 Ratio [b]	p [c]
Solyc06g063370	chlorophyll a-b binding protein (LHCB)	AT4G10340	3	−0.061	0.659	3	0.391	0.212	3	−0.662	0.014
Solyc12g005630	cytochrome b6-f complex iron-sulfur subunit	AT4G03280	3	−0.368	0.077	3	−0.001	0.998	3	−0.739	0.001
Solyc06g009940	photosystem I (PSI) P700 chlorophyll a apoprotein	ATCG00350	3	0.029	0.810	3	−0.121	0.869	3	−0.836	0.015
Solyc02g069450	photosystem I (PSI) reaction center subunit III	AT1G31330	3	0.107	0.605	3	0.048	0.930	3	−0.661	0.038
Solyc08g013670	photosystem I (PSI) reaction center subunit N	AT5G64040	3	0.416	0.244	3	0.412	0.498	3	−0.854	0.033
Solyc05g150152	photosystem I (PSI) reaction centre subunit N protein	AT1G49975	2	0.073	NQ	3	−0.279	0.587	3	−0.599	0.048
Solyc06g060340	photosystem II (PSII) 22 kDa protein	AT1G44575	3	−0.091	0.339	3	0.234	0.063	3	−0.910	0.008
Solyc05g052600	sedoheptulose-1,7-bisphosphatase (SBPase)	AT3G55800	3	0.077	0.001	3	−0.292	0.626	3	−0.758	0.039

[a] Number of biological replicates in which the protein was quantifiable. [b] The average log_2 ratio of protein quantity representing (inoculated/mock) from three biological replicates. [c] The p-value was calculated by Student's t-test if the protein could be quantified from three biological replicates. Color codes: red, significant quantity change with greater than 0.58 of log_2 ratio ($p < 0.05$); blue, significant quantity change less than −0.58 of log_2 ratio; grey, no significant change between the inoculated and mock group ($p \geq 0.05$) or not quantified in all three replicates (NQ); white, the quantity change is between 0.58 and −0.58 of log_2 ratio ($p < 0.05$).

Table 7. Proteins involved in secondary metabolism with a significant change in abundance at 24, 48 and 96 h post-inoculation with *P. infestans*.

			24 h			48 h			96 h		
Gene Accession	Protein Description	Arabidopsis Homolog	# Rep [a]	log_2 Ratio [b]	p [c]	# Rep [a]	log_2 Ratio [b]	p [c]	# Rep [a]	log_2 Ratio [b]	p [c]
Solyc06g150137	cinnamate 4-hydroxylase (C4H)	AT2G30490	3	−0.075	0.505	3	0.164	0.736	3	1.390	0.004
Solyc03g117870	4-coumarate:CoA ligase (4CL)	AT3G21240	3	0.185	0.366	3	0.622	0.178	3	2.293	0.004
Solyc01g091190	5-enolpyruvylshikimate-3-phosphate synthase (ESPS)	AT2G45300	3	−0.124	0.369	3	0.208	0.229	3	1.165	0.022

[a] Number of biological replicates in which the protein was quantifiable. [b] The average log_2 ratio of protein quantity representing (inoculated/mock) from three biological replicates. [c] The p-value was calculated by Student's t-test if the protein could be quantified from three biological replicates. Color codes: red, significant quantity change with greater than 0.58 of log_2 ratio ($p < 0.05$); blue, significant quantity change less than −0.58 of log_2 ratio; grey, no significant change between the inoculated and mock group ($p \geq 0.05$) or not quantified in all three replicates (NQ); white, the quantity change is between 0.58 and −0.58 of log_2 ratio ($p < 0.05$).

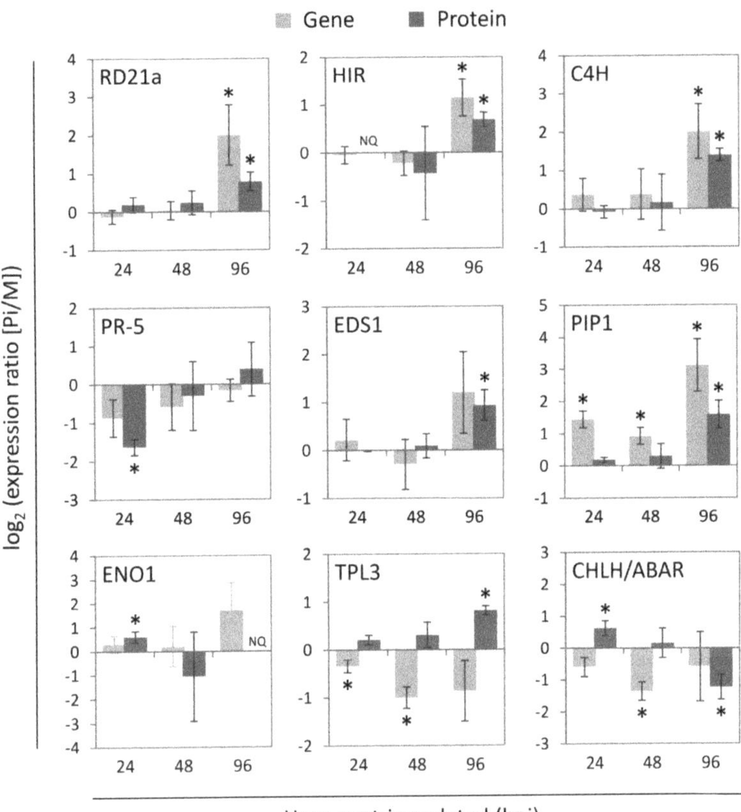

Figure 4. Levels of tomato transcripts encoding proteins that were differentially regulated in response to different pathogenesis stages of *P. infestans*. The transcript levels of the selected *P. infestans*-regulated proteins (RD21a, HIR, C4H, PR-5, EDS1, PIP1, ENO1, TPL3, and CHLH/ABAR) in tomato were determined by qRT-PCR in the same samples prepared for the proteomics analysis in this study. The mean values from three biological repeats are shown (n = 3). All statistically significant differences between the *P. infestans* (Pi) and mock (M)-treated samples are indicated with asterisks (*, $p < 0.05$), based on the Student's *t*-test using the $\log_2$-ratio of the gene or protein expression levels between these two conditions. The internal control *Ubi-3* gene was used for normalization. Error bars are means ± SD. NQ, non-quantifiable as the protein quantification was only measurable in one or two biological replicates.

When examining how these proteins and their gene levels were regulated in the three pathogenesis phases, we found that although PR-5 protein was downregulated at 24 hpi, its gene expression level was not significantly changed. In addition, at this early stage, although ENO1 and CHLH/ABAR protein levels were increased, their coding genes appear to be non-differentially expressed. On the other hand, the protein levels of neither PIP1 nor TPL3 were regulated although the transcript levels of *PIP1* and *TPL3* were enhanced and suppressed at 24 hpi, respectively. Later at 48 hpi, the transcript levels of *PIP1* gene showed upregulation, *TPL3* and *CHLH/ABAR* genes showed downregulation whereas none of their protein levels was significantly altered from the mock-treated group. At 96 hpi, EDS1 and TPL3 were both upregulated, and CHLH/ABAR protein was downregulated but neither of their gene expressions was differentially regulated. On the other hand, RD21a, HIR, C4H, and PIPI1 were regulated at both the transcript and protein levels at 96 hpi.

Since TPL3 functions as the transcription corepressor of JA-responsive genes, the reduction of TPL3 protein level may lead to the activation of JA signaling, which may suppress the production of SA-mediated immune responses. In this study, the suppression of *TPL3* gene expression may be part of the infection strategy of *P. infestans*, which deactivates the SA-mediated immunity to establish a biotrophic interaction with the host plant at the early infection stage. However, the suppressed *TPL3* transcript did not result in suppression of the protein level, possibly due to the translational activity or protein stability being enhanced to compensate for the repression of *TPL3* gene. This may facilitate plants to activate the SA signaling to fight against the biotrophic pathogen. In addition, PR-5 protein, which could directly sabotage the pathogen cell wall assembly, may be suppressed by *P. infestans* specifically at the protein level but not at the transcript level. Also, the increased *PIP1* gene expression did not result in significant protein abundance change in the biotrophic phase. The PIP protease activity has been reported to be a target of *P. infestans* effectors; therefore, this inhibition may cause PIP1 protein to become more unstable, thus generating a lower induction fold at the level of PIP1 protein in comparison with its transcript. Another possible defense strategy may include CHLH/ABAR protein for its role of controlling stomatal closure [66] and relieving the expression of ABA-responsive genes [57], hence the increased CHLH/ABAR protein in the biotrophic phase may represent an early PTI response. On the other hand, since ENO1 is a glycolytic enzyme, the upregulated ENO1 protein could be interpreted as one of the plant early responses to gain more energy or as a pathogen strategy to obtain more carbohydrate nutrients during the biotrophic phase. This translational/post-translational but not transcriptional regulation of ENO1 indicates an urgent response to facilitate the glycolysis pathway.

In the transition phase, the suppressed *TPL3* transcript and enhanced *PIP1* transcript without significant changes in protein levels suggest the continuous action of the pathogen in repressing these immune regulators from the early biotrophic phase. In the meantime, the plant may still try to enhance its immune responses by activating the transcriptional or translational regulation of these genes. In addition to controlling stomatal closure, CHLH/ABAR also influences ABA sensitivity [66–70] and contributes to chloroplast development [71]. Therefore, the suppressed expression of the *CHLH/ABAR* gene without the protein abundance being regulated in the transition phase suggests *P. infestans* may have started to hijack the ABA signaling pathway and disrupt plant growth to prepare itself to initiate necrosis, as well as inhibit stomatal closure, for the next pathogenesis stage.

Although the gene expression of *TPL3* was repressed from the early to the late pathogenesis stages, TPL3 protein was significantly upregulated in the necrotrophic phase. This regulation of TPL3 protein level was able to suppress the JA-responsive genes, which is favorable for *P. infestans* to facilitate the necrotrophic lifestyle. The downregulated CHLH/ABAR protein level but not the transcript level in the necrotrophic phase may reflect a delay effect from the transcriptional regulation in the transition phase. In view of the function of CHLH/ABAR as mentioned above, in the necrotrophic phase, this filamentous pathogen may downregulate CHLH/ABAR protein level to inhibit the stomatal closure to help the new sporangiophore emerge from the stomata and produce sporangia which are spread by wind or water to infect new plants [72]. As PIP1 and RD21a are both proteases for triggering PCD, their upregulated transcription and protein levels in the necrotrophic phase may have different causes: (1) a delay in protein translation production, as there is a time delay from transcriptional induction to protein level increase following an induced state change; (2) the plant eliminating the suppression from pathogen effectors which target these defense proteins starting in the early infection stage; (3) the pathogen acting to induce necrosis by secreting various cell-death inducing effector proteins to thrive on dead host tissues [73]. As both EDS1 and HIR contribute to the R protein (especially NB-LRR)-induced immunity, the upregulated EDS1 and HIR proteins in response to the necrotrophic phase of *P. infestans* suggests a plant action to increase the local concentration of NB-LRR proteins, and possibly other R proteins to enhance the detection sensitivity of pathogen effectors.

3. Materials and Methods

3.1. Plant Material, Growth Conditions and P. infestans Inoculation

Tomato seeds (*Solanum lycopersicum* cv CL5915, originally provided by AVRDC—The World Vegetable Center, Tainan, Taiwan) were planted in soil and grown in a growth chamber for 5 weeks under a 12 h/12 h light-dark cycle, with 50%/70% (light/dark) humidity, 25 °C/20 °C (light/dark) temperature and a light source providing photosynthetic photon flux density (PPFD) of 100 µmole/m^2·s. Fully expanded leaflets with similar size from the 3rd or 4th pair of true leaves were collected for the following experiment. Three leaflets from three individual plants were pooled as one biological replicate at each time point for the pathogenicity assays. *P. infestans* strain was sub-cultured every two weeks and grown on fresh Rye A agar plates [74] in the forever dark incubator at 20 °C for 2–3 weeks before collecting sporangia. By using a cell spreader, sporangia were washed in sterilized distilled water, and after counting under the optical microscope at a magnification of 200X, the concentration was adjusted to 2.0×10^4 sporangia per ml for inoculations. One leaflet from each pair was separately placed in the mock- or pathogen-inoculated group and placed on the distilled water-saturated tissue paper in square Petri dishes before inoculation. The abaxial surface of a tomato leaf was inoculated with 8 droplets of 20 µL sporangial suspension. The mock-treated groups had the same treatment except that the sporangial suspension was replaced with distilled water. After treatment, the dishes containing the leaflets were sealed with Parafilm and incubated in the dark at 20 °C. Samples were collected at 6, 16, and 24 hpi for microscope observation, 24, 48, and 96 hpi for RNA and protein extraction and 24, 48, 96, and 120 hpi for disease phenotype observation.

3.2. Observation of P. infestans Growth by Trypan Blue Staining and Microscope

To visualize the *P. infestans* infection at the early time points, leaves at 6, 16 or 24 hpi were transferred to a beaker and immersed in trypan blue solution (10 g phenol, 10 mL glycerol, 10 mL lactic acid, 10 mL water and 10 mg of trypan blue) diluted with 2-fold of ethanol [75]. The beaker was then heated in boiled water until the staining solution boiled for about 2 min and the leaves turned color to light blue. After being placed at room temperature for about 8 h for complete dyeing, the leaves were destained by replacing the staining buffer with chloral hydrate solution (5 g of chloral hydrate dissolved into 2 mL of distilled water) and shaken at a slow speed of 30 rpm for 24 h [76]. Lastly, the leaf samples were observed under a Zeiss AxioImager Z1 high-performance research microscope (Carl Zeiss AG, Werk Göttingen, Germany) with differential interference contrast optics.

3.3. RNA Extraction and Marker Gene Expression Analysis of P. infestans-Inoculated Tomato Leaf

Total RNA was extracted from each sample by using the Total RNA Mini Kit Plant (Geneaid, New Taipei City, Taiwan), according to the manufacturer's protocol. RNA was quantified by using Nanodrop ND-1000 (NanoDrop Technologies/Thermo Scientific, Wilmington, DE, USA), and 4 µg of total RNA was employed for cDNA synthesis in a 20-µL volume by using SuperScript™ III Reverse Transcriptase (Thermo Scientific, Bellefonte, PA, USA), following to the manufacturer's protocol. PCR was carried out with 20 ng of the cDNA in a 20-µL volume containing 2 µM of each specific primer (Supplemental Table S3) and 2X SuperRed PCR Master Mix (Biotools, New Taipei City, Taiwan), following the manufacturer's protocol.

For RT-PCR analysis, the PCR condition consisted of one cycle of 95 °C for 5 min, followed by 35 cycles of a three-step loop (94 °C for 1 min, 55 °C for 1 min and 72 °C for 1 min) and a final step of 72 °C for 5 min. After gel electrophoresis, the relative intensities of the product bands were assessed using ImageJ software [77]. The relative intensities of gel bands were calculated by the band with the maximal intensity as the value of 1.

The qRT-PCR was used to validate the expression of specific tomato genes. Three biological replicates were used for qRT-PCR analysis, which was performed using SYBR Green reagent (Sigma-Aldrich) and ABI 7500 Fast Real Time PCR system (ThermoFisher

Scientific, Inc.). The PCR cycling steps were 50 °C for 2 min and 94 °C for 10 min for the initial steps followed by 95 °C for 15 s and 60 °C for 1 min for 40 cycles. The gene expressions across different samples were normalized with the internal control Ubi3. The primers used are listed in Supplemental Table S4. The melting curve was used to verify the specificity of the PCR product.

3.4. Sample Preparation for Proteome Analysis

Three biological replicates of mock-treated (24, 48, 96 hpt) and *P. infestans* inoculated (24, 48, 96 hpi) leaves were prepared for the proteomics experiments. For each sample, leaves were ground into powder in a chilled pestle and mortar with liquid N_2 then 0.2 g of the powder was collected. The protein extraction protocol was modified from previous studies [78–80]. The sample was then homogenized with 0.9 mL of ice-cold pH 8.0 homogenization buffer (0.5 M Tris-HCl, pH 7.5; 0.7 M sucrose; 0.1 M KCl; 50 mM EDTA; 1% polyvinylpolypyrrolidone (PVPP); 5 mM dithiothreitol, 1 mM phenylmethylsulfonyl fluoride and 1× Roche protease inhibitor cocktail) by vortexing for at least 3 min or until completely homogenized. The sample was further incubated in the buffer by Intelli Mixer RM-2L (ELMI Ltd., Riga, Latvia) in the cold room for 30 min. The homogenate was filtered through two layers of Mira cloth. After collecting the supernatant of the filtrate by centrifugation at 16,000× *g* for 10 min under 4 °C, an equal volume of phenol (pH 7.8–8.0, Tris-buffered) was added to the supernatant and homogenized by Intelli Mixer RM-2L for 30 min at 4 °C. The upper phenol phase was collected after the solution was centrifuged at 16,000× *g* for 20 min then the back-extraction was performed one more time using an equal volume of phenol. Proteins were precipitated from the final collected phenol phase by mixing with a 5-fold volume of cold 0.1 M ammonium acetate/methanol and incubating at −20 °C overnight. The precipitated proteins were washed once with 0.1 M ammonium acetate/methanol and twice with 80% acetone/10 mM DTT. The air-dried protein pellet was then solubilized in 8 M urea with 50 mM ammonium bicarbonate (ABC) and 5 mM tris (2-carboxyethyl) phosphine hydrochloride (TCEP). The extracted total protein concentration of each sample was measured by the Bradford assay and checked by SDS-PAGE.

For each sample, 100 µg of the solubilized tomato total protein was reduced, alkylated and proteolyzed with lysyl endopeptidase and trypsin following the steps previously described [81]. The digested sample was then desalted using the 100-mg tC18 SepPak cartridge (Waters Corporation, Milford, MA, USA). After being quantified using the Peirce BCA assay kit (Thermo Fisher Scientific, Waltham, MA, USA), the tryptic peptides from the individual sample were dissolved by deionized water containing 2% acetonitrile and 0.1% (*v*/*v*) formic acid to a concentration of 500 ng/µL. The pooled peptide sample of the mock and *P. infestans*-inoculated samples at each time point (as QC sample) was analyzed by LC-MS/MS with the DDA mode in order to construct the DIA spectral library. For the purpose of retention time calibration, the iRT-standard peptides (Biognosys, Schlieren, Switzerland) were mixed with the pooled sample or each individual sample with a ratio of 1:10 by volume.

3.5. Liquid Chromatography-Mass Spectrometry Analysis

The nanoLC−MS/MS was equipped with a self-packed tunnel-frit [82] analytical column (ID 75 µm × 50 cm length) packed with ReproSil-Pur 120A C18-AQ 1.9 µm (Dr. Maisch GmbH, Germany) at 40 °C on a nanoACQUITY UPLC System (Waters Corporation, Milford, MA, USA) connected to a Q Exactive HF Hybrid Quadrupole-Orbitrap mass spectrometer (Thermo Scientific, Bellefonte, PA, USA). The peptides were separated by a 135-min gradient using the mobile phases, including Solvent A (0.1% (*v*/*v*) formic acid) and Solvent B (acetonitrile with 0.1% formic acid). With a flow rate of 250 nL/min, the gradient started with a 40 min equilibration maintained at 2% of B and set as the following segments: 2% to 8% of B in 8 min, 8% to 25% of B in 90 min, then 25% to 48% of B in 5 min, 48% to

80% of B in another 5 min followed by 80% of B wash 10 min and the last equilibrium to 2% B in the last 15 min.

The instrumentation and parameters for DDA and DIA analysis followed previous studies using the Q Exactive HF Hybrid Quadrupole-Orbitrap mass spectrometer [83,84]. Two micrograms of the pooled and individual tryptic peptide samples were analyzed by DDA and DIA mode, respectively. For DDA analysis, the MS instrument was operated in the positive ion mode and two different DDA methods at the MS2 level. In the first DDA method, full-MS was acquired with high resolution (R = 60,000 at m/z 200 at an automatic gain control target of 3.0×10^6) and broadband mass spectra (m/z 350–1650 Da) with a maximum IT of 15 ms, and MS/MS events (R = 15,000 at an automatic gain control target of 1.0×10^5) with a dd-MS^2 IT of 45 ms when a precursor ion charge was 2+, 3+, 4+ and 5+ and minimum AGC target as 4.5×10^3, isolation window was set to 1.2 Th, was detected. The 20 most abundant peptide molecular ions, dynamically determined from the MS1 scan, were selected for MS/MS using stepped normalized collision energies (NCE) as 26.5%, 28%, and 29.5%, with the MS/MS settings of dynamic exclusion as 25 s, peptide matched as preferred, isotopes excluded and apex triggered. The other DDA method was using gas-phase separation with the m/z range of 350–510, 500–710, 700–1060 and 1040-1650 Th and fixed NCE as 28%; the rest of settings were the same as the first DDA method.

For DIA analysis, MS/MS proteome profiling was analyzed by the same LC−MS/MS system. The instrument was operated in the positive ion mode and configured to collect high resolution (R = 120,000 at m/z 200 at an automatic gain control target of 3.0×10^6) broadband mass spectra (m/z 350–1650 Da) with a maximum IT of 60 ms, and MS/MS events (R = 30,000 at an automatic gain control target of 3.0×10^6) with an auto MS^2 IT, isolation window was set to 52.0 m/z, fixed first mass was set to 200 m/z. The 25 segments were selected for MS/MS using a higher energy collisional dissociation (HCD) energy of 28%. The acquisition window covered a mass range from 350 to 1650 m/z through 25 consecutive isolation windows. In each set of DDA or DIA data, the iRT calibration was required with a minimum R^2 of 0.9 by a manual check. During the DIA LC-MS/MS runs across the sequence of the instrument in the study, the coefficient of variation (CV) of iRT peptide retention time was monitored to be less than 3% and the CV of iRT peptide peak intensity to be less than 20% in the QC samples run in-between the DIA analysis of each sample.

3.6. MS Data Analysis

With all the DDA data, Mascot (ver. 2.3, http://www.matrixscience.com/, accessed on 10 April 2021) and X!Tandem (ver. 2013.06.15.1) [85] were used to do a protein database search against a combined database of *Solanum lycopersicum* proteome (ver. 4.1 accessed 16 July 2020; downloaded from the website ftp://ftp.solgenomics.net/tomato_genome/annotation/ITAG4.1_release; 34688 entries; the reverse sequences generated as the decoy database), *P. infestans* proteome (released in April 2020; accessed 17 August 2020; downloaded from the NCBI website ftp://ftp.ncbi.nlm.nih.gov/genomes/genbank/protozoa/Phytophthora_infestans/latest_assembly_versions/GCA_012295175.1_ASM1229517v1, accessed on 10 April 2021; 20172 entries; the reverse sequences generated as the decoy database), sequences of PR-1b protein (Solyc00g174340) which is missing from the ITAG 4.1 database, the iRT standard peptides and BSA (SwissProt Accession: P02769). Search parameters were set as follows: MS tolerance, 10 ppm; precursor monoisotopic mass isotope (M + 1), included; the number of trypsin missed cleavage, 2; fragment mass tolerance, 0.1 Da; enzyme, trypsin/P; fixed modifications, carbamidomethyl cysteine (+57.021 Da); variable modifications, oxidized methionine (+15.995 Da) and acetyl protein N-terminus (+42.016 Da). Next, the search identifications from different search engines and different repeats were combined and statistically scored using PeptideProphet [86] and iProphet [87] within the Trans-Proteomic Pipeline (TPP, ver. 5.2) [88]. MAYU [89] was used to select an iProphet cutoff of 0.913764, resulting in a protein FDR of 1%. SpectraST [90] was used in library generation mode with HCD settings. In the constructed spectral library, there were a total of 65,763 transitions, 53,085 peptides and 11,563 proteins. OpenMS (ver. 2.5) [91] was utilized for decoyed spectral library construction with the workflow adapted from the previously published research [92]. To be included in the final spectral library, product ions were required to contain a minimum of seven amino acids in peptide length and only the best six product ions were selected for each peptide.

The DIA data were analyzed using OpenSWATH (ver. 2.4.0) [93] software against the constructed spectral libraries to identify and quantify peptides and proteins. The retention time alignment used the information of iRT transitions (RT tolerance was set as 7 min for DIA data). In addition to the chromatogram alignment, the spike-in iRT peptide standards were also used for the quality control of the DDA and DIA analyses. Parameters used for feature alignment were set as follows: peptide false-discovery rate (FDR), 0.05; protein FDR, 0.01; alignment method, Local MST; re-alignment method, lowest; retention time (RT) difference, 30 s; alignment score, 0.05. The ratios of protein quantitation between the *P. infestans*-inoculated and mock-treated samples in each replicate were normalized by the most-likely ration normalization principle as previously applied in a DIA study [94]. The mass spectrometry proteomics data have been deposited to the ProteomeXchange Consortium via the PRIDE [95] partner repository with the dataset identifier PXD022266.

3.7. Quantitation Data Analysis

Peptide quantity was exported from the analysis result from OpenSWATH using the measurement by the sum of six best precursors and the peptide ratio was calculated by the peptide quantity of the *P. infestans*-inoculated and mock sample. The protein quantity ratio was calculated by the weighted geometric mean of the unique and/or shared peptide ratios using the peptide quantity as the weighting factor. The criteria for selecting unique and shared peptides for protein or protein group quantification were as follows: (1) if a protein is identified with unique and shared peptides, only the unique peptides will be used; (2) if more than 2 proteins are grouped based on shared identified peptides, both unique and shared peptides will be used as this protein group quantity and any protein without unique peptide identified in this group will be listed in the subset; and (3) if more than 2 proteins are grouped but no unique peptide is identified, then all the shared peptides will be used as this protein group quantity. As the mock and pathogen inoculated detached leaves were collected from paired leaflets of the same plant and samples of three individual plants were pooled in each condition, the initial abundances of proteins (if without treatment) for all conditions at the same growth stage were considered as identical. Thus, a paired Student's t-test (one sample, null hypothesis, no change, mean $\mu = 0$) was performed to uncover differential expression between control and *P. infestans*-inoculated sample for the same growth stage of plants. The t-test was performed based on the protein $\log_2$ ratios of the *P. infestans*- and mock-inoculated sample from three biological replicates. Proteins with a $\log_2$ fold-change of higher than 0.58 or lower than -0.58 and a p-value of less than 0.05 were considered as significant up- or downregulated proteins. The candidate proteins were searched against the *Arabidopsis thaliana* genome assembly *TAIR10* from The Arabidopsis Information Resource (TAIR) database (http://arabidopsis.org) using Protein Basic Local Alignment Search Tool (BLASTP) with E-value $< 1.0 \times 10^{-5}$ (https://blast.ncbi.nlm.nih.gov/Blast.cgi; ver. 2.7.1, accessed on 18 December 2018) first, and these Arabidopsis homologs were then submitted to the web-based platform of the Database for Annotation, Visualization and Integrated Discovery (DAVID; http://david.abcc.ncifcrf.gov, v6.8, accessed on 10 April 2021) for Gene Ontology (GO) enrichment and function analysis.

4. Conclusions

In this study, we investigated the dynamics of the tomato proteome regulation under *P. infestans* biotrophic, biotrophic-necrotrophic transition, and necrotrophic infection stages using a DIA-based quantitative proteomics approach. By constructing a tomato peptide spectral library containing over 11,000 tomato proteins using 2D-LC-MS/MS, a total of 6631 tomato proteins were identified by DIA analysis. At the stage of biotrophic infection, tomato proteins involved in glycolysis, fatty acid/lipid biosynthesis, and ABA signaling were shown to be upregulated whereas proteins involved in direct defense, redox homeostasis, and JA signaling were repressed. The regulated mechanisms may help the pathogen to establish a compatible interaction with the host and obtain more nutrients from the plant cells. From the biotrophic to transition pathogenesis, proteins involved in JA/ET biosynthesis, JA/SA immune signaling, regulation of HR, and defense function were upregulated, whereas proteins involved in ROS production were suppressed. This implies that the immune responses were activated at this stage, but the ROS production may be controlled. When proceeding to the stage of necrotrophic infection, the plant proteome was dramatically affected with regard to most of the resistance-related proteins, including the ones with direct defense functions, immune regulators mediating HR and PCD, and the biosynthetic enzymes of defensive secondary metabolites, which were all upregulated. At this stage, photosynthesis and protein synthesis machinery were ubiquitously downregulated, whereas energy-generation metabolisms were increased, suggesting plants may transfer to the production of energy and regulation as a trade-off between the growth and defense response. Throughout these different pathogenesis stages, the regulation of the defensive phytohormones, including SA, JA, ET, and ABA, possibly caused a synergistic or antagonist effect, and the regulation of the ROS/redox homeostasis appeared to be dynamic during the pathogenesis progression. Several proteins with differentially regulated protein levels across different pathogenesis phases were selected to examine their gene expression levels. Among them, we found that several candidates like PR-5, ENO1, TPL3, PIP1, and CHLH/ABAR could be key targets of *P. infestans* to suppress the host defense/immunity via the transcriptional or translational/post-translational regulation during different infection stage.

It is worth noting that in the early infection stage, it may be essential for *P. infestans* to regulate the protein levels of TPL3, PR-5, and PIP1 to establish a compatible interaction. While in the later infection stages, the regulation of translation and protein stability of TPL3 may be crucial for the pathogen to establish its necrotrophic lifestyle. Hence, this study provides comprehensive information about the regulation in tomato proteome under different pathogenesis stages of *P. infestans*. The translational or post-translational regulation mechanisms of these proteins may also inform the development of a better strategy to control late blight disease in this crop.

Supplementary Materials: Supplementary materials can be found at https://www.mdpi.com/article/10.3390/ijms22084174/s1. Supplemental Figure S1, the expression profile of the marker genes used to clarify *P. infestans* pathogenesis stage. Supplemental Table S1, list of quantified tomato protein with a significant change in abundance at 24, 48, or 96 hpi of *P. infestans* compared to mock. Supplemental Table S2, full list of quantified proteins and peptides in three biological replicates of the infected (24, 48 and 96 hpi) and mock (24, 48 and 96 hpt) experiments. Supplemental Table S3, list of primer sequences for RT-PCR analysis of the *P. infestans* marker genes. Supplemental Table S4, list of primer sequences for qRT-PCR analysis of the selected tomato genes.

Author Contributions: Conceptualization, K.-T.F. and Y.-R.C.; Data curation, W.-H.C.; Funding acquisition, Y.-R.C.; Methodology, K.-T.F., Y.H., C.-H.C. and C.-F.Y.; Project administration, Y.-R.C.; Resources, Y.-R.C.; Software, W.-H.C.; Supervision, Y.-R.C.; Visualization, K.-T.F. and Y.H.; Writing—original draft, K.-T.F., Y.H. and Y.-R.C.; Writing—review & editing, K.-T.F., Y.H. and Y.-R.C. All authors have read and agreed to the published version of the manuscript.

Funding: This work was funded by Academia Sinica, Taiwan (Project ID CDA-105-L03) and the Ministry of Science and Technology, Taiwan (Project ID MOST 107-2113-M-001-006).

Institutional Review Board Statement: Not applicable.

Informed Consent Statement: Not applicable.

Data Availability Statement: The mass spectrometry proteomics data have been deposited to the ProteomeXchange Consortium via the PRIDE partner repository (http://proteomecentral.proteomexchange.org, accessed on 30 October 2020) with the dataset identifier PXD022266.

Acknowledgments: Ruey-Fen Liou at National Taiwan University provided the *Phytophthora infestans* strain and the culturing protocol. The mass spectrometry analysis was supported by the Academia Sinica Metabolomics Core Facility at the Agricultural Biotechnology Research Center of Academia Sinica, funded by Academia Sinica Core Facility and Innovative Instrument Project (AS-CFII-108-108). Miranda Loney at the Agricultural Biotechnology Research Center English Editing Core, Academia Sinica, provided editorial assistance.

Conflicts of Interest: The authors declare no conflict of interest. The funders had no role in the design of the study; in the collection, analyses or interpretation of data; in the writing of the manuscript or in the decision to publish the results.

Abbreviations

ABA	abscisic acid
DIA	data-independent acquisition
DDA	data-dependent acquisition
ET	ethylene
ETI	effector-triggered immunity
ETS	effector-triggered susceptibility
HR	hypersensitive response
JA	jasmonic acid
LB	late blight
PAMP	pathogen-associated molecular pattern
PCD	programmed cell death
PTI	PAMP-triggered immunity
ROS	Reactive-oxidative species
SA	salicylic acid

References

1. Schoina, C.; Govers, F. The Oomycete Phytophthora Infestans, the Irish Potato Famine Pathogen. In *Principles of Plant-Microbe Interactions*; Springer: Cham, Switzerland, 2015; pp. 371–378.
2. Nowicki, M.; Fooled, M.R.; Nowakowska, M.; Kozik, E.U. Potato and Tomato Late Blight Caused by Phytophthora Infestans: An Overview of Pathology and Resistance Breeding. *Plant Dis.* **2012**, *96*, 4–17. [CrossRef] [PubMed]
3. Elsayed, A.Y.; da Silva, D.J.H.; Carneiro, P.C.S.; Mizubuti, E.S.G. The Inheritance of Late Blight Resistance Derived from Solanum Habrochaites. *Crop. Breed. Appl. Biot.* **2012**, *12*, 199–205. [CrossRef]
4. Kamoun, S.; Smart, C.D. Late Blight of Potato and Tomato in the Genomics Era. *Plant Dis.* **2005**, *89*, 692–699. [CrossRef] [PubMed]
5. Hausbeck, M.K.; Lamour, K.H. Phytophthora Capsici on Vegetable Crops: Research Progress and Management Challenges. *Plant Dis.* **2004**, *88*, 1292–1303. [CrossRef]
6. Glazebrook, J. Contrasting Mechanisms of Defense against Biotrophic and Necrotrophic Pathogens. *Annu. Rev. Phytopathol.* **2005**, *43*, 205–227. [CrossRef]
7. Panstruga, R.; Dodds, P.N. Terrific Protein Traffic: The Mystery of Effector Protein Delivery by Filamentous Plant Pathogens. *Science* **2009**, *324*, 748–750. [CrossRef]
8. Koeck, M.; Hardham, A.R.; Dodds, P.N. The Role of Effectors of Biotrophic and Hemibiotrophic Fungi in Infection. *Cell. Microbiol.* **2011**, *13*, 1849–1857. [CrossRef]
9. Wang, B.L.; Liu, J.; Tian, Z.D.; Song, B.T.; Xie, C.H. Monitoring the Expression Patterns of Potato Genes Associated with Quantitative Resistance to Late Blight During Phytophthora Infestans Infection Using Cdna Microarrays. *Plant Sci.* **2005**, *169*, 1155–1167. [CrossRef]
10. Cai, G.; Restrepo, S.; Myers, K.; Zuluaga, P.; Danies, G.; Smart, C.; Fry, W. Gene Profiling in Partially Resistant and Susceptible near-Isogenic Tomatoes in Response to Late Blight in the Field. *Mol. Plant Pathol.* **2013**, *14*, 171–184. [CrossRef]
11. Zuluaga, A.P.; Vega-Arreguin, J.C.; Fei, Z.; Matas, A.J.; Patev, S.; Fry, W.E.; Rose, J.K. Analysis of the Tomato Leaf Transcriptome During Successive Hemibiotrophic Stages of a Compatible Interaction with the Oomycete Pathogen Phytophthora Infestans. *Mol. Plant Pathol.* **2016**, *17*, 42–54. [CrossRef]
12. Larsen, M.K.; Jorgensen, M.M.; Bennike, T.B.; Stensballe, A. Time-Course Investigation of Phytophthora Infestans Infection of Potato Leaf from Three Cultivars by Quantitative Proteomics. *Data Brief.* **2016**, *6*, 238–248. [CrossRef] [PubMed]
13. Ali, A.; Alexandersson, E.; Sandin, M.; Resjo, S.; Lenman, M.; Hedley, P.; Levander, F.; Andreasson, E. Quantitative Proteomics and Transcriptomics of Potato in Response to Phytophthora Infestans in Compatible and Incompatible Interactions. *BMC Genom.* **2014**, *15*, 497. [CrossRef]

14. Xiao, C.; Gao, J.; Zhang, Y.; Wang, Z.; Zhang, D.; Chen, Q.; Ye, X.; Xu, Y.; Yang, G.; Yan, L.; et al. Quantitative Proteomics of Potato Leaves Infected with Phytophthora Infestans Provides Insights into Coordinated and Altered Protein Expression During Early and Late Disease Stages. *Int. J. Mol. Sci.* **2019**, *20*, 136. [CrossRef]
15. Laurindo, B.; Laurindo, R.; Fontes, P.; Vital, C.; Delazari, F.; Baracat-Pereira, M.; da Silva, D. Comparative Proteomics Reveals Set of Oxidative Stress and Thaumatin-Like Proteins Associated with Resistance to Late Blight of Tomato. *Am. J. Plant Sci.* **2018**, *9*, 789–816. [CrossRef]
16. Kelley, B.S.; Lee, S.J.; Damasceno, C.M.; Chakravarthy, S.; Kim, B.D.; Martin, G.B.; Rose, J.K. A Secreted Effector Protein (Sne1) from Phytophthora Infestans Is a Broadly Acting Suppressor of Programmed Cell Death. *Plant J.* **2010**, *62*, 357–366. [CrossRef] [PubMed]
17. van West, P.; de Jong, A.J.; Judelson, H.S.; Emons, A.M.C.; Govers, F. The Ipio Gene of Phytophthora Infestans Is Highly Expressed in Invading Hyphae During Infection. *Fungal Genet. Biol.* **1998**, *23*, 126–138. [CrossRef]
18. Kanneganti, T.-D.; Huitema, E.; Cakir, C.; Kamoun, S. Synergistic Interactions of the Plant Cell Death Pathways Induced by Phytophthora Infestans Nep1-Like Protein Pinpp1. 1 and Inf1 Elicitin. *Mol. Plant-Microbe Interact.* **2006**, *19*, 854–863. [CrossRef] [PubMed]
19. Bartnicki-Garcia, S. Cell Wall Chemistry, Morphogenesis, and Taxonomy of Fungi. *Annu. Rev. Microbiol.* **1968**, *22*, 87–108. [CrossRef]
20. Salzman, R.A.; Koiwa, H.; Ibeas, J.I.; Pardo, J.M.; Hasegawa, P.M.; Bressan, R.A. Inorganic Cations Mediate Plant Pr5 Protein Antifungal Activity through Fungal Mnn1- and Mnn4-Regulated Cell Surface Glycans. *Mol. Plant Microbe Interact.* **2004**, *17*, 780–788. [CrossRef]
21. Batalia, M.A.; Monzingo, A.F.; Ernst, S.; Roberts, W.; Robertus, J.D. The Crystal Structure of the Antifungal Protein Zeamatin, a Member of the Thaumatin-Like, PR-5 Protein Family. *Nat. Struct. Biol.* **1996**, *3*, 19–23. [CrossRef]
22. Ali, S.; Ganai, B.A.; Kamili, A.N.; Bhat, A.A.; Mir, Z.A.; Bhat, J.A.; Tyagi, A.; Islam, S.T.; Mushtaq, M.; Yadav, P.; et al. Pathogenesis-Related Proteins and Peptides as Promising Tools for Engineering Plants with Multiple Stress Tolerance. *Microbiol. Res.* **2018**, *212*, 29–37. [CrossRef]
23. Bantignies, B.; Seguin, J.; Muzac, I.; Dedaldechamp, F.; Gulick, P.; Ibrahim, R. Direct Evidence for Ribonucleolytic Activity of a PR-10-Like Protein from White Lupin Roots. *Plant Mol. Biol.* **2000**, *42*, 871–881. [CrossRef] [PubMed]
24. Zhou, X.J.; Lu, S.; Xu, Y.H.; Wang, J.W.; Chen, X.Y. A Cotton Cdna (Gapr-10) Encoding a Pathogenesis-Related 10 Protein with in Vitro Ribonuclease Activity. *Plant Sci.* **2002**, *162*, 629–636. [CrossRef]
25. Andrade, L.B.; Oliveira, A.S.; Ribeiro, J.K.; Kiyota, S.; Vasconcelos, I.M.; de Oliveira, J.T.; de Sales, M.P. Effects of a Novel Pathogenesis-Related Class 10 (PR-10) Protein from Crotalaria Pallida Roots with Papain Inhibitory Activity against Root-Knot Nematode Meloidogyne Incognita. *J. Agric. Food Chem.* **2010**, *58*, 4145–4152. [CrossRef]
26. Guevara-Morato, M.A.; de Lacoba, M.G.; Garcia-Luque, I.; Serra, M.T. Characterization of a Pathogenesis-Related Protein 4 (PR-4) Induced in Capsicum Chinense L3 Plants with Dual Rnase and Dnase Activities. *J. Exp. Bot.* **2010**, *61*, 3259–3271. [CrossRef]
27. Klarzynski, O.; Plesse, B.; Joubert, J.M.; Yvin, J.C.; Kopp, M.; Kloareg, B.; Fritig, B. Linear Beta-1,3 Glucans Are Elicitors of Defense Responses in Tobacco. *Plant Physiol.* **2000**, *124*, 1027–1038. [CrossRef]
28. Chen, Y.L.; Lee, C.Y.; Cheng, K.T.; Chang, W.H.; Huang, R.N.; Nam, H.G.; Chen, Y.R. Quantitative Peptidomics Study Reveals That a Wound-Induced Peptide from PR-1 Regulates Immune Signaling in Tomato. *Plant Cell* **2014**, *26*, 4135–4148. [CrossRef] [PubMed]
29. Chen, Y.L.; Fan, K.T.; Hung, S.C.; Chen, Y.R. The Role of Peptides Cleaved from Protein Precursors in Eliciting Plant Stress Reactions. *New Phytol.* **2020**, *225*, 2267–2282. [CrossRef] [PubMed]
30. Gamir, J.; Darwiche, R.; Van't Hof, P.; Choudhary, V.; Stumpe, M.; Schneiter, R.; Mauch, F. The Sterol-Binding Activity of Pathogenesis-Related Protein 1 Reveals the Mode of Action of an Antimicrobial Protein. *Plant J.* **2017**, *89*, 502–509. [CrossRef] [PubMed]
31. Kruger, J.; Thomas, C.M.; Golstein, C.; Dixon, M.S.; Smoker, M.; Tang, S.K.; Mulder, L.; Jones, J.D.G. A Tomato Cysteine Protease Required for Cf-2-Dependent Disease Resistance and Suppression of Autonecrosis. *Science* **2002**, *296*, 744–747. [CrossRef]
32. Tian, M.; Win, J.; Song, J.; van der Hoorn, R.; van der Knaap, E.; Kamoun, S. A Phytophthora Infestans Cystatin-Like Protein Targets a Novel Tomato Papain-Like Apoplastic Protease. *Plant Physiol.* **2007**, *143*, 364–377. [CrossRef] [PubMed]
33. Zhou, L.; Cheung, M.-Y.; Li, M.-W.; Fu, Y.; Sun, Z.; Sun, S.-M.; Lam, H.-M. Rice Hypersensitive Induced Reaction Protein 1 (Oshir1) Associates with Plasma Membrane and Triggers Hypersensitive Cell Death. *BMC Plant Biol.* **2010**, *10*, 290. [CrossRef] [PubMed]
34. Qi, Y.; Tsuda, K.; Nguyen le, V.; Wang, X.; Lin, J.; Murphy, A.S.; Glazebrook, J.; Thordal-Christensen, H.; Katagiri, F. Physical Association of Arabidopsis Hypersensitive Induced Reaction Proteins (Hirs) with the Immune Receptor Rps2. *J. Biol. Chem.* **2011**, *286*, 31297–31307. [CrossRef] [PubMed]
35. Lampl, N.; Alkan, N.; Davydov, O.; Fluhr, R. Set-Point Control of Rd21 Protease Activity by Atserpin1 Controls Cell Death in Arabidopsis. *Plant J.* **2013**, *74*, 498–510. [CrossRef]
36. Liu, Y.; Wang, K.; Cheng, Q.; Kong, D.; Zhang, X.; Wang, Z.; Wang, Q.; Xie, Q.; Yan, J.; Chu, J.; et al. Cysteine Protease Rd21a Regulated by E3 Ligase Sinat4 Is Required for Drought-Induced Resistance to Pseudomonas syringae in Arabidopsis. *J. Exp. Bot.* **2020**, *71*, 5562–5576. [CrossRef]
37. DiMario, R.J.; Clayton, H.; Mukherjee, A.; Ludwig, M.; Moroney, J.V. Plant Carbonic Anhydrases: Structures, Locations, Evolution, and Physiological Roles. *Mol. Plant* **2017**, *10*, 30–46. [CrossRef]

38. Zhou, Y.; Vroegop-Vos, I.A.; Van Dijken, A.J.H.; Van der Does, D.; Zipfel, C.; Pieterse, C.M.J.; Van Wees, S.C.M. Carbonic Anhydrases Ca1 and Ca4 Function in Atmospheric Co2-Modulated Disease Resistance. *Planta* **2020**, *251*, 75. [CrossRef]
39. Slaymaker, D.H.; Navarre, D.A.; Clark, D.; del Pozo, O.; Martin, G.B.; Klessig, D.F. The Tobacco Salicylic Acid-Binding Protein 3 (Sabp3) Is the Chloroplast Carbonic Anhydrase, Which Exhibits Antioxidant Activity and Plays a Role in the Hypersensitive Defense Response. *Proc. Natl. Acad. Sci. USA* **2002**, *99*, 11640–11645. [CrossRef]
40. Restrepo, S.; Myers, K.L.; del Pozo, O.; Martin, G.B.; Hart, A.L.; Buell, C.R.; Fry, W.E.; Smart, C.D. Gene Profiling of a Compatible Interaction between Phytophthora Infestans and Solanum Tuberosum Suggests a Role for Carbonic Anhydrase. *Mol. Plant-Microbe Interact.* **2005**, *18*, 913–922. [CrossRef]
41. Wirthmueller, L.; Zhang, Y.; Jones, J.D.; Parker, J.E. Nuclear Accumulation of the Arabidopsis Immune Receptor Rps4 Is Necessary for Triggering Eds1-Dependent Defense. *Curr. Biol.* **2007**, *17*, 2023–2029. [CrossRef]
42. Liu, Y.; Schiff, M.; Marathe, R.; Dinesh-Kumar, S.P. Tobacco Rar1, Eds1 and NPR1/Nim1 Like Genes Are Required for N-Mediated Resistance to Tobacco Mosaic Virus. *Plant J.* **2002**, *30*, 415–429. [CrossRef]
43. Hu, G.S.; deHart, A.K.A.; Li, Y.S.; Ustach, C.; Handley, V.; Navarre, R.; Hwang, C.F.; Aegerter, B.J.; Williamson, V.M.; Baker, B. Eds1 in Tomato Is Required for Resistance Mediated by Tir-Class R Genes and the Receptor-Like R Gene Ve. *Plant J.* **2005**, *42*, 376–391. [CrossRef] [PubMed]
44. Parker, J.E.; Holub, E.B.; Frost, L.N.; Falk, A.; Gunn, N.D.; Daniels, M.J. Characterization of Eds1, a Mutation in Arabidopsis Suppressing Resistance to Peronospora Parasitica Specified by Several Different Rpp Genes. *Plant Cell* **1996**, *8*, 2033–2046. [PubMed]
45. Heidrich, K.; Wirthmueller, L.; Tasset, C.; Pouzet, C.; Deslandes, L.; Parker, J.E. Arabidopsis Eds1 Connects Pathogen Effector Recognition to Cell Compartment-Specific Immune Responses. *Science* **2011**, *334*, 1401–1404. [CrossRef]
46. Glazebrook, J.; Rogers, E.E.; Ausubel, F.M. Isolation of Arabidopsis Mutants with Enhanced Disease Susceptibility by Direct Screening. *Genetics* **1996**, *143*, 973–982. [CrossRef]
47. Cui, H.; Gobbato, E.; Kracher, B.; Qiu, J.; Bautor, J.; Parker, J.E. A Core Function of Eds1 with Pad4 Is to Protect the Salicylic Acid Defense Sector in Arabidopsis Immunity. *New Phytol.* **2017**, *213*, 1802–1817. [CrossRef] [PubMed]
48. Li, J.; Brader, G.; Palva, E.T. Kunitz Trypsin Inhibitor: An Antagonist of Cell Death Triggered by Phytopathogens and Fumonisin B1 in Arabidopsis. *Mol. Plant* **2008**, *1*, 482–495. [CrossRef]
49. Arnaiz, A.; Talavera-Mateo, L.; Gonzalez-Melendi, P.; Martinez, M.; Diaz, I.; Santamaria, M.E. Arabidopsis Kunitz Trypsin Inhibitors in Defense against Spider Mites. *Front. Plant Sci.* **2018**, *9*, 986. [CrossRef]
50. Heitz, T.; Bergey, D.R.; Ryan, C.A. A Gene Encoding a Chloroplast-Targeted Lipoxygenase in Tomato Leaves Is Transiently Induced by Wounding, Systemin, and Methyl Jasmonate. *Plant Physiol.* **1997**, *114*, 1085–1093. [CrossRef] [PubMed]
51. Zhao, Y.; Thilmony, R.; Bender, C.L.; Schaller, A.; He, S.Y.; Howe, G.A. Virulence Systems of Pseudomonas syringae pv. Tomato Promote Bacterial Speck Disease in Tomato by Targeting the Jasmonate Signaling Pathway. *Plant J.* **2003**, *36*, 485–499. [CrossRef]
52. Pauwels, L.; Barbero, G.F.; Geerinck, J.; Tilleman, S.; Grunewald, W.; Perez, A.C.; Chico, J.M.; Vanden Bossche, R.; Sewell, J.; Gil, E.; et al. Ninja Connects the Co-Repressor Topless to Jasmonate Signalling. *Nature* **2010**, *464*, 788–U169. [CrossRef]
53. Causier, B.; Ashworth, M.; Guo, W.; Davies, B. The Topless Interactome: A Framework for Gene Repression in Arabidopsis. *Plant Physiol.* **2012**, *158*, 423–438. [CrossRef]
54. Harvey, S.; Kumari, P.; Lapin, D.; Griebel, T.; Hickman, R.; Guo, W.; Zhang, R.; Parker, J.E.; Beynon, J.; Denby, K. Downy Mildew Effector Harxl21 Interacts with the Transcriptional Repressor Topless to Promote Pathogen Susceptibility. *bioRxiv* **2020**, *16*, e1008835. [CrossRef]
55. Mochizuki, N.; Brusslan, J.A.; Larkin, R.; Nagatani, A.; Chory, J. Arabidopsis Genomes Uncoupled 5 (Gun5) Mutant Reveals the Involvement of Mg-Chelatase H Subunit in Plastid-to-Nucleus Signal Transduction. *Proc. Natl. Acad. Sci. USA* **2001**, *98*, 2053–2058. [CrossRef]
56. Ibata, H.; Nagatani, A.; Mochizuki, N. Chlh/Gun5 Function in Tetrapyrrole Metabolism Is Correlated with Plastid Signaling but Not Aba Responses in Guard Cells. *Front. Plant Sci.* **2016**, *7*, 1650. [CrossRef]
57. Shang, Y.; Yan, L.; Liu, Z.Q.; Cao, Z.; Mei, C.; Xin, Q.; Wu, F.Q.; Wang, X.F.; Du, S.Y.; Jiang, T.; et al. The Mg-Chelatase H Subunit of Arabidopsis Antagonizes a Group of Wrky Transcription Repressors to Relieve Aba-Responsive Genes of Inhibition. *Plant Cell* **2010**, *22*, 1909–1935. [CrossRef] [PubMed]
58. Berens, M.L.; Berry, H.M.; Mine, A.; Argueso, C.T.; Tsuda, K. Evolution of Hormone Signaling Networks in Plant Defense. *Annu. Rev. Phytopathol.* **2017**, *55*, 401–425. [CrossRef] [PubMed]
59. Sagi, M.; Davydov, O.; Orazova, S.; Yesbergenova, Z.; Ophir, R.; Stratmann, J.W.; Fluhr, R. Plant Respiratory Burst Oxidase Homologs Impinge on Wound Responsiveness and Development in Lycopersicon Esculentum. *Plant Cell* **2004**, *16*, 616–628. [CrossRef] [PubMed]
60. Kumar, S.; Kaur, A.; Chattopadhyay, B.; Bachhawat, A.K. Defining the Cytosolic Pathway of Glutathione Degradation in Arabidopsis Thaliana: Role of the Chac/Gcg Family of Gamma-Glutamyl Cyclotransferases as Glutathione-Degrading Enzymes and Atlap1 as the Cys-Gly Peptidase. *Biochem. J.* **2015**, *468*, 73–85. [CrossRef] [PubMed]
61. Rojas, C.M.; Senthil-Kumar, M.; Tzin, V.; Mysore, K. Regulation of Primary Plant Metabolism during Plant-Pathogen Interactions and Its Contribution to Plant Defense. *Front. Plant Sci.* **2014**, *5*, 17. [CrossRef] [PubMed]
62. Dixon, R.A.; Achnine, L.; Kota, P.; Liu, C.J.; Reddy, M.S.; Wang, L. The Phenylpropanoid Pathway and Plant Defence—A Genomics Perspective. *Mol. Plant Pathol.* **2002**, *3*, 371–390. [CrossRef]
63. Vogt, T. Phenylpropanoid Biosynthesis. *Mol. Plant* **2010**, *3*, 2–20. [CrossRef] [PubMed]

64. McManus, J.; Cheng, Z.; Vogel, C. Next-Generation Analysis of Gene Expression Regulation–Comparing the Roles of Synthesis and Degradation. *Mol. Biosyst.* **2015**, *11*, 2680–2689. [CrossRef] [PubMed]
65. Liu, Y.; Beyer, A.; Aebersold, R. On the Dependency of Cellular Protein Levels on Mrna Abundance. *Cell* **2016**, *165*, 535–550. [CrossRef] [PubMed]
66. Tsuzuki, T.; Takahashi, K.; Inoue, S.; Okigaki, Y.; Tomiyama, M.; Hossain, M.A.; Shimazaki, K.; Murata, Y.; Kinoshita, T. Mg-Chelatase H Subunit Affects Aba Signaling in Stomatal Guard Cells, but Is Not an Aba Receptor in Arabidopsis Thaliana. *J. Plant Res.* **2011**, *124*, 527–538. [CrossRef]
67. Zhang, D.-P.; Wu, Z.-Y.; Li, X.-Y.; Zhao, Z.-X. Purification and Identification of a 42-Kilodalton Abscisic Acid-Specific-Binding Protein from Epidermis of Broad Bean Leaves. *Plant Physiol.* **2002**, *128*, 714–725. [CrossRef] [PubMed]
68. Wu, F.Q.; Xin, Q.; Cao, Z.; Liu, Z.Q.; Du, S.Y.; Mei, C.; Zhao, C.X.; Wang, X.F.; Shang, Y.; Jiang, T.; et al. The Magnesium-Chelatase H Subunit Binds Abscisic Acid and Functions in Abscisic Acid Signaling: New Evidence in Arabidopsis. *Plant Physiol.* **2009**, *150*, 1940–1954. [CrossRef] [PubMed]
69. Shen, Y.Y.; Wang, X.F.; Wu, F.Q.; Du, S.Y.; Cao, Z.; Shang, Y.; Wang, X.L.; Peng, C.C.; Yu, X.C.; Zhu, S.Y.; et al. The Mg-Chelatase H Subunit Is an Abscisic Acid Receptor. *Nature* **2006**, *443*, 823–826. [CrossRef]
70. Muller, A.H.; Hansson, M. The Barley Magnesium Chelatase 150-Kd Subunit Is Not an Abscisic Acid Receptor. *Plant Physiol.* **2009**, *150*, 157–166. [CrossRef]
71. Chan, K.X.; Phua, S.Y.; Crisp, P.; McQuinn, R.; Pogson, B.J. Learning the Languages of the Chloroplast: Retrograde Signaling and Beyond. *Annu. Rev. Plant Biol.* **2016**, *67*, 25–53. [CrossRef]
72. Agrios, G.N. *Plant Pathology*, 5th ed.; Elsevier Academic Press: Amsterdam, The Netherlands; Boston, MA, USA, 2005; p. xxiii. 922p.
73. Li, Y.; Han, Y.; Qu, M.; Chen, J.; Chen, X.; Geng, X.; Wang, Z.; Chen, S. Apoplastic Cell Death-Inducing Proteins of Filamentous Plant Pathogens: Roles in Plant-Pathogen Interactions. *Front. Genet.* **2020**, *11*, 661. [CrossRef]
74. Caten, C.; Jinks, J. Spontaneous Variability of Single Isolates of Phytophthora Infestans. I. Cultural Variation. *Can. J. Bot.* **1968**, *46*, 329–348. [CrossRef]
75. Wilson, U.; Coffey, M. Cytological Evaluation of General Resistance to Phytophthora Infestans in Potato Foliage. *Ann. Bot.* **1980**, *45*, 81–90.
76. Yin, J.; Gu, B.; Huang, G.; Tian, Y.; Quan, J.; Lindqvist-Kreuze, H.; Shan, W. Conserved Rxlr Effector Genes of Phytophthora Infestans Expressed at the Early Stage of Potato Infection Are Suppressive to Host Defense. *Front. Plant Sci.* **2017**, *8*, 2155. [CrossRef]
77. Schneider, C.A.; Rasband, W.S.; Eliceiri, K.W. Nih Image to Imagej: 25 Years of Image Analysis. *Nat. Methods* **2012**, *9*, 671–675. [CrossRef] [PubMed]
78. Faurobert, M.; Pelpoir, E.; Chaib, J. Phenol Extraction of Proteins for Proteomic Studies of Recalcitrant Plant Tissues. *Methods Mol. Biol.* **2007**, *355*, 9–14.
79. Wu, X.; Xiong, E.; Wang, W.; Scali, M.; Cresti, M. Universal Sample Preparation Method Integrating Trichloroacetic Acid/Acetone Precipitation with Phenol Extraction for Crop Proteomic Analysis. *Nat. Protoc.* **2014**, *9*, 362–374. [CrossRef] [PubMed]
80. Saravanan, R.S.; Rose, J.K. A Critical Evaluation of Sample Extraction Techniques for Enhanced Proteomic Analysis of Recalcitrant Plant Tissues. *Proteomics* **2004**, *4*, 2522–2532. [CrossRef] [PubMed]
81. Fan, K.T.; Wang, K.H.; Chang, W.H.; Yang, J.C.; Yeh, C.F.; Cheng, K.T.; Hung, S.C.; Chen, Y.R. Application of Data-Independent Acquisition Approach to Study the Proteome Change from Early to Later Phases of Tomato Pathogenesis Responses. *Int. J. Mol. Sci.* **2019**, *20*, 863. [CrossRef] [PubMed]
82. Chen, C.J.; Chen, W.Y.; Tseng, M.C.; Chen, Y.R. Tunnel Frit: A Nonmetallic in-Capillary Frit for Nanoflow Ultra High-Performance Liquid Chromatography-Mass Spectrometryapplications. *Anal. Chem.* **2012**, *84*, 297–303. [CrossRef] [PubMed]
83. Scheltema, R.A.; Hauschild, J.P.; Lange, O.; Hornburg, D.; Denisov, E.; Damoc, E.; Kuehn, A.; Makarov, A.; Mann, M. The Q Exactive HF, a Benchtop Mass Spectrometer with a Pre-Filter, High-Performance Quadrupole and an Ultra-High-Field Orbitrap Analyzer. *Mol. Cell. Proteom. MCP* **2014**, *13*, 3698–3708. [CrossRef]
84. Bruderer, R.; Bernhardt, O.M.; Gandhi, T.; Xuan, Y.; Sondermann, J.; Schmidt, M.; Gomez-Varela, D.; Reiter, L. Optimization of Experimental Parameters in Data-Independent Mass Spectrometry Significantly Increases Depth and Reproducibility of Results. *Mol. Cell. Proteom. MCP* **2017**, *16*, 2296–2309. [CrossRef]
85. Craig, R.; Beavis, R.C. Tandem: Matching Proteins with Tandem Mass Spectra. *Bioinformatics* **2004**, *20*, 1466–1467. [CrossRef] [PubMed]
86. Keller, A.; Nesvizhskii, A.I.; Kolker, E.; Aebersold, R. Empirical Statistical Model to Estimate the Accuracy of Peptide Identifications Made by MS/MS and Database Search. *Anal. Chem.* **2002**, *74*, 5383–5392. [CrossRef] [PubMed]
87. Shteynberg, D.; Deutsch, E.W.; Lam, H.; Eng, J.K.; Sun, Z.; Tasman, N.; Mendoza, L.; Moritz, R.L.; Aebersold, R.; Nesvizhskii, A.I. Iprophet: Multi-Level Integrative Analysis of Shotgun Proteomic Data Improves Peptide and Protein Identification Rates and Error Estimates. *Mol. Cell. Proteom. MCP* **2011**, *10*, M111-007690. [CrossRef] [PubMed]
88. Deutsch, E.W.; Mendoza, L.; Shteynberg, D.; Slagel, J.; Sun, Z.; Moritz, R.L. Trans-Proteomic Pipeline, a Standardized Data Processing Pipeline for Large-Scale Reproducible Proteomics Informatics. *Proteom. Clin. Appl.* **2015**, *9*, 745–754. [CrossRef]

89. Reiter, L.; Claassen, M.; Schrimpf, S.P.; Jovanovic, M.; Schmidt, A.; Buhmann, J.M.; Hengartner, M.O.; Aebersold, R. Protein Identification False Discovery Rates for Very Large Proteomics Data Sets Generated by Tandem Mass Spectrometry. *Mol. Cell. Proteom. MCP* **2009**, *8*, 2405–2417. [CrossRef]
90. Lam, H.; Deutsch, E.W.; Eddes, J.S.; Eng, J.K.; Stein, S.E.; Aebersold, R. Building Consensus Spectral Libraries for Peptide Identification in Proteomics. *Nat. Methods* **2008**, *5*, 873–875. [CrossRef]
91. Röst, H.L.; Sachsenberg, T.; Aiche, S.; Bielow, C.; Weisser, H.; Aicheler, F.; Andreotti, S.; Ehrlich, H.C.; Gutenbrunner, P.; Kenar, E.; et al. OpenMS: A Flexible Open-Source Software Platform for Mass Spectrometry Data Analysis. *Nat. Methods* **2016**, *13*, 741–748. [CrossRef]
92. Schubert, O.T.; Gillet, L.C.; Collins, B.C.; Navarro, P.; Rosenberger, G.; Wolski, W.E.; Lam, H.; Amodei, D.; Mallick, P.; MacLean, B.; et al. Building High-Quality Assay Libraries for Targeted Analysis of Swath MS Data. *Nat. Protoc.* **2015**, *10*, 426–441. [CrossRef]
93. Röst, H.L.; Rosenberger, G.; Navarro, P.; Gillet, L.; Miladinović, S.M.; Schubert, O.T.; Wolski, W.; Collins, B.C.; Malmström, J.; Malmström, L.; et al. OpenSWATH Enables Automated, Targeted Analysis of Data- Independent Acquisition MS Data. *Nat. Biotechnol.* **2014**, *32*, 219–223. [CrossRef] [PubMed]
94. Lambert, J.P.; Ivosev, G.; Couzens, A.L.; Larsen, B.; Taipale, M.; Lin, Z.Y.; Zhong, Q.; Lindquist, S.; Vidal, M.; Aebersold, R.; et al. Mapping Differential Interactomes by Affinity Purification Coupled with Data-Independent Mass Spectrometry Acquisition. *Nat. Methods* **2013**, *10*, 1239–1245. [CrossRef] [PubMed]
95. Perez-Riverol, Y.; Csordas, A.; Bai, J.; Bernal-Llinares, M.; Hewapathirana, S.; Kundu, D.J.; Inuganti, A.; Griss, J.; Mayer, G.; Eisenacher, M.; et al. The PRIDE Database and Related Tools and Resources in 2019: Improving Support for Quantification Data. *Nucleic Acids Res.* **2019**, *47*, D442–D450. [CrossRef] [PubMed]

International Journal of
Molecular Sciences

MDPI

Article

Leaf Apoplast of Field-Grown Potato Analyzed by Quantitative Proteomics and Activity-Based Protein Profiling

Kibrom B. Abreha [1,*,†], Erik Alexandersson [1], Svante Resjö [1], Åsa Lankinen [1], Daniela Sueldo [2,‡], Farnusch Kaschani [3], Markus Kaiser [3], Renier A. L. van der Hoorn [2], Fredrik Levander [4,5] and Erik Andreasson [1]

1 Department of Plant Protection Biology, Swedish University of Agricultural Sciences, SE-234 22 Lomma, Sweden; Erik.Alexandersson@slu.se (E.A.); Svante.Resjo@slu.se (S.R.); Asa.Lankinen@slu.se (Å.L.); Erik.Andreasson@slu.se (E.A.)
2 Plant Chemetics Laboratory, Department of Plant Sciences, University of Oxford, South Parks Road, Oxford OX1 3RB, UK; daniela.sueldo@ntnu.no (D.S.); renier.vanderhoorn@plants.ox.ac.uk (R.A.L.v.d.H.)
3 Chemische Biologie, Zentrum für Medizinische Biotechnologie, Fakultät für Biologie, Universität Duisburg-Essen, Universitätsstr. 2, 45117 Essen, Germany; farnusch.kaschani@uni-due.de (F.K.); markus.kaiser@uni-due.de (M.K.)
4 Department of Immunotechnology, Lund University, SE-221 00 Lund, Sweden; fredrik.levander@immun.lth.se
5 National Bioinformatics Infrastructure Sweden (NBIS), Science for Life Laboratory, Lund University, SE-221 00 Lund, Sweden
* Correspondence: kibrom.abreha@slu.se; Tel.: +4-640-415-000
† Current address: Department of Plant Breeding, Swedish University of Agricultural Sciences, SE-234 22 Lomma, Sweden.
‡ Current address: Department of Biology, Norwegian University of Science and Technology, 5 Hogskoleringen, N-7491 Trondheim, Norway.

Citation: Abreha, K.B.; Alexandersson, E.; Resjö, S.; Lankinen, Å.; Sueldo, D.; Kaschani, F.; Kaiser, M.; van der Hoorn, R.A.L.; Levander, F.; Andreasson, E. Leaf Apoplast of Field-Grown Potato Analyzed by Quantitative Proteomics and Activity-Based Protein Profiling. *Int. J. Mol. Sci.* **2021**, *22*, 12033. https://doi.org/10.3390/ijms222112033

Academic Editors: Setsuko Komatsu and Sixue Chen

Received: 30 September 2021
Accepted: 2 November 2021
Published: 6 November 2021

Publisher's Note: MDPI stays neutral with regard to jurisdictional claims in published maps and institutional affiliations.

Abstract: Multiple biotic and abiotic stresses challenge plants growing in agricultural fields. Most molecular studies have aimed to understand plant responses to challenges under controlled conditions. However, studies on field-grown plants are scarce, limiting application of the findings in agricultural conditions. In this study, we investigated the composition of apoplastic proteomes of potato cultivar Bintje grown under field conditions, i.e., two field sites in June–August across two years and fungicide treated and untreated, using quantitative proteomics, as well as its activity using activity-based protein profiling (ABPP). Samples were clustered and some proteins showed significant intensity and activity differences, based on their field site and sampling time (June–August), indicating differential regulation of certain proteins in response to environmental or developmental factors. Peroxidases, class II chitinases, pectinesterases, and osmotins were among the proteins more abundant later in the growing season (July–August) as compared to early in the season (June). We did not detect significant differences between fungicide Shirlan treated and untreated field samples in two growing seasons. Using ABPP, we showed differential activity of serine hydrolases and β-glycosidases under greenhouse and field conditions and across a growing season. Furthermore, the activity of serine hydrolases and β-glycosidases, including proteins related to biotic stress tolerance, decreased as the season progressed. The generated proteomics data would facilitate further studies aiming at understanding mechanisms of molecular plant physiology in agricultural fields and help applying effective strategies to mitigate biotic and abiotic stresses.

Keywords: ABPP; apoplast; proteomics; serine hydrolases; β-glycosidases; potato; field-omics

1. Introduction

In an agricultural field, plants are continuously exposed to varying climate conditions and challenged by mu below- and above-ground microbes, which can trigger morphological and molecular changes. For example, compared to plants grown under greenhouse conditions, Arabidopsis grown in a field displays a different leaf morphology [1] and apoplastic proteome profile [2] indicating changes in response to the stresses or variable environmental condition. Therefore, molecular studies performed under laboratory-based conditions investigating response to single stress might not translate directly to field conditions where there are multiple confounding stresses [2,3]. Field experiments are an integral component to test applicability of laboratory-based findings, and they are necessary to solve agricultural challenges (for example, in relation to plant breeding and plant protection). Therefore, field experiments are important to gain basic knowledge about the biology of plant field performance and ecology and to understand the differences in physiology and molecular function of plants growing under field conditions. This is crucial to facilitate applicability of field-based studies (for instance, creating a basis for decision support systems in agriculture based on molecular knowledge) [4].

Potatoes (*Solanum tuberosum* L.; $2n = 4x = 48$), the third most important food crop in the world, are exposed to multiple biotic and abiotic stresses [5]. The plant apoplast is an important arena in plant–microbe interactions [6–8] and plays a crucial role in the plant's response to abiotic stresses [9], as proteins from the plant and the attacking pathogens are secreted into the apoplast [7,8]. Apoplastic proteomic studies are crucial to understand defense response against pathogenic microbes [10,11]. In potatoes, many proteins in the leaf apoplast change in abundance following the application of host defense inducers [12,13] and *P. infestans* inoculation [14]. An apoplastic study has also been used for predicting activation of defense response in *Solanum* species growing under natural and agricultural conditions [15].

Omics techniques have been used in laboratory-based studies to better understand potato response to biotic [14,16,17] and abiotic stresses [18–20]. However, application of these techniques to elucidate the molecular processes in field-grown plants, the so called field-omics approach, is scarce [21]. Label-free quantitative proteomics, based on sensitive and reliable mass spectrometry, has emerged as a powerful tool to investigate plant responses to biotic stress and made it possible to investigate differential abundance of proteins in many different systems [14,22,23]. Nevertheless, a change in abundance of a protein identified by quantitative proteomics does not necessarily translate into a change in activity. Recent developments in activity-based protein profiling (ABPP), a method that uses chemical probes that irreversibly bind to the active residue of distinct protein classes in a complex proteome sample, can be used to investigate the functional status of the proteins [24]. This powerful tool has been successfully applied to enhance molecular understanding of different plant–microbe interactions [25,26] and the plant's response to biotic stresses [27]. ABPP is used to determine the activity of certain protein families, including serine hydrolases and β-glucosidases, in plants challenged by biotic and abiotic stresses [26,27]. Serine hydrolases, carrying an activated serine residue in the catalytic triad, are a large superfamily of enzymes that includes proteases, esterases, lipases, and peptidases [24,26]. β-Glucosidases are glycoside hydrolases that carry a glutamate and aspartate residue at the catalytic site [28]. These enzyme groups are important in cellular processes such as cell wall remodeling, development, biotic and abiotic stress responses. However, to the best of our knowledge, ABPP has not been applied in studies conducted in plants under field conditions.

The objective of this study was to investigate the apoplastic proteome using label-free quantitative proteomics and ABPP to identify apoplastic proteins under field conditions. Therefore, we investigated the apoplastic proteome of field grown potato plants from two different years and followed the change in the abundance of apoplastic proteins during the course of the growing season. Moreover, we performed ABPP on serine hydrolases and β-glucosidases, to understand the functional state of proteins involved in some biological processes.

2. Results and Discussion

2.1. Analysis of Apoplastic Peptides

To get insights into the proteome of field-grown plants, we collected the leaf apoplastic fluid from fully expanded potato leaves of field-grown plants from two field sites over a two–year period. The apoplast is in many cases the first plant compartment that directly interact with the environment and does not contain any dominating protein such as rubisco. Therefore, we chose to analyze this part of the potato proteins under field conditions. The leaf apoplastic fluid was collected using our mobile laboratory [21,29,30] that allows us to isolate, aliquot, and freeze the samples directly on the site, thus reducing the probability of sample degradation, which can be a problem in proteome sampling [31].

In total, we identified 3960 peptides, of which 501 were identical except for the charge differences of the intact peptide precursor ion detected by MS. Average abundance was computed for the identical peptides resulting in 3459 unique peptides, corresponding to 1257 proteins, which were subjected to quantitative analysis. Of the 3459 peptides, 2335 peptides were uniquely identified for a protein in the sequence database. Therefore, they were designated as diagnostic peptides. The use of our assembled RNA-seq data obtained from potato Desirée (cv.), SW93-1015 (breeding line), and Sarpo Mira (cv.) improved peptide identification by 6% as estimated by the fraction of peptides mapped uniquely to accessions in the RNA-seq data. This increase in peptide identification was however less than the previously reported 17% [14], maybe due to the fact that we used a different potato cultivar Bintje, not represented in our RNA-seq database used for peptide identification. We identified 85 of the 104 peptides previously reported to be have prediction potential for *P. infestans* resistance in leaf and tuber, as well as for tuber yield, in the potato clones SW93-1015 and Desirée [32].

Of the 1257 proteins, 419 were predicted to contain a signal peptide as determined by SignalP version 4.1 [33]. This is in line with the previous studies [6,26], where 30–60% of proteins identified in the apoplast contained a classical signal peptide. Furthermore, Pfam analysis of proteins identified in samples from field-grown plants identified 461 protein families. Peptidases, peroxidases, glycosyl hydrolases, protein GDSL lipases, thaumatin, heat shock proteins, Leucine rich repeat proteins, and epimerases being the predominant protein families in the apoplast of field-grown potato cv. Bintje (Figure 1).

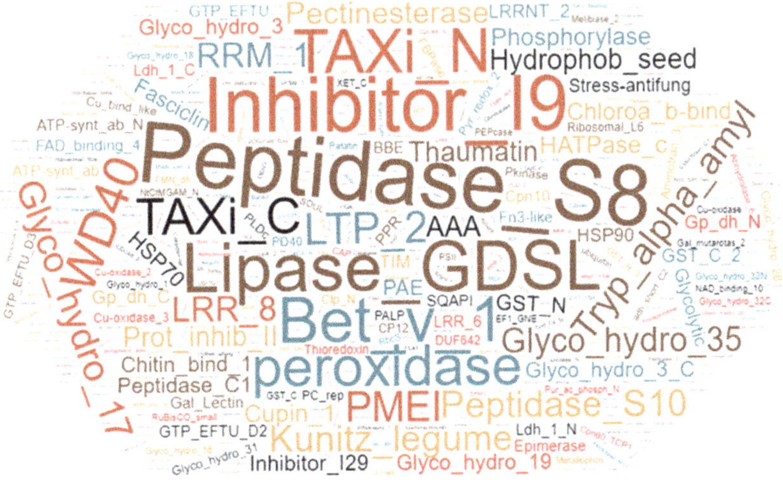

Figure 1. Word-cloud representation of the leaf apoplastic proteome from potato cultivar Bintje. Identified proteins were classified into families using Pfam analysis. Scale of the fonts and colors in the word cloud represents relative abundance of the protein family in the apoplast samples from field-grown plants.

In our dataset, the correlation between peptide abundance and proportion of samples with missing values was $r^2 = -0.31$ (Supplementary Figure S1). Such a negative correlation between peptide abundance and missing values is expected and has previously been reported [34].

2.2. Effect of Fungicide Application on Potato Apoplastic Proteome

In order to investigate whether the application of a fungicide to control a disease change the abundance of proteins in the apoplast, we compared fungicide treated samples with untreated ones. The fungicide application may directly or indirectly, by changing the microbial communities [35], affect the apoplastic proteome of the plants. Some of the field plants were treated with fungicide Shirlan (classified as non-systemic), targeting foliar and tuber late blight infections in potato. According to the Public Release Summary on the fungicide [36], spraying with C^{14}-tagged fluazinam (500 g L^{-1}), the active ingredient in Shirlan, revealed small traces of the label in the potato pulp, showing some translocation of the fungicide into the plant. We performed separately two-group comparison (t-test) analysis in Qlucore for 12 samples from 2011 and 18 samples from 2012 in Mosslunda; however, no differentially abundant protein was detected between fungicide-treated and untreated plants in both years in Mosslunda ($q < 0.1$) (data not shown). Therefore, we conclude that the application of fungicide Shirlan did not significantly change the abundance of the apoplastic proteins in potato leaves. Further studies investigating levels and timing of fungicide application, and its translocation and stability in plant tissues would provide crucial insights into its effects on apoplastic proteome.

2.3. Apoplastic Proteome Differences between Growing Sites and between Years

To further understand and describe the dynamics of apoplastic proteome in potato leaves growing in field conditions, we compare the data set between the two sites Borgeby in 2010 and in Mosslunda both in 2011 and 2012 growing seasons. The PCA clustered the samples from the two growing sites together (Figure 2A).

Similarly, ref. [2] found an overlap in apoplastic proteome composition of *Arabidopsis* plants collected from two field sites. It has been shown that the effect of weather conditions on tuber proteome of potato grown at two different fields is minimal [37]. The overlap between field samples observed in the PCA plot indicates the similarity between the apoplastic protein compositions of the potato plants grown in Mosslunda and those grown in Borgeby (Figure 2A). This indicates similar apoplastic proteome profile among the plants grown in these sites. However, a transcriptome analysis of grapevine berries grown in different sites identified more than 8000 differentially expressed genes [38].

To describe individual proteins that were differentially regulated between potato plants grown at these two sites, we performed a two-group comparison (t-test) in Qlucore ($q < 0.001$) and found 314 peptides (9%) from 234 proteins (Figure 2C). These results indicate that, despite the overlap of Mosslunda and Borgeby samples in the PCA plot, hence similarity in apoplastic proteome, not all of the proteins had a similar pattern of abundance at both growing sites. The abundance of subtilisin-like proteases (Q9LWA4), endochitinase (DMP400046624), Kunitz trypsin inhibitor (DMP400046980), pectinesterase (DMP400055021), and peroxidases (DMP400052953, Q9SD46) were higher in Borgeby plants compared to those grown in Mosslunda and those of glucan endo-1,3-β-D-glucosidase (DMP400051976), glyceraldehyde-3-phosphate dehydrogenase (DMP400017652), and methionine synthase (DMP400048869) were lower (Supplementary Table S2).

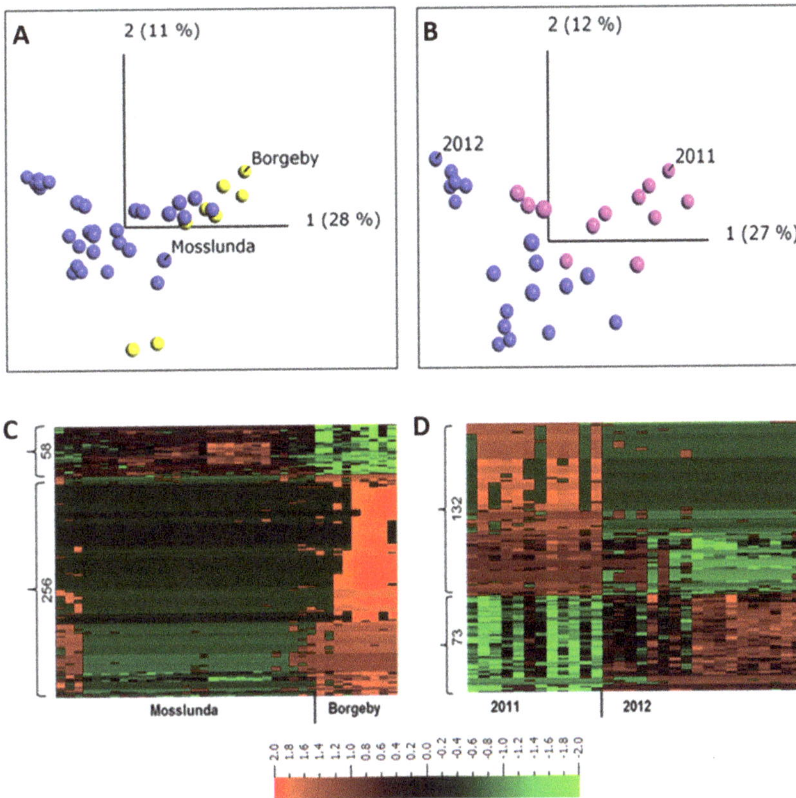

Figure 2. Principal components and heat map analyses of apoplastic proteome samples isolated from potato cultivar Bintje grown at two experimental sites (Borgeby and Mosslunda) and in two different growing seasons (2011 and 2012) in Mosslunda. (**A**) Unsupervised principal component analysis plot of the samples in Mosslunda and Borgeby; (**B**) Unsupervised principal component analysis plot of the samples in Mosslunda in 2011 and 2012; (**C**) Abundance of peptides from Borgeby respectively compared to those in Mosslunda; (**D**) Abundance of peptides collected in 2011 respectively compared to those collected in 2012 in Mosslunda. Two-group comparisons (t-test) were performed in Qlucore with false discovery rate using Benjamini–Hochberg correction ($q < 0.001$). Heat maps are sorted using hierarchal clustering and red represents higher abundance (Fold change, log2).

Similarly, to investigate the differences between the years, apoplastic samples were collected in 2011 and 2012 from plants grown in Mosslunda. The PCA analysis grouped the samples from the same year together, but there was an overlap between the samples from both years (Figure 2B). A two-group comparison (t-test) analysis in Qlucore ($q \leq 0.001$) found 205 peptides (6% of total peptides, corresponding to 156 proteins) differentially abundant in 2011 and 2012 in Mosslunda (Figure 2D). This indicates that relatively few apoplastic proteins differ in abundance between the years. Compared to in 2012, osmotin-like protein (Q41350), pectinesterase (DMP400031280), PAE (DMP400041742), polygalacturonase (DMP400021809), and β-galactosidase (DMP400026688) showed increased abundance in 2011 (Supplementary Table S3). Dal Santo et al. [38] found 625 grapevine genes differentially expressed in at least one of the three growing seasons. Although investigation across more years and locations is required to draw solid conclusions, the result indicates stability of apoplastic proteome profile in potato. It is also possible that, given the sites are in the same climate conditions; the variation in weather and biotic condition was minimal thus did not significantly alter the apoplastic proteome.

2.4. Abundance of Apoplastic Proteins across a Growing Season

Under field conditions, microbial populations and abiotic stresses change throughout the growing season [39,40] which might accordingly alter the apoplastic proteome profile. To understand the possible changes in the apoplastic proteome within the same growing season, plant samples without disease symptoms collected in June, July, and August in Mosslunda were investigated. The samples from the same month were clustered together, and a multi-group comparison ($q \leq 0.001$) identified 320 peptides (9.3%) from 240 proteins that were differentially regulated in samples collected in at least one of these months (Figure 3A,B).

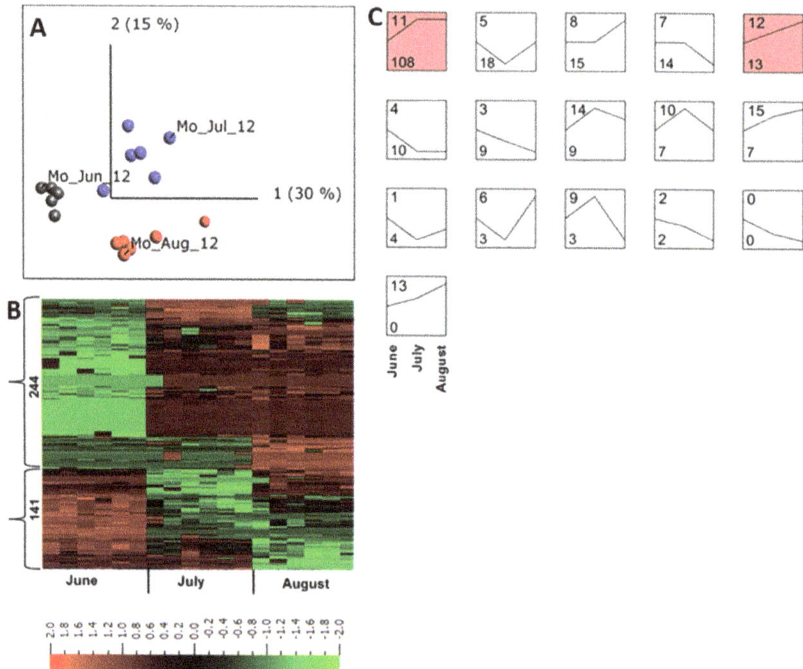

Figure 3. Quantitative analysis of apoplastic proteome samples isolated from potato cultivar Bintje grown in Mosslunda in 2012. (**A**) Unsupervised principal component analysis plot of all the samples collected in June (Mo_Jun_12), July (Mo_Jul_12), and August (Mo_Jul_12). Each circle represents one biological replicate. (**B**) Heat maps and the number of peptides up- and down-regulated in plants grown under field conditions in Mosslunda. We performed a multi-group comparison with false discovery rate < 0.001 (according to the Benjamini–Hochberg procedure for determining q). Heat map of the differentially regulated peptides ($q < 0.001$) was sorted using hierarchal clustering and red represents higher abundance (Fold change, log2). (**C**) STEM clustering analysis of apoplastic peptides in June, July, August of 2012 in Mosslunda. Proteins that were significantly ($q \leq 0.001$) increased or decreased in at least one of the months across the growing season were used for the STEM clustering analysis. Top left of each box is the profile number and bottom left of each box indicates the number of peptides that fit the defined abundance pattern in June, July, and August. The STEM analysis identified 16 profiles, of which profiles 11 and 12 contains statistically significant number of proteins ($p < 0.05$).

STEM clustering [41] of differentially regulated proteins identified 16 abundance pattern profiles for proteins that were co-regulated throughout the growing season (Figure 3C). Of those, only two were significant profiles ($p < 0.05$). Profile 11 represented proteins with lower abundance in June and their abundance increased at a similar degree in July and August, whereas profile 13 comprised proteins with continued increase in abundance during the season (Figure 3C).

In July and August most of the proteins with increased abundance in profile 11 were involved in plant response to specific biotic and abiotic stresses (Table 1, Figure 3C, profile 11).

Table 1. Differentially abundant proteins in plants collected in June, July, or August in 2012 in fields in Mosslunda at false discovery rate < 0.001 (according to Benjamini−Hochberg), corresponding to the abundance pattern identified in STEM clustering profile 11 (Figure 3). Only unique peptides were used for the analysis. Shown are peptides with $\log_2$ fold change ≥ 4 in July and August compared to their abundance in June.

Peptide Sequence	Protein IDs	Protein Name	Genome	Signal P	Log2 Fold Change	
			Location		July	August
TDPNQNTGIVIQK	DMP400016183	Pectinesterase	chr03	Yes	4.72	4.73
DGQPSEQHFGLFYPDQR	Q70BW9	1,3-beta-glucan glucanohydrolase			4.71	4.81
GQTWVIDAPR	DMP400005465	Osmotin	chr08	Yes	4.68	4.83
GLTWSVPTGR	DMP400022299	Peroxidase	chr01	Yes	4.68	4.63
RLDPGQTWVIDAPR	Q5XUH0	Osmotin-like protein			4.65	4.84
MLNEGFVPDDVSLK	Q9FHR3	Putative pentatricopeptide repeat-containing protein At5g37570			4.65	4.56
NIQNAISGAGLGNQIK	DMP400051976	Glucan endo-1,3-beta-D-glucosidase	chr10	Yes	4.64	4.64
TSNLYAIGEMEIEENKK	DMP400023312	DUF26 domain-containing protein 2	chr12	Yes	4.62	4.67
LLALSDTPYK	DMP400046980	Kunitz trypsin inhibitor	chr06	Yes	4.62	4.57
VCWPVPNK	DMP400033260	Xylem serine proteinase 1	chr10	No	4.61	4.65
SPSAYLNNPAGER	DMP400007784	Ceramidase	chr03	Yes	4.61	4.24
RYCGMLNVPTGEN-LDCNNQR	DMP400002757	Class II chitinase	chr02	Yes	4.6	4.72
QRCPDAYSYPQDD-PTSTFTCPSDSTNYR	DMP400005463	Osmotin OSML13	chr08	Yes	4.59	4.34
GVIFFGDSPYVFLPGMDVSK	DMP400015799	Xyloglucan-specific endoglucanase inhibitor 4	chr01	Yes	4.58	4.49
IFESCSTDTFQIR	DMP400041178	Embryo-specific 3	chr01	Yes	4.57	4.51
YCGICCEECK	DMP400037307	Snakin-1	chr04	Yes	4.57	4.45
ALPTYTPESPADATR	DMP400038185	Transketolase, chloroplastic	chr10	No	4.56	4.62
VITSSTEAQAYTPGR	Q43143	Pectinesterase/pectinesterase inhibitor U1			4.53	4.54
GFEAAPSVSFTVDGEEK	DMP400000884	Serine carboxypeptidase III	chr11	No	4.52	4.62
FVVVVDDSK	M1BPR5	Uncharacterized protein (Solanum tuberosum)			4.52	4.58
AETWVQEETRALISLR	Q43326	Box II Factor			4.52	4.56
KFGLTVDNVLDAR	DMP400031346	Reticuline oxidase	chr02	Yes	4.52	4.52
LCPQGGDGGTFANLDK	DMP400055305	Peroxidase	chr01	Yes	4.51	4.61
CLCGSPLPDCK	DMP400038422	Polygalacturonase inhibitor protein	chr07	Yes	4.51	4.49
TVTNLGDGQSTYTAK	DMP400027005	Subtilisin-like protease preproenzyme	chr12	Yes	4.48	4.51
LCGEIPKGEYMK	DMP400014905	Polygalacturonase inhibiting protein	chr09	Yes	4.45	4.17
ADNLDTCYR	DMP400025990	41 kD chloroplast nucleoid DNA binding protein (CND41)	chr08	Yes	4.43	4.23

Table 1. *Cont.*

Peptide Sequence	Protein IDs	Protein Name	Genome Location	Signal P	Log2 Fold Change July	Log2 Fold Change August
GTGDFTGR	SW_g323.t1	Pathogenesis-related protein 1b (Solanum tuberosum)			4.41	4.49
RIVDIPAGAFSFNSNT-GAGTIIDSGTVFTR	DMP400009572	Aspartic proteinase nepenthesin-1	chr01	Yes	4.38	4.55
VIIADIQNDLGNSLVK	DMP400032777	Short chain alcohol dehydrogenase	chr12	No	4.37	4.56
TLPESTTNENK	K7WVA0	Acyl-CoA-binding protein (Solanum tuberosum)			4.37	4.42
CHAVQCTANINGECPGQLK	DMP400023388	Osmotin		Yes	4.35	4.68
TNCNFDGDGR	Q01591	Osmotin-like protein TPM-1			4.35	4.41
LSEDGQVLEVLEDVEGK	DMP400030201	Strictosidine synthase	chr07	Yes	4.31	4.59
SMVGTPLMPGISVDTYIF-ALYDEDLKPGPGSER	DMP400001406	Glucan endo-1,3-beta-glucosidase	chr01	Yes	4.3	4.65
GNLDIFSGR	DMP400035839	Wound/stress protein	chr04	Yes	4.27	4.6
ITGNDYSSGVR	DMP400007118	Citrate binding protein	chr11	Yes	4.26	4.54
AVGEAGLGNDIK	DMP400062364	Glucan endo-1,3-beta-glucosidase, basic isoform 2	chr01	No	4.24	4.58
HAGPQFDYLEK	DMP400019521	Glutathione S-transferase omega	chr10	No	4.23	4.58
SSSTDVFGR	DMP400043338	Subtilisin-like protease	chr02	Yes	4.21	4.61
YLVTIGGVEGNPGR	DMP400017956	Miraculin	chr03	Yes	4.21	4.52
MYQLSFK	DMP400050666	Unidentified	chr08	Yes	4.21	4.48
ADAGHVLVEK	DMP400022826	MRNA binding protein	chr09	No	4.15	4.49
GQGTVGTEINR	DMP400023006	Threonine dehydratase biosynthetic, chloroplastic	chr09	No	4.14	4.52
WQPSGADQAANR	P52405	Endochitinase 3			4.1	4.45

Proteins with increased abundance in July and August included peroxidases (DMP400022299, DMP400055305), serine carboxypeptidase III (DMP400000884), class II chitinase (DMP400002757), pectinesterases (DMP400016183, Q43143), and osmotins (DMP400005465, DMP400005463, Q5XUH0). In addition, ceramidase (DMP400007784), 1,3-β-glucan glucanohydrolase (Q70BW9), glucan endo-1,3-β-D-glucosidase (DMP400051976), and Kunitz trypsin inhibitor (DMP400046980) also showed increased abundance ($\log_2$ fold change ≥ 4) in July and August compared to that in June (Table 1; Figure 3C, profile 11). This was corroborated by our finding that the number of potato plants with activated immunity increased (as measured as PR protein accumulation) at the end of the growing season, which might be associated with increased presence of pathogens as the season progressed [15].

2.5. Difference in Protein Abundance under Field and Greenhouse Conditions

To investigate the apoplastic proteome differences between plants grown under greenhouse and those grown under field conditions, we first conducted an unsupervised PCA. The resulting PCA plot showed a clear clustering of the field- and greenhouse-grown samples into different groups (Figure 4A), regardless of the sampling year and growing site, suggesting distinct apoplastic proteome profiles for field- and greenhouse-grown plants.

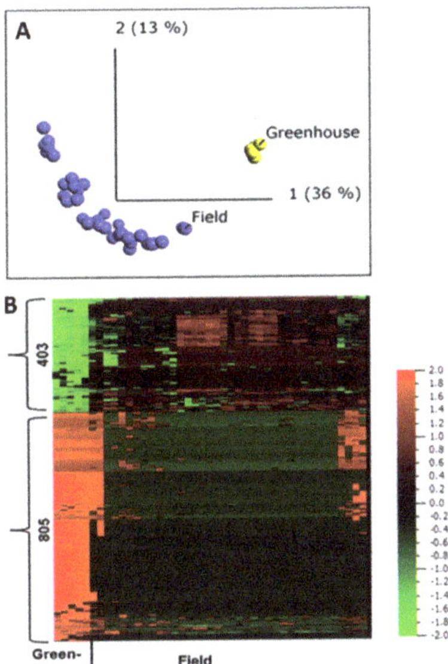

Figure 4. Quantitative analysis of apoplastic proteome samples isolated from potato cultivar Bintje grown under greenhouse and field conditions. (**A**) Unsupervised principal component analysis plot of all the samples. Each circle represents one biological replicate; (**B**) heat maps and numbers of peptides up- and down-regulated under greenhouse and field conditions according to a two-group comparison in Qlucore with a false discovery rate < 0.001 (according to the Benjamini–Hochberg procedure for determining q). The heat map of the differentially regulated peptides ($q < 0.001$) was sorted using hierarchal clustering and red represents higher abundance (Fold change, log2).

This is in agreement with a study that compared apoplastic proteome profiles between field- and laboratory-grown *Arabidopsis* using a limited number of samples [2]. Our data also strengthens the notion that peptide biomarkers developed using apoplastic proteome of field-grown potato can be a powerful tool for trait prediction [32].

A two-group comparison (*t*-test) in Qlucore, field vs. greenhouse, identified 1208 peptides that belong to 606 proteins, making up to 48% of all the proteins identified in the apoplast, to be differentially abundant in plants grown in the field and greenhouse ($q \leq 0.001$). Of those, 781 peptides were diagnostic, assigned to only one protein in our database. The abundance of 805 peptides was lower in field-grown samples compared to those grown in the greenhouse (Figure 4B). This difference in abundance of most of the proteins in greenhouse samples might be due to differences in extraction efficiency associated with variation in leaf anatomy between plants grown in greenhouse and field conditions [1], or a limited sampling of the greenhouse samples.

To describe the identified proteins that were differentially abundant under field and greenhouse conditions, we carried out a MapMan analysis [42] using 350 differentially abundant proteins with PGSC identity numbers (Supplementary Figure S2). The analysis identified many stress-related proteins such as proteins associated with proteolysis, proteins involved in cell wall synthesis or degradation, proteins classified as pathogenesis-related proteins (PR-proteins), peroxidases, and proteins involved in signaling; most of these proteins were at lower abundance in field-grown plants than in plants grown in the greenhouse (Supplementary Figure S2A). However, most of the redox and heat shock proteins were more abundant in field-grown plants (Supplementary Figure S2B).

2.6. ABPP Reveals Seasonal Effects

Serine and glycosyl hydrolase protein families are commonly found in the apoplast [6]; both families were among the most abundant in our samples (Figure 1). Similarly, we also found that most of the differentially regulated proteins in field-grown plants vs. greenhouse-grown plants were associated with catalytic and hydrolytic activities, as shown using MapMan pathway analysis (Figure 5, Supplementary Figure S2). Therefore, we studied the activity profile of serine hydrolase and β-glucosidase proteins in the apoplast using ABPP. The labeling of the apoplast for active serine hydrolases identified a signal (#1) at 100 kDa only in the field-grown potato plants (Figure 5A). In contrast, a signal (#3) at 40 kDa had higher intensity in greenhouse- than in field-grown plants.

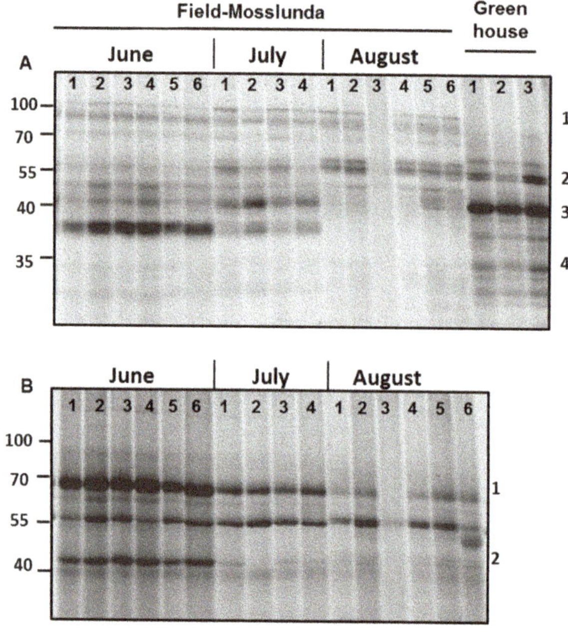

Figure 5. Serine hydrolase and β-glycosidase activity profiling of potato cultivar Bintje grown under greenhouse and field conditions in Mosslunda in June, July, and August 2012. Apoplastic proteins were labeled by 2 μM probe for (**A**) serine hydrolase and (**B**) β-glycosidase. The probe-labelled proteins were separated on 12% sodium dodecyl sulfate-polyacrylamide electrophoresis gels and detected using a fluorescence scanner.

Surprisingly, we found that the activity of serine hydrolases and β-glycosidases was generally decreasing as the growing season progressed (Figure 5). The intensity of the signal (#4) below 40 kDa for active serine hydrolases decreased from June to August in field-grown plants (Figure 5A). Using a probe for β-glycosidases, we identified that intensity signal (#1) at 70 kDa and signal (#2) at 40 kDa decreased later in the season (Figure 5B), showing decreasing activity of these proteins as the season progressed.

2.7. Serine Hydrolases and β-Glycosidases Identified by ABPP and MS

ABPP has been applied in a limited number of protein families (van der Hoorn et al., 2011). In this study, to investigate the proteins identified in the activity profile of the potato apoplastic fluid, the samples were purified after labelling with a mix of two biotinylated probes targeting serine hydrolases and β-glycosidases. The detected protein bands in ABPP were excised and analyzed using MS. The results of the MS analysis of the apoplastic proteome, which was used to identify the composition of protein signals, revealed that serine hydrolase signal (#1) at 100 kDa corresponded to subtilisin-like proteins, such as P69B (DMP400056894), P69E (DMP400007008), P69F (DMP400006964), subtilase (DMP400011990), and serine protease (DMP400006965) (Figure 6).

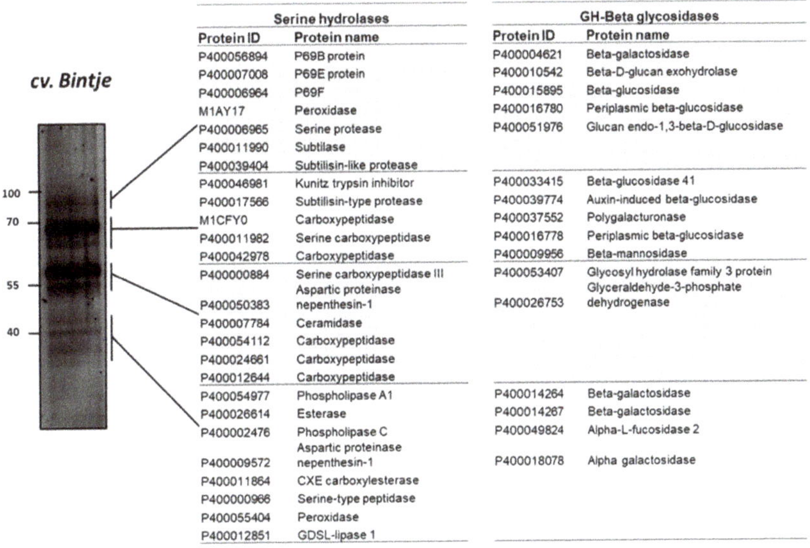

Serine hydrolases	
Protein ID	**Protein name**
P400056894	P69B protein
P400007008	P69E protein
P400006964	P69F
M1AY17	Peroxidase
P400006965	Serine protease
P400011990	Subtilase
P400039404	Subtilisin-like protease
P400046981	Kunitz trypsin inhibitor
P400017566	Subtilisin-type protease
M1CFY0	Carboxypeptidase
P400011982	Serine carboxypeptidase
P400042978	Carboxypeptidase
P400000884	Serine carboxypeptidase III
P400050383	Aspartic proteinase nepenthesin-1
P400007784	Ceramidase
P400054112	Carboxypeptidase
P400024661	Carboxypeptidase
P400012644	Carboxypeptidase
P400054977	Phospholipase A1
P400026614	Esterase
P400002476	Phospholipase C
P400009572	Aspartic proteinase nepenthesin-1
P400011864	CXE carboxylesterase
P400000966	Serine-type peptidase
P400055404	Peroxidase
P400012851	GDSL-lipase 1

GH-Beta glycosidases	
Protein ID	**Protein name**
P400004621	Beta-galactosidase
P400010542	Beta-D-glucan exohydrolase
P400015895	Beta-glucosidase
P400016780	Periplasmic beta-glucosidase
P400051976	Glucan endo-1,3-beta-D-glucosidase
P400033415	Beta-glucosidase 41
P400039774	Auxin-induced beta-glucosidase
P400037552	Polygalacturonase
P400016778	Periplasmic beta-glucosidase
P400009956	Beta-mannosidase
P400053407	Glycosyl hydrolase family 3 protein
P400026753	Glyceraldehyde-3-phosphate dehydrogenase
P400014264	Beta-galactosidase
P400014267	Beta-galactosidase
P400049824	Alpha-L-fucosidase 2
P400018078	Alpha galactosidase

Figure 6. Identification of serine hydrolases and β-glycosidases proteins that were captured by activity-based probes. Leaf apoplastic proteome of the potato sample was co-labelled by 5 μM biotinylated probes for β-glucosidase (JJB111) and serine hydrolases (FP-biotin). Biotinylated proteins were then affinity-purified with streptavidin beads and separated on 12% sodium dodecyl sulfate-polyacrylamide electrophoresis gel. The gel was stained by SYPRO Ruby staining.

Proteins corresponding to signals #3 and #4 included carboxylesterase (DMP400011864), esterase (DMP400026614), serine-type peptidase (DMP400000966), and GDSL-lipase 1 (DMP4000-12851) (Figure 6); GDSL-lipase 1 is involved in plant defense against pathogens such as *Pseudomonas syringae* [43]. Moreover, β-glycosidases were identified at 100 kDa (β-galactosidases [DMP400004621, DMP4000-15895, and DMP400016780]) and at 70 kDa (polygalacturonase [DMP400037552], β-glucosidase [DMP400033415], β-mannosidase [DMP400009956]), and β-galactosidase (DMP400014264 and DMP400014267) and α-galactosidase (DMP400018078) were detected at 40 kDa (signal #4) (Figure 6).

3. Materials and Methods

3.1. Plant Material and Field Sites

Potato cultivar Bintje was used in this study, and field experiments were conducted at two experimental sites in southern Sweden: Mosslunda (55°58′ N, 14°6.3′ E) in 2011 and 2012 and Borgeby in 2010 (55°45′ N, 13°23′ E). The experiment site in Borgeby had a sandy clay soil with 2.8 % humus, 16 % clay content, and 55% fine sand, with pH 7.1. The nutrient concent were Phosphorus (9.9 mg), Potassium (9.9 mg), Magnesium (10 mg), and Calcium (310 mg) per 100 g soil. In Mosslunda in 2011, the soil was sandy (79%), with a low clay content (7%) and humus (4.2%), and the chemical property was pH was 7.2, and Phosphorus (31 mg), Potassium (12 mg), Magnesium (13 mg), Calcium (490 mg) per 100 g of soil. In 2012 in Mosslunda, the soil was sandy (84%), with a low clay content (3%) and humus (4.1%), the pH was 6.5, and Phosphorus (21 mg/100 g), Potassium (11 mg), Magnesium (10 mg), Calcium (200 mg) per 100 g of soil. The average temperature, relative humidity and precipitation during this study was in Borgeby (17.6 °C, 73.8% RH, and 357 mm), and in Mosslunda in 2011 (16.9 °C, 79.9% RH, and 293 mm) and in 2012 (16.0 °C, 78,0% RH, and 213 mm). Furthermore, the monthly weather condition in Mosslunda in 2012 was in June (13.0 °C, 76 % RH, and 190 mm), July (17.1, 79% RH, and 242 mm) and August (17.0 °C, 80.0% RH, and 265 mm). Plants were grown following good experimental practice in accordance with EU directive 93/71, KIFS 2004:4, STAFS 2001:1, and standard operative procedures, SLU 2004 [44]. Plants in the greenhouse were grown at SLU Alnarp (55°65′ N, 13°07′ E), under conditions described previously [44]. Some plants were treated with a fungicide Shirlan (ISK Biosciences Europe S.A., Machelen, Belgium; active ingredient: fluazinam 500 g L^{-1}) according to manufacturer's recommendation (0.3 L ha^{-1}). No pathogen-related symptoms were visible on any of the analyzed leaves.

3.2. Apoplast Isolation and Protein Digestion

Apoplastic fluid was isolated, between 10 a.m. and 3 p.m., from five fully expanded leaves of a single plant by vacuum infiltration-centrifugation using a mobile field laboratory as previously described [21,29,30]. Isolated samples were aliquoted on site, frozen in liquid nitrogen, and kept at −80 °C until used for either label-free quantitative proteomics analysis or activity-based protein profiling (ABPP). For quantitative proteomics analysis, a 30 µL aliquot of each apoplastic sample was cleaned and trypsin-digested as described in [32] before analysis by mass spectrometry.

3.3. Mass Spectrometry

For high-performance liquid chromatography-tandem mass spectrometry (HPLC-MS/MS) analysis, 5 µL of the peptide solution was injected into Eksigent nanoLC-2D HPLC system coupled to a LTQ Orbitrap XL ETD (Eksigent Technologies, Dublin, CA, USA). Peptides were separated using a linear gradient for 90 min at a flow rate of 300 nL min^{-1}. The eluted peptide spectra were acquired, analyzed, and the four most intense ions were selected for fragmentation in linear trap quadrupole (LTQ) [45]. The raw data were converted to mzML [46] and Mascot generic format (MGF) files using ProteoWizard [47] and uploaded to the Proteios software environment (ProSE) [48], where the MGF and mzML files were used for MS/MS identification and feature detection, respectively. Protein identification was performed in Mascot (www.matrixscience.com) (accessed on 1 November 2021) and X!Tandem (www.thegpm.org/tandem) (accessed on 1 November 2021) by searching a protein sequence database containing all *Solanum* proteins in Uniprot (http://www.uniprot.org/) (accessed on 1 November 2021) and potato genome project (https://solgenomics.net/) (accessed on 17 June 2015). Protein sequences based on de novo assembled transcripts of potato clones Desirée, Sarpo Mira, and SW93-1015 were included in the database [14]. Reversed sequences of all the proteins in the database (449,968 protein entries) were included as decoys. For MS/MS searches, MS mass tolerance was set to 5 ppm and MS/MS fragment tolerance to 0.5 Da. One missed cleavage, fixed cysteine carbamidomethylation, and variable oxidation of methionine were allowed. All search results at the peptide spectrum level were subsequently filtered at a false discovery rate (FDR) of 1% [48]. Feature detection to quantify peptides was performed on the mzML files from Proteios using Dinosaur [49]. An alignment algorithm was used in Proteios to propagate peptide identities between LC-MS/MS runs [50] and the report was exported for further analysis. After feature matching across all runs with a recall of 0.99 and a precision of 0.80 based on common identifications, the peptide feature level FDR could be estimated to 4% from the fraction of decoy identifications. The proteomics data used for quantitative analysis have been deposited to the ProteomeXchange Consortium (http://proteomecentral.proteomexchange.org) (accessed on 1 November 2021) via the PRIDE partner repository with the dataset identifier PXD006392.

3.4. Quantitative Analysis of Peptides

The quantitative dataset with peptide precursor intensities was analyzed using Normalyzer v1.1.1 [51], and after comparing 12 normalization methods the Loess-G was used for normalization [52]. The log_2-transformed normalized data were used for visualization and statistical tests for identification of differentially abundant proteins; missing values were treated as described in [12]. Qlucore Omics Explorer v3.2 software (http://www.qlucore.se/) (accessed on 1 November 2021) was used to generate principal components analysis (PCA) plots and perform comparative analysis. Unsupervised PCA plots were generated to show similarities or differences between the samples. To identify peptide and proteins with differential abundance, comparative analyses were performed two group comparison (t-test) in Qlucore with the Benjamini−Hochberg false discovery ratio (FDR) procedure ($q \leq 0.001$). Heat maps sorted with hierarchical clustering were generated and the list of peptide and proteins differentially regulated among the samples was exported for bioinformatics analysis. If the protein was represented by two or more peptides, a median abundance value was calculated.

3.5. Labelling of Apoplastic Proteome Activity

For the activity-based profiling, 5 µL of 500 mM sodium acetate (NaAc, pH = 5) and 1 µL of 250 mM dithiothreitol (DTT) were added to 43 µL of ice-thawed apoplastic fluid. Two micromoles of each probe for serine hydrolases (FP-Rh) [53] and Glycosyl hydrolases (JJB70) [28] were added to the labeling reaction. For the inhibition tests, equal amounts of labeling reactions were pre-incubated for 30 min at room temperature (RT) with 100 µM of inhibitor 3,4 dichloroisocoumarin (DCI) against each probe. In samples without the inhibitor, DMSO was added instead and all labeling reactions were incubated for 1 h in darkness at RT. The reactions were stopped by adding 15 µL of 4X gel loading buffer, and the whole samples were boiled at 95 °C for 5 min. Proteins were separated on 12% sodium dodecyl sulfate-polyacrylamide electrophoresis (SDS-PAGE) gels and probe-labelled proteins were detected using a Typhoon 9400 fluorescence scanner (GE Healthcare, Bio-Sciences AB, Uppsala, Sweden).

3.6. Affinity Purification and Identification of Serine Hydrolases and β-glycosidases

Apoplastic fluid of Bintje (2.5 mL) was incubated with 5 µM biotinylated probes for β-glycosidase (JJB111) [28] and serine hydrolases (FP-biotin) [53], 50 mM NaAc, and 1 mM DTT for 1 h in the dark at RT. The labelled proteins were affinity-purified as described previously [26,28], eluted from streptavidin beads by adding 30 µL of 4X gel loading buffer and boiling at 95 °C for 10 min. The samples were centrifuged and 15 µL of the supernatant was loaded and separated on 12% SDS-PAGE gel. The gel was incubated with SYPRO fix for 30 min and stained overnight in the dark with SYPRO Ruby Protein Gel Stain (Thermo Fischer Scientific, Waltham, MA, USA). Protein bands were detected and excised from the gel using scalpel, and gel pieces were subjected to in-gel tryptic digestion [26,28].

Peptide and protein identification was performed using LC-MS/MS of the in-gel digests on an Orbitrap Elite instrument (Thermo, [54]) that was coupled to an EASY-nLC 1000 liquid chromatography (LC) system (Thermo). The LC was operated in the one-column mode. The analytical column was a fused silica capillary (75 µm × 36 cm) with an integrated PicoFrit emitter (New Objective) packed in-house with Reprosil-Pur 120 C18-AQ 1.9 µm resin (Dr. Maisch). The analytical column was encased by a column oven (Sonation) attached to a nanospray flex ion source (Thermo). The column oven temperature was adjusted to 45 °C during data acquisition. The LC was equipped with two mobile phases: solvent A (0.1% formic acid, FA, in water) and solvent B (0.1% FA in acetonitrile, ACN). All solvents were of UPLC grade (Sigma-Aldrich, St. Louis, MO, USA). Peptides were directly loaded onto the analytical column with a maximum flow rate that would not exceed the set pressure limit of 980 bar (usually around 0.4–0.6 µL/min). Peptides were subsequently separated on the analytical column by running a 50 min gradient of solvent A and solvent B (start with 7% B; gradient 7 to 35% B for 40 min; gradient 35 to 100% B for 5 min and 100% B for 5 min) at a flow rate of 300 nl/min. The mass spectrometer was operated using Xcalibur software version 2.2 SP1.48. The mass spectrometer was set in the positive ion mode. Precursor ion scanning was performed in the Orbitrap analyzer (FTMS; Fourier Transform Mass Spectrometry) in the scan range of m/z 300–1500 and at a resolution of 60,000 with the internal lock mass option turned on (lock mass was 445.120025 m/z, polysiloxane) [55]. Product ion spectra were recorded in a data dependent fashion in the ion trap (ITMS) in a variable scan range and at a rapid scan rate. The ionization potential (spray voltage) was set to 1.8 kV. Peptides were analyzed using a repeating cycle consisting of a full precursor ion scan (1.0×10^6 ions or 50 ms) followed by 12 product ion scans (1.0×10^4 ions or 100 ms) where peptides are isolated based on their intensity in the full survey scan (threshold of 500 counts) for tandem mass spectrum (MS2) generation. The MS2 permits peptide sequencing and identification. Collision induced dissociation (CID) energy was set to 35% for the generation of MS2 spectra. During MS2 data acquisition dynamic ion exclusion was set to 60 s with a maximum list of excluded ions consisting of 500 members and a repeat count of one. Ion injection time prediction, preview mode for the FTMS, monoisotopic precursor selection and charge state screening were enabled. Only charge states higher than 1 were considered for fragmentation.

Peptide and Protein Identification was performed using MaxQuant. RAW spectra were submitted to an Andromeda [56] search in MaxQuant version 1.5.3.30 using the default settings. [57] Label-free quantification was activated. MS/MS spectra data were searched against the Uniprot *S. tuberosum* (UP000011115_4113.fasta; 53104 entries) [58], as well as assembled RNA-seq datasets of *Solanum dulcamara* (DUL.fasta; 26392 entries) and *Solanum tuberosum* cv. Desiree (DES.fasta; 24703 entries) [14,16]. Further analysis and filtering of the results was carried out in Perseus version 1.5.5.3 [59]. Proteins unique or with higher spectral count in a specific protein band in the SDS-PAGE are reported herein. The identified proteins with their spectral counts are listed in Supplementary Table S1.

3.7. *Bioinformatics Analysis*

We used Pfam enrichment analysis [60] for investigation of protein families in the apoplastic proteome. SignalP 4.1 was used to predict the presence of secretion signals in identified proteins [33]. In order to identify proteins with a similar differential regulation pattern, we performed STEM structure analysis in STEM v1.3.8 with default parameters [41]. Location of the identified proteins in potato genome was predicted using SPUD DB Genome browser version 1.70 (http://solanaceae.plantbiology.msu.edu/) (accessed on 1 November 2021). Functional categories of identified proteins were determined by gene ontology (GO) enrichment analysis in agriGO version 2.0 [61]. MapMan version 3.6.0 [42] was used for pathway analysis based on the potato mapping file obtained from GoMapMan [62]. To establish a correlation between the quantitative profiling and ABPP, proteins and peptides found by both approaches were identified based on the sequence similarity analysis. The Potato Genome Sequencing Consortium (PGSC) protein numbers throughout this report have been abbreviated for readability; for example, protein PGSCDMP400044750 is abbreviated to DMP400044750.

4. Conclusions

Proteomics can be useful for understanding the physiology of field-grown plants, and we found similarities between years and sites. We show that fungicide may not affect the apoplastic proteome significantly, but this requires a more detailed investigation on the translocation and stability of the fungicide applied and changes in phyllosphere microbial community as well as induction of plant defense responses. Variation in proteomics profile including presence of proteins with differential abundance and activity among samples grown in different sites, growing season, as well as across the months within the growing season indicates dynamic regulation of parts of the apoplastic proteome in response to biotic and abiotic factors. Indeed, most of the differentially regulated proteins were associated with stress related processes. Nevertheless, this warrants further investigation to identify proteins that can be useful for molecular-assisted decision making in management strategies of these stresses. The study combines quantitative analysis with ABPP to gain insights into the actual activity of certain protein classes. This study also shows the importance of collecting apoplastic proteomes in field conditions and that understanding the proteome in agricultural fields would be a new dimension in order to understand the physiological state of field-grown plants (field-omics) and support biotic and abiotic stress mitigation strategies.

Supplementary Materials: The following are available online at https://www.mdpi.com/article/10.3390/ijms222112033/s1.

Author Contributions: K.B.A., E.A. (Erik Andreasson), Å.L. and F.L. conceived the experiment. K.B.A. collected apoplastic samples, performed ABPP. assays and quantitate proteomics analysis, and drafted the manuscript. S.R. and E.A. (Erik Alexandersson) helped with the quantitative analysis. D.S., F.K., M.K. and R.A.L.v.d.H. supervised the ABPP experiments. All authors have read and agreed to the published version of the manuscript.

Funding: This work was financially supported by Novo Nordisk Foundation, Mistra Biotech, SLF (R-19-25-282), Formas (2020-01211) and European Molecular Biology Organization (EMBO) short-term fellowship, the ERC starting grant (M.K.; grant No. 258413), ERC Consolidator grant (D.S. and R.A.L.v.d.H., grant No. 616449) and the Deutsche Forschungsgemeinschaft (M.K.; grant No. INST 20876/127-1 FUGG).

Data Availability Statement: The data presented in this study are openly available in ProteomeXchange Consortium at [http://proteomecentral.proteomexchange.org] (accessed on 1 November 2021) with Project DOI: [10.6019/PXD006392] and reference number [PXD006392].

Acknowledgments: We thank Sofia Hydbom and Mia Mogren for their assistance with Apoplast isolation and protein digestion, and Karin Hansson for performing the LC-MS/MS acquisition.

Conflicts of Interest: The authors declare no conflict of interest.

References

1. Mishra, Y.; Jankanpaa, H.J.; Kiss, A.Z.; Funk, C.; Schroder, W.P.; Jansson, S. Arabidopsis plants grown in the field and climate chambers significantly differ in leaf morphology and photosystem components. *BMC Plant Biol.* **2012**, *12*, 6. [CrossRef]
2. Ruhe, J.; Agler, M.T.; Placzek, A.; Kramer, K.; Finkemeier, I.; Kemen, E.M. Obligate Biotroph Pathogens of the Genus Albugo Are Better Adapted to Active Host Defense Compared to Niche Competitors. *Front. Plant Sci.* **2016**, *7*, 820. [CrossRef]
3. Poorter, H.; Fiorani, F.; Pieruschka, R.; Wojciechowski, T.; van der Putten, W.H.; Kleyer, M.; Schurr, U.; Postma, J. Pampered inside, pestered outside? Differences and similarities between plants growing in controlled conditions and in the field. *New Phytol.* **2016**, *212*, 838–855. [CrossRef] [PubMed]
4. Jones, J.W.; Antle, J.M.; Basso, B.; Boote, K.J.; Conant, R.T.; Foster, I.; Godfray, H.C.J.; Herrero, M.; Howitt, R.E.; Janssen, S.; et al. Brief history of agricultural systems modeling. *Agric. Syst.* **2016**, *155*, 240–254. [CrossRef]
5. Oerke, E.C. Crop losses to pests. *J. Agric. Sci.* **2005**, *144*, 31. [CrossRef]
6. Alexandersson, E.; Ali, A.; Resjo, S.; Andreasson, E. Plant secretome proteomics. *Front. Plant Sci.* **2013**, *4*, 9. [CrossRef]
7. De Wit, P.J. Apoplastic fungal effectors in historic perspective; a personal view. *New Phytol.* **2016**, *212*, 805–813. [CrossRef]
8. Jashni, M.K.; Mehrabi, R.; Collemare, J.; Mesarich, C.H.; de Wit, P.J.G.M. The battle in the apoplast: Further insights into the roles of proteases and their inhibitors in plant-pathogen interactions. *Front. Plant Sci.* **2015**, *6*, 584. [CrossRef] [PubMed]
9. Song, Y.; Zhang, C.J.; Ge, W.N.; Zhang, Y.F.; Burlingame, A.L.; Guo, Y. Identification of NaCl stress-responsive apoplastic proteins in rice shoot stems by 2D-DIGE. *J. Proteom.* **2011**, *74*, 1045–1067. [CrossRef]
10. Delaunois, B.; Jeandet, P.; Clement, C.; Baillieul, F.; Dorey, S.; Cordelier, S. Uncovering plant-pathogen crosstalk through apoplastic proteomic studies. *Front. Plant Sci.* **2014**, *5*, 249. [CrossRef]
11. Wang, Y.; Wang, Y.C.; Wang, Y.M. Apoplastic Proteases: Powerful Weapons against Pathogen Infection in Plants. *Plant Commun.* **2020**, *1*, 100085. [CrossRef]
12. Bengtsson, T.; Weighill, D.; Proux-Wera, E.; Levander, F.; Resjo, S.; Burra, D.D.; Moushib, L.I.; Hedley, P.E.; Liljeroth, E.; Jacobson, D.; et al. Proteomics and transcriptomics of the BABA-induced resistance response in potato using a novel functional annotation approach. *BMC Genom.* **2014**, *15*, 315. [CrossRef] [PubMed]
13. Burra, D.D.; Berkowitz, O.; Hedley, P.E.; Morris, J.; Resjo, S.; Levander, F.; Liljeroth, E.; Andreasson, E.; Alexandersson, E. Phosphite-induced changes of the transcriptome and secretome in Solanum tuberosum leading to resistance against Phytophthora infestans. *BMC Plant Biol.* **2014**, *14*, 254. [CrossRef]

14. Ali, A.; Alexandersson, E.; Sandin, M.; Resjo, S.; Lenman, M.; Hedley, P.; Levander, F.; Andreasson, E. Quantitative proteomics and transcriptomics of potato in response to Phytophthora infestans in compatible and incompatible interactions. *BMC Genom.* **2014**, *15*, 497. [CrossRef]
15. Lankinen, A.; Abreha, K.B.; Masini, L.; Ali, A.; Resjo, S.; Andreasson, E. Plant immunity in natural populations and agricultural fields: Low presence of pathogenesis-related proteins in Solanum leaves. *PLoS ONE* **2018**, *13*, e0207253. [CrossRef]
16. Frades, I.; Abreha, K.B.; Proux-Wéra, E.; Lankinen, Å.; Andreasson, E.; Alexandersson, E. A novel workflow correlating RNA-seq data to resistance levels of wild Solanum species and potato clones to Phytophthora infestans. *Front. Plant Sci.* **2015**, *6*, 718. [CrossRef] [PubMed]
17. Yogendra, K.N.; Kushalappa, A.C. Integrated transcriptomics and metabolomics reveal induction of hierarchies of resistance genes in potato against late blight. *Funct Plant Biol.* **2016**, *43*, 766–782. [CrossRef]
18. Evers, D.; Legay, S.; Lamoureux, D.; Hausman, J.F.; Hoffmann, L.; Renaut, J. Towards a synthetic view of potato cold and salt stress response by transcriptomic and proteomic analyses. *Plant Mol. Biol.* **2012**, *78*, 503–514. [CrossRef] [PubMed]
19. Gong, L.; Zhang, H.X.; Gan, X.Y.; Zhang, L.; Chen, Y.C.; Nie, F.J.; Shi, L.; Li, M.; Guo, Z.Q.; Zhang, G.H.; et al. Transcriptome Profiling of the Potato (*Solanum tuberosum* L.) Plant under Drought Stress and Water-Stimulus Conditions. *PLoS ONE* **2015**, *10*, e0128041.
20. Boguszewska-Mankowska, D.; Gietler, M.; Nykiel, M. Comparative proteomic analysis of drought and high temperature response in roots of two potato cultivars. *Plant Growth Regul.* **2020**, *92*, 345–363. [CrossRef]
21. Alexandersson, E.; Jacobson, D.; Vivier, M.A.; Weckwerth, W.; Andreasson, E. Field-omics-understanding large-scale molecular data from field crops. *Front. Plant Sci.* **2014**, *5*, 286. [CrossRef] [PubMed]
22. Zhu, W.H.; Smith, J.W.; Huang, C.M. Mass Spectrometry-Based Label-Free Quantitative Proteomics. *J. Biomed. Biotechnol.* **2010**, 840518. [CrossRef] [PubMed]
23. Liu, Y.H.; Lu, S.; Liu, K.F.; Wang, S.; Huang, L.Q.; Guo, L.P. Proteomics: A powerful tool to study plant responses to biotic stress. *Plant Methods* **2019**, *15*, 1–20. [CrossRef] [PubMed]
24. Van der Hoorn, R.A.; Colby, T.; Nickel, S.; Richau, K.H.; Schmidt, J.; Kaiser, M. Mining the Active Proteome of Arabidopsis thaliana. *Front. Plant Sci.* **2011**, *2*, 89. [CrossRef] [PubMed]
25. Bozkurt, T.O.; Schornack, S.; Win, J.; Shindo, T.; Ilyas, M.; Oliva, R.; Cano, L.M.; Jones, A.M.E.; Huitema, E.; van der Hoorn, R.A.L.; et al. Phytophthora infestans effector AVRblb2 prevents secretion of a plant immune protease at the haustorial interface. *Proc. Natl. Acad. Sci. USA* **2011**, *108*, 20832–20837. [CrossRef]
26. Kaschani, F.; Gu, C.; Niessen, S.; Hoover, H.; Cravatt, B.F.; van der Hoorn, R.A. Diversity of serine hydrolase activities of unchallenged and botrytis-infected Arabidopsis thaliana. *Mol. Cell. Proteom. MCP* **2009**, *8*, 1082–1093. [CrossRef]
27. Sueldo, D.; Ahmed, A.; Misas-Villamil, J.; Colby, T.; Tameling, W.; Joosten, M.H.A.J.; van der Hoorn, R.A.L. Dynamic hydrolase activities precede hypersensitive tissue collapse in tomato seedlings. *New Phytol.* **2014**, *203*, 913–925. [CrossRef]
28. Chandrasekar, B.; Colby, T.; Emran Khan Emon, A.; Jiang, J.; Hong, T.N.; Villamor, J.G.; Harzen, A.; Overkleeft, H.S.; van der Hoorn, R.A. Broad-range glycosidase activity profiling. *Mol. Cell. Proteom. MCP* **2014**, *13*, 2787–2800. [CrossRef]
29. Ali, A.; Moushib, L.I.; Lenman, M.; Levander, F.; Olsson, K.; Carlson-Nilson, U.; Zoteyeva, N.; Liljeroth, E.; Andreasson, E. Paranoid potato: Phytophthora-resistant genotype shows constitutively activated defense. *Plant Signal. Behav.* **2012**, *7*, 400–408. [CrossRef]
30. Andreasson, E.; Abreha, K.B.; Resjö, S. Isolation of Apoplast. In *Isolation of Plant Organelles and Structures: Methods and Protocols*; Taylor, N.L., Millar, A.H., Eds.; Springer: New York, NY, USA, 2017; pp. 233–240.
31. Tyers, M.; Mann, M. From genomics to proteomics. *Nature* **2003**, *422*, 193–197. [CrossRef]
32. Chawade, A.; Alexandersson, E.; Bengtsson, T.; Andreasson, E.; Levander, F. Targeted Proteomics Approach for Precision Plant Breeding. *J. Proteome Res.* **2016**, *15*, 638–646. [CrossRef]
33. Petersen, T.N.; Brunak, S.; von Heijne, G.; Nielsen, H. SignalP 4.0: Discriminating signal peptides from transmembrane regions. *Nat. Methods* **2011**, *8*, 785–786. [CrossRef] [PubMed]
34. Webb-Robertson, B.J.M.; Wiberg, H.K.; Matzke, M.M.; Brown, J.N.; Wang, J.; McDermott, J.E.; Smith, R.D.; Rodland, K.D.; Metz, T.O.; Pounds, J.G.; et al. Review, Evaluation, and Discussion of the Challenges of Missing Value Imputation for Mass Spectrometry-Based Label-Free Global Proteomics. *J. Proteome Res.* **2015**, *14*, 1993–2001. [CrossRef]
35. Doherty, J.R.; Botti-Marino, M.; Kerns, J.P.; Ritchie, D.F.; Roberts, J.A. Response of Microbial Populations on the Creeping Bentgrass Phyllosphere to Periodic Fungicide Applications. *Plant Health Prog.* **2017**, *18*, 44–49. [CrossRef]
36. Byrnes, C. Public release summary of the evaluation by the NRA of the new active constituent: Fluazinam in the product: SHIRLAN FUNGICIDE. *Natl. Regist. Auth. Agric. Vet. Chem.* **2011**, 1–28. Available online: https://apvma.gov.au/node/13751 (accessed on 1 November 2021).
37. Hoehenwarter, W.; Larhlimi, A.; Hummel, J.; Egelhofer, V.; Selbig, J.; van Dongen, J.T.; Wienkoop, S.; Weckwerth, W. MAPA Distinguishes Genotype-Specific Variability of Highly Similar Regulatory Protein Isoforms in Potato Tuber. *J. Proteome Res.* **2011**, *10*, 2979–2991. [CrossRef] [PubMed]
38. Dal Santo, S.; Tornielli, G.B.; Zenoni, S.; Fasoli, M.; Farina, L.; Anesi, A.; Guzzo, F.; Delledonne, M.; Pezzotti, M. The plasticity of the grapevine berry transcriptome. *Genome Biol.* **2013**, *14*, r54. [CrossRef]
39. Wiik, L. Potato Late Blight and Tuber Yield: Results from 30 Years of Field Trials. *Potato Res.* **2014**, *57*, 77–98. [CrossRef]

40. Zhang, B.; He, H.B.; Ding, X.L.; Zhang, X.D.; Zhang, X.P.; Yang, X.M.; Filley, T.R. Soil microbial community dynamics over a maize (*Zea mays* L.) growing season under conventional- and no-tillage practices in a rainfed agroecosystem. *Soil Till Res.* **2012**, *124*, 153–160. [CrossRef]
41. Ernst, J.; Bar-Joseph, Z. STEM: A tool for the analysis of short time series gene expression data. *BMC Bioinform.* **2006**, *7*, 191. [CrossRef]
42. Thimm, O.; Blasing, O.; Gibon, Y.; Nagel, A.; Meyer, S.; Kruger, P.; Selbig, J.; Muller, L.A.; Rhee, S.Y.; Stitt, M. MAPMAN: A user-driven tool to display genomics data sets onto diagrams of metabolic pathways and other biological processes. *Plant J.* **2004**, *37*, 914–939. [CrossRef]
43. Kwon, S.J.; Jin, H.C.; Lee, S.; Nam, M.H.; Chung, J.H.; Kwon, S.I.; Ryu, C.M.; Park, O.K. GDSL lipase-like 1 regulates systemic resistance associated with ethylene signaling in Arabidopsis. *Plant J.* **2009**, *58*, 235–245. [CrossRef] [PubMed]
44. Liljeroth, E.; Bengtsson, T.; Wiik, L.; Andreasson, E. Induced resistance in potato to Phytphthora infestans-effects of BABA in greenhouse and field tests with different potato varieties. *Eur. J. Plant Pathol.* **2010**, *127*, 171–183. [CrossRef]
45. Resjo, S.; Ali, A.; Meijer, H.J.G.; Seidl, M.F.; Snel, B.; Sandin, M.; Levander, F.; Govers, F.; Andreasson, E. Quantitative Label-Free Phosphoproteomics of Six Different Life Stages of the Late Blight Pathogen Phytophthora infestans Reveals Abundant Phosphorylation of Members of the CRN Effector Family. *J. Proteome Res.* **2014**, *13*, 1848–1859. [CrossRef]
46. Martens, L.; Chambers, M.; Sturm, M.; Kessner, D.; Levander, F.; Shofstahl, J.; Tang, W.H.; Ropp, A.; Neumann, S.; Pizarro, A.D.; et al. mzML-a Community Standard for Mass Spectrometry Data. *Mol. Cell. Proteom.* **2011**, *10*, R110.000133. [CrossRef]
47. French, W.R.; Zimmerman, L.J.; Schilling, B.; Gibson, B.W.; Miller, C.A.; Townsend, R.R.; Sherrod, S.D.; Goodwin, C.R.; McLean, J.A.; Tabb, D.L. Wavelet-Based Peak Detection and a New Charge Inference Procedure for MS/MS Implemented in ProteoWizard's msConvert. *J. Proteome Res.* **2015**, *14*, 1299–1307. [CrossRef] [PubMed]
48. Hakkinen, J.; Vincic, G.; Mansson, O.; Warell, K.; Levander, F. The Proteios Software Environment: An Extensible Multiuser Platform for Management and Analysis of Proteomics Data. *J. Proteome Res.* **2009**, *8*, 3037–3043. [CrossRef]
49. Teleman, J.; Chawade, A.; Sandin, M.; Levander, F.; Malmstrom, J. Dinosaur: A Refined Open-Source Peptide MS Feature Detector. *J. Proteome Res.* **2016**, *15*, 2143–2151. [CrossRef]
50. Sandin, M.; Ali, A.; Hansson, K.; Mansson, O.; Andreasson, E.; Resjo, S.; Levander, F. An Adaptive Alignment Algorithm for Quality-controlled Label-free LC-MS. *Mol. Cell. Proteom.* **2013**, *12*, 1407–1420. [CrossRef] [PubMed]
51. Chawade, A.; Alexandersson, E.; Levander, F. Normalyzer: A Tool for Rapid Evaluation of Normalization Methods for Omics Data Sets. *J. Proteome Res.* **2014**, *13*, 3114–3120. [CrossRef]
52. Smyth, G.K.; Ritchie, M.; Thorne, N.; Wettenhall, J. LIMMA: Linear models for microarray data. In Bioinformatics and Computational Biology Solutions Using R and Bioconductor. *Stat. Biol. Health* **2005**, 397–420.
53. Kaschani, F.; Nickel, S.; Pandey, B.; Cravatt, B.F.; Kaiser, M.; van der Hoorn, R.A.L. Selective inhibition of plant serine hydrolases by agrochemicals revealed by competitive ABPP. *Bioorganic Med. Chem.* **2012**, *20*, 597–600. [CrossRef] [PubMed]
54. Michalski, A.; Damoc, E.; Lange, O.; Denisov, E.; Nolting, D.; Muller, M.; Viner, R.; Schwartz, J.; Remes, P.; Belford, M.; et al. Ultra High Resolution Linear Ion Trap Orbitrap Mass Spectrometer (Orbitrap Elite) Facilitates Top Down LC MS/MS and Versatile Peptide Fragmentation Modes. *Mol. Cell. Proteom.* **2012**, *11*. [CrossRef]
55. Olsen, J.V.; de Godoy, L.M.F.; Li, G.Q.; Macek, B.; Mortensen, P.; Pesch, R.; Makarov, A.; Lange, O.; Horning, S.; Mann, M. Parts per million mass accuracy on an orbitrap mass spectrometer via lock mass injection into a C-trap. *Mol. Cell. Proteom.* **2005**, *4*, 2010–2021. [CrossRef]
56. Cox, J.; Neuhauser, N.; Michalski, A.; Scheltema, R.A.; Olsen, J.V.; Mann, M. Andromeda: A Peptide Search Engine Integrated into the MaxQuant Environment. *J. Proteome Res.* **2011**, *10*, 1794–1805. [CrossRef] [PubMed]
57. Cox, J.; Mann, M. MaxQuant enables high peptide identification rates, individualized p.p.b.-range mass accuracies and proteome-wide protein quantification. *Nat. Biotechnol.* **2008**, *26*, 1367–1372. [CrossRef]
58. Cox, J.; Hein, M.Y.; Luber, C.A.; Paron, I.; Nagaraj, N.; Mann, M. Accurate Proteome-wide Label-free Quantification by Delayed Normalization and Maximal Peptide Ratio Extraction, Termed MaxLFQ. *Mol. Cell. Proteom.* **2014**, *13*, 2513–2526. [CrossRef]
59. Tyanova, S.; Temu, T.; Sinitcyn, P.; Carlson, A.; Hein, M.Y.; Geiger, T.; Mann, M.; Cox, J. The Perseus computational platform for comprehensive analysis of (prote)omics data. *Nat. Methods* **2016**, *13*, 731–740. [CrossRef] [PubMed]
60. Finn, R.D.; Coggill, P.; Eberhardt, R.Y.; Eddy, S.R.; Mistry, J.; Mitchell, A.L.; Potter, S.C.; Punta, M.; Qureshi, M.; Sangrador-Vegas, A.; et al. The Pfam protein families database: Towards a more sustainable future. *Nucleic Acids Res.* **2016**, *44*, D279–D285. [CrossRef]
61. Du, Z.; Zhou, X.; Ling, Y.; Zhang, Z.; Su, Z. agriGO: A GO analysis toolkit for the agricultural community. *Nucleic Acids Res.* **2010**, *38*, W64–W70. [CrossRef]
62. Ramsak, Z.; Baebler, S.; Rotter, A.; Korbar, M.; Mozetic, I.; Usadel, B.; Gruden, K. GoMapMan: Integration, consolidation and visualization of plant gene annotations within the MapMan ontology. *Nucleic Acids Res.* **2014**, *42*, D1167–D1175. [CrossRef]

MDPI
St. Alban-Anlage 66
4052 Basel
Switzerland
Tel. +41 61 683 77 34
Fax +41 61 302 89 18
www.mdpi.com

International Journal of Molecular Sciences Editorial Office
E-mail: ijms@mdpi.com
www.mdpi.com/journal/ijms

www.ingramcontent.com/pod-product-compliance
Lightning Source LLC
LaVergne TN
LVHW070137170726
843515LV00007B/1813

* 9 7 8 3 0 3 6 5 2 6 6 2 1 *